SOCIAL LEARNING

Opinion Formation and Decision-Making over Graphs

Vincenzo Matta
Virginia Bordignon
Ali H. Sayed

now
the essence of knowledge

Published, sold and distributed by:
now Publishers Inc.
PO Box 1024
Hanover, MA 02339
United States
Tel. +1-781-985-4510
www.nowpublishers.com
sales@nowpublishers.com

Outside North America:
now Publishers Inc.
PO Box 179
2600 AD Delft
The Netherlands
Tel. +31-6-51115274

EURASIP–Now Publishers Open Access Book Series on Information and Learning Sciences

ISBN: 978-1-63828-472-7
E-ISBN: 978-1-63828-473-4
DOI: 10.1561/9781638284734

Suggested citation: Vincenzo Matta, Virginia Bordignon, and Ali H. Sayed (2025). *Social Learning: Opinion Formation and Decision-Making over Graphs*. Boston–Delft: Now Publishers.

Dedication

To Chiara and Sara, for their smiling eyes that inspire me. To my mother, and in memory of my father, for the sacrifices I will never repay (V. Matta)

To Cláudia, Inete, and in memory of Assis Bordignon (V. Bordignon)

To Thomas Kailath, for his unwavering guidance and support (A. H. Sayed)

Contents

Preface

Social learning is a timely and highly relevant topic that addresses themes such as the study of opinion formation and propagation over networks, or how cooperating agents (e.g., humans, robots, or sensors) affect one another and make decisions based on decentralized observations.

Many complex cognitive systems are made up of individual agents whose activities are the result of sophisticated "social" interactions with other agents. Consider how people build their opinions about a particular phenomenon. The opinions form through repeated interactions with other people, whether in person or virtually (e.g., over a social network). A diffusion mechanism occurs, by which ideas, information, and even false news spread throughout the network. Nature provides many other examples of cooperative learning in the form of biological networks.

Social learning occurs in man-made systems as well, in the form of multi-agent decision-making procedures. One example is a robotic swarm deployed over a hazardous area for a rescue operation. Multi-agent decision-making can be critical in this scenario, as some robots operating in adverse conditions (e.g., with limited visibility or partial information) would only be able to complete their task by cooperating with other robots that have better access to critical information.

The primary focus of this text is on techniques for information diffusion and decision-making over graphs, as well as the examination of how agents' decisions evolve dynamically in response to interactions with neighbors and the environment. There are at least two reasons why research on social learning is important. On one hand, it provides for a more in-depth explanation of the fundamental cognitive mechanisms that enable opinion formation and the dissemination of knowledge (or disinformation) across graphs. On the other hand, the study of social learning is important for the design of reliable distributed decision-making strategies, which encounter applications in a range of settings involving highly dynamic environments, nonstationary data and uncertain models, untrustworthy or malicious agents, sparsely connected graphs, and restricted communication.

The text provides a unifying framework and a comprehensive presentation for understanding and developing social learning strategies. The treatment starts from the theory of optimal single-agent learning, to arrive gradually at the foundations of social learning by multiple agents connected through a graph, whose structure can induce interesting and diversified phenomena. For example, we will see how connected graphs enable agreement across the agents, whereas a "mind control" mechanism emerges over weakly connected graphs, where the network is split into influencers and influenced agents.

After a detailed illustration of the traditional techniques, the focus is shifted to recent advances and trends in social learning. For example, we will show that traditional strategies produce stubborn agents, which oppose new states of information and are reluctant to respond to changes in the environment. We then explain how to endow social networks with adaptation and learning capabilities to detect these drifting situations. We also present methodologies to deal with the sharing of incomplete or partial information, and we explain how to design social machine learning solutions where the agents rely exclusively on data.

The text relies on various powerful tools, such as stochastic convergence, large deviation analysis, martingales, and the Rademacher complexity. The necessary elements to understand and use these tools are collected in the appendices.

Acknowledgments

We are indebted to many Ph.D. students and post-doctoral associates for their contributions to social learning, as well as for their input at various stages of this project, either through feedback on earlier drafts or through conversations that deepened our understanding of the topics. In particular, we are grateful to (in alphabetical order): Marco Carpentiero, Michele Cirillo, Ping Hu, Mert Kayaalp, Malek Khammassi, Kostas Ntemos, Hawraa Salami, Augusto Santos, Valentina Shumovskaia, Stefan Vlaski, Bicheng Ying, and Xiaochuan Zhao.

Vincenzo Matta, *Salerno, Italy.*
Virginia Bordignon, *Lausanne, Switzerland.*
Ali H. Sayed, *Lausanne, Switzerland.*

April 2025

Chapter 1

Introduction

By *social learning*, in this book we refer to an ensemble of mathematical models and inferential strategies for opinion formation and decision-making over graphs [27].

To motivate the use of the term "learning," let us consider a situation where there exist some possible choices, called the *hypotheses* or *classes*. These choices could correspond to the weather condition (such as sunny or rainy), to the outcome of a soccer match (such as a victory, draw, or loss), or to the type of restaurant that a group of friends would like to book. Some agents collect data related to the phenomenon of interest and their *learning* objective is to assign probability scores, called *beliefs*, to all possible hypotheses. These scores would quantify the levels of confidence or "opinions" of each agent about each of the potential hypotheses. For example, in the weather forecasting problem, the data sensed by the agents can be measurements of humidity, atmospheric pressure, or temperature, and the opinions formed by one agent in relation to the sunny or rainy condition could be in the form: "Tomorrow will likely be sunny with 90% confidence and rainy with 10% confidence." In other words, each agent will form a belief vector, which happens to be a probability vector with nonnegative entries adding up to 1. Each entry of this vector will represent the credit that the agent assigns to the corresponding hypothesis being the truth. This process of belief formation enables automatic *decision-making*, since an agent can select as the most plausible hypothesis the one corresponding to the highest belief. When the agents continuously collect streams of data supporting increasing evidence in favor of one particular hypothesis, it is expected that they will ultimately place all the probability mass on that hypothesis.

The qualification "social," on the other hand, refers to the networked or graphical structure (i.e., the *graph*) that links multiple agents together, as happens in the context of social networks, biological networks, or robotic swarms. These networks (also referred to as *multi-agent networks* [151, 152]) consist of multiple communicating agents, equipped with sensing and *cognitive* abilities that allow them to cooperate and extract meaningful information from measurements. Nature itself provides numerous examples of cooperative learning through sophisticated dynamics arising, e.g., over biological networks [11, 39, 98], in animal behavior [5, 39, 51, 64, 138, 156], and in brain science [15, 38, 162].

In social learning, the decentralized interaction between dispersed agents takes place through repeated *local consultation* steps, where neighboring agents, i.e., agents linked by edges over a graph, are allowed to exchange their beliefs over these edges. Consider, for instance, the manner in which humans form their opinions about a certain phenomenon of interest. In this case, the opinions of an individual take shape via repeated interactions with other people they can consult with (i.e., their neighbors), whether through direct contact or virtually over a social platform. A diffusion mechanism emerges through which opinions, information, or even fake news propagate. Social learning strategies also arise over man-made systems, e.g., over distributed networks of sensors that collect measurements and exchange information to solve a decision-making problem. Compared with standalone learning strategies, *social* (i.e., networked or decentralized) strategies yield improved performance and robustness. They also enable agents to overcome their individual limitations by leveraging collaboration during the learning process.

1.1 Examples of Social Learning

The social learning problem is encountered across a range of disciplines and applications, including cognitive sciences (e.g., psychology), social sciences (e.g., economics), statistics, biology, engineering design, and others. Depending on the context, the term "social learning" might emphasize different aspects. For example, in [10] the topic of social learning is addressed from the perspective of psychology, whereas in [43] the focus is on learning dynamics that arise in economics. In our treatment, social learning will be useful to examine how the beliefs assigned to some hypotheses of interest evolve through interactions over a graph. It will also be a driver to enable decision-making by networked agents. In other words, the framework adopted herein is general enough to allow applications across different fields, such as the study of opinion formation over graphs, the dissemination of misinformation over these same topologies, as well as the ability to perform decision-making by robotic swarms, meteorological stations, or by communication and control networks in engineering design.

The manner in which a group of individuals is able to aggregate dispersed information has been the subject of several studies before. As early as the 18th century [48], scientists have been studying how multiple individuals can combine their information to learn some underlying truth. In [76], the following social experiment was described. People at a fair were asked to guess the weight of an ox, and 787 guesses were collected. The interesting result was that, while the individual guesses varied, their median value approached the true weight of the animal. The success of aggregating estimates in this experiment reinforces the idea of the "wisdom of the crowd," according to which a collective of agents could combine opinions to improve the reliability of the conclusions reached by a single individual.

Example 1.1 (Brazil-Italy soccer match). Assume the World Cup final is between Brazil and Italy. Three friends want to predict the winner of the match. Friend 1 is Italian, friends 2 and 3 are Brazilian. Figure 1.1 shows four possible scenarios arising from the opinion formation process, with focus on the belief of friend 1.

In cases (a) and (b) the three friends do not communicate with each other (in the graph shown in the top part of the figure, we see each node connected only to itself), i.e., they form their individual opinions only based on their own private information. In case (a) the data collected by friend 1 and the model they use to interpret the data support victory by Brazil. The belief of friend 1 accordingly places more mass on that hypothesis. In contrast, in case (b) friend 1 is biased by being a supporter of the Italian team, resulting in an opinion favoring their victory.

Let us now consider how the situation changes if friend 1 interacts with their Brazilian friends, according to the communication graph displayed in the bottom part of the figure. We assume the data and models of friends 2 and 3 always support the hypothesis of a Brazilian victory. In case (c) the belief of friend 1 in favor of Brazil is reinforced by the interaction with friends 2 and 3, thus leading to a higher mass concentration on that hypothesis. The most interesting situation occurs in case (d). Here, owing to cooperation and the sharing of information, friend 1 ends up changing their mind and is driven to believe that Brazil will win.[1]

Despite its simplicity, the previous example illustrates the fundamental interplay that arises between data, models, and network, and how this

[1]The perspective given in this particular example might be biased by the birth nationality of the authors, especially by the fact that one of the authors is a minority.

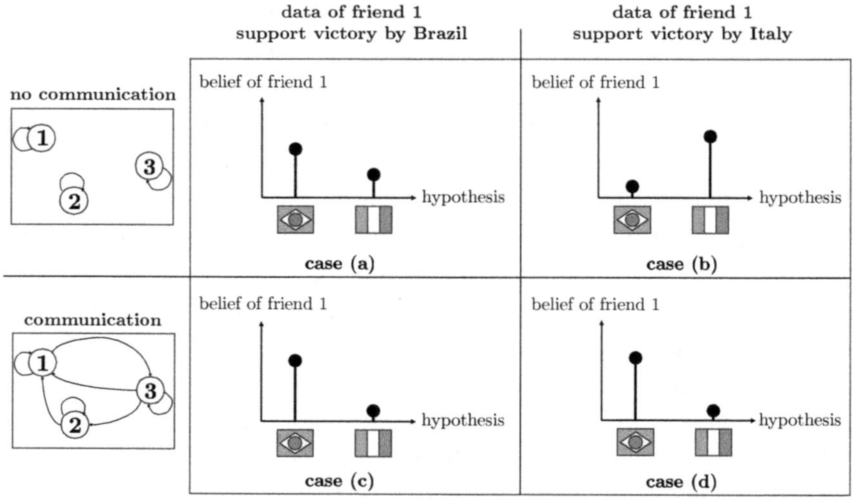

Figure 1.1: Illustration of belief formation for Example 1.1.

interplay can influence the final beliefs. This type of behavior is observed in real opinion formation processes, and it will be well captured by the social learning strategies derived in the forthcoming chapters. Specific instances will be discussed in some detail in Chapter 5.

In recent years, there have been many useful works devoted to the study of the social learning problem, such as [1, 2, 25, 42, 83, 96, 106, 118, 132, 135, 147, 175]. These studies have two main ramifications. From a *behavioral* perspective, the focus is on proposing and examining mathematical models for social learning that are able to capture the collective behavior of groups of cognitive agents. From a *design-oriented* or *engineering* perspective, the focus is on devising powerful social learning algorithms to accomplish specific tasks, and on assessing their quality, for example, their capacity to infer the right hypothesis from the evolving beliefs or the speed of convergence of the decision process.

Example 1.2 (Distributed sensing and decision-making). An example of an engineering system whose design is inspired by social learning is a collection of sensors recording data from a common region. These could be, for example, meteorological stations measuring different attributes such as air humidity, atmospheric pressure, or temperature. The goal of the network is to predict the state of the weather in the region under observation. Fusing information from multiple sensors can be useful to deliver superior learning performance. This is particularly relevant since in many situations the information at each individual sensor can be insufficient to allow it to make a correct weather forecast on its own. For instance, some sensors might be able to collect only humidity data, while others collect only pressure data. However, through mutual interactions, all sensors could be able to arrive at more informed predictions.

Another example is a robotic swarm deployed over a hazardous area for a rescue operation. Assume the robots patrol different portions of the area under control. Only neighboring robots can communicate with each other. All robots must make a decision and consequently take a coordinated action. Some robots operating under disadvantageous conditions (e.g., with limited visibility or partial information) would only be able to perform their assigned task (such as saving a life during the rescue operation) by leveraging cooperation with other robots that have better access to critical information. For example, the more informed robots might be closer to the origin of a fire. In this case, cooperation is critical since some robots might be "blind" to the fire event or detect it with some great delay.

We will discover in future chapters that several interesting phenomena arise in the context of social learning. It is often the case that the data observed by the agents are ruled by some *common truth* (i.e., one and the same hypothesis), giving rise to a scenario that we will refer to as *objective evidence*. Under this model, we will identify meaningful situations where the agents are able to learn the common hypothesis. However, other cases are encountered over real-world networks. For example, we can have *multiple individual truths* (leading to a *subjective evidence* scenario), where the observations of distinct agents in the network are ruled by distinct hypotheses; *fake news*, where some agents purposely inject artificial data to steer the agents' opinions toward some wrong hypothesis; or situations where the data distributions do not match perfectly any of the hypotheses postulated by the agents. We will identify situations where the agents are subject to manipulation, can become stubborn and be slow in accepting new truths, and can even end up following a herding behavior. Understanding the learning mechanisms arising under these different possibilities is useful to dissect the social learning dynamics and to answer interesting questions, such as: Do the agents agree on some hypothesis? If so, which one? Can some agents influence other agents in their choice?

Example 1.3 (Find the best restaurant). There are cases where it is difficult to define a common truth for all agents. For example, consider a group of friends that exchange opinions to choose an Italian restaurant in New York City. Some of the friends mostly care about food quality, while others care about price. The data available to the friends interested in quality relate to information such as ingredients or recipes, and the learning models they use to interpret the data enable them to pinpoint restaurants with higher food quality. In contrast, the data available to the remaining friends relate to prices, and their learning models give priority to restaurants with lower prices. This situation is one instance of the *subjective evidence* model treated in Chapter 5.

Depending on the relative number of friends interested in quality over price, on the data types and models used, and on the graph of interactions that describes who talks to whom, diversified outcomes are possible, as we will discover later in Chapter 5. For instance, if the graph is sufficiently connected, then all friends will be able to agree on one and the same restaurant, which would somehow optimize the quality-price ratio. However, the choice would be more or less unbalanced in favor of quality or price depending on different factors, such as the number of friends interested in quality over price, and the graph dictating the friends' interactions. If these interactions are sparse (i.e., if the communication graph is not sufficiently connected) we can also have disagreement, with different restaurants chosen by different friends.

1.2 Building Opinions

Several factors drive the process of opinion formation over graphs. These factors will be quantified in future chapters, and their roles will appear explicitly in the expressions defining the social learning strategies. Here we provide a brief overview. We identify seven main elements, which are described below.

Prior convictions. At any given time epoch, each cognitive agent will have its own personal opinion regarding the plausible states of nature, summarized in a probability vector that constitutes the *prior* belief. This opinion arises from different mechanisms, also depending on the particular application or context. For example, the prior belief can be completely flat (i.e., equal mass is assigned to all hypotheses) because the agent is completely ignorant about the hypotheses, or it can be biased because the agent has some preferences, or it can also originate from the agent's experience accumulated as the outcome of a previous learning process. The aim of the social learning process is to update these prior convictions by exploiting: *i)* new information or knowledge coming from private data observed by the agent; and *ii)* interaction with other (neighboring) agents.

Data. The effect of the world on the agents occurs through the *private* observations or measurements arriving from the environment at the individual agents. The quality of these measurements, the way they are distributed across the agents and over time are of utmost importance for the learning outcome.

Likelihood models. In order to update the prior convictions with the new information contained in the data, each agent will need to quantify how the data are related to the possible hypotheses. To do so, the agent will employ a *likelihood model* that describes the probabilistic mechanism by which the data are generated given a particular hypothesis. For example, the model would describe which humidity values are more likely to occur if the state of nature happens to be "rainy." In Figure 1.2 we show an example with three models (namely, three probability density functions) associated with the possible hypotheses.

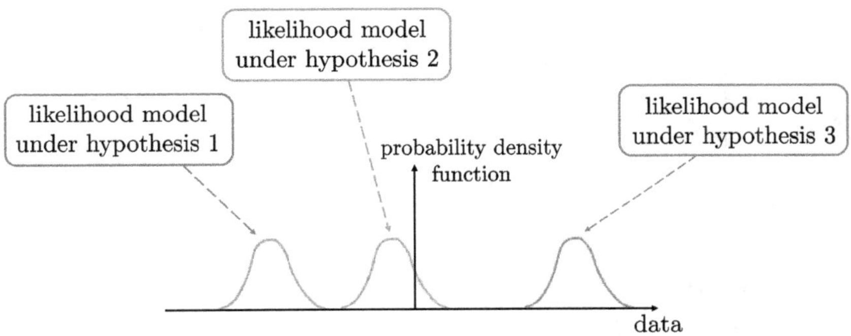

Figure 1.2: Example of probabilistic mechanisms linking the data to the hypotheses. Here we have three hypotheses corresponding to three probability density functions.

The agents will adopt some likelihood models depending on their knowledge about the specific learning task. For example, consider a decision-making network deployed to detect which symbol has been transmitted over a communication channel. The measurements would correspond to received signals corrupted by Gaussian noise. The statistical models linking the received measurement to the transmitted symbol will take the form of Gaussian distributions with different means corresponding to the different symbols. In other examples, the agents will need to learn their models directly from data during a training phase, by using some clues available to describe the relations between the hypotheses and the data. We will see specific examples of this training process in Chapter 12, in the context of *social machine learning*.

Update rule. In order to update the prior convictions using the knowledge extracted from the observed data and the assumed likelihood models, each agent will implement an update rule, whose general flow diagram is shown in Figure 1.3. The agent computes an updated belief by blending the prior belief with the data, whose information content is evaluated through the available likelihood models. In our treatment, the update rule will often be Bayes' rule, but other choices are useful, as we will see in Chapters 8 and 13.

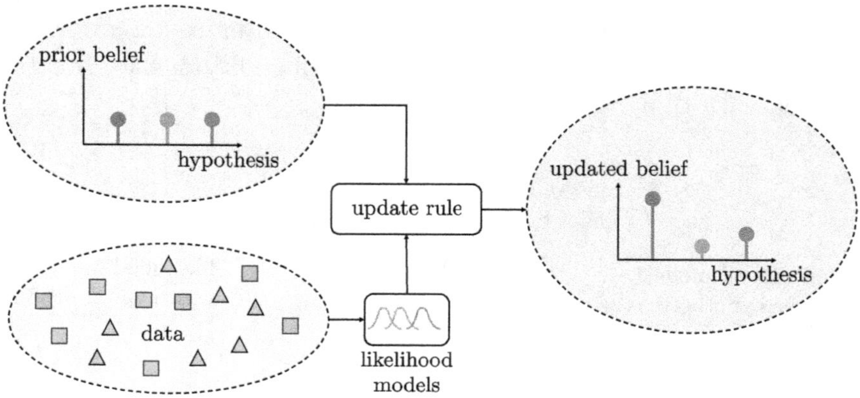

Figure 1.3: Schematic illustration of the belief update process.

Belief diffusion. The exchange of information between neighbors enables agents to solve the inference problem collaboratively, which might bring significant improvements over the noncooperative case. Through proper cooperation, the agents exploit the knowledge distributed across the entire network to deliver superior performance and to overcome the limitations that might exist in their individual data or models. Due to various physical constraints, the agents are not generally allowed to exchange their raw data. One such constraint is usually privacy; another one is complexity. For example, over a distributed cloud storage system it is seldom the case that one can share the (huge) datasets. It is more likely to share summary information, such as beliefs in a social learning context. Also in human learning, usually we would not share with our friends the entire set of information (i.e., the data) that led us to form our personal opinions, but we would rather share opinions or impressions.

Moreover, it is often the case that we share only *part* of our opinions. For instance, assume two friends are interested in ranking some commercial brands. In many cases, they talk specifically of a single commercial brand and then automatically update their opinions regarding other commercial brands. This particular learning mechanism will be examined in Chapter 11, in the context of social learning under *partial information.*

Network. When the agents exchange information, they do it according to some communication graph, i.e., over a network. The network structure determines the communication paths and the flow of the exchanged information across the agents, as well as the weights given by each agent to the information received from its neighbors. Different connectivity and weight patterns give rise to influence dynamics and rich belief formation scenarios.

Pooling. Once an agent receives the beliefs from its neighbors, it has to blend them suitably. In other words, it is necessary to devise a pooling rule to construct the final belief arising from the *social* learning mechanism. Popular pooling rules are the *geometric* and *arithmetic* averaging rules — see Chapter 3.

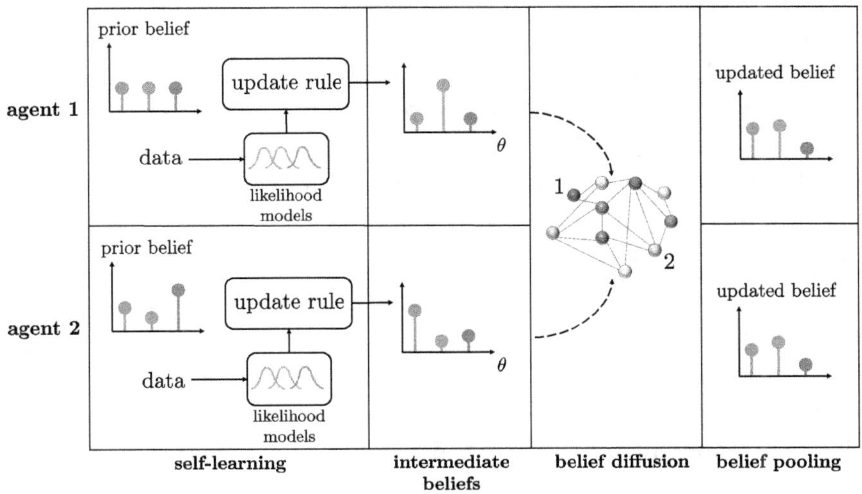

Figure 1.4: Schematic illustration of social learning.

The combined interaction of the seven elements described here is represented in Figure 1.4, where we show the behavior of two agents, denoted by 1 and 2. In a *self-learning* step, the agents implement a *local* rule to update their prior convictions using the knowledge extracted from the new data, based on the assumed likelihood models. In the considered example, we see that agent 1 starts with a flat belief, while agent 2 starts from a belief biased in favor of the red hypothesis. Interestingly, after incorporating evidence from the data, agent 1 departs from its initial agnostic assignment and gains confidence in favor of the green hypothesis. In contrast, agent 2 abandons its initial bias toward the red hypothesis in favor of the blue one.

During the belief-diffusion stage, each agent shares over the network (i.e., with its neighbors) the intermediate beliefs produced during the self-learning stage. Then, each agent processes the received beliefs through a suitable pooling rule. In the considered example, the beliefs of agents 1 and 2 after pooling become more similar to each other. This is a direct effect of incorporating the opinions from other agents in the network.

1.3 Book Organization

Social learning is a timely research topic with applications in several domains, and there are of course several works on the subject. It is therefore useful to describe the main distinguishing features of the present work.

We bring together into a *unifying treatment* the fundamentals of social learning and the most recent advances in the field. In particular, these advances consider important features encountered in many applications, such as adaptation under nonstationary conditions, the exchange of incomplete information, or the necessity for agents to build their private models from scratch relying on some clues available before social learning takes place. We derive a number of versatile social learning strategies that are well-suited to highly dynamic and uncertain environments where real-world networks usually operate.

The theoretical analysis of these social learning methodologies relies on advanced probability and mathematical tools, such as convergence of random series, unconventional central limit theorems, large deviation analysis, and advanced statistical learning tools. These tools are different from those traditionally used in similar contexts, e.g., in multi-agent distributed optimization or regression problems. For this reason, one added value of the book is to present these tools in an organic manner (with the help of some appendices) to introduce the reader gradually to the necessary background. In broad outline, the work is organized as follows:

Bayesian learning. One pillar of belief formation is *Bayes' rule*, which solves optimally the single-agent learning problem. Behavioral studies also reveal that standalone agents form their opinions in a "Bayesian way," i.e., their beliefs evolve according to Bayes' rule when they learn in isolation. We explain in Chapter 2 how belief vectors can be updated by means of Bayes' rule, especially in response to streaming observations. We examine the convergence behavior of this rule and provide useful information-theoretic interpretations for its optimality.

Non-Bayesian learning. In contrast to the single-agent case, when distributed agents are organized into a network structure, they aggregate their individual beliefs in a non-Bayesian way dictated by the physical constraints that the network imposes. In Chapter 3 we introduce these fundamental constraints, and derive the corresponding pooling policies that combine the agents' opinions and activate a belief diffusion mechanism over the network graph.

Graphs and network models. Chapter 4 illustrates the network models relevant to the treatment, emphasizing the role of network descriptors that are useful for the learning process, such as graphs, nodes, edges, neighborhoods, combination policies, and connectedness regimes.

Opinion formation over graphs. After having introduced, in the first chapters, the background on the necessary statistical and graph tools, in Chapters 5, 6, and 7 we examine carefully the behavior of the derived social learning strategies. The analysis provides a detailed characterization of the opinion formation mechanism and reveals how interesting and diversified phenomena emerge, depending on the data, models, and network structure. For example, under subjective evidence when different agents promote different hypotheses (say, hypotheses a or b), a "truth-is-somewhere-in-between" effect can arise, where *all* agents end up choosing a third option c. We will also see that a "mind-control" effect can arise over weakly connected graphs, where some agents can exert a domineering role over other agents, with the network split into influencers and influenced agents.

Adaptive social learning. We will explain that traditional social learning implementations cause the agents to become stubborn and to react slowly to drifts in the environment conditions. These traditional strategies are inherently nonadaptive and, hence, not suited to applications where continual learning must be guaranteed in the midst of nonstationary phenomena.

Adaptation in social learning is critical because in most situations the agents must be ready to change their mind and adapt their opinions. In order to address this issue, we introduce an adaptive social learning (ASL) strategy in Chapter 8 by showing how to modify the Bayesian update to embed into it the ability for continuous adaptation and learning. We introduce advanced mathematical tools in Chapters 9 and 10 to provide an accurate performance assessment of adaptive social learning and to ascertain the fundamental laws governing it (e.g., the weak law of small adaptation parameters, asymptotic normality, and large deviations).

Partial information sharing. We examine social learning under partial information sharing in Chapter 11, which arises when the agents exchange only a subset of their opinions about the hypotheses under consideration. For example, the agents may be interested in forming opinions about the candidates in an election process, but they would limit their interactions to discussing only one of the candidates. We will explain how the opinion formation process is affected by such partial information sharing mechanism.

Social machine learning. In most studies, social learning algorithms rely on predefined likelihood models that are assumed to be perfectly known beforehand. However, this is not always the case. In Chapter 12 we develop a social *machine* learning framework, where we examine the process governing the formation of the individual agent's memory, i.e., we focus on *how* the agents build their private models from some empirical clues observed prior to the social learning phase. The models obtained during this preliminary phase of training are then deployed to run the social learning algorithms. We show that the resulting fully data-driven strategy achieves consistent learning despite the challenges introduced by the lack of exact likelihood models.

Extensions and future directions. Chapter 13 is devoted to the presentation of possible research lines, extensions, and open questions. In particular, we discuss the effect of non-Bayesian updates, alternative adaptive rules based on censored beliefs, and the inverse problem where an inferential engine estimates the graph structure after monitoring the agents' beliefs.

1.4 Notation, Symbols, and Conventions

Table 1.1 collects the main conventions used in our exposition. More local definitions will be introduced in the individual chapters at the necessary moment.

In our treatment we often work with vectors and matrices. Vectors are denoted by small letters. The notation

$$x = [a, b, c] \tag{1.1}$$

defines a vector with entries a, b, c. When we need to perform linear algebra operations, we must specify if we deal with row or column vectors. Unless otherwise indicated, all vectors will be column vectors. Therefore, when we write $x \in \mathbb{R}^d$, we implicitly imply that x is a $d \times 1$ vector. Matrices are denoted by capital letters, and their entries by the corresponding small letter, to which we append two subscripts to pinpoint the particular matrix entry. The notation $A = [a_{jk}]$ specifies that the matrix A collects entries denoted by a_{jk}, where j is the row index and k the column index. Likewise, the notation $x = [x_k]$ indicates that the vector x collects entries denoted by x_k, where k is the entry index. Sometimes we employ the alternative notation $[A]_{jk}$ to extract the (j, k) entry of a matrix. This is particularly convenient when we work with products or powers of matrices. For example, $[A^2]_{jk}$ denotes the (j, k) entry of the matrix A^2. A matrix or vector with all null entries will be denoted by 0.

For infinite sequences of the form

$$x_1, x_2, \ldots \tag{1.2}$$

we use the notation $\{x_t\}_{t\in\mathbb{N}}$, $\{x_t\}_{t=1}^{\infty}$, or simply $\{x_t\}$ when the indexing is clear from the context. This notation is also used for finite collections of objects. Moreover, for objects with multiple indices, the notation $\{x_{k,t}\}_{k=1}^{K}$ refers to the collection of K values $x_{1,t}, x_{2,t}, \ldots, x_{K,t}$ for a fixed t.

Several arguments employed in this text rely on probability theory. The exposition is not focused on a measure-theoretic approach, so that readers will be able to follow most of the arguments without a background in measure theory. We use **bold** font for random quantities and normal font for their realizations or for deterministic quantities.

When we introduce random quantities that describe a particular setting (e.g., the agents' data in a social learning problem) we implicitly assume that they live in a common probability space $(\Omega, \mathscr{F}, \mathbb{P})$, where Ω is the

Table 1.1: List of the main notational conventions used in this book.

$\mathbb{R}$	Field of real numbers
$\mathbb{C}$	Field of complex numbers
$\mathbb{N}$	Set of natural numbers $1, 2, \ldots$
$\mathbb{1}$	Column vector with all its entries equal to 1
$\mathbb{1}_d$	$d \times 1$ vector with all its entries equal to 1
I	Identity matrix
I_d	$d \times d$ identity matrix
$\mathbb{I}[\mathcal{C}]$	Indicator function $\mathbb{I}[\mathcal{C}] = \begin{cases} 1 & \text{if condition } \mathcal{C} \text{ is true,} \\ 0 & \text{if condition } \mathcal{C} \text{ is false.} \end{cases}$
$\boldsymbol{x}$	**Bold** font denotes random quantities
x	Normal font denotes deterministic quantities or realizations of random quantities
$\mathbb{E}\boldsymbol{x}$	Expected value of $\boldsymbol{x}$
$\mathsf{VAR}[\boldsymbol{x}]$	Variance of $\boldsymbol{x}$
$\mathbb{P}[\mathcal{E}]$	Probability of event $\mathcal{E}$
A	Matrices are denoted by capital letters
A^T	Transpose of matrix A
$\mathrm{col}\{a_1, a_2, \ldots, a_N\}$	Column vector obtained by stacking the entries (or vectors) $a_1, a_2, \ldots, a_N$
$x_{k,t}$	Vector quantity relative to agent k at time t
$x_{k,t}(\theta)$	θth entry of vector $x_{k,t}$
$\|x\|$	Euclidean norm of x
$f(x) = o(g(x))$ as $x \to x_0$	$f(x)/g(x) \to 0$ as $x \to x_0$
$f(x) = O(g(x))$ as $x \to x_0$	$f(x)/g(x)$ remains bounded as $x \to x_0$
$\boldsymbol{x}_n \xrightarrow[n\to\infty]{\text{a.s.}} \boldsymbol{x}$	$\boldsymbol{x}_n$ converges to $\boldsymbol{x}$ almost surely as $n \to \infty$
$\boldsymbol{x}_n \xrightarrow[n\to\infty]{\text{p}} \boldsymbol{x}$	$\boldsymbol{x}_n$ converges to $\boldsymbol{x}$ in probability as $n \to \infty$
$\boldsymbol{x}_n \xrightarrow[n\to\infty]{\text{d}} \boldsymbol{x}$	$\boldsymbol{x}_n$ converges to $\boldsymbol{x}$ in distribution as $n \to \infty$

sample space, $\mathscr{F}$ (also called the event space) is a σ-field of subsets of Ω, and $\mathbb{P}$ a probability measure on $\mathscr{F}$. When we refer to sets and functions, we implicitly assume that they are measurable. Likewise, to avoid measurability issues, we assume that the probability spaces are complete.[2]

We reserve the term "random variable" to scalar real-valued quantities and use "random vector" to denote vectors whose entries are random variables. Discrete, a.k.a. categorical random variables, belong to discrete alphabets, such as a discrete random variable $\boldsymbol{x}$ taking on values in the set $\mathcal{X} = \{a, b, c\}$. These variables are described in terms of a probability mass function (pmf), e.g.,

$$\mathbb{P}[\boldsymbol{x} = a] = p(a). \tag{1.3}$$

A pmf can be equivalently regarded as a probability vector $p \in \Delta_{|\mathcal{X}|}$, where $|\mathcal{X}|$ is the cardinality of $\mathcal{X}$, and $\Delta_{|\mathcal{X}|}$ denotes the probability simplex in $\mathbb{R}^{|\mathcal{X}|}$. For example, if $x \in \mathcal{X} = \{a, b, c\}$, we can write

$$p = [p(a), p(b), p(c)]. \tag{1.4}$$

A continuous random vector $\boldsymbol{x}$ is defined on $\mathcal{X} = \mathbb{R}^d$, for some $d \in \mathbb{N}$. If the random vector admits a probability density function (pdf) $p(x)$ with respect to the Lebesgue measure on $\mathbb{R}^d$, for a set $\mathcal{S} \subseteq \mathbb{R}^d$ we have

$$\mathbb{P}[\boldsymbol{x} \in \mathcal{S}] = \int_{\mathcal{S}} p(x) dx. \tag{1.5}$$

As done for pmfs, we write p in place of $p(x)$ to refer to the entire pdf, not to a particular value x. When, for two pdfs $p(x)$ and $q(x)$, we write $p = q$ or $p(x) = q(x)$, we imply that the equality $p(x) = q(x)$ holds for all $x \in \mathcal{X}$, possibly excluding a set of zero Lebesgue measure. When for two random vectors $\boldsymbol{x}$ and $\boldsymbol{y}$ we write $\boldsymbol{x} = \boldsymbol{y}$, the equality is intended to hold with probability 1. Likewise, when we say that a random variable $\boldsymbol{x}$ is positive or write $\boldsymbol{x} > 0$, we mean that the inequality holds with probability 1.

The expectation of a random variable $\boldsymbol{x}$ is denoted by $\mathbb{E}\boldsymbol{x}$. The same symbol is used for vectors, where expectation is meant to be computed for each entry of the vector. When we evaluate the expectation of more involved functions, we use parentheses, e.g., $\mathbb{E}[(\boldsymbol{x} - 3)^2]$. Sometimes we write $\mathbb{E}_p\boldsymbol{x}$ to emphasize that the expectation is computed by assuming that the random variable $\boldsymbol{x}$ is distributed according to some pmf or pdf p. When we write $\mathbb{E}$ (and $\mathbb{P}$) without subscripts, the underlying distribution should be clear from the context.

[2]In a complete probability space, all subsets of zero-measure sets are measurable, and it is known that every measure can be completed [145].

With reference to equalities, inequalities, and more general relations, the acronyms LHS (left-hand side) and RHS (right-hand side) will be used to indicate a specific side of the relation. For example, the LHS of

$$a \to b \tag{1.6}$$

is a. When a formula contains multiple relations, the LHS (resp., RHS) will indicate the leftmost (resp., rightmost) side. For example, in

$$a = b = c, \tag{1.7}$$

c is the RHS.

Chapter 2

Bayesian Learning

A central quantity in this book is the *belief vector*, a probability vector whose entries quantify the credibility that a cognitive agent assigns to different hypotheses of interest. For example, assume we are interested in predicting the outcome of a soccer match, which can be represented by a hypothesis $\theta \in \{\text{victory, draw, loss}\}$. We start from some *prior* convictions arising from personal impressions or previous evidence, such as statistics on the recent performance of the involved teams. Then, we can progressively update our initial opinion about the possible outcome by collecting new pieces of information, which can originate from different sources. We can access this information both *individually* (e.g., by hearing the latest news about players' conditions) or *socially* (e.g., from discussions with friends). The ultimate belief arising from this process can be represented by a probability vector μ, such as

$$\mu = [\mu(\text{victory}), \mu(\text{draw}), \mu(\text{loss})] = [0.6, 0.3, 0.1]. \tag{2.1}$$

Establishing how the belief is formed is a problem of paramount importance, with applications in several disciplines. From a *design-oriented* perspective, belief formation is a critical tool to solve a number of inference and learning tasks. For example, a classification problem can be solved by choosing the most credited hypothesis. From a *behavioral* perspective, there is enormous interest across various communities in establishing formal rules that govern the mechanism of belief formation within many cognitive systems, such as biological systems, the brain, self-organizing systems, and social networks. For instance, interesting studies [74] have shown that inference and learning processes in the brain evolve in a "Bayesian" way, according to a *free-energy minimization principle* — see Section 2.3.

Remarkably, the design-oriented and behavioral perspectives enjoy fruitful cross-fertilization: Methods established for design inspire behavioral models and, conversely, understanding of human cognition improves design by creating or perfecting tools in several fields, such as signal processing, data analysis, machine learning, or artificial intelligence.

The mechanism of opinion formation becomes even more interesting in *distributed* systems, where it becomes important to understand how spatially separated agents should blend their own private information and the beliefs of their neighbors to construct opinions. For various reasons, the belief vector emerges as a key player in the theory of social learning.

From a more technical perspective, it is worth noting that the opinion formation process lies somewhere in between estimation and classification problems [90, 155]. It is not simply a classification problem, since in classification we are often mainly interested in the final decision. For example, a belief vector equal to $[0.55, 0.45]$ would yield the same decision as $[0.99, 0.01]$, but the meaning and reliability of the two decisions are different. The "analog" value of the belief, namely, the mass assigned to each hypothesis is important and helps explain decisions. Likewise, opinion formation is not simply an estimation problem, since, as we will see, the peculiarities of the belief (e.g., it defines a probability distribution over a discrete set) require specific mathematical tools and lead to results that are different from those traditionally employed in estimation theory.

2.1 The Bayesian Way

In order to introduce the concept of belief it is convenient to start with the single-agent setting. Let Θ be a set of cardinality H, which collects the hypotheses the agent is interested in. The particular elements contained in Θ depend on the application. Without loss of generality, we take $\Theta = \{1, 2, \ldots, H\}$ unless otherwise specified. *Before* starting the learning process, the agent has some convictions as regards each hypothesis $\theta \in \Theta$, which are summarized in the *prior* belief $\pi(\theta)$:

$$\pi(\theta) \geq 0, \qquad \sum_{\theta \in \Theta} \pi(\theta) = 1. \tag{2.2}$$

Let $x \in \mathcal{X}$ represent the data available for learning. The term "learning" means that the agent aims at *updating* its prior belief based on the observation x, thus building the *posterior* belief $\mu(\theta|x)$:

$$\mu(\theta|x) \geq 0, \qquad \sum_{\theta\in\Theta} \mu(\theta|x) = 1. \tag{2.3}$$

In order to build the posterior belief, the agent relies on generative models linking the data to the hypotheses, and encoded in the likelihood models[1] $\ell(x|\theta)$. As a function of x for a given θ, $\ell(x|\theta)$ is a probability function. For example, the observations can be modeled as continuous random vectors in the space $\mathcal{X} = \mathbb{R}^d$, with $\ell(x|\theta)$ being a probability density function; or by random variables in a discrete space $\mathcal{X} = \{a, b, c\}$, with $\ell(x|\theta)$ being a probability mass function. The nature of the likelihood models is assumed to be the same for all hypotheses, i.e., the functions $\ell(x|\theta)$ correspond either to pdfs or pmfs for all $\theta \in \Theta$. They are accordingly subject to the following normalization conditions:

$$\begin{cases} \displaystyle\int_{\mathcal{X}} \ell(x|\theta)dx = 1 & \forall\theta \in \Theta \qquad \text{(for pdf)}, \\[2ex] \displaystyle\sum_{x\in\mathcal{X}} \ell(x|\theta) = 1 & \forall\theta \in \Theta \qquad \text{(for pmf)}. \end{cases} \tag{2.4}$$

In some applications, it is useful to deal with mixed-type data, i.e., data that are not represented only by continuous or discrete random variables. For example, we might represent x as a vector with different entries having different characteristics. Some entries in x can be described by continuous variables, e.g., the price of a commercial product, while other entries are better described by categorical attributes, e.g., the product brand. To avoid added complexity in the presentation, in our treatment we will mostly focus on the case where $\ell(x|\theta)$ is a pdf or a pmf. However, we remark that the results presented in this text apply to more heterogeneous cases, provided that $\ell(x|\theta)$ can be meaningfully defined in terms of the so-called Radon-Nikodym derivative [21].

2.1.1 From Priors and Likelihoods to Beliefs

The product $\pi(\theta)\ell(x|\theta)$ identifies a joint probability distribution for the hypothesis/data pair (θ, x). Under this distribution, the posterior belief can be computed as the conditional probability that θ is true given x,

[1] In this text we call "likelihood model" or simply "likelihood" the pdf/pmf $\ell(x|\theta)$ for a given $\theta \in \Theta$. We remark that, in statistics, it is more frequent to use the terms "likelihood function" and "likelihood" when $\ell(x|\theta)$ is regarded as a function of θ for a given x [110, 155].

through *Bayes' rule*:[2]

$$\mu(\theta|x) = \frac{\pi(\theta)\ell(x|\theta)}{m(x)}, \qquad m(x) = \sum_{\theta\in\Theta} \pi(\theta)\ell(x|\theta), \tag{2.5}$$

where $m(x)$ is the marginal pdf or pmf of x, a.k.a. *evidence* in Bayesian theory [20]. As is typical in the Bayesian framework, expressions like (2.5) are often conveniently abbreviated as

$$\mu(\theta|x) \propto \pi(\theta)\ell(x|\theta). \tag{2.6}$$

The proportionality sign $\propto$ is used because we regard the belief $\mu(\theta|x)$ as a function of θ, while the normalization term $m(x)$ that makes $\mu(\theta|x)$ a probability vector depends only on x, and, hence, is a proportionality constant that is independent of θ.

Bayes' rule is reassuring under several viewpoints. First of all, if the postulated joint model $\pi(\theta)\ell(x|\theta)$ is true, Bayes' rule computes exactly the conditional probability of hypothesis θ given data x, which identifies naturally the best candidate to quantify the agent's credibility on the different hypotheses, and is therefore the building block to solve many inference problems. For example, if we want to maximize the probability of guessing the hypothesis correctly, we should seek the value of θ that yields the *maximum a posteriori probability* (MAP), namely,

$$\widehat{\theta}_{\mathsf{MAP}} = \arg\max_{\theta\in\Theta} \mu(\theta|x). \tag{2.7}$$

Notably, Bayes' rule works well even under *mismatched* models, i.e., when the observed data are not obeying the postulated likelihood models employed to perform the belief update. As we will see in Lemma 2.3, under this more challenging setting, Bayes' rule is able to provide the best fit to the true underlying data model.

Example 2.1 (Bayes' rule with Bernoulli likelihoods). Consider a data sample $x \in \{0,1\}$ and a Bernoulli likelihood model

$$\ell(x|\theta) = q_\theta\, \mathbb{I}[x=0] + (1-q_\theta)\, \mathbb{I}[x=1], \tag{2.8}$$

for certain probabilities q_θ, parametrized by $\theta \in \Theta$. We recall from Table 1.1 that $\mathbb{I}$ is the indicator function, which assumes the value 1 when the condition identified by its

[2] $\mu(\theta|x)$ is defined only when $m(x) \neq 0$. When the data are distributed according to $m(x)$, this is immaterial since the set $\{x : m(x) = 0\}$ has zero probability under $m(x)$ [7]. More generally, the set $\{x : m(x) = 0\}$ has zero probability when the support of the true data distribution is contained in the support of the distribution identified by $m(x)$ — see Definition E.1. We will find instances of the latter case in our analysis.

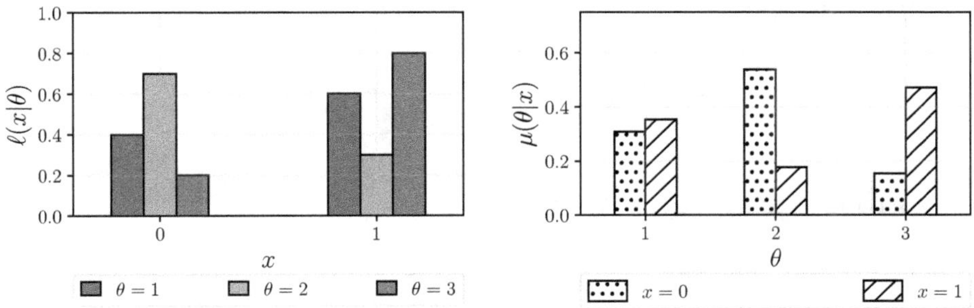

Figure 2.1: (*Left*) Bernoulli likelihood models in Example 2.1, with hypotheses $\theta = 1, 2, 3$ displayed with different colors. (*Right*) Posterior beliefs given the observation of $x = 0$ (in dotted hatching), or $x = 1$ (in parallel hatching).

argument is true and the value 0 otherwise. Starting from the prior belief $\pi(\theta)$, we are interested in evaluating the posterior belief $\mu(\theta|x)$ following Bayes' rule seen in (2.6), which yields

$$\mu(\theta|x) \propto \pi(\theta)\ell(x|\theta) = \pi(\theta)\Big(q_\theta\, \mathbb{I}[x=0] + (1-q_\theta)\, \mathbb{I}[x=1]\Big). \tag{2.9}$$

Accounting for the normalization term, we obtain

$$\mu(\theta|x) = \frac{\pi(\theta)q_\theta}{\sum\limits_{\theta' \in \Theta} \pi(\theta')q_{\theta'}}\, \mathbb{I}[x=0] + \frac{\pi(\theta)(1-q_\theta)}{\sum\limits_{\theta' \in \Theta} \pi(\theta')(1-q_{\theta'})}\, \mathbb{I}[x=1]. \tag{2.10}$$

Note that a likelihood can sometimes be uninformative about the hypotheses, which in this example happens when we have a uniform Bernoulli likelihood. In fact, if $q_\theta = 1/2$ for a certain hypothesis θ, from (2.8) we have $\ell(x|\theta) = 1/2$ for all $x \in \{0, 1\}$, which means that the data sample x bears no information about θ. Accordingly, for θ such that $q_\theta = 1/2$, Eq. (2.10) reveals that the Bayesian update remains equal to the prior belief, namely we get $\mu(\theta|x) = \pi(\theta)$, corroborating the absence of information in the likelihood.

Let us now consider a numerical example, with a set of three hypotheses $\Theta = \{1, 2, 3\}$, a flat prior belief, i.e., $\pi(\theta) = 1/3$ for $\theta = 1, 2, 3$, and the following likelihood (see the left panel of Figure 2.1):

$$\begin{aligned} \ell(x|1) &= 0.4\, \mathbb{I}[x=0] + 0.6\, \mathbb{I}[x=1], \\ \ell(x|2) &= 0.7\, \mathbb{I}[x=0] + 0.3\, \mathbb{I}[x=1], \\ \ell(x|3) &= 0.2\, \mathbb{I}[x=0] + 0.8\, \mathbb{I}[x=1]. \end{aligned} \tag{2.11}$$

We evaluate the posterior belief $\mu(\theta|x)$ using (2.10):

$$\begin{aligned} \mu(1|x) &= \frac{0.4}{0.4+0.7+0.2}\, \mathbb{I}[x=0] + \frac{0.6}{0.6+0.3+0.8}\, \mathbb{I}[x=1], \\ \mu(2|x) &= \frac{0.7}{0.4+0.7+0.2}\, \mathbb{I}[x=0] + \frac{0.3}{0.6+0.3+0.8}\, \mathbb{I}[x=1], \\ \mu(3|x) &= \frac{0.2}{0.4+0.7+0.2}\, \mathbb{I}[x=0] + \frac{0.8}{0.6+0.3+0.8}\, \mathbb{I}[x=1]. \end{aligned} \tag{2.12}$$

In the right panel of Figure 2.1, we see the posterior belief $\mu(\theta|x)$ given the observation of $x = 0$ or $x = 1$. Note that the case $x = 0$ results in a belief vector whose largest entry is $\theta = 2$, that is, $x = 0$ is perceived as a relatively strong evidence supporting hypothesis $\theta = 2$. On the other hand, the case $x = 1$ reinforces hypothesis $\theta = 3$.

2.2 Properties of Bayes' Rule

In many inference and learning problems, it is necessary to deal with *streams* of data:

$$x_1, x_2, \ldots, x_t, \tag{2.13}$$

with each data sample belonging to some space $\mathcal{X}$. The "time" index t need not correspond to a physical time instant. Depending on the application, it might denote a time instant, but also the amount of observed data, or the number of iterations of an algorithm. Applying (2.6) to the product space $\mathcal{X}^t$, we can write

$$\mu(\theta|x_1, x_2, \ldots, x_t) \propto \pi(\theta)\, \ell(x_1, x_2, \ldots, x_t|\theta). \tag{2.14}$$

It is convenient to introduce a more compact notation to work with streaming data. The belief about hypothesis θ at time t will be denoted by

$$\mu_t(\theta) \triangleq \mu(\theta|x_1, x_2, \ldots, x_t), \tag{2.15}$$

yielding the belief vector

$$\mu_t \triangleq [\mu_t(1), \mu_t(2), \ldots, \mu_t(H)]. \tag{2.16}$$

We adopt the convention that index $t = 0$ corresponds to the prior belief vector, namely, the initial belief vector is $\mu_0 = \pi$. Note that in (2.15) and (2.16), the subscript t denotes dependence on a stream with t samples, while the explicit dependence on the data $\{x_1, x_2, \ldots, x_t\}$ has been suppressed. When the belief is evaluated on *random* data, we will write more explicitly

$$\boldsymbol{\mu}_t(\theta) = \mu(\theta|\boldsymbol{x}_1, \boldsymbol{x}_2, \ldots, \boldsymbol{x}_t), \tag{2.17}$$

where the belief vector is written in bold to reflect the randomness of the data. In order to facilitate the reading, we make an important remark as regards notation. Throughout the treatment, we will use bold font when the stochastic nature of the pertinent variables is important (e.g., when we deal with stochastic convergence). Otherwise, we will stick to normal font whenever randomness is not relevant to illustrate the results.

If the *joint* likelihood needed in (2.14) is built assuming that the streaming data are independent and identically distributed (iid) conditioned on θ, then it can be written in product form as

$$\ell(x_1, x_2, \ldots, x_t|\theta) = \prod_{\tau=1}^{t} \ell(x_\tau|\theta). \tag{2.18}$$

One fundamental property of the Bayesian update (2.14) is that, under (2.18), it can be implemented *sequentially*,[3] meaning that the overall rule that updates the prior belief μ_0 to a posterior belief based on the entire stream of data $x_1, x_2, \ldots, x_t$, can be equivalently obtained in an online manner as follows. First, perform a Bayesian update of the prior belief μ_0 by using data x_1, yielding the posterior belief μ_1. Then, take μ_1 as the prior and perform a Bayesian update through x_2 obtaining the posterior μ_2, and so on. The belief μ_t obtained through this sequential procedure is equivalent to the Bayesian posterior with prior μ_0 and data $x_1, x_2, \ldots, x_t$.

Lemma 2.1 (Sequential Bayesian updates). Let $\mu_0(\theta)$ be the prior belief and consider, for $t \in \mathbb{N}$, a joint likelihood in the form

$$\ell(x_1, x_2, \ldots, x_t|\theta) = \prod_{\tau=1}^{t} \ell(x_\tau|\theta). \tag{2.19}$$

Assume that

$$\sum_{\theta\in\Theta} \mu_0(\theta) \prod_{\tau=1}^{t} \ell(x_\tau|\theta) > 0. \tag{2.20}$$

Then,

$$\mu_t(\theta) \propto \mu_{t-1}(\theta)\, \ell(x_t|\theta), \tag{2.21}$$

where $\mu_t(\theta)$ and $\mu_{t-1}(\theta)$ are the Bayesian updates until time instants t and $t-1$, respectively.

Proof. Using (2.14), we can write (observe that in the following applications of Bayes' rule, the denominator hidden by the proportionality sign is always nonzero in view of (2.20))

$$\mu_t(\theta) = \mu(\theta|x_1, x_2, \ldots, x_t) \propto \mu_0(\theta)\ell(x_1, x_2, \ldots, x_t|\theta), \tag{2.22}$$

$$\mu_{t-1}(\theta) = \mu(\theta|x_1, x_2, \ldots, x_{t-1}) \propto \mu_0(\theta)\ell(x_1, x_2, \ldots, x_{t-1}|\theta). \tag{2.23}$$

[3] The sequential nature of the Bayesian update is preserved if we relax the condition of identical distribution over time, namely, even if at time τ we have $\ell^{(\tau)}(x_\tau|\theta)$.

Substituting (2.19) into (2.22) yields

$$\begin{aligned}\mu_t(\theta) &\propto \pi(\theta)\prod_{\tau=1}^{t}\ell(x_\tau|\theta)\\ &= \mu_0(\theta)\left(\prod_{\tau=1}^{t-1}\ell(x_\tau|\theta)\right)\ell(x_t|\theta)\\ &= \mu_0(\theta)\ell(x_1,x_2,\ldots,x_{t-1}|\theta)\ell(x_t|\theta)\\ &\propto \mu_{t-1}(\theta)\ell(x_t|\theta),\end{aligned} \tag{2.24}$$

where in the last step we used (2.23), thus concluding the proof.

■

Lemma 2.1 ensures that, given an additional piece of information x_t, we can update the knowledge summarized in the previous-step belief $\mu_{t-1}(\theta)$ by taking into account the knowledge contained in the likelihood $\ell(x_t|\theta)$ *corresponding to time* t. Notably, this update gives the same result that we would have obtained by updating the initial belief $\mu_0(\theta)$ through the overall likelihood $\prod_{\tau=1}^{t}\ell(x_\tau|\theta)$.

The sequential nature of the Bayesian update (2.21) is a compelling property, both from a theoretical and practical perspective. From the practical standpoint, it is critical for *online* applications where processing the entire bulk of data in a single shot is unfeasible, or when it is necessary to incorporate streaming pieces of information as soon as they arrive. From the theoretical standpoint, the sequential construction confirms that the Bayesian update is logically coherent. In fact, we see that at a certain epoch $t-1$ all knowledge relative to previous epochs is summarized in the belief vector μ_{t-1}. According to Lemma 2.1, this summary is all we need to incorporate future data.

Example 2.2 (Bayes' rule with streaming data and Gaussian likelihoods). Consider the joint likelihood in (2.18) evaluated when the *single-sample* likelihood is Gaussian:

$$\ell(x|\theta) = \frac{1}{\sqrt{2\pi\sigma^2}}\exp\left\{-\frac{(x-\nu_\theta)^2}{2\sigma^2}\right\}, \tag{2.25}$$

with θ−dependent means ν_θ, and variance σ^2 common to all hypotheses. Starting from the initial belief vector μ_0, we are interested in evaluating the posterior belief vector μ_t corresponding to the stream of t data samples.

From (2.14) and (2.25) we can write

$$\mu_t(\theta) \propto \mu_0(\theta)\,\ell(x_1,x_2,\ldots,x_t|\theta) \propto \mu_0(\theta)\prod_{\tau=1}^{t}\exp\left\{-\frac{(x_\tau-\nu_\theta)^2}{2\sigma^2}\right\}. \tag{2.26}$$

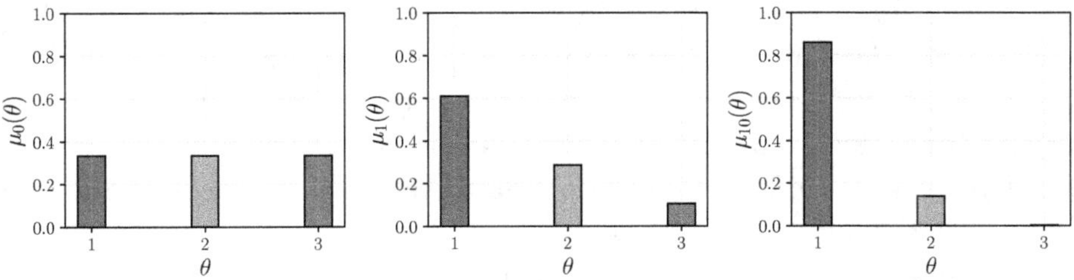

Figure 2.2: Belief evolution over a Gaussian data stream under the setting described in Example 2.2. The data are generated according to the likelihood model corresponding to $\theta = 1$. (*Left*) Prior belief. (*Center*) Belief given the first sample ($t = 1$). (*Right*) Belief given the first 10 samples ($t = 10$).

Accounting for the normalization term, we obtain

$$\mu_t(\theta) = \frac{\mu_0(\theta) \exp\left\{-\sum_{\tau=1}^{t} \frac{(x_\tau - \nu_\theta)^2}{2\sigma^2}\right\}}{\sum_{\theta' \in \Theta} \mu_0(\theta') \exp\left\{-\sum_{\tau=1}^{t} \frac{(x_\tau - \nu_{\theta'})^2}{2\sigma^2}\right\}}. \tag{2.27}$$

It is also useful to notice that, by splitting the product appearing in (2.26), we can write

$$\mu_t(\theta) \propto \underbrace{\mu_0(\theta) \prod_{\tau=1}^{t-1} \exp\left\{-\frac{(x_\tau - \nu_\theta)^2}{2\sigma^2}\right\}}_{\propto \mu_{t-1}(\theta)} \underbrace{\exp\left\{-\frac{(x_t - \nu_\theta)^2}{2\sigma^2}\right\}}_{\propto \ell(x_t|\theta)}, \tag{2.28}$$

which is in agreement with Lemma 2.1. Exploiting (2.28) and accounting for the normalization term, we obtain the following formula, which is suited to a *sequential* evaluation of the belief under Gaussian likelihoods:

$$\mu_t(\theta) = \frac{\mu_{t-1}(\theta) \exp\left\{-\frac{(x_t - \nu_\theta)^2}{2\sigma^2}\right\}}{\sum_{\theta' \in \Theta} \mu_{t-1}(\theta') \exp\left\{-\frac{(x_t - \nu_{\theta'})^2}{2\sigma^2}\right\}}. \tag{2.29}$$

Let us now illustrate the numerical example considered in Figure 2.2. We generate a random data stream

$$\boldsymbol{x}_1, \boldsymbol{x}_2, \ldots, \boldsymbol{x}_t, \tag{2.30}$$

made of independent samples drawn from the Gaussian model $\ell(x|1)$. Then, we evaluate the sequence of beliefs $\boldsymbol{\mu}_t(\theta)$ starting from flat prior beliefs. In Figure 2.2 we display the prior beliefs and the beliefs $\boldsymbol{\mu}_t(\theta)$ corresponding to 1 and 10 samples. We see how, starting from an initial state of ignorance (flat prior), the increase of information from $t = 1$ to $t = 10$ leads the belief vector to place most of the mass on $\theta = 1$, namely, on the model from which the stream is actually generated.

Another useful property of Bayes' rule is *consistency*, i.e., the ability to guess the right model as the number of collected samples increases [71, 72]. Specifically, assuming that an infinite stream of iid data $\boldsymbol{x}_1, \boldsymbol{x}_2, \ldots$ arising from the *same* model $\ell(x|\vartheta^o)$ is observed, then the sequence of belief vectors $\boldsymbol{\mu}_t$ converges to a probability vector that places all its mass on the correct hypothesis $\vartheta^o \in \Theta$.

Before stating the result in a formal way, we need to introduce the Kullback-Leibler (KL) divergence between two pdfs or two pmfs $f(x)$ and $g(x)$ [52] — see Definition B.4:

$$D(f||g) = \mathbb{E}_f \log \frac{f(\boldsymbol{x})}{g(\boldsymbol{x})}, \tag{2.31}$$

where we recall that the symbol $\mathbb{E}_f$ means that $\boldsymbol{x}$ is distributed according to $f(x)$. As explained in Section 1.4, we drop the argument x in $f(x)$ and $g(x)$ and write simply f and g to globally denote the pertinent pdf or pmf. Similarly, we will write ℓ_θ to denote the pdf or pmf $\ell(x|\theta)$ (regarded as a function of x for a fixed θ), where we add the subscript θ to emphasize the dependence on θ.

Lemma 2.2 (Consistency of Bayes' rule under correct models). Let $\{\ell_\theta\}$ be likelihood models fulfilling the following conditions:

$$0 < D(\ell_\theta||\ell_{\theta'}) < \infty \qquad \forall \theta, \theta' \in \Theta, \quad \theta \neq \theta', \tag{2.32}$$

namely, the pdfs or pmfs ℓ_θ corresponding to different hypotheses are all distinct and with finite KL divergences. Consider an infinite stream of iid data samples $\boldsymbol{x}_1, \boldsymbol{x}_2, \ldots$, each one distributed according to ℓ_{ϑ^o}, with $\vartheta^o \in \Theta$, and let $\boldsymbol{\mu}_t$ be the belief vector obtained through Bayes' rule (2.21), based on models $\{\ell_\theta\}$ and on a prior μ_0 placing nonzero mass on all $\theta \in \Theta$. Then, for any choice of $\vartheta^o \in \Theta$,

$$\boldsymbol{\mu}_t(\vartheta^o) \xrightarrow[t\to\infty]{\text{a.s.}} 1. \tag{2.33}$$

Proof. We observe preliminarily that $\ell(\boldsymbol{x}_t|\theta) > 0$ almost surely. This condition holds for $\theta = \vartheta^o$ since the true model of the data samples is $\ell(x|\vartheta^o)$, and it holds for all $\theta \neq \vartheta^o$ since in view of (2.32) we have

$$D(\ell_{\vartheta^o}||\ell_\theta) < \infty. \tag{2.34}$$

Since $\ell(\boldsymbol{x}_t|\theta) > 0$ almost surely, from Lemma 2.1 we can write the belief about any $\theta \in \Theta$ as

$$\boldsymbol{\mu}_t(\theta) = \frac{\boldsymbol{\mu}_{t-1}(\theta)\ell(\boldsymbol{x}_t|\theta)}{\sum_{\theta' \in \Theta} \boldsymbol{\mu}_{t-1}(\theta')\ell(\boldsymbol{x}_t|\theta')}, \tag{2.35}$$

but for an ensemble of realizations with zero probability.

Next, we show that $\boldsymbol{\mu}_t(\theta) > 0$ almost surely, for all t and θ. This property can be established by induction. First, we observe that the property $\boldsymbol{\mu}_t(\theta) > 0$ is true for $t = 0$, since the prior belief vector μ_0 is assumed to have positive entries. Second, we consider the induction step, and show that the property holds for t if it holds for $t-1$. To this end, it suffices to use (2.35), along with the fact that $\ell(\boldsymbol{x}_t|\theta) > 0$ almost surely.

We can now focus on establishing the claim of the theorem. To this end, we work in terms of belief ratios, which are well defined since $\boldsymbol{\mu}_t(\theta) > 0$ almost surely. Let $\theta \neq \vartheta^o$. In view of (2.35), the belief ratio $\boldsymbol{\mu}_t(\vartheta^o)/\boldsymbol{\mu}_t(\theta)$ is given by

$$\frac{\boldsymbol{\mu}_t(\vartheta^o)}{\boldsymbol{\mu}_t(\theta)} = \frac{\boldsymbol{\mu}_{t-1}(\vartheta^o)\ell(\boldsymbol{x}_t|\vartheta^o)}{\boldsymbol{\mu}_{t-1}(\theta)\ell(\boldsymbol{x}_t|\theta)}. \tag{2.36}$$

Taking the logarithm we obtain

$$\log\frac{\boldsymbol{\mu}_t(\vartheta^o)}{\boldsymbol{\mu}_t(\theta)} = \log\frac{\boldsymbol{\mu}_{t-1}(\vartheta^o)}{\boldsymbol{\mu}_{t-1}(\theta)} + \log\frac{\ell(\boldsymbol{x}_t|\vartheta^o)}{\ell(\boldsymbol{x}_t|\theta)}. \tag{2.37}$$

Developing the recursion over time and dividing by t gives

$$\frac{1}{t}\log\frac{\boldsymbol{\mu}_t(\vartheta^o)}{\boldsymbol{\mu}_t(\theta)} = \frac{1}{t}\log\frac{\mu_0(\vartheta^o)}{\mu_0(\theta)} + \frac{1}{t}\sum_{\tau=1}^{t}\log\frac{\ell(\boldsymbol{x}_\tau|\vartheta^o)}{\ell(\boldsymbol{x}_\tau|\theta)}. \tag{2.38}$$

The iid property of $\boldsymbol{x}_1, \boldsymbol{x}_2, \ldots$ and the finiteness condition in (2.32) allow us to use the strong law of large numbers (Theorem D.7) to establish the convergence of the second term on the RHS of (2.38) in the following manner:

$$\frac{1}{t}\sum_{\tau=1}^{t}\log\frac{\ell(\boldsymbol{x}_\tau|\vartheta^o)}{\ell(\boldsymbol{x}_\tau|\theta)} \xrightarrow[t\to\infty]{\text{a.s.}} \mathbb{E}_{\ell_{\vartheta^o}}\log\frac{\ell(\boldsymbol{x}_\tau|\vartheta^o)}{\ell(\boldsymbol{x}_\tau|\theta)} = D(\ell_{\vartheta^o}||\ell_\theta). \tag{2.39}$$

Since the first term on the RHS of (2.38) tends to 0, from (2.38) we conclude that, for all $\theta \neq \vartheta^o$,

$$\frac{1}{t}\log\frac{\boldsymbol{\mu}_t(\vartheta^o)}{\boldsymbol{\mu}_t(\theta)} \xrightarrow[t\to\infty]{\text{a.s.}} D(\ell_{\vartheta^o}||\ell_\theta) > 0, \tag{2.40}$$

where we also used the positivity condition in (2.32). Equation (2.40) implies that

$$\log\frac{\boldsymbol{\mu}_t(\vartheta^o)}{\boldsymbol{\mu}_t(\theta)} \xrightarrow[t\to\infty]{\text{a.s.}} \infty \quad \forall\theta \neq \vartheta^o. \tag{2.41}$$

Since the entries of the belief vector are bounded, Eq. (2.41) is equivalent to

$$\boldsymbol{\mu}_t(\theta) \xrightarrow[t\to\infty]{\text{a.s.}} 0 \quad \forall\theta \neq \vartheta^o. \tag{2.42}$$

Moreover, since the entries of the belief vector add up to 1, Eq. (2.42) is equivalent to stating that $\boldsymbol{\mu}_t(\vartheta^o)$ converges almost surely to 1, thus establishing the claim.

■

Note that if we want the claim in (2.33) to hold for a particular ϑ^o, then it is not necessary to require condition (2.32) to hold for any pair (θ, θ'), but only the following relaxed condition would suffice:

$$0 < D(\ell_{\vartheta^o}||\ell_\theta) < \infty, \qquad \forall\theta \neq \vartheta^o. \tag{2.43}$$

However, since ϑ^o is unknown, we enforce condition (2.32) because we want to ensure that the claim holds *for any* possible choice of ϑ^o.

It is useful to illustrate Lemma 2.2 by means of an example.

Example 2.3 (Consistency under Gaussian likelihoods). Consider a problem with 3 hypotheses, namely $\Theta = \{1, 2, 3\}$, and a family of Gaussian likelihoods, as introduced in Example 2.2, with variance $\sigma^2 = 1$ and means

$$\nu_1 = 0.5, \quad \nu_2 = 1, \quad \nu_3 = 1.5. \tag{2.44}$$

Let $\boldsymbol{x}_1, \boldsymbol{x}_2, \ldots$ be a stream of samples independently drawn from the same model $\ell(x|1)$, that is, the true underlying hypothesis is $\vartheta^o = 1$. In the left panel of Figure 2.3, we see the shape of the Gaussian likelihoods $\ell(x|\theta)$.

From Lemma 2.2 we expect that, as t grows, the belief vector $\boldsymbol{\mu}_t$ places all its mass on hypothesis ϑ^o, as long as condition (2.32) is satisfied. In order to verify this condition, it is useful to evaluate the KL divergence between two Gaussian distributions (with the same variance σ^2):

$$\begin{aligned} D(\ell_\theta || \ell_{\theta'}) &= \mathbb{E}_{\ell_\theta} \log \frac{\ell(\boldsymbol{x}|\theta)}{\ell(\boldsymbol{x}|\theta')} \\ &= \frac{1}{2\sigma^2} \mathbb{E}_{\ell_\theta} \left[(\boldsymbol{x} - \nu_{\theta'})^2 - (\boldsymbol{x} - \nu_\theta)^2 \right] \\ &= \frac{1}{2\sigma^2} \Big[2(\nu_\theta - \nu_{\theta'}) \underbrace{\mathbb{E}_{\ell_\theta} \boldsymbol{x}}_{=\nu_\theta} + \nu_{\theta'}^2 - \nu_\theta^2 \Big] \\ &= \frac{(\nu_\theta - \nu_{\theta'})^2}{2\sigma^2}. \end{aligned} \tag{2.45}$$

Using (2.44) and (2.45), we can see that (2.32) is satisfied.

We illustrate the result of Lemma 2.2 by simulating the evolution of the belief vector $\boldsymbol{\mu}_t$, updated according to the recursion in (2.29), over 100 iterations. The prior belief vector has uniform entries. The resulting behavior is reported in the right panel of Figure 2.3. We see that, as t grows, all the belief mass tends to be concentrated on the true underlying hypothesis, $\vartheta^o = 1$.

In practice, it is seldom the case that the true distribution that generated the observations is exactly equal to one of the postulated likelihood models. What one can hope for is to have some reasonable approximation for the true distribution through one of the likelihood models. This problem was originally addressed in [19], with reference to a more general setting involving also a continuous parameter θ. Specifically, assuming that an infinite stream of iid data $\boldsymbol{x}_1, \boldsymbol{x}_2, \ldots$ arising from a certain pdf or pmf f is observed, if there exists a model $\ell_{\vartheta^\star}$ that is the closest (in terms of KL divergence) to f, then the sequence of belief vectors $\boldsymbol{\mu}_t$ will converge to a probability vector that places all its mass on hypothesis $\vartheta^\star$. This means

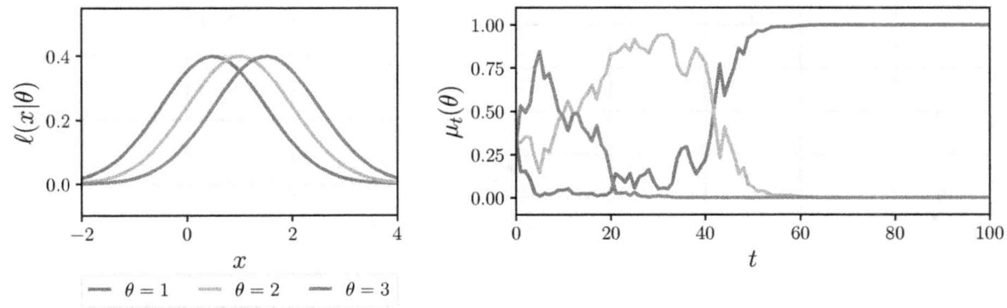

Figure 2.3: (*Left*) Gaussian likelihood models in Example 2.3. (*Right*) Belief evolution over 100 iterations. We see that, as t grows, all the belief mass is concentrated on the true underlying hypothesis, $\vartheta^o = 1$.

that the Bayesian learning approach is able to indicate which model fits best the true underlying distribution.

Lemma 2.3 (Convergence of Bayes' rule under mismatched models). Consider an infinite stream of iid data samples $\boldsymbol{x}_1, \boldsymbol{x}_2, \ldots$, each one distributed according to a probability (density or mass) function f. Let $\{\ell_\theta\}$ be likelihood models of the same nature as f (namely, for all $\theta \in \Theta$, ℓ_θ is a pdf if f is a pdf, and a pmf otherwise) and let $\boldsymbol{\mu}_t$ be the belief vector obtained through Bayes' rule (2.21), based on models $\{\ell_\theta\}$ and on a prior μ_0 placing nonzero mass on all $\theta \in \Theta$. If

$$D(f||\ell_\theta) < \infty \quad \forall \theta \in \Theta \tag{2.46}$$

and if the minimization problem

$$\min_{\theta \in \Theta} D(f||\ell_\theta) \tag{2.47}$$

admits a unique minimizer $\vartheta^\star$, then

$$\boldsymbol{\mu}_t(\vartheta^\star) \xrightarrow[t\to\infty]{\text{a.s.}} 1. \tag{2.48}$$

Proof. We reuse the arguments of Lemma 2.2 until (2.38) with $\vartheta^\star \neq \theta$ in place of ϑ^o to write

$$\frac{1}{t}\log\frac{\boldsymbol{\mu}_t(\vartheta^\star)}{\boldsymbol{\mu}_t(\theta)} = \frac{1}{t}\log\frac{\mu_0(\vartheta^\star)}{\mu_0(\theta)} + \frac{1}{t}\sum_{\tau=1}^{t}\log\frac{\ell(\boldsymbol{x}_\tau|\vartheta^\star)}{\ell(\boldsymbol{x}_\tau|\theta)}. \tag{2.49}$$

Again, the first term on the RHS of (2.49) tends to 0, while the second term can be rewritten as

$$\frac{1}{t}\sum_{\tau=1}^{t}\log\frac{\ell(\boldsymbol{x}_\tau|\vartheta^\star)}{\ell(\boldsymbol{x}_\tau|\theta)} = \frac{1}{t}\sum_{\tau=1}^{t}\log\frac{f(\boldsymbol{x}_\tau)}{\ell(\boldsymbol{x}_\tau|\theta)} - \frac{1}{t}\sum_{\tau=1}^{t}\log\frac{f(\boldsymbol{x}_\tau)}{\ell(\boldsymbol{x}_\tau|\vartheta^\star)}. \tag{2.50}$$

Given the iid property of $\boldsymbol{x}_1, \boldsymbol{x}_2, \ldots$ and the finiteness condition in (2.46), we can once more appeal to the strong law of large numbers (Theorem D.7) to note that

$$\begin{aligned}\frac{1}{t}\sum_{\tau=1}^{t}\log\frac{\ell(\boldsymbol{x}_\tau|\vartheta^\star)}{\ell(\boldsymbol{x}_\tau|\theta)} &\xrightarrow[t\to\infty]{\text{a.s.}} \mathbb{E}_f\log\frac{f(\boldsymbol{x}_\tau)}{\ell(\boldsymbol{x}_\tau|\theta)} - \mathbb{E}_f\log\frac{f(\boldsymbol{x}_\tau)}{\ell(\boldsymbol{x}_\tau|\vartheta^\star)}\\ &= D(f||\ell_\theta) - D(f||\ell_{\vartheta^\star}).\end{aligned}\tag{2.51}$$

Since $\vartheta^\star$ is the unique minimizer of (2.47), it follows that, for all $\theta \neq \vartheta^\star$,

$$D(f||\ell_\theta) - D(f||\ell_{\vartheta^\star}) > 0.\tag{2.52}$$

From (2.52) and (2.51) we conclude that

$$\log\frac{\boldsymbol{\mu}_t(\vartheta^\star)}{\boldsymbol{\mu}_t(\theta)} \xrightarrow[t\to\infty]{\text{a.s.}} \infty \quad \forall\theta\neq\vartheta^\star,\tag{2.53}$$

which, since the beliefs are bounded, implies that

$$\boldsymbol{\mu}_t(\theta) \xrightarrow[t\to\infty]{\text{a.s.}} 0 \quad \forall\theta\neq\vartheta^\star.\tag{2.54}$$

This is equivalent to (2.48) because the entries of the belief vector must add up to 1, and the proof is complete.

■

We remark that Lemma 2.2 can be regarded as a special case of Lemma 2.3 corresponding to $f = \ell_{\vartheta^o}$. In fact, with this particular choice we have

$$\vartheta^\star = \arg\min_{\theta\in\Theta} D(f||\ell_\theta) = \arg\min_{\theta\in\Theta} D(\ell_{\vartheta^o}||\ell_\theta) = \vartheta^o,\tag{2.55}$$

since $D(\ell_{\vartheta^o}||\ell_{\vartheta^o}) = 0$ and, in view of (2.32), the KL divergences corresponding to $\theta \neq \vartheta^o$ are all positive.

Example 2.4 (Convergence under mismatched Gaussian likelihoods). Consider a problem with 3 hypotheses, namely, $\Theta = \{1, 2, 3\}$, and a family of Gaussian likelihoods with unit variance and means

$$\nu_1 = 0.5, \quad \nu_2 = 1, \quad \nu_3 = 1.5.\tag{2.56}$$

Let $\boldsymbol{x}_1, \boldsymbol{x}_2, \ldots$ be a stream of samples independently drawn from a Gaussian distribution that does not belong to the family of models $\{\ell(x|\theta)\}_{\theta\in\Theta}$. We denote by $f(x)$ the true Gaussian pdf of the data samples, with mean $\nu_f = 0.6$ and unit variance. In the left panel of Figure 2.4, we display the true pdf $f(x)$ and the mismatched pdfs $\ell(x|\theta)$ for $\theta = 1, 2, 3$.

From Lemma 2.3 we expect that, as t grows, the belief vector $\boldsymbol{\mu}_t$ places all its mass on hypothesis $\vartheta^\star$, which is the unique minimizer of $D(f||\ell_\theta)$. As seen in Example 2.3, we can compute $D(f||\ell_\theta)$ as

$$D(f||\ell_\theta) = \mathbb{E}_f\log\frac{f(\boldsymbol{x})}{\ell(\boldsymbol{x}|\theta)} = \frac{(\nu_f - \nu_\theta)^2}{2},\tag{2.57}$$

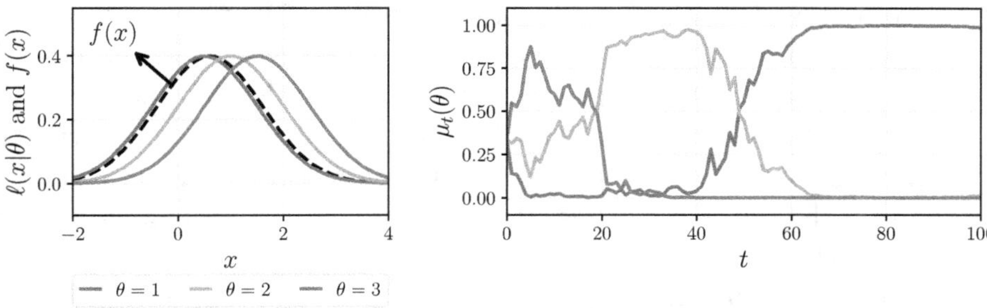

Figure 2.4: (*Left*) Mismatched Gaussian models $\ell(x|\theta)$ (solid line) and true Gaussian model $f(x)$ (dashed line) corresponding to Example 2.4. (*Right*) Belief evolution over 100 iterations. We see that, as t grows, all the belief mass is concentrated on the unique minimizer $\vartheta^\star = 1$, namely, on the hypothesis that provides the best fit to the true model (see the left panel).

resulting in the KL divergence values

$$D(f||\ell_1) = 0.005, \quad D(f||\ell_2) = 0.080, \quad D(f||\ell_3) = 0.405. \tag{2.58}$$

In this example, the minimizer of $D(f||\ell_\theta)$ is given by $\vartheta^\star = 1$. As a matter of fact, in the left panel of Figure 2.4 we see that the likelihood providing the best fit to the true model (dashed line) corresponds to hypothesis 1 (blue solid line). In addition, we simulate the evolution of the belief vector $\boldsymbol{\mu}_t$, updated according to the recursion in (2.29), over 100 iterations. The resulting behavior is displayed in the right panel of Figure 2.4. We see that, as t grows, the belief mass becomes progressively concentrated on the unique minimizer $\vartheta^\star = 1$, corresponding to the hypothesis that provides the best fit to the true model.

2.3 Information-Theoretic Interpretations

An interesting problem in Bayesian theory is to determine the optimal belief update rule relative to a suitable criterion. In other words, the posterior belief is no longer treated as a *fait accompli* forced by the rules of conditional probability, but should arise instead as the solution to a meaningful optimization problem. A relevant class of optimization problems emerging in this context addresses the following question. Given the prior knowledge embodied in $\pi(\theta)$, and the likelihood model $\ell(x|\theta)$, which is the posterior belief $\mu(\theta)$ that provides the best rule to process the available information? In order to find an optimized rule, it is necessary to define a suitable function to measure the cost associated with a particular posterior belief. We will see that typical cost functions involve information-theoretic quantities.

One of the earliest works following this path is [174], where an information conservation principle is formulated to construct the cost function, and the corresponding minimization problem is shown to lead to the Bayesian posterior as the optimal solution. This study stimulated new formulations and led to a debate on the interpretation of Bayes' rule in terms of information-theoretic quantities [105]. Since then, the cost function has been remastered and modified in different guises. A commonly accepted formulation is the *free-energy minimization criterion.* This optimization principle, originally enunciated at the end of the 19th century, can explain several inference and learning techniques, including Bayesian inference, maximum likelihood learning with latent variables, variational approximate Bayesian theory, mirror descent optimization, maximum entropy methods, as well as brain modeling and cognition in self-organizing systems [97, 155]. We illustrate the free-energy minimization principle applied to the learning problem of our interest.

Let Δ_H be the probability simplex of dimension H. Denoting by $p \in \Delta_H$ the (unknown) belief vector we must optimize over, the free-energy function pertinent to our problem is defined as [74, 97, 155]:[4]

$$F(p) = \sum_{\theta \in \Theta} p(\theta) \log \frac{1}{\pi(\theta)\ell(x|\theta)} - H(p), \tag{2.59}$$

where

$$H(p) = \sum_{\theta \in \Theta} p(\theta) \log \frac{1}{p(\theta)} \tag{2.60}$$

is the entropy of the pmf p [52, 158] — see Definition B.1. The first term on the RHS of (2.59) can be interpreted as a cost value for selecting p based on the available information reflected by the joint model $\pi(\theta)\ell(x|\theta)$, while the entropy serves as a measure for the complexity of p. The free-energy function is usually rearranged in the following form:

$$F(p) = D(p||\pi) - \sum_{\theta \in \Theta} p(\theta) \log \ell(x|\theta), \tag{2.61}$$

where

$$D(p||\pi) = \sum_{\theta \in \Theta} p(\theta) \log \frac{p(\theta)}{\pi(\theta)} \tag{2.62}$$

is the KL divergence between pmfs p and π [52] — see Definition B.4.

[4]We should have written $p(\theta|x)$, since we are actually evaluating a *posterior* belief. We omit the explicit dependence on x for notational simplicity.

Using (2.62) and the marginal pdf or pmf $m(x)$ defined in (2.5), Eq. (2.61) can be manipulated as follows:

$$F(p) = \sum_{\theta\in\Theta} p(\theta) \log \frac{p(\theta)}{\pi(\theta)} + \sum_{\theta\in\Theta} p(\theta) \log \frac{1}{\ell(x|\theta)} \tag{2.63a}$$

$$= \sum_{\theta\in\Theta} p(\theta) \log \frac{p(\theta)}{\frac{\pi(\theta)\ell(x|\theta)}{m(x)}} - \log m(x) \tag{2.63b}$$

$$= D(p||\mu^{\mathsf{Bu}}) - \log m(x), \tag{2.63c}$$

where we denote by μ^{Bu} the belief arising from the Bayesian update (2.5), namely,

$$\mu^{\mathsf{Bu}}(\theta|x) = \frac{\pi(\theta)\ell(x|\theta)}{m(x)}, \qquad m(x) = \sum_{\theta\in\Theta} \pi(\theta)\ell(x|\theta). \tag{2.64}$$

Since $D(p||\mu^{\mathsf{Bu}}) \geq 0$, with equality if, and only if, $p = \mu^{\mathsf{Bu}}$, we obtain the following remarkable result:

$$\mu^{\mathsf{Bu}} = \arg\min_{p\in\Delta_H} F(p), \tag{2.65}$$

namely, *free energy is minimized by the Bayesian posterior* μ^{Bu}. It is worth mentioning that, in the context of variational Bayesian inference, the negative free energy is also known as ELBO *(evidence lower bound)*, because in view of (2.63c) we have [155]

$$\log m(x) = D(p||\mu^{\mathsf{Bu}}) - F(p) \geq -F(p), \tag{2.66}$$

showing that the negative free energy is a lower bound on the logarithm of the evidence $m(x)$.

We next describe another useful information-theoretic interpretation of Bayes' rule. We start by constructing a belief vector μ^{lik} that uses only the information contained in the likelihood $\ell(x|\theta)$ and disregards the information contained in the prior π. This construction can be done by scaling the likelihood to transform it into a probability (i.e., belief) vector

$$\mu^{\mathsf{lik}}(\theta|x) \triangleq \frac{\ell(x|\theta)}{\sum_{\theta'\in\Theta} \ell(x|\theta')}. \tag{2.67}$$

We refer to μ^{lik} as the "likelihood" posterior. Observe that (2.67) can be interpreted as a Bayesian update with likelihood $\ell(x|\theta)$ and *uniform prior* ($\pi(\theta) = 1/H$ for all $\theta \in \Theta$), namely,

$$\mu^{\mathsf{lik}}(\theta|x) = \frac{(1/H)\ell(x|\theta)}{m_u(x)}, \qquad m_u(x) = \frac{1}{H} \sum_{\theta\in\Theta} \ell(x|\theta). \tag{2.68}$$

Using (2.68) in (2.61), the free energy can be rewritten as

$$F(p) = D(p||\pi) + \underbrace{\sum_{\theta\in\Theta} p(\theta) \log \frac{1}{\mu^{\mathsf{lik}}(\theta|x)}}_{\text{cross-entropy } H(p,\mu^{\mathsf{lik}})} - \underbrace{\log\left(\sum_{\theta'\in\Theta} \ell(x|\theta')\right)}_{\text{independent of } p}, \tag{2.69}$$

where we see the appearance of the *cross-entropy* between the target belief p and the "likelihood" posterior μ^{lik} — see Definition B.2. Since the last term in (2.69) does not contain the target belief p, minimizing the free energy is tantamount to minimizing the following modified cost function:

$$\widetilde{F}(p) \triangleq D(p||\pi) + H(p, \mu^{\mathsf{lik}}). \tag{2.70}$$

We see that the cost function in (2.70) adds to the KL divergence between p and the prior the cross-entropy between p and the "likelihood" posterior. Moreover, by writing explicitly the KL divergence and the cross-entropy, we see that

$$\begin{aligned}\widetilde{F}(p) &= \sum_{\theta\in\Theta} p(\theta) \log \frac{p(\theta)}{\pi(\theta)} + \sum_{\theta\in\Theta} p(\theta) \log \frac{1}{\mu^{\mathsf{lik}}(\theta)} \\ &= \sum_{\theta\in\Theta} p(\theta) \log \frac{1}{\pi(\theta)} + \sum_{\theta\in\Theta} p(\theta) \log \frac{p(\theta)}{\mu^{\mathsf{lik}}(\theta)},\end{aligned} \tag{2.71}$$

and we conclude that the cost function $\widetilde{F}(p)$ can be equivalently written as

$$\widetilde{F}(p) = H(p, \pi) + D(p||\mu^{\mathsf{lik}}). \tag{2.72}$$

That is, the roles of the KL divergence and the cross-entropy in (2.70) and (2.72) can be interchanged without altering the cost function. In summary, we find that the Bayesian posterior μ^{Bu} minimizes the free energy $F(p)$. Since we showed that minimizing $F(p)$ is equivalent to minimizing $\widetilde{F}(p)$, from (2.70) and (2.72) we conclude that the Bayesian posterior also minimizes the sum of a KL divergence term and a cross-entropy term involving the prior π and the "likelihood" posterior μ^{lik}.

2.4 Stochastic-Optimization Interpretation

In this section we provide a third interpretation and show how Bayes' rule can arise from a *stochastic-optimization* problem solved by means of a stochastic mirror descent (SMD) algorithm [17, 36, 136, 155]. Assume that we observe a stream of iid data $\boldsymbol{x}_1, \boldsymbol{x}_2, \ldots$ drawn from an unknown probability (density or mass) function f. As usual, a prior belief vector π and the likelihood models $\{\ell_\theta\}$ are available. The goal is to learn from the data stream which model $\ell_{\vartheta^\star}$ provides the best fit to the true model f, a question that can be formulated in terms of the following optimization problem:

$$\vartheta^\star = \arg\min_{\theta\in\Theta} D(f||\ell_\theta). \tag{2.73}$$

We work under the same assumptions used in Lemma 2.3 and, in particular, we are assuming in (2.73) that $D(f||\ell_\theta)$ is minimized at a unique value $\vartheta^\star$. Problem (2.73) can be reformulated in terms of belief vectors p, namely, we can solve the following equivalent problem over p belonging to the simplex Δ_H:

$$e_{\vartheta^\star} = \arg\min_{p\in\Delta_H} \sum_{\theta\in\Theta} p(\theta) D(f||\ell_\theta). \tag{2.74}$$

In (2.74), we are denoting by $e_{\vartheta^\star} \in \mathbb{R}^H$ the basis vector that has all zero entries, except for the $\vartheta^\star$th entry that is equal to 1. In order to justify why (2.74) is equivalent to (2.73), we observe that, assuming a unique minimizer $\vartheta^\star$, we can write

$$\begin{aligned}\sum_{\theta\in\Theta} p(\theta)D(f||\ell_\theta) - D(f||\ell_{\vartheta^\star}) &= \sum_{\theta\in\Theta} p(\theta)\Big(D(f||\ell_\theta) - D(f||\ell_{\vartheta^\star})\Big)\\ &= \sum_{\theta\neq\vartheta^\star} p(\theta) \underbrace{\Big(D(f||\ell_\theta) - D(f||\ell_{\vartheta^\star})\Big)}_{>0} \geq 0.\end{aligned} \tag{2.75}$$

Accordingly, the LHS of (2.75) is minimized if we set $p(\theta) = 0$ for all $\theta \neq \vartheta^\star$. Thus, the minimum of the cost function in (2.74) is $D(f||\ell_{\vartheta^\star})$, and the unique minimizer is a probability vector placing unit mass on $\vartheta^\star$, namely, the vector $e_{\vartheta^\star}$. Expanding the cost function in (2.74) we can write

$$\sum_{\theta\in\Theta} p(\theta)D(f||\ell_\theta) = \mathbb{E}_f\left[\sum_{\theta\in\Theta} p(\theta)\log\frac{1}{\ell(\boldsymbol{x}|\theta)} + \log f(\boldsymbol{x})\right], \tag{2.76}$$

which, since the term $\log f(\boldsymbol{x})$ does not depend on p, can be replaced by the cost function

$$\mathbb{E}_f\left[\sum_{\theta\in\Theta} p(\theta)\log\frac{1}{\ell(\boldsymbol{x}|\theta)}\right]. \tag{2.77}$$

Actually, the function $f(x)$ relative to which the expectation is evaluated is unknown. Therefore, the cost function in (2.77) cannot be computed. In the theory of optimization, one way to circumvent this type of difficulty is to implement a *stochastic* gradient descent (SGD) algorithm, where the gradient of (2.77) is replaced by a stochastic instantaneous approximation, yielding, for $t = 1, 2, \ldots$ (with p_0 being the initial or prior belief),

$$\begin{aligned} p_t &= p_{t-1} - \gamma \nabla Q_t(p_{t-1}) \\ &= \arg\min_{p \in \mathbb{R}^H} \|p_{t-1} - \gamma \nabla Q_t(p_{t-1}) - p\|^2, \end{aligned} \tag{2.78}$$

where $\gamma > 0$ is the step-size or learning rate parameter, and where we introduced the instantaneous loss function

$$Q_t(p) \triangleq \sum_{\theta \in \Theta} p(\theta) \log \frac{1}{\ell(x_t|\theta)}. \tag{2.79}$$

The first equality in (2.78) is the standard SGD formulation, whereas the second equality is a straightforward identity that will be useful soon [155]. Unfortunately, algorithm (2.78) does not account for the fact that p must belong to the probability simplex (i.e., its entries must be nonnegative and add up to 1). In order to incorporate this constraint, we can resort to a *projected* stochastic gradient algorithm, by restricting to Δ_H the search space appearing in the second formulation of (2.78):

$$\begin{aligned} p_t &= \arg\min_{p \in \Delta_H} \|p_{t-1} - \gamma \nabla Q_t(p_{t-1}) - p\|^2 \\ &= \arg\min_{p \in \Delta_H} \left\{ \left(\nabla Q_t(p_{t-1}) \right)^{\mathsf{T}} p + \frac{1}{2\gamma} \|p - p_{t-1}\|^2 \right\}, \end{aligned} \tag{2.80}$$

where the last equality follows by expanding the squared norm and ignoring terms that are independent of p. Exploiting the form of the loss function $Q_t(p)$ in (2.79), we observe that

$$\frac{\partial Q_t(p)}{\partial p(\theta)} = \log \frac{1}{\ell(x_t|\theta)}, \tag{2.81}$$

which implies

$$\left(\nabla Q_t(p_{t-1}) \right)^{\mathsf{T}} p = \sum_{\theta \in \Theta} p(\theta) \log \frac{1}{\ell(x_t|\theta)} = Q_t(p). \tag{2.82}$$

Accordingly, Eq. (2.80) can be rewritten as

$$p_t = \arg\min_{p \in \Delta_H} \left\{ Q_t(p) + \frac{1}{2\gamma} \|p - p_{t-1}\|^2 \right\}. \tag{2.83}$$

The update in (2.83) can be interpreted as minimizing $Q_t(p)$ while keeping under control the Euclidean distance of p from the previous iterate p_{t-1}. The *mirror descent* method replaces the Euclidean distance term appearing in (2.83) with a more general similarity measure [17, 36, 136, 155]. Specifically, this measure is chosen from the family of *Bregman divergences* [34]. A Bregman divergence $B_g(p, p')$ is constructed as follows [36, 155]:

$$B_g(p, p') \triangleq g(p) - g(p') + \left(\nabla g(p')\right)^{\mathsf{T}} (p' - p), \qquad p, p' \in \Delta_H, \tag{2.84}$$

where $g : \Delta_H \mapsto \mathbb{R}$ is a continuously differentiable and strictly convex function (see Definition A.2),[5] a.k.a. *mirror function* in the context of mirror descent methods. Replacing the term $(1/2)\|p - p_{t-1}\|^2$ in (2.83) with a Bregman divergence results in the stochastic mirror descent algorithm [17, 36, 136, 155]

$$p_t = \arg\min_{p\in\Delta_H} \left\{ Q_t(p) + \frac{1}{\gamma} B_g(p, p_{t-1}) \right\}. \tag{2.85}$$

One choice of the Bregman divergence that fits our problem where we need to compare probability distributions is the KL divergence, which is obtained when the mirror function is chosen as the *negative entropy*

$$g(p) = \sum_{\theta\in\Theta} p(\theta) \log p(\theta) = -H(p). \tag{2.86}$$

With this choice, Eq. (2.85) becomes

$$\begin{aligned} p_t &= \arg\min_{p\in\Delta_H} \left\{ Q_t(p) + \frac{1}{\gamma} D(p\|p_{t-1}) \right\} \\ &= \arg\min_{p\in\Delta_H} \left\{ \sum_{\theta\in\Theta} p(\theta) \log \frac{1}{\ell(x_t|\theta)} + \frac{1}{\gamma} D(p\|p_{t-1}) \right\}, \end{aligned} \tag{2.87}$$

where in the last equality we used (2.79). Notably, for the case $\gamma = 1$, the representation in (2.87) coincides with the minimization of the free energy in the form (2.61), with π replaced by p_{t-1}. This is a remarkable conclusion, since it implies that, with $\gamma = 1$, the individual iterate of the SMD algorithm is nothing but a Bayesian update rule!

To solve (2.87) for general values of γ, we multiply the quantity within brackets by γ and write the KL divergence explicitly, obtaining

$$p_t = \arg\min_{p\in\Delta_H} \left\{ \sum_{\theta\in\Theta} p(\theta) \log \frac{p(\theta)}{p_{t-1}(\theta)\ell^{\gamma}(x_t|\theta)} \right\}. \tag{2.88}$$

[5] In general, to define a Bregman divergence, the domain of g must be a closed convex set (see Definition A.1), not necessarily a probability simplex.

Note that the function $p_{t-1}(\theta)\ell^{\gamma}(x_t|\theta)$ can be turned into a pmf by normalization, namely,

$$p'(\theta) = \frac{p_{t-1}(\theta)\ell^{\gamma}(x_t|\theta)}{\sum\limits_{\theta'\in\Theta} p_{t-1}(\theta')\ell^{\gamma}(x_t|\theta')}. \tag{2.89}$$

Since the normalization term does not depend on p, the minimization problem in (2.88) can be turned into minimization of the KL divergence between p and p', yielding the solution

$$p_t(\theta) \propto p_{t-1}(\theta)\ell^{\gamma}(x_t|\theta). \tag{2.90}$$

Observe now that the goal of the considered stochastic-optimization framework is to approximate, for a sufficiently large number of iterations, the solution $e_{\vartheta^\star}$ to problem (2.74). It is therefore useful to examine the asymptotic behavior, over an infinite stream of random data $\boldsymbol{x}_1, \boldsymbol{x}_2, \ldots$, of the belief generated by (2.90). By inspecting the proof of Lemma 2.3, it is easily seen that the parameter γ does not affect the conclusion of the lemma. Accordingly, if the prior belief p_0 places nonzero mass on all $\theta \in \Theta$, we have that

$$\boldsymbol{p}_t(\vartheta^\star) \xrightarrow[t\to\infty]{\text{a.s.}} 1, \tag{2.91}$$

where the bold notation $\boldsymbol{p}_t$ is now necessary since, as done in Lemma 2.3, we are focusing on the limiting, almost-sure behavior of the belief when evaluated over an infinite stream of *random* data. Since $\boldsymbol{p}_t$ is a probability vector, we conclude from (2.91) that the sequence of SMD iterates converges almost surely to $e_{\vartheta^\star}$, a belief vector that places all its mass on hypothesis $\vartheta^\star$ in (2.73).

In principle, the fact that the algorithm converges does not reveal anything special as regards the *instantaneous* beliefs p_t, which carry information about *how* the agent is progressively learning to reach the final conclusion. These running beliefs are critical for the learning process, since ideally we would like to guarantee that the agent is able to make the best possible choice at any time instant, in a manner that is compatible with the data observed up to that time. The remarkable conclusion stemming from the above analysis is that the stochastic mirror descent algorithm with the similarity measure equal to the KL divergence and the step-size γ equal to 1, actually provides the *best instantaneous belief, corresponding to the Bayesian update.* This conclusion is not obvious, since the rationale behind the stochastic-optimization approach is to solve (2.73) or (2.74) using a sufficiently large number of iterations, with *no guarantees of optimized solutions for the individual iteration t.*

Before concluding this section, there are two useful observations regarding the choice of the step-size γ in (2.85). First, in the general theory of SMD (and SGD), it is common to select a step-size that vanishes (with suitable decay rate) as $t \to \infty$ to guarantee convergence to the true solution. In our particular case, *constant* step-sizes are sufficient to guarantee convergence.

Second, we see that in the modified posterior (2.90), the step-size γ can be used to tune the relative importance of the likelihood. The modified posterior arose from the cost function in (2.87) which, for $\gamma \neq 1$, is a modification of the free energy where the KL divergence term is weighted by $1/\gamma$. This is only one possibility for deriving posterior beliefs based on specific constraints [174]. One could consider variations of the cost functions in (2.61), (2.70), or (2.72) by weighting the individual terms in a different way, so as to unbalance the relative importance of past information (encoded in the prior) and fresh data (encoded in the likelihood). We will revisit this approach in Chapter 8 when we introduce *adaptive* social learning and when we show the advantages of having *non-Bayesian updates* in Chapter 13.

Chapter 3

From Single-Agent to Social Learning

We are now ready to formulate the *social* learning problem, namely, the multi-agent, decentralized version of the single-agent problem addressed in the previous chapter. Differently from what was assumed there, now we allow each agent to exploit information received from some other agents, called *neighbors*. We denote by $\mathcal{N}_k$ the set of agents whose information is exploited by agent k. This set can include agent k, but this is not mandatory in our model. Technically, $\mathcal{N}_k$ represents the *in-neighborhood* of agent k, as explained later in Chapter 4.

The total number of agents in the graph will be denoted by K. Each agent $k = 1, 2, \ldots, K$ observes a stream of data

$$x_{k,1}, x_{k,2}, \ldots, x_{k,t} \tag{3.1}$$

belonging to a space $\mathcal{X}_k$, where the subscript k highlights that the observation spaces are allowed to be heterogeneous across the agents. Each agent k attempts to construct a belief vector $\mu_{k,t}$, relying on a prior $\mu_{k,0}$ and on *private* likelihood models $\{\ell_k(x|\theta)\}_{\theta\in\Theta}$. The nature of $\ell_k(x|\theta)$, regarded as a function of x, may vary across the agents as well. For example, it can be a pdf for certain agents, and a pmf for other agents. However, and as was also assumed in the single-agent setting, the nature of $\ell_k(x|\theta)$ does not vary across θ.

Moreover, note that each likelihood model $\ell_k(x|\theta)$ describes a *marginal* distribution for the data of agent k, which means that no joint model encompassing the dependence across the agents' data (i.e., over space) is used. This is because, as we will discuss more thoroughly in the next section, we will focus on *non-Bayesian* social learning, where the inter-agent dependence is not known or too complex to be accounted for.

3.1 Bayesian versus Non-Bayesian Learning

In Chapter 2 we saw that Bayes' rule is optimal under several paradigms. Therefore, an agent acting rationally should perform *Bayesian learning*, which means that its belief should be the Bayesian posterior. We showed in Lemma 2.1 that a standalone agent can perform Bayesian learning by using an online algorithm where the belief at each iteration t, updated sequentially by taking the previous belief and the likelihood of the new data, corresponds exactly to the Bayesian posterior computed over the amount of data available up to t.

The scenario changes dramatically when we move from single-agent to *social* learning. Under the latter setting, spatially distributed agents are linked by a graph, introducing nontrivial communication dynamics and spatial dependence into the learning process. To see why, let us focus on the perspective of a single agent from the group of agents. In order to be fully rational (i.e., fully Bayesian) this agent would need to know a *joint* model encompassing all agents, and should use it to compute a posterior based on the *entirety of data observed across the network*. These requirements are far from being satisfied in practice. First, each agent usually possesses only local (i.e., marginal) generative models of the form $\ell_k(x|\theta)$ to link its private data to the hypotheses, while it has no information about the generative models of the other agents. And even if an agent had full knowledge of the marginal models of the other agents, such knowledge would be generally insufficient to determine the joint model. Second, the data $x_{k,t}$ at each agent is usually private, and cannot be shared with other agents. More commonly, the agents are only allowed (or inclined) to share summary information, such as opinions or decisions, rather than their raw data.

In other words, in a distributed setting, the agents possess limited resources to deliberate, and they have access to incomplete information about their environment. The only information available to an agent is contained in its private data, its local likelihood model, and also in the opinions received from its immediate neighbors. This sharing of opinions also results in some redundancy and nontrivial correlations among different information sources over the graph. In addition, the agents face the challenge of not having knowledge of the full graph structure. Even when the agents have global knowledge of the network topology and the agents' data structure, retrieving fully Bayesian knowledge from summary information collected from neighboring agents is in general NP-hard [91]. The next example illustrates one simple scenario that shows the complexity of a fully Bayesian solution in the decentralized setting.

Example 3.1 (Multi-Agent Bayesian processing). Consider 4 agents organized into a network. Agent k, for $k = 1, 2, 3, 4$, observes a data sample ξ_k.[1] Specifically, the agents observe iid Bernoulli data, which are related to a hypothesis $\theta \in \Theta = \{\theta_1, \theta_2\}$. The likelihood model of agent k is, for $\xi \in \{0, 1\}$,

$$\ell_k(\xi|\theta) = q_\theta^{(k)} \, \mathbb{I}[\xi = 0] + \left(1 - q_\theta^{(k)}\right) \mathbb{I}[\xi = 1], \tag{3.2}$$

with $q_{\theta_1}^{(k)} \neq q_{\theta_2}^{(k)}$. All agents assume uniform prior beliefs, so that the exact Bayesian posterior (2.64) given all data samples $\{\xi_1, \xi_2, \xi_3, \xi_4\}$ is

$$\mu^{\mathsf{Bu}}(\theta|\xi_1, \xi_2, \xi_3, \xi_4) \propto \ell_1(\xi_1|\theta)\, \ell_2(\xi_2|\theta)\, \ell_3(\xi_3|\theta)\, \ell_4(\xi_4|\theta). \tag{3.3}$$

We want to illustrate the complexity associated with the decentralized computation of (3.3) when the agents cannot share the data. The agents are instead allowed to share the beliefs with their neighbors, according to the directed graph in Figure 3.1. We describe next a procedure that will enable agent 4 to obtain (3.3).

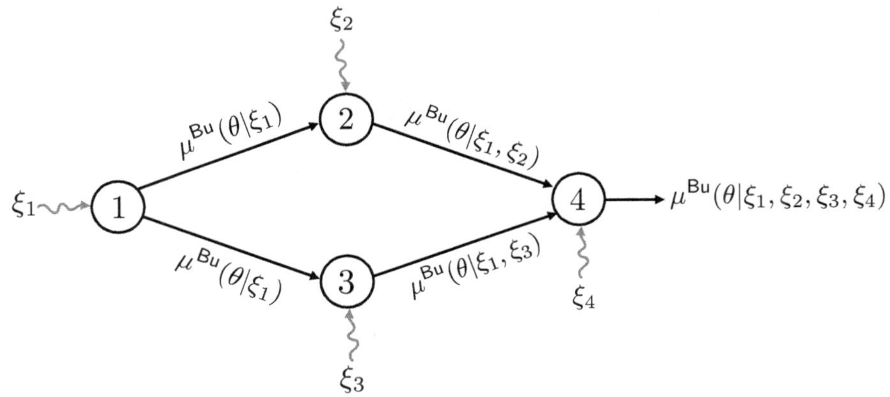

Figure 3.1: Diagram showing the flow of information across the network of four agents in Example 3.1.

Agent 1 updates its belief. When agent 1 receives observation ξ_1, it performs a Bayesian update yielding the belief

$$\mu^{\mathsf{Bu}}(\theta|\xi_1) \propto \ell_1(\xi_1|\theta). \tag{3.4}$$

[1] In this example we denote the data by ξ_k in place of x_k. This is done to avoid confusion, since in the previous chapter the subscript on x referred to time, whereas here it refers to the agent.

Agent 1 then sends the updated belief vector to agents 2 and 3.

Agents 2 and 3 update their beliefs. We first focus on agent 2 and show that, from its data ξ_2 and the belief (3.4) received from agent 1, it can compute the exact Bayesian posterior corresponding to the subset of data $\{\xi_1, \xi_2\}$. In fact, agent 2 can perform a Bayesian update by using (3.4) as the prior, yielding

$$\mu^{\mathsf{Bu}}(\theta|\xi_1,\xi_2) \propto \mu^{\mathsf{Bu}}(\theta|\xi_1)\,\ell(\xi_2|\theta) \overset{(3.4)}{\propto} \ell_1(\xi_1|\theta)\,\ell_2(\xi_2|\theta), \tag{3.5}$$

which is the exact Bayesian posterior given $\{\xi_1, \xi_2\}$. Agent 3 could perform a similar procedure and get

$$\mu^{\mathsf{Bu}}(\theta|\xi_1,\xi_3) \propto \ell_1(\xi_1|\theta)\,\ell_3(\xi_3|\theta). \tag{3.6}$$

Agents 2 and 3 send to agent 4 beliefs (3.5) and (3.6), respectively.

Agent 4 updates its belief. Note that the beliefs received by agent 4 contain redundant information about observation ξ_1. In order to compute the exact Bayesian posterior (3.3), agent 4 needs to disentangle the observations ξ_1, ξ_2, and ξ_3 from the received beliefs. To this end, agent 4 uses beliefs (3.5) and (3.6) (received from agents 2 and 3, respectively) to compute the quantity

$$a \triangleq \frac{\mu^{\mathsf{Bu}}(\theta_1|\xi_1,\xi_2)}{\mu^{\mathsf{Bu}}(\theta_2|\xi_1,\xi_2)} = \frac{\ell_1(\xi_1|\theta_1)\,\ell_2(\xi_2|\theta_1)}{\ell_1(\xi_1|\theta_2)\,\ell_2(\xi_2|\theta_2)}. \tag{3.7}$$

If agent 4 knows the likelihood models of agents 2 and 3, it can recover ξ_1 and ξ_2 by checking the possible values of a as follows:

$$\begin{cases} \xi_1 = 0,\ \xi_2 = 0 & \text{if } a = \dfrac{q_{\theta_1}^{(1)}}{q_{\theta_2}^{(1)}}\dfrac{q_{\theta_1}^{(2)}}{q_{\theta_2}^{(2)}}, \\[2ex] \xi_1 = 0,\ \xi_2 = 1 & \text{if } a = \dfrac{q_{\theta_1}^{(1)}}{q_{\theta_2}^{(1)}}\dfrac{1-q_{\theta_1}^{(2)}}{1-q_{\theta_2}^{(2)}}, \\[2ex] \xi_1 = 1,\ \xi_2 = 0 & \text{if } a = \dfrac{1-q_{\theta_1}^{(1)}}{1-q_{\theta_2}^{(1)}}\dfrac{q_{\theta_1}^{(2)}}{q_{\theta_2}^{(2)}}, \\[2ex] \xi_1 = 1,\ \xi_2 = 1 & \text{if } a = \dfrac{1-q_{\theta_1}^{(1)}}{1-q_{\theta_2}^{(1)}}\dfrac{1-q_{\theta_1}^{(2)}}{1-q_{\theta_2}^{(2)}}, \end{cases} \tag{3.8}$$

assuming that the parameters of the Bernoulli distributions are such that the above four values of a are distinct. A similar procedure, with four different comparisons, can be applied to the beliefs received from agent 3 to recover ξ_3. Upon recovery of ξ_1, ξ_2, and ξ_3, agent 4 can finally compute (3.3).

The analysis in this example explains why obtaining a fully Bayesian solution in a decentralized setting becomes soon unfeasible as the number of time steps increases. Moreover, the complexity also increases when the observation model is less simple (e.g., discrete random variables with more than two values or continuous random variables) or when we have more hypotheses to classify. In addition, note that we made the assumption that agent 4 knows the likelihood models of agents 2 and 3, while in typical applications each agent knows only its own private models. Note also that, as we will see later, in traditional social learning all agents update and share their beliefs in parallel, which introduces additional complexity with respect to the simplified sequential scheme considered in this example.

The issues encountered in the distributed setting motivated many investigators to move away from insisting on the fully Bayesian perspective. The departure from Bayesian thinking is endorsed by the theory of *bounded rationality*. The qualification "bounded" highlights the fact that, due to cognitive and knowledge constraints, the agents are unable to implement fully rational rules and must implement instead *non-Bayesian* rules [43, 161]. Under sophisticated learning tasks, psychological experiments have supported the theory that non-Bayesian decision-making can take place. It has been observed that the subjects of these experiments were not fully rational in the face of new information and deviated from a fully Bayesian approach toward some more attainable decision *heuristics* [49]. This behavior can be justified in terms of a trade-off between *decision accuracy* and *cognitive effort*. The subject is not only seeking a learning strategy that yields good performance, but also keeps deliberation cost under control.

3.2 Non-Bayesian Social Learning

The experimental evidence of bounded rationality in humans and the intractability of Bayesian computation within groups motivate the development of *non-Bayesian* social learning. Within this paradigm, several useful methods have been proposed [3, 25, 84, 96, 106, 131, 135, 157, 175]. All these methods share the following common structure, in a manner similar to strategies used for optimization and learning over graphs [151, 152, 155]: *i)* a *self-learning* step, where each agent learns individually from its private data; followed by a *ii) cooperation* step, where the individual knowledge is shared among agents according to a communication structure dictated by the network graph.[2] In summary, we arrive at the "equation"

$$\textsf{social learning} = \textsf{self-learning} + \textsf{cooperation} \tag{3.9}$$

In the self-learning step we assume each agent acts *individually* in a canonical Bayesian way. That is, each agent k at time t performs *locally* a Bayesian update by blending its prior belief vector $\mu_{k,t-1}$ and the likelihood

[2]We opt to focus on the adapt-then-combine form, where the cooperation step comes after the self-learning step. It is also possible to consider the combine-then-adapt form, where the order is reversed [131, 135, 152].

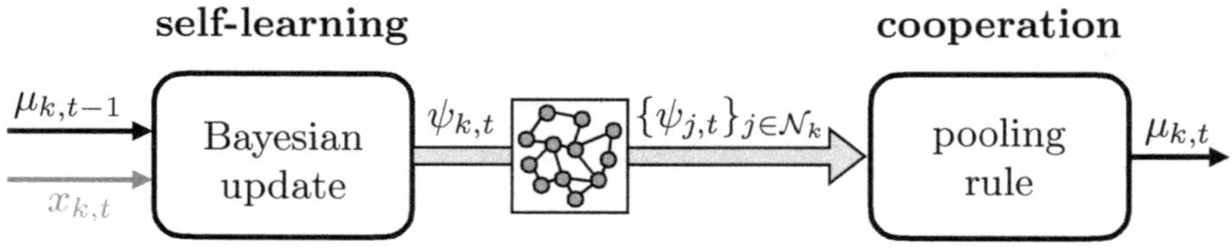

Figure 3.2: Non-Bayesian social learning. In the self-learning step, each agent k performs individually a *Bayesian update* given the prior belief vector $\mu_{k,t-1}$ and the private data $x_{k,t}$. The resulting intermediate belief vector $\psi_{k,t}$ is then diffused across the network. In the cooperation step, agent k aggregates the intermediate beliefs $\{\psi_{j,t}\}_{j\in\mathcal{N}_k}$ received from neighbors by using a pooling rule.

$\ell_k(x_{k,t}|\theta)$ computed from the locally available data $x_{k,t}$. The output of the Bayesian update is an intermediate belief vector $\psi_{k,t}$ to be shared with neighboring agents.

Then, during the cooperation step, agent k forms its final belief vector $\mu_{k,t}$ by using a certain *pooling* rule C_k to combine the intermediate belief vectors $\{\psi_{j,t}\}_{j\in\mathcal{N}_k}$ received from its neighbors. These steps can be formally written according to the following recursion:

$$\psi_{k,t}(\theta) \propto \mu_{k,t-1}(\theta)\ell_k(x_{k,t}|\theta) \qquad \text{(self-learning)}, \tag{3.10a}$$

$$\mu_{k,t} = \mathsf{C}_k\Big(\{\psi_{j,t}\}_{j\in\mathcal{N}_k}\Big) \qquad \text{(cooperation)}. \tag{3.10b}$$

Figure 3.2 summarizes the essential features of non-Bayesian social learning.

While time-varying combination rules can be considered, in our presentation it is sufficient to focus on time-invariant rules, C_k. Note also that the pooling operator in (3.10b) aggregates only the *current* updated belief vectors $\{\psi_{j,t}\}_{j\in\mathcal{N}_k}$, and does not account for the entire history of beliefs received up to time t. This property, sometimes referred to as "imperfect recall," is an instance of the principle of *bounded rationality*, aimed at reducing computational and memory complexity [131]. In this connection, we observe that the intermediate belief vector $\psi_{j,t}$ depends on the previous-lag belief vector $\mu_{j,t-1}$. In view of the sequential nature of Bayes' rule (see Lemma 2.1), previous-lag beliefs are sufficient to build the optimal posterior in the single-agent case. This property is in general lost in the distributed setting due to the reasons mentioned in the previous section.

We see from (3.10a) and (3.10b) that the structure of the self-learning step is determined by Bayes' rule. Therefore, the critical part is to select a combination rule, i.e., how the intermediate neighboring beliefs should be processed. To this end, we will go through the following path. First, in the next section we introduce an approach that derives the combination rule from the optimization of suitable information-theoretic measures. We will see how the two most popular social learning strategies will arise naturally from this approach. Later, in Section 3.4, the very same strategies will arise as the unique solutions to a formulation with meaningful physical constraints placed on the agents' behavior.

3.3 Information-Theoretic Viewpoint

A general principle to build an aggregate belief vector $\mu_{k,t}$ from the ensemble of belief vectors $\{\psi_{j,t}\}_{j\in\mathcal{N}_k}$ is to minimize some measure of discrepancy between the new belief and the ensemble of beliefs. The purpose is to make the aggregate belief as close as possible to all beliefs in the ensemble, i.e., to fuse the different viewpoints brought by the different agents. In the next two sections we illustrate two choices for the discrepancy measures. Similar choices are considered in [101], albeit for probability density functions as opposed to probability mass functions.

For the sake of simplicity, the derivations in the forthcoming Sections 3.3.1 and 3.3.2 are carried out under the assumption that all the entries of the belief vectors $\psi_{j,t}$ are nonzero. The case where some of them are zero can be obtained from continuity arguments — see the expressions in (B.1).

3.3.1 Geometric-Averaging Rule

One meaningful way to quantify the discrepancy between a candidate belief vector p at agent k and the received belief vectors $\{\psi_{j,t}\}_{j\in\mathcal{N}_k}$, is a convex combination of the KL divergences between p and the received belief vectors, namely,

$$\sum_{j\in\mathcal{N}_k} a_{jk} D(p||\psi_{j,t}), \tag{3.11}$$

where the scalars a_{jk} are defined for $j \in \mathcal{N}_k$ and obey the following convexity conditions:

$$a_{jk} > 0, \qquad \sum_{j\in\mathcal{N}_k} a_{jk} = 1. \tag{3.12}$$

We now want to compute the aggregate belief vector $\mu_{k,t}$ as the one that minimizes the combination of KL divergences in (3.11), namely,

$$\mu_{k,t} = \arg\min_{p \in \Delta_H} \left\{ \sum_{j \in \mathcal{N}_k} a_{jk} D(p || \psi_{j,t}) \right\}. \tag{3.13}$$

Note that the condition $\sum_{j \in \mathcal{N}_k} a_{jk} = 1$ is not a limitation, since if we formulate (3.13) with a generic set of positive weights, we can always scale these weights to reduce the problem to one with convex weights, without altering the solution. The objective function in (3.13) can be manipulated as follows:

$$\begin{aligned} \sum_{j \in \mathcal{N}_k} a_{jk} D(p || \psi_{j,t}) &= \sum_{j \in \mathcal{N}_k} a_{jk} \sum_{\theta \in \Theta} p(\theta) \log \frac{p(\theta)}{\psi_{j,t}(\theta)} \\ &= \sum_{\theta \in \Theta} p(\theta) \log \frac{p(\theta)}{\prod_{j \in \mathcal{N}_k} [\psi_{j,t}(\theta)]^{a_{jk}}}. \end{aligned} \tag{3.14}$$

Proceeding as done to manage (2.88), we see that the denominator on the RHS of (3.14) is equivalent to a pmf up to a scaling term independent of p (i.e., the RHS is equivalent to a KL divergence up to an additive constant), implying that the solution to (3.13) is

$$\mu_{k,t}(\theta) \propto \prod_{j \in \mathcal{N}_k} [\psi_{j,t}(\theta)]^{a_{jk}}. \tag{3.15}$$

In summary, the pooling rule in (3.15), resulting from the optimization problem in (3.13), prescribes that each agent computes a *weighted geometric average* of the Bayesian updates gathered from its own neighborhood, up to a normalization factor necessary to yield a vector belonging to the probability simplex Δ_H. The overall social learning strategy obtained by using the Bayesian update (in the self-learning step) followed by the geometric-average pooling rule (in the cooperation step) is summarized in listing (3.16).

Social learning with geometric averaging

start from the prior belief vectors $\mu_{k,0}$ for $k = 1, 2, \ldots, K$
for $t = 1, 2, \ldots$
  for $k = 1, 2, \ldots, K$
    agent k observes $x_{k,t}$
    for $\theta = 1, 2, \ldots, H$

$$\psi_{k,t}(\theta) = \frac{\mu_{k,t-1}(\theta)\ell_k(x_{k,t}|\theta)}{\sum\limits_{\theta' \in \Theta} \mu_{k,t-1}(\theta')\ell_k(x_{k,t}|\theta')} \qquad \text{(self-learning)}$$

    end
  end

  for $k = 1, 2, \ldots, K$
    for $\theta = 1, 2, \ldots, H$

$$\mu_{k,t}(\theta) = \frac{\prod_{j \in \mathcal{N}_k} [\psi_{j,t}(\theta)]^{a_{jk}}}{\sum\limits_{\theta' \in \Theta} \prod_{j \in \mathcal{N}_k} [\psi_{j,t}(\theta')]^{a_{jk}}} \qquad \text{(cooperation)}$$

    end
  end
end

(3.16)

Equation (3.15) implies that

$$\log \mu_{k,t}(\theta) = \sum_{j \in \mathcal{N}_k} a_{jk} \log \psi_{j,t}(\theta) + \text{const.}, \tag{3.17}$$

which explains why the geometric-average rule is also referred to as *log-linear* or *logarithmic* pooling.

Examining (3.16), we see that the social learning algorithm with geometric averaging involves two normalization operations. The first normalization is implemented by each agent k during the self-learning stage, to compute the intermediate belief vector $\psi_{k,t}$ that must be shared over the network. The second normalization is implemented by each agent during the cooperation stage to compute the final belief vector $\mu_{k,t}$. It is clear that we can combine both steps into a single update written as

$$\mu_{k,t}(\theta) \propto \prod_{j \in \mathcal{N}_k} \Big[\mu_{j,t-1}(\theta)\ell_j(x_{j,t}|\theta)\Big]^{a_{jk}}. \tag{3.18}$$

This form does not require double normalization and is appealing in situations where the intermediate beliefs are not required.

3.3.2 Arithmetic-Averaging Rule

A second way to quantify the discrepancy between a belief vector p at agent k and the received belief vectors is the following:

$$\sum_{j\in\mathcal{N}_k} a_{jk} D(\psi_{j,t}||p), \tag{3.19}$$

where the roles of p and $\psi_{j,t}$ are reversed in comparison with (3.11). We again impose the conditions

$$a_{jk} > 0, \quad \sum_{j\in\mathcal{N}_k} a_{jk} = 1. \tag{3.20}$$

The aggregate belief vector $\mu_{k,t}$ is then obtained as

$$\mu_{k,t} = \arg\min_{p\in\Delta_H} \left\{ \sum_{j\in\mathcal{N}_k} a_{jk} D(\psi_{j,t}||p) \right\}. \tag{3.21}$$

The objective function in (3.21) can be manipulated as follows:

$$\begin{aligned}
\sum_{j\in\mathcal{N}_k} a_{jk} D(\psi_{j,t}||p) &= \sum_{j\in\mathcal{N}_k} a_{jk} \sum_{\theta\in\Theta} \psi_{j,t}(\theta) \log \frac{\psi_{j,t}(\theta)}{p(\theta)} \\
&= \sum_{\theta\in\Theta} \sum_{j\in\mathcal{N}_k} a_{jk}\psi_{j,t}(\theta) \log \frac{\psi_{j,t}(\theta)}{p(\theta)} \\
&= \sum_{\theta\in\Theta} \sum_{j\in\mathcal{N}_k} a_{jk}\psi_{j,t}(\theta) \log \frac{\sum_{j\in\mathcal{N}_k} a_{jk}\psi_{j,t}(\theta)}{p(\theta)} \\
&\quad \underbrace{+ \sum_{\theta\in\Theta} \sum_{j\in\mathcal{N}_k} a_{jk}\psi_{j,t}(\theta) \log \frac{\psi_{j,t}(\theta)}{\sum_{j\in\mathcal{N}_k} a_{jk}\psi_{j,t}(\theta)}}_{\text{independent of } p} \\
&= D\left(\sum_{j\in\mathcal{N}_k} a_{jk}\psi_{j,t} \middle|\middle| p \right) + \text{const.}
\end{aligned} \tag{3.22}$$

Therefore, we see that the objective function in (3.21) is minimized when the KL divergence on the last line in (3.22) is equal to 0, which occurs for the choice

$$\mu_{k,t}(\theta) = \sum_{j\in\mathcal{N}_k} a_{jk}\psi_{j,t}(\theta), \tag{3.23}$$

namely, a *weighted arithmetic* average of the intermediate beliefs. Such linear rule is among the first combination policies used in social learning [175].

One of the earliest appearances of arithmetic averaging is in the context of the *consensus* algorithm [58], which dealt with a static case without streaming data. The objective there was to pool different beliefs belonging to spatially distributed agents, so that they can agree on a common belief. In more recent years, arithmetic averaging has been successfully applied in the context of diffusion strategies used for optimization and learning over graphs [151, 152, 155]. The overall social learning strategy corresponding to the linear combination rule is summarized in listing (3.24).

Social learning with arithmetic averaging

start from the prior belief vectors $\mu_{k,0}$ for $k = 1, 2, \ldots, K$
for $t = 1, 2, \ldots$
  for $k = 1, 2, \ldots, K$
    agent k observes $x_{k,t}$
    for $\theta = 1, 2, \ldots, H$

$$\psi_{k,t}(\theta) = \frac{\mu_{k,t-1}(\theta)\ell_k(x_{k,t}|\theta)}{\sum\limits_{\theta' \in \Theta} \mu_{k,t-1}(\theta')\ell_k(x_{k,t}|\theta')} \quad \text{(self-learning)}$$

    end
  end

  for $k = 1, 2, \ldots, K$
    for $\theta = 1, 2, \ldots, H$

$$\mu_{k,t}(\theta) = \sum_{j \in \mathcal{N}_k} a_{jk}\psi_{j,t}(\theta) \quad \text{(cooperation)}$$

    end
  end
end

(3.24)

Comparing (3.13) against (3.21), an interesting interpretation emerges. Recall that the KL divergence is not symmetric, and that the distribution appearing as the first argument is the one under which the expectation is evaluated — see Definition B.4. In other words, the first argument is taken as the true underlying distribution. In (3.13), the KL divergences are computed by assuming that the true underlying pmf is p, which is the belief obtained by aggregating all beliefs received from neighboring agents. In contrast, in (3.21), each individual KL divergence corresponding to agent j is computed by assuming a different underlying truth, which is represented by the intermediate belief vector $\psi_{j,t}$. This difference ultimately results in two different pooling rules, the geometric and arithmetic averaging rules, respectively.

3.4 Behavioral Viewpoint

In the previous section we obtained two possible forms for the pooling operator C_k in (3.10b) (namely, the geometric and arithmetic averaging rules) by optimizing suitable information-theoretic metrics. A completely different route is followed in [131], where the pooling rule is derived from a set of axioms that represent relevant *behavioral* attributes of the agents. We now illustrate this alternative construction.

In preparation for the forthcoming technical analysis, it is convenient to simplify the notation (we will omit the dependence on time and agent indices) and formulate the problem in the following general manner. Given a collection of *input* belief vectors $q_1, q_2, \ldots, q_K$, we want to aggregate them into an *output* belief vector q through a *continuous* mapping $\mathsf{C} : \Delta_H^K \mapsto \Delta_H$, namely,

$$q = \mathsf{C}(q_1, q_2, \ldots, q_K). \tag{3.25}$$

We will now introduce a set of behavioral assumptions that the pooling operator C should fulfill.

The first assumption is *label neutrality*, which ensures that the way the input beliefs are processed cannot depend on the particular labeling chosen for the hypotheses.

Assumption 3.1 (Label neutrality). Let $\Pi : \Theta \mapsto \Theta$ be a permutation. For any $p \in \Delta_H$, the symbol $p^{(\Pi)}$ denotes a permuted vector whose θth entry is

$$p^{(\Pi)}(\theta) = p\Big(\Pi(\theta)\Big). \tag{3.26}$$

A pooling operator C fulfills label neutrality when permuting the output belief vector is equivalent to applying C to the permuted input vectors. Formally, label neutrality holds when, for any $\Pi : \Theta \mapsto \Theta$ and any collection $\{q_1, q_2, \ldots, q_K\}$ of input belief vectors,

$$q^{(\Pi)} = \mathsf{C}\left(q_1^{(\Pi)}, q_2^{(\Pi)}, \ldots, q_K^{(\Pi)}\right), \tag{3.27}$$

with the output belief vector q being defined by (3.25).

Second, consider the case where all input beliefs are equal, namely,

$$q_1 = q_2 = \ldots = q_K = p. \tag{3.28}$$

It is natural to request that the pooling operator should return p. The second assumption, called *unanimity*, summarizes this requirement.

Assumption 3.2 (Unanimity). The pooling operator C is unanimous when $\mathsf{C}(p, p, \ldots, p) = p$ for all $p \in \Delta_H$.

Third, consider a collection of belief vectors $q_1, q_2, \ldots, q_K$ having all positive entries, and a belief vector q formed by using (3.25). Assume that a single belief vector q_i changes into another belief q_i' with an increase in confidence about some hypothesis $\bar{\theta}$ and a decrease for the remaining hypotheses, namely,

$$q_i'(\bar{\theta}) > q_i(\bar{\theta}), \qquad q_i'(\theta) \le q_i(\theta) \quad \forall \theta \ne \bar{\theta}. \tag{3.29}$$

It would be natural to expect that also the pooling operator reflects the increase in confidence. This is formalized by the next assumption.

Assumption 3.3 (Monotonicity). Consider two collections of input belief vectors, $\{q_1, q_2, \ldots, q_K\}$ and $\{q_1', q_2', \ldots, q_K'\}$, both placing nonzero[3] mass on all $\theta \in \Theta$ and fulfilling, for some $i \in \{1, 2, \ldots, K\}$, the conditions

$$\text{i)} \quad q_j = q_j' \quad \forall j \in \{1, 2, \ldots, K\} \setminus \{i\}, \tag{3.30}$$

$$\text{ii)} \quad \begin{cases} q_i'(\bar{\theta}) > q_i(\bar{\theta}), \\ q_i'(\theta) \le q_i(\theta) \quad \forall \theta \ne \bar{\theta}. \end{cases} \tag{3.31}$$

Let

$$q = \mathsf{C}(q_1, q_2, \ldots, q_K), \qquad q' = \mathsf{C}(q_1', q_2', \ldots, q_K'). \tag{3.32}$$

Monotonicity holds when

$$q'(\bar{\theta}) > q(\bar{\theta}). \tag{3.33}$$

The previous assumptions (label neutrality, unanimity, and monotonicity) are basic requirements that a meaningful combination rule is expected to fulfill. We now complete the set of behavioral axioms by specifying how each entry of the output belief vector is influenced by the various entries of the input belief vectors. We present two possibilities, namely, the assumption of *independence of irrelevant alternatives*, and the assumption of *separability*. We will see later how these different assumptions lead to different pooling rules.

[3]Actually, one might want to impose monotonicity even when some entries of the input belief vectors are zero. However, this extended notion of monotonicity would not be fulfilled simultaneously with the other behavioral assumptions considered later in Theorem 3.1.

Independence of irrelevant alternatives is based on the following rationale. Assume we know that the hypothesis of interest belongs to a subset $\mathcal{S} \subset \Theta$ of all possible hypotheses, and assume we want to construct from the available input belief vectors $\{q_j\}$, a *conditional* belief vector given that $\theta \in \mathcal{S}$. In performing this pooling, one meaningful criterion is to take into account only the input beliefs conditioned on the same subset $\mathcal{S}$. In this sense, *when focusing on a conditional belief given a specific subset* $\mathcal{S}$, the hypotheses that do not belong to $\mathcal{S}$ are deemed as irrelevant alternatives. The assumption of independence of irrelevant alternatives prescribes that the pooling operator C does automatically guarantee this type of construction for any subset $\mathcal{S}$, namely, that the output q of the pooling rule, conditioned on a subset $\mathcal{S}$ of the hypotheses, is equivalent to the output of the same pooling operator applied only to the input belief vectors $\{q_j\}$ conditioned on $\mathcal{S}$.

Assumption 3.4 (Independence of irrelevant alternatives). For any belief vector $p \in \Delta_H$ that places nonzero mass on a subset of hypotheses $\mathcal{S} \subset \Theta$, we introduce the belief *conditioned* on $\mathcal{S}$:

$$p^{|\mathcal{S}}(\theta) = \begin{cases} \dfrac{p(\theta)}{\sum\limits_{\theta' \in \mathcal{S}} p(\theta')} & \text{for } \theta \in \mathcal{S}, \\ 0 & \text{otherwise.} \end{cases} \tag{3.34}$$

Let $q = \mathsf{C}(q_1, q_2, \dots, q_K)$. Independence of irrelevant alternatives holds when, for any subset $\mathcal{S} \subset \Theta$ and any collection $\{q_1, q_2, \dots, q_K\}$ of input belief vectors that place nonzero mass in $\mathcal{S}$,

$$q^{|\mathcal{S}} = \mathsf{C}\Big(q_1^{|\mathcal{S}}, q_2^{|\mathcal{S}}, \dots, q_K^{|\mathcal{S}}\Big). \tag{3.35}$$

In Assumptions 3.4, since the belief vectors $\{q_j\}$ have nonzero mass in $\mathcal{S}$ (i.e., each q_j has at least one nonzero entry in $\mathcal{S}$), each *conditional* belief vector $q_j^{|\mathcal{S}}$ is well-posed. Therefore, in (3.35) it is legitimate to apply the combination rule to the conditional belief vectors $\left\{q_j^{|\mathcal{S}}\right\}$. As a result, under Assumptions 3.4 also the (conditional) output belief vector $q^{|\mathcal{S}}$ is well-posed, which means that the (unconditional) output belief vector q has automatically nonzero mass in $\mathcal{S}$. In particular, taking $\mathcal{S} = \{\theta\}$, we conclude that if $q_j(\theta) > 0$ for all j, then $q(\theta) > 0$.

Another possibility to complete the set of behavioral assumptions is *separability*, which establishes that the θth entry of the output belief vector depends solely on the θth entry of the input belief vectors.

Assumption 3.5 (Separability). Let $q = \mathsf{C}(q_1, q_2, \ldots, q_K)$. Separability holds when $q(\theta)$ does not depend on $q_1(\theta'), q_2(\theta'), \ldots, q_K(\theta')$ for $\theta' \neq \theta$.

3.4.1 Geometric-Averaging Rule, Revisited

The next theorem, proved in [131], reveals the unique form that the pooling operator can take under label neutrality, unanimity, monotonicity, and independence of irrelevant alternatives.

Theorem 3.1 (Geometric averaging). Let Assumptions 3.1, 3.2, 3.3, and 3.4 be satisfied, and let $q = \mathsf{C}(q_1, q_2, \ldots, q_K)$. If $|\Theta| > 2$, we must have, for all $\theta \in \Theta$,

$$q(\theta) \propto \prod_{j=1}^{K} [q_j(\theta)]^{a_j}, \tag{3.36}$$

with $a_j > 0$ and $\sum_{j=1}^{K} a_j = 1$.

Proof. It suffices to examine the case where $q_j(\theta) > 0$ for $j = 1, 2, \ldots, K$ and for all $\theta \in \Theta$. The case where some entries of the belief vector are zero will follow automatically from the continuity of the pooling operator. Consider two arbitrary hypotheses θ', θ'', and the set $\mathcal{S} = \{\theta', \theta''\}$. We focus on the logarithmic ratio $\log(q(\theta')/q(\theta''))$, which is well-posed since the belief vectors $\{q_j\}$ have all positive entries and the same holds for q in view of the observation following Assumption 3.4. We can write (the notation $[v]_\theta$ extracts the θth entry of the vector v)

$$\log \frac{q(\theta')}{q(\theta'')} \overset{\text{(a)}}{=} \log \frac{q^{|\mathcal{S}}(\theta')}{q^{|\mathcal{S}}(\theta'')} \overset{\text{(b)}}{=} \log \frac{\left[\mathsf{C}\left(q_1^{|\mathcal{S}}, q_2^{|\mathcal{S}}, \ldots, q_K^{|\mathcal{S}}\right)\right]_{\theta'}}{\left[\mathsf{C}\left(q_1^{|\mathcal{S}}, q_2^{|\mathcal{S}}, \ldots, q_K^{|\mathcal{S}}\right)\right]_{\theta''}}, \tag{3.37}$$

where (a) follows from definition (3.34), and (b) from the independence of irrelevant alternatives. The jth conditional belief vector can be represented as

$$\begin{aligned} q_j^{|\mathcal{S}} &= \Bigg[0, \ldots, 0, \underbrace{\frac{q_j(\theta')}{q_j(\theta') + q_j(\theta'')}}_{\text{label } \theta'}, 0, \ldots, 0, \underbrace{\frac{q_j(\theta'')}{q_j(\theta') + q_j(\theta'')}}_{\text{label } \theta''}, 0, \ldots, 0\Bigg] \\ &= \Bigg[0, \ldots, 0, \underbrace{\frac{q_j(\theta')/q_j(\theta'')}{1 + q_j(\theta')/q_j(\theta'')}}_{\text{label } \theta'}, 0, \ldots, 0, \underbrace{\frac{1}{1 + q_j(\theta')/q_j(\theta'')}}_{\text{label } \theta''}, 0, \ldots, 0\Bigg]. \end{aligned} \tag{3.38}$$

Let us denote by z_j the value of the ratio $q_j(\theta')/q_j(\theta'')$, namely,

$$\frac{q_j(\theta')}{q_j(\theta'')} = z_j \quad \text{for } j = 1, 2, \ldots, K. \tag{3.39}$$

The RHS of (3.37) is in principle a function of the values $\{z_j\}$ and of the labels θ', θ''. However, we now show that label neutrality implies that the dependence on the particular labels disappears. From (3.38) and (3.39) we have (for the sake of presentation we consider a vector of length $H = 8$)

$$q_j^{|S} = \Big[0, 0, 0, \underbrace{\frac{z_j}{1+z_j}}_{\text{label } \theta'}, 0, 0, \underbrace{\frac{1}{1+z_j}}_{\text{label } \theta''}, 0\Big]. \tag{3.40}$$

Assume instead that (3.39) is verified for a different pair of labels $(\dot{\theta}', \dot{\theta}'')$. Then, the *location* of the values changes, namely, we obtain another belief vector, say,

$$\dot{q}_j^{|S} = \Big[\underbrace{\frac{z_j}{1+z_j}}_{\text{label } \dot{\theta}'}, 0, 0, 0, 0, 0, 0, \underbrace{\frac{1}{1+z_j}}_{\text{label } \dot{\theta}''}\Big], \tag{3.41}$$

which is a permuted version of $q_j^{|S}$ in (3.40). Now, in (3.37) we apply the pooling operator C to the conditional belief vectors $q_j^{|S}$ to compute the output belief vector q. Then, we compute the ratio between $q(\theta')$ and $q(\theta'')$, namely, the θ'th and θ''th entries of this vector. Assume now that we apply C to the *permuted* belief $\dot{q}_j^{|S}$ to compute another output belief vector $\dot{q}(\theta')$. By label neutrality, we must have

$$\dot{q}\left(\dot{\theta}'\right) = q\left(\theta'\right), \qquad \dot{q}\left(\dot{\theta}''\right) = q\left(\theta''\right). \tag{3.42}$$

This implies that the ratio in (3.37) does not depend on the particular pair of labels, but only on the values $\{z_j\}$, which allows us to write

$$\log\frac{q(\theta')}{q(\theta'')} = g\left(\log\frac{q_1(\theta')}{q_1(\theta'')}, \log\frac{q_2(\theta')}{q_2(\theta'')}, \ldots, \log\frac{q_K(\theta')}{q_K(\theta'')}\right), \tag{3.43}$$

where the function g does not depend on the labels θ', θ''. When we say that g does not depend on the labels, we mean that its functional form remains the same if we vary the labels, namely, we do *not* have different functions $g_{\theta'\theta''}$ for different pairs of labels. Moreover, g is continuous because, in the RHS of (3.37), the pooling operator C is continuous by assumption, and, in view of (3.38), each conditional belief $q_j^{|S}$ can be regarded as a continuous function of $\log\frac{q_j(\theta')}{q_j(\theta'')}$.

Consider now a third hypothesis θ'''. We can write

$$\begin{aligned}\log\frac{q(\theta')}{q(\theta''')} &= \log\frac{q(\theta')}{q(\theta'')} + \log\frac{q(\theta'')}{q(\theta''')}\\ &= g\Big(\underbrace{\log\frac{q_1(\theta')}{q_1(\theta'')}, \log\frac{q_2(\theta')}{q_2(\theta'')}, \ldots, \log\frac{q_K(\theta')}{q_K(\theta'')}}_{x}\Big)\\ &\quad + g\Big(\underbrace{\log\frac{q_1(\theta'')}{q_1(\theta''')}, \log\frac{q_2(\theta'')}{q_2(\theta''')}, \ldots, \log\frac{q_K(\theta'')}{q_K(\theta''')}}_{y}\Big)\end{aligned} \tag{3.44}$$

and also

$$\log \frac{q(\theta')}{q(\theta''')} = g\Bigg(\underbrace{\log \frac{q_1(\theta')}{q_1(\theta''')}, \log \frac{q_2(\theta')}{q_2(\theta''')}, \ldots, \log \frac{q_K(\theta')}{q_K(\theta''')}}_{x+y}\Bigg). \tag{3.45}$$

Grouping (3.44) and (3.45) we find that g satisfies

$$g(x+y) = g(x) + g(y), \qquad x, y \in \mathbb{R}^K, \tag{3.46}$$

which is a multidimensional Cauchy functional equation, to be solved by seeking a function $g : \mathbb{R}^K \mapsto \mathbb{R}$ [66]. It is known that the unique continuous solutions to (3.46) are in the following form [66]:

$$g(x) = \sum_{j=1}^{K} a_j x_j, \tag{3.47}$$

where x_j is the jth entry of x, and where the solution space is spanned by $a_j \in \mathbb{R}$, for $j = 1, 2, \ldots, K$. Using the definition of g from (3.43), Eq. (3.47) becomes

$$\log \frac{q(\theta')}{q(\theta'')} = \sum_{j=1}^{K} a_j \log \frac{q_j(\theta')}{q_j(\theta'')}. \tag{3.48}$$

By exponentiation and normalization (since q must be a belief vector), Eq. (3.48) leads to (3.36). It is easily verified that (3.36) fulfills Assumptions 3.1 and 3.4 for all choices of $\{q_j\}$ and $\{a_j\}$. To conclude the proof, we show that the combination weights $\{a_j\}$ must add up to 1 and be positive. Consider $q_j = p$ for all j, with p having positive entries and $p(\theta') \neq p(\theta'')$ for two hypotheses θ' and θ'' (the case where all entries of p are equal is trivial). By unanimity we have

$$\log \frac{p(\theta')}{p(\theta'')} = \sum_{j=1}^{K} a_j \log \frac{p(\theta')}{p(\theta'')} \iff \sum_{j=1}^{K} a_j = 1. \tag{3.49}$$

Finally, we show that positivity of the weighting coefficients follows from monotonicity. To see why, consider, for the input belief vectors, two assignments $\{q_j\}$ and $\{q_j'\}$ as defined in Assumption 3.3, along with the corresponding output belief vectors q and q'. Recall that the input belief vectors of the two assignments are equal to each other, but for the ith input belief vectors q_i and q_i', which are different and satisfy the following inequalities:

$$q_i'(\bar{\theta}) > q_i(\bar{\theta}), \qquad q_i'(\theta) \leq q_i(\theta) \quad \forall \theta \neq \bar{\theta} \tag{3.50}$$

for some hypothesis $\bar{\theta} \in \Theta$. From (3.36) we obtain the following representation for the output belief $q(\bar{\theta})$:

$$q(\bar{\theta}) = \frac{\prod_{j=1}^{K} q_j^{a_j}(\bar{\theta})}{\prod_{j=1}^{K} q_j^{a_j}(\bar{\theta}) + \sum_{\theta \neq \bar{\theta}} \prod_{j=1}^{K} q_j^{a_j}(\theta)} = \left(1 + \sum_{\theta \neq \bar{\theta}} \prod_{j=1}^{K} \left(\frac{q_j(\theta)}{q_j(\bar{\theta})}\right)^{a_j}\right)^{-1}, \tag{3.51}$$

and a similar expression holds for $q'(\bar{\theta})$. Therefore, in view of (3.51) we can write

$$q(\bar{\theta}) = \left(1 + \sum_{\theta \neq \bar{\theta}} c_\theta \left(\frac{q_i(\theta)}{q_i(\bar{\theta})}\right)^{a_i}\right)^{-1},$$
$$q'(\bar{\theta}) = \left(1 + \sum_{\theta \neq \bar{\theta}} c_\theta \left(\frac{q'_i(\theta)}{q'_i(\bar{\theta})}\right)^{a_i}\right)^{-1}, \tag{3.52}$$

where we defined

$$c_\theta \triangleq \prod_{j \neq i} \left(\frac{q_j(\theta)}{q_j(\bar{\theta})}\right)^{a_j} = \prod_{j \neq i} \left(\frac{q'_j(\theta)}{q'_j(\bar{\theta})}\right)^{a_j}, \tag{3.53}$$

with the equality following from (3.30). To prove monotonicity we must show that under the considered assignments for the input belief vectors, the output belief vectors satisfy the inequality $q'(\bar{\theta}) > q(\bar{\theta})$. To this end, observe from (3.52) that we have the following equivalence:

$$q'(\bar{\theta}) > q(\bar{\theta}) \Longleftrightarrow \sum_{\theta \neq \bar{\theta}} c_\theta \left(\frac{q'_i(\theta)}{q'_i(\bar{\theta})}\right)^{a_i} < \sum_{\theta \neq \bar{\theta}} c_\theta \left(\frac{q_i(\theta)}{q_i(\bar{\theta})}\right)^{a_i}. \tag{3.54}$$

In view of (3.50), the inequality on the RHS of the implication holds if, and only if, $a_i > 0$. This implies that the inequality on the LHS holds if, and only if, $a_i > 0$, and the proof is complete.

■

In summary, Theorem 3.1 reveals that under the considered behavioral assumptions, the pooling rule is a *weighted geometric average* of the beliefs, up to a normalization factor necessary to yield a probability vector. Note that the theorem requires Θ to contain at least three elements. This is because, for $|\Theta| = 2$, the independence of irrelevant alternatives is trivially satisfied, thus imposing no restrictions on the combination rule. In other words, this behavioral approach does not help deduce the form of the social learning algorithm for the case of two hypotheses.

We are now ready to exploit the general result in (3.36) in our social learning algorithm, specifically, in (3.10b). In order to use (3.36) in (3.10b), we need to take into account two facts. First, each agent k receives the intermediate beliefs only from its neighbors $j \in \mathcal{N}_k$. Thus, the ensemble of *input* belief vectors on which the pooling operator acts is given by the collection $\{\psi_{j,t}\}_{j \in \mathcal{N}_k}$. Second, the pooling operator C_k can be dependent on the particular agent k, which implies that the weight $a_j > 0$ appearing in (3.36) is replaced by a weight $a_{jk} > 0$, defined for $j \in \mathcal{N}_k$, and fulfilling the condition $\sum_{j\in\mathcal{N}_k} a_{jk} = 1$. In summary, the belief vector $\mu_{k,t}$ corresponding to agent k at time t is obtained through the following combination rule, for all $\theta \in \Theta$:

$$\mu_{k,t}(\theta) \propto \prod_{j\in\mathcal{N}_k} [\psi_{j,t}(\theta)]^{a_{jk}} \propto \prod_{j\in\mathcal{N}_k} [\mu_{j,t-1}(\theta)\ell_j(x_{j,t}|\theta)]^{a_{jk}}, \tag{3.55}$$

where in the last step we used the Bayesian update from (3.10a).

We see that rule (3.55) is identical to rule (3.18), namely, to the rule obtained in Section 3.3 based on information-theoretic principles. We arrive at the remarkable conclusion that social learning with *geometric averaging* is optimal both under the minimization problem in (3.13) and under the behavioral Assumptions 3.1, 3.2, 3.3, and 3.4. In addition, note that the information-theoretic approach adopted in Section 3.3 is less restrictive in that it does *not* require the condition $|\Theta| > 2$.

Before proceeding further, it is useful to examine another property of the pooling operator in (3.36), known as *external Bayesianity* [79, 80, 101]. It can be stated as follows. Assume *all* agents observe the *same data* x and have the *same likelihood* $\ell(x|\theta)$. Under these conditions, it is meaningful to expect that the belief obtained by first updating all the K beliefs with the common likelihood $\ell(x|\theta)$ and then combining the results, is equivalent to the belief obtained by first combining the K beliefs and then updating the result. A pooling rule satisfies external Bayesianity if this interchangeability property holds when all agents have the same data and likelihood.

We now verify that the geometric pooling operator in (3.36) is externally Bayesian [135]. If we first compute the Bayesian updates of the individual belief vectors $\{q_j\}$:

$$q_j^{\mathsf{Bu}}(\theta) \propto q_j(\theta)\ell(x|\theta), \tag{3.56}$$

and then combine them by using the geometric-averaging rule, we obtain

$$\begin{aligned} q(\theta) &\propto \prod_{j=1}^{K} \left[q_j^{\mathsf{Bu}}(\theta)\right]^{a_j} \propto \prod_{j=1}^{K} [q_j(\theta)\ell(x|\theta)]^{a_j} \\ &= [\ell(x|\theta)]^{\sum_{j=1}^{K} a_j} \times \prod_{j=1}^{K} q_j^{a_j}(\theta) = \ell(x|\theta) \prod_{j=1}^{K} q_j^{a_j}(\theta), \end{aligned} \tag{3.57}$$

where in the last step we used the condition $\sum_{j=1}^{K} a_j = 1$. Now note that the term $\ell(x|\theta)\prod_{j=1}^{K} q_j^{a_j}(\theta)$ in (3.57) can be interpreted as resulting from the following process. First combine the individual beliefs $q_j(\theta)$ with geometric averaging to obtain an aggregate belief $\propto \prod_{j=1}^{K} q_j^{a_j}(\theta)$, and then perform a Bayesian update, using this aggregate belief along with the likelihood $\ell(x|\theta)$, to obtain $q(\theta)$. We have thus shown that the geometric-averaging rule is externally Bayesian.

It is worth mentioning that there exists in the literature an axiomatic approach, different from the one we have illustrated here, which aims at finding the general functional form of pooling operators that are externally Bayesian. In these alternative approaches, external Bayesianity is a requirement rather than a consequence. Remarkably, it can be shown that the weighted geometric average is the only rule that satisfies external Bayesianity under certain additional constraints — see [79, 80].

3.4.2 Arithmetic-Averaging Rule, Revisited

The next theorem, proved in [131], reveals how the pooling rule changes if independence of irrelevant alternatives is replaced by separability.

Theorem 3.2 (Arithmetic averaging). Let Assumptions 3.1, 3.2, 3.3, and 3.5 be satisfied, and let $q = \mathsf{C}(q_1, q_2, \ldots, q_K)$. If $|\Theta| > 2$, we must have, for all $\theta \in \Theta$,

$$q(\theta) = \sum_{j=1}^{K} a_j q_j(\theta), \tag{3.58}$$

with $a_j > 0$ and $\sum_{j=1}^{K} a_j = 1$.

Proof. Let us focus on $q(\theta)$, namely, the θth entry of the belief vector q resulting from the pooling rule. In view of the separability assumption, $q(\theta)$ depends only on the values $\{q_j(\theta)\}_{j=1}^{K}$, which allows us to write, for a certain function g_θ,

$$q(\theta) = g_\theta(q_1(\theta), q_2(\theta), \ldots, q_K(\theta)). \tag{3.59}$$

Note that in principle separability allows the functional form of g_θ to depend on θ (which justifies the subscript θ). However, we now show that label neutrality implies that this dependence on θ disappears. In fact, consider a set of values $z_1, z_2, \ldots, z_K$, with

$$q_j(\theta') = z_j \quad \text{for } j = 1, 2, \ldots, K \tag{3.60}$$

and with $q(\theta') = z$ for some value z. Assume instead that (3.60) is verified for a different label θ''. Then, by label neutrality, the θ''th entry of the output belief vector must take on the same value, i.e., we must have $q(\theta'') = z$. In other words, the mapping by which the function g_θ associates the input values $\{z_j\}$ with the output value z does not depend on the particular label θ, but only on the values $\{z_j\}$. Therefore, it is legitimate to write

$$q(\theta) = g(q_1(\theta), q_2(\theta), \ldots, q_K(\theta)), \tag{3.61}$$

where we do not have different functions g_θ corresponding to different labels, i.e., the functional form of g does not depend on θ. Moreover, g is continuous by continuity of the pooling operator.

Consider now two hypotheses θ' and θ''. Since the pooling operator produces a belief vector whose entries add up to 1, we can write

$$\begin{aligned} g(q_1(\theta'), q_2(\theta'), \ldots, q_K(\theta')) + g(q_1(\theta''), q_2(\theta''), \ldots, q_K(\theta'')) \\ = 1 - \sum_{\theta \notin \{\theta', \theta''\}} g(q_1(\theta), q_2(\theta), \ldots, q_K(\theta)). \end{aligned} \tag{3.62}$$

Consider next another ensemble of beliefs $\{\widetilde{q}_j\}$, with the following assignment for $j = 1, 2, \ldots, K$:

$$\widetilde{q}_j(\theta') = q_j(\theta') + q_j(\theta''), \tag{3.63}$$
$$\widetilde{q}_j(\theta'') = 0, \tag{3.64}$$
$$\widetilde{q}_j(\theta) = q_j(\theta) \quad \forall \theta \notin \{\theta', \theta''\}. \tag{3.65}$$

We can apply (3.62) to $\{\widetilde{q}_j\}$ and get

$$\begin{aligned} g(\widetilde{q}_1(\theta'), \widetilde{q}_2(\theta'), \ldots, \widetilde{q}_K(\theta')) + g(\widetilde{q}_1(\theta''), \widetilde{q}_2(\theta''), \ldots, \widetilde{q}_K(\theta'')) \\ = 1 - \sum_{\theta \notin \{\theta', \theta''\}} g(q_1(\theta), q_2(\theta), \ldots, q_K(\theta)). \end{aligned} \tag{3.66}$$

Grouping (3.62) and (3.66), we can write

$$\begin{aligned} & g\underbrace{(q_1(\theta'), q_2(\theta'), \ldots, q_K(\theta'))}_{x} + g\underbrace{(q_1(\theta''), q_2(\theta''), \ldots, q_K(\theta''))}_{y} \\ = \;& g\underbrace{(q_1(\theta') + q_1(\theta''), q_2(\theta') + q_2(\theta''), \ldots, q_K(\theta') + q_K(\theta''))}_{x+y} + \underbrace{g(0, 0, \ldots, 0)}_{g_0}. \end{aligned} \tag{3.67}$$

Introducing, for $x \in [0, 1]^K$, the function

$$h(x) = g(x) - g_0, \tag{3.68}$$

we can represent (3.67) as

$$h(x + y) = h(x) + h(y), \tag{3.69}$$

where the entries x_j and y_j of the K-dimensional vectors x and y obey the condition $0 \le x_j + y_j \le 1$ for $j = 1, 2, \ldots, K$. Therefore, Eq. (3.69) is a *conditional* multidimensional Cauchy equation, with the qualification "conditional" indicating the fact that the admissible domain for the vectors x and y is restricted [104]. For our particular restricted domain $0 \le x_j + y_j \le 1$ for $j = 1, 2, \ldots, K$, it is known that the unique family of continuous solutions is the same as in the unrestricted case (3.46) examined before, namely we have that [4, 57, 104][4]

$$h(x) = \sum_{j=1}^{K} a_j x_j. \tag{3.71}$$

Returning to the definitions of $h(x)$ in (3.68) and $g(x)$ in (3.61), we obtain

$$q(\theta) = \sum_{j=1}^{K} a_j q_j(\theta) + g_0. \tag{3.72}$$

Now, applying unanimity to a common input belief vector p we get

$$p = p \sum_{j=1}^{K} a_j + g_0, \tag{3.73}$$

which, considering a vector p with one entry equal to 0, yields $g_0 = 0$. As a result, Eq. (3.72) coincides with (3.58). Using (3.73) with $g_0 = 0$ we also see that the combination weights must add up to 1. Moreover, it is readily seen that the pooling rule in (3.58) satisfies monotonicity if, and only if, the combination weights are all positive. We complete the proof by observing that, for all choices of input belief vectors $\{q_j\}$ and positive combination weights $\{a_j\}$ that add up to 1, the pooling rule (3.58) returns a valid belief vector q (i.e., a probability vector) and fulfills Assumptions 3.1 and 3.5.

■

By following a similar argument to the one applied to the geometric rule in the previous section, we can now incorporate (3.58) into our social learning algorithm, i.e., into (3.10b), to get

$$\mu_{k,t}(\theta) = \sum_{j \in \mathcal{N}_k} a_{jk} \psi_{j,t}(\theta), \tag{3.74}$$

with the weights a_{jk}, defined for $j \in \mathcal{N}_k$, fulfilling the conditions $a_{jk} > 0$ and $\sum_{j \in \mathcal{N}_k} a_{jk} = 1$. Rule (3.74) is identical to (3.23), which arose in Section 3.3 from information-theoretic principles. Remarkably again, we conclude that social learning with *arithmetic averaging* is optimal both under the minimization problem in (3.21) and under Assumptions 3.1, 3.2, 3.3, and 3.5, that is, under behavioral axioms where *independence of irrelevant alternatives* is replaced by *separability*. As was the case for geometric averaging, for arithmetic averaging the behavioral approach requires at least three hypotheses, while the information-theoretic approach adopted in Section 3.3 covers even the case $|\Theta| = 2$.

[4]Actually, the fundamental result proved in [57] refers to the case $K = 1$. However, it can be straightforwardly extended to an arbitrary K by noting that, in view of (3.69), we can also represent the function h as the sum of K functions

$$h(x) = h(x_1, 0, \ldots, 0) + h(0, x_2, 0, \ldots, 0) + \ldots + h(0, 0, \ldots, x_K) \tag{3.70}$$

and then apply the result available for $K = 1$ to any of these functions.

Table 3.1: Summary of pooling rules and pertinent criteria.

Pooling rule	Information-theoretic	Behavioral
Geometric avg.	$\sum_{j \in \mathcal{N}_k} a_{jk} D(p \| \psi_{j,t})$	Indep. irrelevant alternatives
Arithmetic avg.	$\sum_{j \in \mathcal{N}_k} a_{jk} D(\psi_{j,t} \| p)$	Separability

In summary, from the analysis conducted in this section and Section 3.3, we learned that the most popular combination rules, the geometric (a.k.a. log-linear) and arithmetic (a.k.a. linear) rules, are the optimal solutions to two different formulations: One based on information-theoretic arguments and another based on behavioral arguments. In one problem, we focus on minimizing an information-theoretic measure that quantifies the discrepancy between the target belief and the beliefs of neighboring agents. In the other problem, we place a set of axiomatic constraints on the agents' admissible behavior. Table 3.1 summarizes the results. In the column referred to behavioral assumptions, we report only the distinguishing behavioral assumption for the two combination rules, implicitly implying that the other three assumptions (label neutrality, unanimity, and monotonicity) are fulfilled in both cases.

3.5 Unifying Framework

The earlier Figure 3.2 illustrates the structure of the social learning framework. In the figure, the first block performs a Bayesian update from a previous-lag belief to an intermediate belief using the fresh data sample. The second block implements a suitable combination rule, such as geometric or arithmetic pooling.

However, in several applications other requirements emerge, under which the criteria adopted to design the scheme in Figure 3.2 will need to be revisited. We provide two relevant examples, which will help motivate a unifying framework for non-Bayesian social learning.

Example 3.2 (Adaptive social learning). The first example pertains to a modification of Bayes' rule, motivated by a strong need for *adaptation* when the agents need to be responsive to drifts in the environmental conditions, e.g., the state of nature, the statistical properties of the streaming data, the likelihood models, or the network topology. We will explain later in Chapter 8 that the social learning scheme of Figure 3.2 is not capable of adapting well to changes in the environment and that the agents exhibit significant stubbornness in their behavior.

In order to modify the Bayesian update to infuse adaptation, we return to the information-theoretic approach from Section 2.3. We explained there that Bayes' rule results from minimizing the free-energy $F(p)$ introduced in (2.61) or its variation $\widetilde{F}(p)$ in (2.70). In either of these formulations, the contributions of the prior and likelihood are weighted equally. One principled way to induce adaptation is to modify the cost function so as to give more relative importance to the *new data*, which appears as an argument of the likelihood. To this end, we will formulate in Chapter 8 the following alternative problem to determine the intermediate belief vector

$$\psi_{k,t} = \arg\min_{p\in\Delta_H} \left\{(1-\delta)D(p||\mu_{k,t}^{\mathsf{Bu}}) + \delta D(p||\mu_{k,t}^{\mathsf{lik}})\right\}, \quad 0<\delta<1, \tag{3.75}$$

where $\mu_{k,t}^{\mathsf{Bu}}$ denotes the traditional Bayesian update defined in (2.64), specifically computed by using as the prior the previous-lag belief vector $\mu_{k,t-1}$, while $\mu_{k,t}^{\mathsf{lik}}$ denotes the posterior defined in (2.67), which assumes a uniform prior. In both cases we use the likelihood $\ell_k(x_{k,t}|\theta)$.

Observe that (3.75) minimizes a convex combination of two KL divergences. The first divergence measures the discrepancy between the target belief vector p and the Bayesian update $\mu_{k,t}^{\mathsf{Bu}}$, which takes into account both the new data and the past belief vector $\mu_{k,t-1}$. The second divergence measures the discrepancy between p and the posterior $\mu_{k,t}^{\mathsf{lik}}$, which ignores the past belief and is based on the new data only. The design parameter $\delta\in(0,1)$ determines the level of adaptation by giving more or less importance to fresh data over past data. As $\delta\to 0$, we recover the traditional Bayesian update. On the other hand, as δ increases, the role of $D(p||\mu_{k,t}^{\mathsf{lik}})$ is enhanced and the minimization in (3.75) will tend to promote belief vectors that are closer to $\mu_{k,t}^{\mathsf{lik}}$. Since $\mu_{k,t}^{\mathsf{lik}}$ does not embody prior information, this mechanism provides a way to depress information accumulated from past data and to emphasize information coming from fresh data.

Example 3.3 (Partial information sharing). It has been assumed so far that each agent k sends over the network its full belief vector $\psi_{k,t}$. This might not always be the case. For instance, due to privacy reasons, or simply for the desire of "discussing" particular opinions, the agents might not be willing to share their entire belief vector. Another relevant requirement concerns the need to reduce communication costs by compressing the information to be transmitted. In either case, the agents share an *encoded* version of their beliefs, yielding an intermediate processing step in the social learning mechanism. This encoding step would be implemented locally, i.e., before communication takes place, but it would then require a *decoding* stage before the social learning phase.

One interesting type of encoding is the sharing of *partial information*. For instance, the agents might exchange the belief about one *hypothesis of interest* $\vartheta^\bullet\in\Theta$. Such model can be conveniently abstracted by saying that each agent k sends an encoded version of its intermediate belief vector $\psi_{k,t}$. In this case, the encoded version amounts to extracting only the entry $\psi_{k,t}(\vartheta^\bullet)$, whereas the entries corresponding to $\theta\neq\vartheta^\bullet$ are not shared. Upon receiving $\psi_{j,t}(\vartheta^\bullet)$ from its neighbors $j\in\mathcal{N}_k\backslash\{k\}$, agent k can perform a decoding operation to fill in the missing entries in the intermediate belief vectors $\psi_{j,t}$. This process gives rise to a *decoded* or *reconstructed* belief vector $\widehat{\psi}_{j,t}^{(k)}$ that agent k

assigns to agent j. One possible decoding rule is as follows:

$$\widehat{\psi}_{j,t}^{(k)}(\theta) = \begin{cases} \psi_{j,t}(\vartheta^{\bullet}) & \text{if } \theta = \vartheta^{\bullet}, \\ \dfrac{1 - \psi_{j,t}(\vartheta^{\bullet})}{H-1} & \text{otherwise,} \end{cases} \tag{3.76}$$

where we note that the remaining probability mass, $1 - \psi_{j,t}(\vartheta^{\bullet})$, is distributed equally across the hypotheses for which no information is shared. The problem of social learning under partial information sharing will be studied in Chapter 11.

Motivated by the aforementioned two examples, we can introduce a unifying framework for non-Bayesian social learning described by the diagram in Figure 3.3 and summarized by the following four steps.

$$(\mu_{k,t-1}, x_{k,t}) \overset{\text{update}}{\longrightarrow} \psi_{k,t}, \tag{3.77a}$$

$$\psi_{k,t} \overset{\text{encode}}{\longrightarrow} e_{k,t}, \tag{3.77b}$$

$$\left(\psi_{k,t}, \{e_{j,t}\}_{j\in\mathcal{N}_k\setminus\{k\}}\right) \overset{\text{decode}}{\longrightarrow} \left\{\widehat{\psi}_{j,t}^{(k)}\right\}_{j\in\mathcal{N}_k}, \tag{3.77c}$$

$$\left\{\widehat{\psi}_{j,t}^{(k)}\right\}_{j\in\mathcal{N}_k} \overset{\text{combine}}{\longrightarrow} \mu_{k,t}. \tag{3.77d}$$

Update. The self-learning step (3.77a) is implemented through a *general* update rule. This is not necessarily a Bayesian update. For example, it could be an adaptive rule as in Example 3.2 and Chapter 8.

Encode. In step (3.77b) we introduce a local encoder that is used by each agent k *before sharing*, to map its locally updated belief vector into some encoded vector $e_{k,t} \in \mathcal{E}_k$. The space $\mathcal{E}_k$ depends on the particular encoding scheme. The encoding operation reflects the necessity of accounting for some constraints relative to the information that can be shared over the network. For instance, $e_{k,t}$ might be a single entry of $\psi_{k,t}$ as in Example 3.3 and Chapter 11.

Decode. Agent k has direct access to *its own* uncoded belief vector $\psi_{k,t}$, whereas it only receives encoded signals $e_{j,t}$ from neighbors $j \in \mathcal{N}_k\setminus\{k\}$. Therefore, in step (3.77c) we introduce a decoder that is employed by agent k to construct a set of belief vectors $\{\widehat{\psi}_{j,t}\}_{j\in\mathcal{N}_k}$ from the available information $(\psi_{k,t}, \{e_{j,t}\}_{j\in\mathcal{N}_k\setminus\{k\}})$. For example, when $e_{j,t}$ is a single entry of the belief vector, one possible decoding rule is given by (3.76).

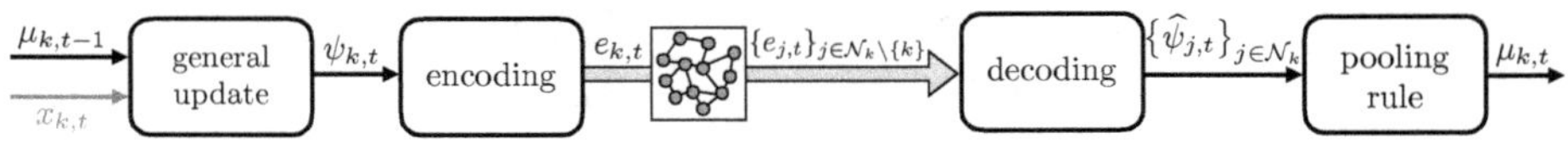

Figure 3.3: Unifying framework for non-Bayesian social learning. In comparison with Figure 3.2, the following distinguishing elements emerge: *i)* a general update rule that each agent k applies to its past belief vector $\mu_{k,t-1}$ and new data $x_{k,t}$; *ii)* an *encoding* step that each agent k applies to the intermediate belief vector before sharing it within the social group; and *iii)* a *decoding* step that each agent k applies to the available information, i.e., to its own belief vector $\psi_{k,t}$ and the encoded information received from neighbors $j \in \mathcal{N}_k\setminus\{k\}$.

Combine. After the above three steps, each agent k possesses a collection of *reconstructed* belief vectors $\{\widehat{\psi}^{(k)}_{j,t}\}_{j\in\mathcal{N}_k}$ associated with its neighbors. According to the analysis conducted in the previous sections, in step (3.77d) agent k will adopt a suitable combination rule to aggregate these beliefs.

Chapter 4

Network Models

Social learning relies on the local exchange of information between spatially dispersed agents that interact according to a certain network topology. We describe in this chapter the network models relevant to decentralized learning and comment on their fundamental properties.

4.1 Network Graphs

We focus in our treatment on networks of agents. Any two agents may be connected directly by an edge if they are neighbors, or they may be connected by a path that passes through other intermediate agents, or they may not be connected at all. The topology of a network can be described in terms of graphs — see, e.g., [22, 151, 155, 171].

Definition 4.1 (Graphs). A network of size K is generally represented by a *directed* graph consisting of K vertices (which we will refer to more frequently as nodes or agents) and a set of *directed* edges connecting the nodes. By "directed" we mean that an edge might exist from node j to node k and not in the opposite direction. When all edges in a graph exist in both directions, the graph is called "undirected".

A relevant concept associated with a graph is the *path.*

Definition 4.2 (Paths). A *directed* path from node j to node k is a sequence of directed edges, where the first edge in the sequence starts at j and the last edge ends at k. When the starting and ending nodes coincide, the path is called a *cycle.* When there is an edge that connects a node to itself, then the cycle has length 1 and is called a *self-loop.*

Another relevant concept is the *neighborhood.*

Definition 4.3 (Neighborhoods). The neighborhood $\mathcal{N}_k$ of node k is the set of nodes j (possibly including node k itself) for which there exists an edge starting at j and ending at k.

Note that we have introduced a *directed* neighborhood, a.k.a. *in-neighborhood.* In fact, the neighborhood $\mathcal{N}_k$ includes only nodes connected to k by an edge *entering* node k. As we have seen in the previous chapter, the neighborhood $\mathcal{N}_k$ identifies the agents from which agent k collects the beliefs to be combined during the pooling stage, and is therefore sufficient to describe the social learning strategies. Obviously, we could define the dual directed neighborhood (a.k.a. *out-neighborhood*) of nodes connected to k by an edge *emanating from* k. In the following, to make the terminology lighter, we will simply use the term *neighborhood,* and it should be clear from the context if we are referring to an in-neighborhood or to an out-neighborhood. Likewise, unless otherwise specified, graphs and related descriptors (e.g., edges) are implicitly intended to be directed.

We observe that there might exist multiple paths starting at node j and ending at node k. However, it is readily seen that the *shortest* path cannot exceed length $K-1$ (because if a node is visited more than once along the path, we can cut redundant sub-paths). This also implies that the shortest *cycle* cannot exceed length K. In fact, a cycle (of length greater than 1) linking k to itself can always be made of a path from k to some other node k', plus one edge from k' to k. By choosing the shortest path from k to k', we see that the length of a cycle cannot exceed K.

In our treatment, it is useful to consider *weighted* graphs, where we associate a nonnegative weight a_{jk} with each pair (j,k), including the case $j = k$. These nonnegative weights can be conveniently collected into a $K \times K$ *combination matrix* $A = [a_{jk}]$. Every such matrix with nonnegative entries will be called a *nonnegative matrix.*

Definition 4.4 (Weighted graphs and combination matrices). We associate with every nonnegative square matrix $A = [a_{jk}]$, also called a combination matrix, a weighted graph constructed as follows. First, one constructs the *support graph* of A, where an edge will exist *from* node j *to* node k whenever $a_{jk} > 0$. In particular, node k would have a self-loop when $a_{kk} > 0$. Then, the graph becomes weighted by associating the matrix entry a_{jk} with the edge emanating from j

toward k. It is useful to note that the neighborhood of node k, introduced in Definition 4.3, can be alternatively written in terms of the combination matrix as

$$\mathcal{N}_k = \{j : a_{jk} > 0\}. \tag{4.1}$$

We will generally be dealing with weighted graphs and their associated combination matrices. To lighten the terminology, we will often simply refer to the "graph" rather than to the "weighted graph."

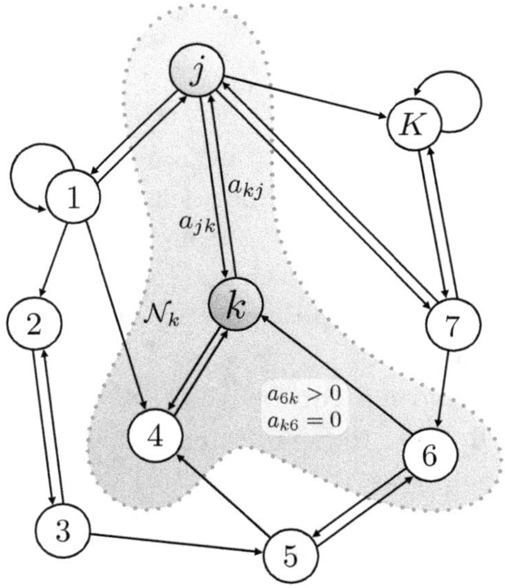

Figure 4.1: Agents that are linked by edges can share information. The neighborhood of agent k is marked by the dotted line and consists of the set $\mathcal{N}_k = \{4, 6, j\}$. Likewise, the neighborhood of agent 1 consists of the set $\mathcal{N}_1 = \{1, j\}$. For emphasis in the figure, we are representing edges between agents by two separate directed arrows. We will continue to use this representation whenever useful. Otherwise, in future network pictures we will represent undirected edges by a single segment with no arrows. We show the combination weights for some agents. For example, the weights a_{jk} and a_{kj} are associated with the directed edges from j to k and from k to j, respectively. We also emphasize that $a_{k6} = 0$ since there is no edge from k to 6, while $a_{6k} > 0$ because there is an edge from 6 to k.

Figure 4.1 shows one example of a network graph, where we emphasize the combination weights and the neighborhood of agent k. An edge between two neighboring agents is represented by a directed arrow to indicate the direction in which information can flow. The absence of an edge signifies that the corresponding combination weight is equal to 0.

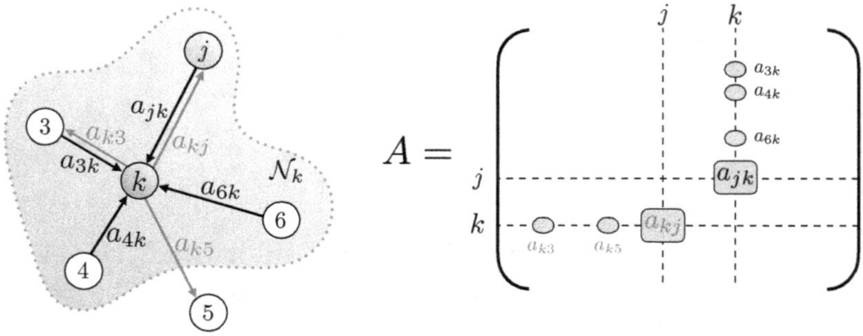

Figure 4.2: We associate a $K \times K$ combination matrix A with every network of K agents. The (j, k) entry of A contains the combination weight a_{jk}, which scales the information arriving at agent k and originating at agent j.

Let us comment on the practical meaning of the combination weights. As we have seen in Chapter 3, in the social learning strategies, agent k combines the information it receives from agents $j \in \mathcal{N}_k$ by using a set of convex positive combination weights $\{a_{jk}\}_{j \in \mathcal{N}_k}$. This scaling can be interpreted as a measure of the confidence that agent k assigns to its interaction with agent j.

The weights a_{jk} and a_{kj} can be different, and one or both weights can also be zero. In fact, the subscripts j and k in a_{jk} have different meanings. This is emphasized in Figure 4.2. The row index j designates the source agent (i.e., the sender) and the column index k designates the sink agent (i.e., the receiver). In other words, the entries on the kth column of A contain the coefficients used by agent k to scale the information corresponding to each agent (i.e., each row) j. If $a_{jk} = 0$, the information from agent j is not used by agent k — see Figure 4.2.

In some cases we need to distinguish between the physical network topology and the topology resulting from the combination matrix A. For example, consider a collection of electronic devices organized into a wireless communication network. The physical topology defines whether one agent can send and/or receive information from another agent. To scale the information exchanged across the network, the agents will need to assign some weights on top of this topology. Obviously, the weight a_{jk} will be zero when agent k cannot receive from agent j. However, this weight can be zero even when agent k is physically connected to agent j and can receive information from it. This happens when agent k deliberately decides to ignore the information from agent j. Thus, there can be a difference between the underlying physical topology and the actual "communication" topology defined by the support graph of A. However, once the combination matrix is defined, the underlying physical topology is "overwritten" by the support graph of A. In other words, if k ignores information received from j, then it is irrelevant to know whether this happens because k is physically unable to receive information from j, or because k chooses to ignore it on purpose. In summary, moving forward, whenever we refer to a network topology we will be in effect referring to the topology defined by the support graph of A.

The next definition lists four useful notions of connectivity.

Definition 4.5 (Network connectivity). We distinguish four types of connectivity.

i) **Connected graph.** Consider the nontrivial case $K > 1$. A graph (or network) is said to be connected when, for $j = 1, 2, \ldots, K$ and $k = 1, 2, \ldots, K$, with $j \neq k$, there exists a path originating at j and ending at k. In other words, given any two nodes, there are paths in both directions linking them (the paths need not be the same). Note that over a connected graph, for any node k there always exists a path originating and ending at k since, even in the absence of a self-loop (i.e., even when $a_{kk} = 0$) we can join two paths from k to j and from j to k. In the trivial case $K = 1$, we will say that the graph is connected when $a_{11} > 0$.

ii) **Primitive graph.** A connected graph is said to be primitive when there exist paths of *common* length $m > 0$ linking any two distinct nodes in both directions and linking any node to itself.

iii) **Strong graph.** A connected graph is said to be strong when it has at least one *self-loop*, meaning that $a_{kk} > 0$ for some node k. As we will show later in Section 4.3, a strong graph is also primitive, but the reverse implication does not hold.

iv) **Weak graph.** The graphs that do not belong to the family described by definition i) are said to be weak.

Note that in the definition of a connected graph, since j and k are arbitrary nodes, we require that any two distinct nodes are linked in *both* directions, either directly when they are neighbors or by passing through intermediate nodes when they are not neighbors. In this way, information can flow in *both* directions between any two distinct agents in the network, although the forward path from a node j to some other node k need not be the same as the backward path from k to j.

Figure 4.3 shows four examples of networks corresponding to the four families of graphs in Definition 4.5. The leftmost graph is *connected*, because

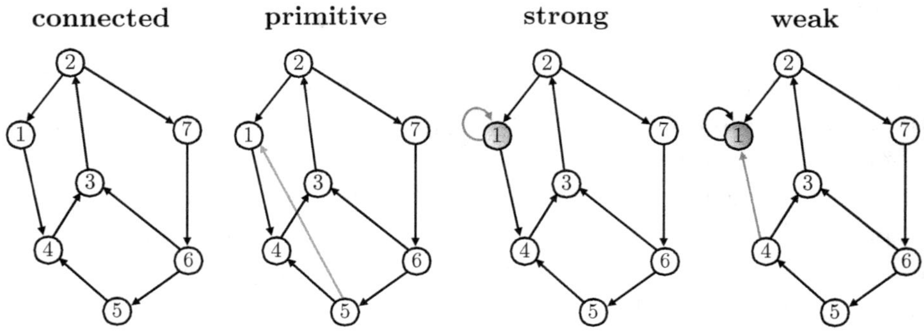

Figure 4.3: (*Leftmost panel*) Connected graph, where there exist paths between any two agents in both directions. (*Second panel*) Primitive graph, where there exist paths of length 16 between any two distinct agents, in both directions, and from any agent to itself. (*Third panel*) Strong graph, obtained from the connected graph in the leftmost panel by simply adding a self-loop at agent 1. (*Rightmost panel*) Weak graph, obtained from the strong graph in the third panel by simply reversing the arrow connecting agents 1 and 4. This slight change makes agent 1 incapable of sending information to any of the other agents in the network, even though information can reach agent 1 from all other agents (directly or indirectly).

if we select any two nodes j and k, we can find paths linking them in both directions. For example, for nodes 2 and 4, one valid path from 2 to 4 goes through 1 and one valid path for the reverse direction from 4 to 2 goes through 3. Similarly, paths can be determined linking all other combinations of nodes in both directions. Note that in the considered example no pair of nodes is *directly* connected by edges in both directions. Nevertheless, we can still find paths linking any pair of nodes in both directions.

Note also that the paths connecting 1 and 2 must have odd length, whereas the paths connecting 1 and 3 must have even length. Therefore, no paths of common length can be found, and the graph is not primitive. In comparison, the graph shown in the second panel of Figure 4.3 is primitive. In fact, it can be verified that, thanks to the addition of the edge represented in green, there exist paths of common length 16 between any two distinct nodes, in both directions, and that there also exist paths of length 16 linking any node to itself. We remark that this graph has no self-loops.

The third panel shows a network with the same structure as the one in the leftmost panel, but for the additional presence of a self-loop at node 1, which makes the graph *strong*. Note that now it is possible to find paths of common length 10 between any two distinct nodes, in both directions, and also from any node to itself. Therefore, the graph is also primitive.

Finally, observe the rightmost graph in Figure 4.3. Compared with the strong graph in the third panel, we simply reversed the direction of the arrow that emanated from 1 toward 4. With this slight modification, the information from agent 1 cannot reach any other agent in the network and agent 1 is only at the receiving end. This graph is accordingly *weak*.

Throughout our treatment, we will use interchangeably the words "network" and "graph." Moreover, to avoid misunderstanding, we hasten to add that the terms "connected," "strong," and "weak," are not uniformly defined in the literature and can refer to slightly different notions. Nevertheless, we will remain faithful to Definition 4.5. We will shortly see that the graph families in this definition entail some useful correspondences with the families of irreducible, primitive, and reducible matrices that arise in matrix theory.

4.2 Combination Matrices

We have explained before how to associate a weighted graph with a nonnegative square matrix, which we called the combination matrix and denoted by A. The particular choice of the weights influences specific properties that will be of interest for the social learning strategies. For example, we explained in Chapter 3 how convex combination weights arise. Moreover, the choice of the weights determines the spectral properties of the matrix (i.e., the structure of its eigenvalues and eigenvectors), which will be seen to be critical for the asymptotic properties of the social learning strategies.

However, some other critical properties relative to the flow of information over the network are more immediately revealed by the mere support graph of A, namely, by the *unweighted* graph that encodes the network "skeleton" and describes only the interconnections between agents. In matrix analysis, there exist powerful tools to characterize the interplay between matrices and their support graphs. In particular, for nonnegative matrices, an elegant theory was developed by Perron and Frobenius [93, 126]. We exploit this theory to great effect in our analysis.

To start with, we show how the nth power of a nonnegative square matrix is related to paths of length n over its support graph.

Lemma 4.1 (Paths and matrix powers). Let A be a nonnegative $K \times K$ matrix, and consider the support graph of A. Over this graph, a path of length n between nodes j and k (including the case $j = k$) exists if, and only if, the (j, k) entry of the matrix power A^n is positive.

Proof. From the rules of matrix multiplication, the (j, k) entry of the nth power of A is given by

$$[A^n]_{jk} = \sum_{m_1=1}^{K} \sum_{m_2=1}^{K} \cdots \sum_{m_{n-1}=1}^{K} a_{jm_1}\, a_{m_1 m_2} \cdots a_{m_{n-1}k}. \tag{4.2}$$

Therefore, $[A^n]_{jk} > 0$ if, and only if, there exists at least one sequence of agent indices $\{m_1, m_2, \ldots, m_{n-1}\}$ associated with nonzero scaling weights $\{a_{jm_1}, a_{m_1m_2}, \ldots, a_{m_{n-1}k}\}$, i.e., if, and only if, we have a path

$$j \xrightarrow{a_{jm_1}} m_1 \xrightarrow{a_{m_1m_2}} m_2 \longrightarrow \cdots \longrightarrow m_{n-1} \xrightarrow{a_{m_{n-1}k}} k \qquad [n \text{ edges}]. \tag{4.3}$$

■

Next, we introduce *irreducible matrices*, which are tightly coupled with *connected networks*. Although the notion of irreducible matrices applies to matrices with entries of arbitrary sign, we restrict our treatment to the case of nonnegative matrices.

Definition 4.6 (Irreducible matrices). A nonnegative $K \times K$ matrix A is irreducible when its support graph is connected. In other words, when for any pair (j, k), including the case $j = k$, there exists a shortest path of some length $n_{jk} \le K$ (depending in general on j and k) that starts at j and ends at k. In view of Lemma 4.1, the following property holds for irreducible matrices: For any pair (j, k), including the case $j = k$, there exists a positive integer $n_{jk} \le K$ such that

$$[A^{n_{jk}}]_{jk} > 0. \tag{4.4}$$

Some of the forthcoming results will examine the spectral properties of useful families of matrices. In preparation for these results, it is useful to recall the basic definitions of the algebraic and geometric multiplicity of an eigenvalue [93, 126].

Definition 4.7 (Eigenvalue multiplicity). Consider a square matrix A (not necessarily nonnegative) and an eigenvalue λ of A. We have the following definitions:

i) The algebraic multiplicity of λ is the number of times it is repeated as a root of the characteristic equation $\det(A-\lambda I)=0$. An eigenvalue occurring only once is called *simple*. When we say "there are h eigenvalues equal to λ" we mean that the algebraic multiplicity of λ is equal to h.

ii) The geometric multiplicity of λ is the maximal number of linearly independent eigenvectors associated with it, namely, the dimension of the null space of $A-\lambda I$. The geometric multiplicity cannot exceed the algebraic multiplicity.

iii) An eigenvalue whose geometric multiplicity equals the algebraic multiplicity is called *semisimple*.

The next theorem establishes the fundamental properties of nonnegative irreducible matrices. Before stating the result, it might be useful to recall that the spectral radius of a square matrix is equal to the largest magnitude of its eigenvalues.

Theorem 4.1 (Perron-Frobenius theorem [126, p. 673]). Let A be a nonnegative irreducible $K \times K$ matrix. Then the following properties hold:

i) The matrix A has a *simple* eigenvalue λ equal to the spectral radius $\rho(A)$ and, moreover, $\rho(A) > 0$. All other eigenvalues of A are not equal to $\rho(A)$, but they can have *magnitude* equal to $\rho(A)$, i.e., λ need not be the only eigenvalue on the spectral circle.

ii) With proper sign scaling, all entries of the eigenvector of A corresponding to the eigenvalue $\lambda = \rho(A)$ can be made *positive*. Let v denote this eigenvector, with its entries $\{v_k\}$ normalized to add up to 1, i.e.,

$$Av = \lambda v, \quad \mathbb{1}^\top v = 1, \quad v_k > 0 \quad \text{for } k = 1, 2, \ldots, K. \tag{4.5}$$

We refer to v as the *Perron vector* of A. All eigenvectors of A associated with the other eigenvalues cannot be made nonnegative (they have entries with varied signs or complex-valued entries).

4.2.1 Convergence of Matrix Powers

As will become apparent in the next chapters, social learning algorithms will rely on repeated exchanges of information between neighboring nodes over a graph. This iterative process will correspond to information traversing longer and longer paths across the network as the number of iterations increases. Technically, such repeated interactions can be represented by matrix powers A^t, with t denoting the number of iterations. It is therefore critical to study the convergence properties of these matrix powers.

One useful notion of convergence for the matrix powers is *Cesàro summability*.

Definition 4.8 (Cesàro summability). A square matrix A (not necessarily nonnegative) is said to be Cesàro-summable when the arithmetic mean of the matrix powers converges, i.e., when there exists a matrix $A^{\bullet}$ such that

$$\lim_{t\to\infty} \frac{1}{t}\sum_{\tau=1}^{t} A^{\tau} = A^{\bullet}. \tag{4.6}$$

Cesàro summability will be useful in the study of social learning. For example, it will be exploited in Chapter 5 to examine the convergence of the beliefs under social learning with geometric averaging.

In some other cases we will appeal to a stronger notion of convergence. This will be the case in Chapters 6 and 9, when we will examine the error probability performance of social learning. The stronger notion we refer to requires that the matrix powers converge. In this case we say that A is a convergent matrix. It is readily seen that any convergent matrix is Cesàro-summable, since convergence of the powers implies convergence of their arithmetic means. The converse is in general not true.

The next theorem establishes the fundamental result on the convergence of matrix powers.

Theorem 4.2 (Convergent matrices [126, p. 630]). A square matrix A (not necessarily nonnegative) is said to be convergent when the limit as $t \to \infty$ of the sequence of powers A^t exists. This situation happens if, and only if, one of the following conditions is verified:

i) $\rho(A) < 1$, in which case A^t converges to a null matrix, i.e., a matrix with all entries equal to 0.

ii) $\rho(A) = 1$ with $\lambda = 1$ being a semisimple eigenvalue and with no other eigenvalues on the unit circle.

Theorem 4.2 is very general; it provides necessary and sufficient conditions for the convergence of any square matrix. We now focus on the family of nonnegative irreducible matrices. For this family, we are going to show in Theorem 4.3 that the convergence of matrix powers is tightly coupled with the notion of *primitive matrices*, which are introduced next.

Definition 4.9 (Primitive matrices). A nonnegative irreducible matrix A is said to be primitive when it has only one eigenvalue on its spectral circle and is called imprimitive otherwise. In view of Theorem 4.1, for primitive matrices the only eigenvalue on the spectral circle is equal to $\rho(A)$, whereas imprimitive matrices have other eigenvalues, all different from $\rho(A)$, but having magnitude equal to $\rho(A)$. Furthermore, the following property, a.k.a. Frobenius' test for primitivity, holds [126, p. 673]: A nonnegative irreducible matrix A is primitive if, and only if, there exists a positive integer m such that the entries of A^m are all positive. In view of Lemma 4.1, we see that primitive matrices are automatically associated with primitive graphs in the sense of Definition 4.5.

Comparing Frobenius' test for primitivity with the property of irreducible matrices in (4.4), we find now that the power m in A^m is uniform across the graph nodes, i.e., it does not change with the indices j and k.

For primitive matrices, the conclusions from the Perron-Frobenius theorem can be strengthened to establish the limiting behavior of the matrix powers, as stated in the next theorem.

Theorem 4.3 (Powers of primitive matrices [126, p. 674]). Consider a nonnegative irreducible matrix A and let $\lambda = \rho(A)$. Then, the matrix A is primitive if, and only if, $\lim_{t\to\infty}(A/\lambda)^t$ exists, in which case the limit will be a rank-one matrix with positive entries according to the following formula:

$$\lim_{t\to\infty}\left(\frac{A}{\lambda}\right)^t = \frac{v\,u^\mathsf{T}}{u^\mathsf{T}v}, \tag{4.7}$$

where v and u are the Perron vectors of A and A^T, respectively.
Furthermore, let λ_2 denote the second largest-magnitude eigenvalue of A, and let r be such that

$$\frac{|\lambda_2|}{\rho(A)} < r < 1. \tag{4.8}$$

Then, there exists a constant, C depending on A and r, such that

$$\left|\left[\left(\frac{A}{\lambda}\right)^t - \frac{v\,u^\mathsf{T}}{u^\mathsf{T}v}\right]_{jk}\right| \le Cr^t \tag{4.9}$$

for all indices j and k and all $t \in \mathbb{N}$.

4.3 Strong and Primitive Graphs

The assumption of a *connected* graph ensures that information will be flowing between any two arbitrary agents and that this flow of information is bidirectional: Information flows from j to k and from k to j, although the paths over which the flows occur need not be the same and the manner in which information is scaled over these paths can also be different. As we will see, e.g., in Chapters 5 and 7, connected graphs will be critical to guarantee full propagation of information across the network and enable successful social learning.

In addition to being connected, a primitive graph features the existence of *common-length* paths between any two distinct nodes, in both directions, and from each node to itself. In view of Theorem 4.3, for combination matrices associated with primitive graphs, the asymptotic behavior of the matrix powers is known. This additional knowledge will be exploited in our analysis to characterize the performance of social learning strategies, e.g., in Chapters 6 and 9.

Furthermore, the assumption of a strong graph requires that the network is connected and, additionally, there exists at least one agent in the network that trusts its own information and will assign some positive weight to it. This is a reasonable condition and is characteristic of many real networks. If $a_{kk} = 0$ for all k, then this means that all agents will be ignoring their individual information and will be relying instead on information received from other agents. The next lemma shows that strong graphs are always primitive.

Lemma 4.2 (Strong graphs are primitive). If the graph associated with a nonnegative $K \times K$ matrix A is strong, then there exists a positive integer m such that all entries of A^m are positive and, hence, A is a primitive matrix.

Proof. Since the graph associated with A is strong, it is also a connected graph. According to Definition 4.6, this means that A is an irreducible matrix. It follows that, for any pair (j, k), including the case $j = k$, there exists an integer $n_{jk} > 0$ such that the (j, k) entry of the matrix power $A^{n_{jk}}$ is positive. Note that this integer is dependent on indices j and k. We now go a step further and show that, over *strong* graphs, a common (i.e., independent of the particular agents j and k) power m exists such that all entries of A^m are positive.

Recall from Definition 4.5 that a strong graph is a connected graph with the additional requirement that there exists at least one agent k_0 with a self-loop, i.e., with $a_{k_0k_0} > 0$. We know from (4.4) that $\left[A^{n_{jk_0}}\right]_{jk_0} > 0$ for any agent j in the network. Then,

$$\begin{aligned}\left[A^{(n_{jk_0}+1)}\right]_{jk_0} &= \left[A^{n_{jk_0}}A\right]_{jk_0} = \sum_{m=1}^{K}\left[A^{n_{jk_0}}\right]_{jm} a_{mk_0}\\ &\geq \left[A^{n_{jk_0}}\right]_{jk_0} a_{k_0k_0} > 0,\end{aligned} \tag{4.10}$$

which implies that the positivity of the (j, k_o) entry is maintained at higher powers of A once it is satisfied at power n_{jk_o}. Let

$$m' \triangleq \max_{j \in \{1,2,\ldots,K\}} n_{jk_o}. \tag{4.11}$$

Note that $m' \leq K$ since index n_{jk} identifies the shortest path between nodes j and k, and we know that the shortest path cannot be longer than K. From (4.10) and (4.11), we can also write

$$\left[A^{m'}\right]_{jk_o} > 0 \tag{4.12}$$

for all j, which means that the entries on the k_oth column of $A^{m'}$ are all positive. Interchanging the roles of k_o and j, we can define an index

$$m'' \triangleq \max_{j \in \{1,2,\ldots,K\}} n_{k_o j}. \tag{4.13}$$

This index is still upper bounded by K and guarantees that

$$\left[A^{m''}\right]_{k_o j} > 0 \tag{4.14}$$

for all j, which means that the entries on the k_oth row of $A^{m''}$ are all positive.

Now, let $m = m' + m''$ and let us examine the entries of the matrix A^m. We can write schematically

$$A^m = A^{m'} A^{m''} = \begin{bmatrix} \times & \times & + & \times \\ \times & \times & + & \times \\ \times & \times & + & \times \\ \times & \times & + & \times \end{bmatrix} \begin{bmatrix} \times & \times & \times & \times \\ \times & \times & \times & \times \\ + & + & + & + \\ \times & \times & \times & \times \end{bmatrix}, \tag{4.15}$$

where the $+$ signs are used to refer to the *positive* entries on the k_oth column of $A^{m'}$ and the k_oth row of $A^{m''}$, whereas the $\times$ signs are used to refer to the remaining entries of $A^{m'}$ and $A^{m''}$, which are nonnegative. It is clear from the above equality that the resulting entries of A^m will all be positive, and we conclude that A is primitive from Definition 4.9.

■

It is useful to summarize the ties between network connectivity and irreducible or primitive matrices:

$$\begin{cases} \text{connected graph} \iff \text{irreducible matrix,} \\ \quad \text{strong graph} \implies \text{primitive matrix.} \end{cases} \tag{4.16}$$

Observe that in the second relation we do not have a double implication, since a primitive matrix can arise even when the graph is not strong. In other words, a primitive matrix is always associated with a connected graph since it is irreducible by definition, but this graph could have no self-loops — see the second panel from the left in Figure 4.3.

4.4 Stochastic Combination Matrices

We explained in Chapter 3 that, in the context of social learning, the combination weights employed by each agent to scale the information received from its neighbors form a convex combination, i.e., they are nonnegative and add up to 1. This property gives rise to *left stochastic* (a.k.a. column stochastic) combination matrices A. The term "stochastic matrix," arising in the theory of Markov chains, does not refer to any randomness in the entries of A; it simply means that the columns of A consist of nonnegative weights that add up to 1.

Definition 4.10 (Left and doubly stochastic matrices). A nonnegative $K \times K$ matrix A is said to be left stochastic when the entries on each of its columns add up to 1, namely, when

$$\sum_{j=1}^{K} a_{jk} = \sum_{j \in \mathcal{N}_k} a_{jk} = 1 \quad \Longleftrightarrow \quad \mathbb{1}^\top A = \mathbb{1}^\top. \tag{4.17}$$

Note that Eq. (4.17) implies that at least one weight a_{jk}, for $j = 1, 2, \ldots, K$, must be nonzero. Recalling definition (4.1), this means that, for a left stochastic matrix, the neighborhood of every node k is nonempty.
In the special case where also the entries on each row of A add up to 1 (which does not necessarily require A to be a symmetric matrix), the matrix is said to be *doubly stochastic*, and we have $A\mathbb{1} = \mathbb{1}$, or

$$A\frac{\mathbb{1}}{K} = \frac{\mathbb{1}}{K}. \tag{4.18}$$

This implies that, if a doubly stochastic matrix is also irreducible, its Perron vector is $v = \mathbb{1}/K$, i.e., the Perron vector has uniform entries.

From now on, we will always assume that the combination matrix is left stochastic. The next lemma shows one property of left stochastic matrices that will be useful in the sequel.

Lemma 4.3 (Spectral radius of left stochastic matrices). For any left stochastic matrix A,

$$\rho(A) = 1. \tag{4.19}$$

That is, the spectral radius of a left stochastic matrix is equal to 1.

Proof. The spectral radius is upper bounded by any matrix norm, and in particular by the maximum absolute column sum norm, yielding [93, 126]

$$\rho(A) \leq \max_{k \in \{1,2,\ldots,K\}} \sum_{j=1}^{K} |a_{jk}| = \max_{k \in \{1,2,\ldots,K\}} \sum_{j=1}^{K} a_{jk} = 1, \tag{4.20}$$

where the first equality holds because A is nonnegative and the second one because it is left stochastic. On the other hand, Eq. (4.17) implies that 1 is an eigenvalue of A^{T}, and since a matrix and its transpose share the same eigenvalues, we conclude that 1 is an eigenvalue of A. The claim then follows from (4.20).

■

Another useful property of left stochastic matrices is that they are always Cesàro-summable.

Theorem 4.4 (All left stochastic matrices are Cesàro-summable [126, p. 697]). Let A be a left stochastic matrix. Then, there exists a left stochastic matrix $A^{\bullet}$ such that

$$\lim_{t \to \infty} \frac{1}{t} \sum_{\tau=1}^{t} A^{\tau} = A^{\bullet}. \tag{4.21}$$

Moreover, if A is irreducible, with Perron vector v, then

$$A^{\bullet} = v \mathbb{1}^{\mathsf{T}}. \tag{4.22}$$

That is, the limiting matrix $A^{\bullet}$ is a rank-one matrix that has all columns equal to the Perron vector of A.

For a connected network, the previous theorem shows that the time-average of the combination-matrix powers converge to a matrix whose columns are all equal to the Perron vector v associated with A. As observed before, raising A to power t corresponds to applying the combination matrix t times, that is, to performing t nested combination steps. In other words, $[A^t]_{jk}$, the (j,k) entry of the matrix A^t, represents the weight that agent k would assign to agent j after t combination steps. According to (4.22), for all k, the time-average of weights $[A^t]_{jk}$ would converge to v_j, the jth entry of the Perron vector. As a result, weight v_j quantifies the importance or *centrality* that agent j assumes in the network. In fact, in graph theory, one useful indicator for the relative importance of the network nodes is the so-called *eigenvector centrality score* [16]. When a weighted graph is connected (see Definition 4.5) and described by a combination matrix A, the centrality score assigned to node j is represented by the jth entry of the Perron vector associated with A. Note that, over weighted graphs, the centrality score assigned to the nodes accounts not only for the network topology, but also for the intensity of interaction between the nodes, represented by the values of the combination weights.

For left stochastic and *primitive* matrices, the conclusions from Theorem 4.3 admit a simpler form that will be repeatedly used in our treatment.

Corollary 4.1 (Powers of left stochastic and primitive matrices). If a $K \times K$ matrix A is left stochastic and primitive, with Perron vector v, then

$$\lim_{t \to \infty} A^t = v \mathbb{1}^\top. \tag{4.23}$$

That is, the sequence of matrix powers converges to a rank-one matrix that has all columns equal to the Perron vector of A. Furthermore, denoting by λ_2 the second largest-magnitude eigenvalue of A, and letting

$$|\lambda_2| < r < 1, \tag{4.24}$$

there exists a constant C depending on A and r, such that

$$\left| \left[A^t - v \mathbb{1}^\top \right]_{jk} \right| \leq C r^t \tag{4.25}$$

for all indices j and k and all $t \in \mathbb{N}$.

Proof. Since $\rho(A) = 1$ in view of Lemma 4.3, Eq. (4.17) implies that $\mathbb{1}/K$ is the Perron vector of $A^\top$ (recall that the Perron vector is scaled so that its entries add up to 1). Therefore, since A is primitive, we can apply (4.7) with the choices $u = \mathbb{1}/K$ and $\lambda = \rho(A) = 1$ — see (4.19). Then, Eq. (4.23) follows from the relation $\mathbb{1}^\top v = 1$, which holds since v is the Perron vector of A. Equation (4.25) then follows from (4.9). ∎

4.5 Weak Graphs

We focused so far on connected graphs, i.e., networks where any two agents are reachable through some paths in both directions. We wish now to characterize the remaining types of networks, where some pairs of agents are connected only in one direction, or they are not even connected through any path. According to Definition 4.5, we refer to these networks as *weak graphs*.

Interestingly, the combination matrices associated with weak graphs can be represented in a canonical form, a.k.a. Gantmacher normal form [77], as detailed in the next theorem.[1]

[1] To avoid confusion, we remark that in [126, Eq. (8.4.6)] the combination matrix is right (instead of left) stochastic (i.e., the rows, and not the columns, add up to 1). As a result, the blocks referring to the sending and receiving networks introduced in Theorem 4.5 are switched.

Theorem 4.5 (Canonical form for reducible left stochastic matrices [126, p. 695]). Let A be a left stochastic $K \times K$ matrix associated with a weak graph. In the trivial case where the graph is made of isolated subnetworks that do not communicate with each other, A has a block-diagonal structure where each block is a left stochastic matrix corresponding to each isolated subnetwork. In this case the graph is said to be *completely reducible.* Otherwise, any weak graph can be partitioned into two groups of subnetworks, namely, $S \geq 1$ *sending* networks and $R \geq 1$ *receiving* networks. Then, A can always be reduced to the following canonical form by a suitable permutation of the agent labels:

$$A = \left[\begin{array}{cccc|cccc} A_1 & 0 & \cdots & 0 & A_{1,S+1} & A_{1,S+2} & \cdots & A_{1,S+R} \\ 0 & A_2 & \cdots & 0 & A_{2,S+1} & A_{2,S+2} & \cdots & A_{2,S+R} \\ \vdots & \vdots & \ddots & \vdots & \vdots & \vdots & \cdots & \vdots \\ 0 & 0 & \cdots & A_S & A_{S,S+1} & A_{S,S+2} & \cdots & A_{S,S+R} \\ \hline 0 & 0 & \cdots & 0 & A_{S+1} & A_{S+1,S+2} & \cdots & A_{S+1,S+R} \\ 0 & 0 & \cdots & 0 & 0 & A_{S+2} & \cdots & A_{S+2,S+R} \\ \vdots & \vdots & \ddots & \vdots & \vdots & \vdots & \ddots & \vdots \\ 0 & 0 & \cdots & 0 & 0 & 0 & \cdots & A_{S+R} \end{array}\right], \tag{4.26}$$

where the individual submatrices have the following properties:

i) All submatrices on the main diagonal are square.

ii) **Top left block.** The submatrix A_s, for $s = 1, 2, \ldots, S$, dictates the inner communication structure relative to agents in the sth sending network. Each submatrix A_s is irreducible (thus corresponding to a connected subgraph), and the entries on each of its columns add up to 1 since A is left stochastic.

iii) **Top right block.** The submatrix $A_{s,S+r}$, for $s = 1, 2, \ldots, S$ and $r = 1, 2, \ldots, R$, dictates the communication from agents in the sth sending network to agents in the rth receiving network.

iv) **Bottom right block.** The submatrix A_{S+r}, for $r = 1, 2, \ldots, R$, dictates the inner communication structure relative to agents in the rth receiving network. Each submatrix A_{S+r} is either irreducible or a 1×1 matrix equal to 0.[2] The submatrices $A_{S+r,S+r'}$, for $r, r' = 1, 2, \ldots, R$, dictate the communication from the rth to the r'th receiving network.

v) For each $r = 1, 2, \ldots, R$, at least one of the submatrices *lying above* A_{S+r} (i.e., $A_{1,S+r}, A_{2,S+r}, \ldots, A_{S+r-1,S+r}$) has at least one nonzero entry.

Theorem 4.5 reveals the structural properties of weak graphs, which are conveniently summarized below and represented in Figure 4.4.

- Recalling that the entry a_{jk} of the combination matrix A is relative to the flow of information *from j to k*, then the null bottom left block in

[2]Even when A_{S+r} is a 1×1 matrix equal to 0, the corresponding agent interacts with the network through the connections described by the submatrices lying above A_{S+r} in view of point v).

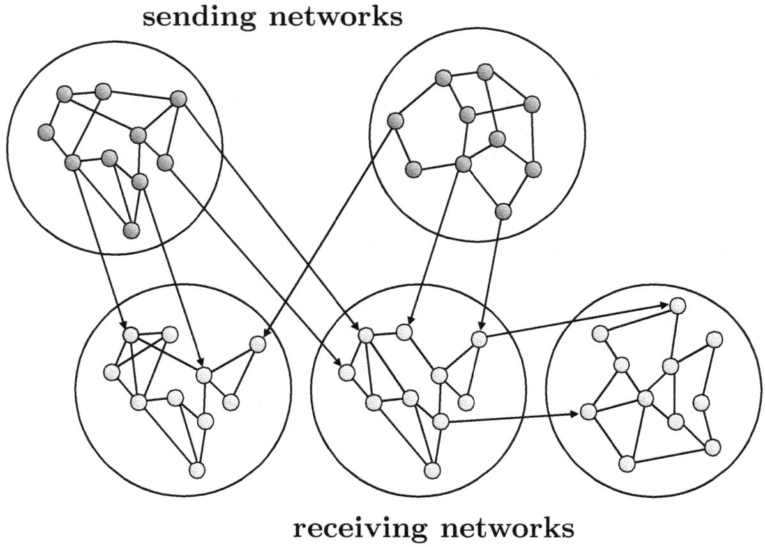

Figure 4.4: One example of a weak graph, reflecting the canonical structure in Theorem 4.5, with two sending networks (blue nodes) and three receiving networks (yellow nodes). As explained before, undirected edges are depicted with no arrows.

(4.26) signifies that the communication between agents in the sending networks and agents in the receiving networks is *one-directional.* That is, a link can exist from an agent in a sending network to an agent in a receiving network, but not in the reverse direction.

- The block-diagonal structure of the top left block in (4.26) signifies that sending networks do not communicate with each other.
- Point v) of Theorem 4.5 implies that a receiving network r *must* necessarily receive information from at least one agent external to r. Note that this agent need not belong to a sending network. It can also belong to a receiving network r' different from r. However, observe that we must have $r' < r$, because the matrix blocks below A_{S+r} are null or absent.
- Another conclusion stemming from point v) of Theorem 4.5 is that each receiving network is reachable through a path that originates at some sending network. Let us explain why this is the case. Point v) implies that there exists at least one nonzero entry in at least

one of the submatrices $A_{1,S+1}, A_{2,S+1}, \ldots, A_{S,S+1}$, which correspond to connections from the sending networks to the receiving network $r = 1$. This means that the receiving network $r = 1$ is connected to a sending network. On the other hand, using again point v), we know that, if the receiving network $r = 2$ is not connected to any sending network, it must necessarily be connected to the receiving network $r = 1$. In this case, the receiving network $r = 2$ can be reached from a sending network by using a path that goes through the receiving network $r = 1$. By iterating this reasoning, we conclude that each receiving network can be reached through a path that originates at some sending network.

- According to the Gantmacher normal form, a receiving network *must* receive, but can also send. In comparison, a sending network *cannot receive* and might also not send (indeed, a sending network s evolves in isolation when the submatrices $A_{s,S+1}, A_{s,S+2}, \ldots, A_{s,S+R}$ are all null). Therefore, a more rigorous (albeit less appealing) classification would have been "non-receiving vs. receiving," in place of "sending vs. receiving."

In summary, the agents in a weak graph can be conveniently partitioned into two groups, $\mathcal{S}$ and $\mathcal{R}$, where the group $\mathcal{S}$ consists of the S *sending networks*, whereas the group $\mathcal{R}$ consists of the R *receiving networks*. To avoid misunderstanding, we remark that S (resp., R) does not denote the total number of agents in $\mathcal{S}$ (resp., $\mathcal{R}$), which is instead given by the cardinality $|\mathcal{S}|$ (resp., $|\mathcal{R}|$).

The representation in (4.26) can be compactly written as follows:

$$A = \left[\begin{array}{c|c} A_{\mathcal{S}} & A_{\mathcal{SR}} \\ \hline 0 & A_{\mathcal{R}} \end{array}\right], \tag{4.27}$$

where the matrix $A_{\mathcal{SR}}$, i.e., the top right block, globally collects the edges from agents in the ensemble of sending networks to agents in the ensemble of receiving networks, while the matrix $A_{\mathcal{R}}$, i.e., the bottom right block, describes the communication structure involving all receiving networks. The matrix $A_{\mathcal{S}}$ pertains to the sending networks and, since we know these networks do not communicate with each other, it takes the block-diagonal form shown in the top left block in (4.26).

The structure highlighted in the canonical form of Theorem 4.5 is not that uncommon in real-world networks. Actually, it is quite frequent over social networks, where some influential agents (e.g., celebrities) have a large number of followers, while the influential agents themselves may not consult information from most of these followers. Another example is that of media networks, which promote the emergence of opinions by feeding information to users without paying attention to feedback from them. A similar effect arises when social networks operate in the presence of stubborn agents, which insist on their opinion regardless of the evidence provided by local observations or by neighboring agents [3, 173].

4.5.1 Convergent Matrices over Weak Graphs

The next theorem arises in the theory of Markov chains, where sending and receiving networks are referred to as *ergodic* and *transient* classes, respectively. This terminology is related to their persistence as $t \to \infty$, in the sense that the limiting distribution of the Markov chain is concentrated only on the states belonging to the ergodic classes. As we will see later in Section 5.6, the physical interpretation in our social learning context is that sending networks are *influential* and determine the limiting behavior of the entire network, whereas receiving networks are *influenced.* The exact asymptotic behavior of the sequence of matrix powers is characterized in the next theorem [126, p. 698].

Theorem 4.6 (Matrix powers over weak graphs). Let A be a left stochastic $K \times K$ matrix associated with a weak graph. For each $s = 1, 2, \dots, S$, let A_s be the $K_s \times K_s$ submatrix associated with the sth sending network according to the canonical form (4.26). Denote by $v^{(s)}$ the Perron vector of A_s, and collect the Perron vectors corresponding to all sending networks into the block-diagonal matrix

$$V \triangleq \begin{bmatrix} v^{(1)}\mathbb{1}_{K_1}^\top & 0 & \cdots & 0 \\ 0 & v^{(2)}\mathbb{1}_{K_2}^\top & \cdots & 0 \\ \vdots & \vdots & \ddots & \vdots \\ 0 & 0 & \cdots & v^{(S)}\mathbb{1}_{K_S}^\top \end{bmatrix}, \tag{4.28}$$

where $\mathbb{1}_{K_s}$ is the $K_s \times 1$ vector with all entries equal to 1. Let also

$$W \triangleq V A_{\mathcal{SR}} \left(I_{|\mathcal{R}|} - A_{\mathcal{R}}\right)^{-1}, \tag{4.29}$$

where $I_{|\mathcal{R}|}$ (recall that $|\mathcal{R}|$ is the number of agents in the receiving networks) is the $|\mathcal{R}| \times |\mathcal{R}|$ identity matrix, and the submatrices $A_{\mathcal{SR}}$ and $A_{\mathcal{R}}$ are defined in the block-triangular representation (4.27). Then, we have the following results:

i) The matrix A (which is Cesàro-summable since it is left stochastic — see

Theorem 4.4) satisfies the condition

$$\lim_{t\to\infty}\frac{1}{t}\sum_{\tau=1}^{t}A^{\tau}=\left[\begin{array}{c|c}V & W\\ \hline 0 & 0\end{array}\right]. \tag{4.30}$$

ii) The sequence of matrix powers A^t converges if, and only if, all the submatrices $\{A_s\}_{s=1}^S$ associated with the sending networks are primitive. Moreover, if the sequence is convergent, the limiting matrix is

$$\lim_{t\to\infty}A^{t}=\left[\begin{array}{c|c}V & W\\ \hline 0 & 0\end{array}\right]. \tag{4.31}$$

Proof. First, we want to establish that for each receiving network $r = 1, 2, \ldots, R$, the spectral radius $\rho(A_{S+r})$ of matrix A_{S+r} is strictly smaller than 1. Observe that $\rho(A_{S+r}) \leq 1$ since the spectral radius of A is equal to 1 from Lemma 4.3, and we recall that the eigenvalues of a block-triangular matrix are the eigenvalues of the block matrices on the main diagonal. We know that A_{S+r} is either a scalar equal to 0, or an irreducible matrix (thus, corresponding to a connected network). In the former case we have $\rho(A_{S+r}) = 0$. Thus, assume that A_{S+r} is irreducible. Reasoning by contradiction, we assume $\rho(A_{S+r}) = 1$, which, in view of Theorem 4.1, would imply

$$A_{S+r}v^{(S+r)}=v^{(S+r)} \tag{4.32}$$

for a Perron vector $v^{(S+r)}$. From (4.32) we can also write

$$\underbrace{\mathbb{1}_{K_{S+r}}^{\top}A_{S+r}}_{u^{\top}}v^{(S+r)}=\mathbb{1}_{K_{S+r}}^{\top}v^{(S+r)}, \tag{4.33}$$

where K_{S+r} denotes the number of agents belonging to the rth receiving network. From point v) in Theorem 4.5 we know that each receiving network receives information from at least one agent in the rest of the network. Since the columns of A add up to 1, we deduce that at least one column of A_{S+r} must have a sum that is strictly smaller than 1, which in turn implies that the vector u defined in (4.33) has at least one entry strictly smaller than 1. In this case, the equality in (4.33) would be impossible. We conclude that $\rho(A_{S+r}) < 1$ and, hence, $\rho(A_{\mathcal{R}}) < 1$.

Now we proceed to prove (4.30). Owing to the block-triangular representation in (4.27), we can write

$$A^{t}=\left[\begin{array}{c|c}A_{\mathcal{S}}^{t} & W_t\\ \hline 0 & A_{\mathcal{R}}^{t}\end{array}\right], \tag{4.34}$$

where W_t is some unknown $|\mathcal{S}| \times |\mathcal{R}|$ matrix. Since A is left stochastic, it is also Cesàro summable in view of Theorem 4.4. Therefore, using the representation in (4.34), it is legitimate to write

$$\lim_{t\to\infty}\frac{1}{t}\sum_{\tau=1}^{t}A^{\tau}=\left[\begin{array}{c|c}\lim\limits_{t\to\infty}\frac{1}{t}\sum\limits_{\tau=1}^{t}A_{\mathcal{S}}^{\tau} & \lim\limits_{t\to\infty}\frac{1}{t}\sum\limits_{\tau=1}^{t}W_{\tau}\\ \hline 0 & \lim\limits_{t\to\infty}\frac{1}{t}\sum\limits_{\tau=1}^{t}A_{\mathcal{R}}^{\tau}\end{array}\right]. \tag{4.35}$$

Since $A_{\mathcal{S}}$ is a block-diagonal matrix collecting the submatrices A_s of the sending networks, for $s = 1, 2, \ldots, S$, when we compute $A_{\mathcal{S}}^{\tau}$, we obtain a block-diagonal matrix whose blocks are given by A_s^{τ}. Since the submatrices A_s are irreducible, from Theorem 4.4 we obtain

$$\lim_{t\to\infty} \frac{1}{t} \sum_{\tau=1}^{t} A_s^{\tau} = v^{(s)} \mathbb{1}_{K_s}^{\top}, \tag{4.36}$$

which, using (4.28), yields

$$\lim_{t\to\infty} \frac{1}{t} \sum_{\tau=1}^{t} A_{\mathcal{S}}^{\tau} = V. \tag{4.37}$$

Regarding the bottom right block in (4.35), it vanishes as $t \to \infty$ because, in particular, we have

$$\lim_{t\to\infty} A_{\mathcal{R}}^{t} = 0 \tag{4.38}$$

since $\rho(A_{\mathcal{R}}) < 1$. Defining

$$W = \lim_{t\to\infty} \frac{1}{t} \sum_{\tau=1}^{t} W_{\tau}, \tag{4.39}$$

and substituting (4.37) and (4.38) into (4.35), we obtain

$$\lim_{t\to\infty} \frac{1}{t} \sum_{\tau=1}^{t} A^{\tau} = \left[\begin{array}{c|c} V & W \\ \hline 0 & 0 \end{array}\right]. \tag{4.40}$$

Now we show how to determine W. We have the identity

$$\lim_{t\to\infty} \frac{1}{t} \sum_{\tau=1}^{t} A^{\tau} = \lim_{t\to\infty} \frac{1}{t} \sum_{\tau=1}^{t-1} A^{\tau} A \tag{4.41}$$

Substituting (4.27) and (4.47) into (4.41), we have

$$\left[\begin{array}{c|c} V & W \\ \hline 0 & 0 \end{array}\right] = \left[\begin{array}{c|c} V & W \\ \hline 0 & 0 \end{array}\right] \left[\begin{array}{c|c} A_{\mathcal{S}} & A_{\mathcal{S}\mathcal{R}} \\ \hline 0 & A_{\mathcal{R}} \end{array}\right]. \tag{4.42}$$

Considering the top right block only, and performing the pertinent matrix-block multiplication, we obtain the following relation:

$$W = V A_{\mathcal{S}\mathcal{R}} + W A_{\mathcal{R}}, \tag{4.43}$$

which implies (4.29). This means that we have established (4.30), and the proof of part i) is complete. We switch to part ii).

We want to establish that the sequence of matrix powers A^t converges if, and only if, all the submatrices $\{A_s\}_{s=1}^{S}$ are primitive, and that if the sequence converges, its limit is given by (4.31). First, observe that the condition $\rho(A_{\mathcal{R}}) < 1$ means that the bottom right block $A_{\mathcal{R}}$ contributes to the spectrum of A with eigenvalues lying strictly inside the unit circle. Consider now the top left block $A_{\mathcal{S}}$, which is a block-diagonal matrix collecting the submatrices A_s of the sending networks, for $s = 1, 2, \ldots, S$. Owing to the block-diagonal structure, if we raise A to a power t, each of the matrices A_s will be raised to t. In view of Theorem 4.2, if at least one of these matrices is imprimitive, then it is not convergent, and A will not be convergent either. Let us then focus on the case where all the submatrices $\{A_s\}_{s=1}^{S}$ are primitive.

Observe that the eigenvalues associated with the top left block $A_{\mathcal{S}}$ are the eigenvalues of the submatrices $\{A_s\}_{s=1}^{S}$ on the main diagonal. Since these submatrices are left stochastic and primitive, each of them has a *simple* eigenvalue equal to 1, and no other eigenvalues on the unit circle. From this observation, and since $A_{\mathcal{R}}$ is associated with eigenvalues lying strictly inside the unit circle, we conclude that the eigenvalue 1 has algebraic multiplicity S. Let us now examine the eigenvectors associated with this eigenvalue. Recalling that we denote by $v^{(s)}$ the Perron vector of the submatrix A_s, for $s = 1$ we have

$$A \times \begin{bmatrix} v^{(1)} \\ 0 \\ \vdots \\ 0 \end{bmatrix} = \begin{bmatrix} A_1 v^{(1)} \\ 0 \\ \vdots \\ 0 \end{bmatrix} = \begin{bmatrix} v^{(1)} \\ 0 \\ \vdots \\ 0 \end{bmatrix}, \tag{4.44}$$

for $s = 2$ we have

$$A \times \begin{bmatrix} 0 \\ v^{(2)} \\ 0 \\ \vdots \\ 0 \end{bmatrix} = \begin{bmatrix} 0 \\ A_2 v^{(2)} \\ 0 \\ \vdots \\ 0 \end{bmatrix} = \begin{bmatrix} 0 \\ v^{(2)} \\ 0 \\ \vdots \\ 0 \end{bmatrix}, \tag{4.45}$$

and so on. Therefore, we can associate with the eigenvalue 1 the following S eigenvectors:

$$\begin{bmatrix} v^{(1)} \\ 0 \\ 0 \\ \vdots \\ 0 \\ \vdots \\ 0 \end{bmatrix}, \begin{bmatrix} 0 \\ v^{(2)} \\ 0 \\ \vdots \\ 0 \\ \vdots \\ 0 \end{bmatrix}, \ldots, \begin{bmatrix} 0 \\ 0 \\ 0 \\ \vdots \\ v^{(S)} \\ \vdots \\ 0 \end{bmatrix}, \tag{4.46}$$

which are mutually orthogonal and, hence, the geometric multiplicity of the eigenvalue 1 is equal to S. This means that the geometric multiplicity of the eigenvalue 1 is equal to its algebraic multiplicity. In other words, the eigenvalue 1 is semisimple and, as observed before, is the only eigenvalue on the unit circle. In view of Theorem 4.2, the matrix A is convergent. This means that the sequence of matrix powers A^t converges. Owing to the block-triangular representation in (4.27), we can write

$$A^t = \left[\begin{array}{c|c} A_{\mathcal{S}}^t & W_t \\ \hline 0 & A_{\mathcal{R}}^t \end{array}\right] \overset{t \to \infty}{\longrightarrow} \left[\begin{array}{c|c} V & W \\ \hline 0 & 0 \end{array}\right], \tag{4.47}$$

where: *i)* $A_{\mathcal{S}}^t$ converges to V in view of Corollary 4.1; *ii)* $A_{\mathcal{R}}^t$ vanishes because $\rho(A_{\mathcal{R}}) < 1$; and *iii)* W_t is some unknown $|\mathcal{S}| \times |\mathcal{R}|$ matrix, whose limit W is known to exist since we have established that A is a convergent matrix. On the other hand, we have the identity

$$\lim_{t\to\infty} A^t = \left(\lim_{t\to\infty} A^{t-1}\right) A. \tag{4.48}$$

Substituting (4.27) and (4.47) into (4.48) we have

$$\left[\begin{array}{c|c} V & W \\ \hline 0 & 0 \end{array}\right] = \left[\begin{array}{c|c} V & W \\ \hline 0 & 0 \end{array}\right] \left[\begin{array}{c|c} A_{\mathcal{S}} & A_{\mathcal{S}\mathcal{R}} \\ \hline 0 & A_{\mathcal{R}} \end{array}\right]. \tag{4.49}$$

Considering the top right block only, and performing the pertinent matrix-block multiplication, we obtain the following relation:

$$W = VA_{\mathcal{SR}} + WA_{\mathcal{R}}, \tag{4.50}$$

which implies (4.29).

■

The matrix power A^t is left stochastic for any t, which implies that the limiting matrix

$$\left[\begin{array}{c|c} V & W \\ \hline 0 & 0 \end{array}\right] \tag{4.51}$$

is left stochastic. Since this limiting matrix has a null bottom right block, the entries on each column of its top right block, $W = [w_{jk}]$ (defined for $j \in \mathcal{S}$ and $k \in \mathcal{R}$), must add up to 1, i.e., for any $k \in \mathcal{R}$ we have

$$\sum_{j \in \mathcal{S}} w_{jk} = 1. \tag{4.52}$$

Furthermore, from (4.29) we can write

$$W = VA_{\mathcal{SR}} \left(I_{|\mathcal{R}|} + A_{\mathcal{R}} + A_{\mathcal{R}}^2 + \dots \right). \tag{4.53}$$

By expanding the matrix products, the (j, k) entry of the matrix W can be represented as follows:

$$w_{jk} = \sum_{h \in \mathcal{S}} [V]_{jh} \sum_{h' \in \mathcal{R}} [A_{\mathcal{SR}}]_{hh'} \left([I_{|\mathcal{R}|}]_{h'k} + [A_{\mathcal{R}}]_{h'k} + [A_{\mathcal{R}}^2]_{h'k} + \dots \right). \tag{4.54}$$

Assume that j belongs to the sth sending network and denote the ensemble of agents in this network by $\mathcal{A}_s$. By exploiting the structure of the matrix V defined by (4.28), from (4.54) we can write

$$w_{jk} = v_j^{(s)} \sum_{h \in \mathcal{A}_s} \sum_{h' \in \mathcal{R}} [A_{\mathcal{SR}}]_{hh'} \left([I_{|\mathcal{R}|}]_{h'k} + [A_{\mathcal{R}}]_{h'k} + [A_{\mathcal{R}}^2]_{h'k} + \dots \right), \tag{4.55}$$

from which we find that w_{jk} aggregates the sum of influences over all paths originating at the sending network $\mathcal{A}_s$ (i.e., the agents belonging to the sending network of agent j) and ending at agent $k \in \mathcal{R}$. Accordingly, $w_{jk} > 0$ if, and only if, there exists a directed path from some $h \in \mathcal{A}_s$ to k. Moreover, Eq. (4.52) implies that w_{jk} must be nonzero for at least one $j \in \mathcal{S}$. This means that each agent k is reachable through a path that originates at a sending network. Note that the latter property is consistent with the comments following Theorem 4.5, in particular, with the implications of point v).

4.6 Combination Policies

In this section we describe some common policies used to build the combination matrix A. We first need to define the desired support graph of A, and then design a combination policy to assign the combination weights on top of this graph. A list of popular combination policies is reported in Table 4.1. In the table, the symbol $\mathsf{deg}_k = |\mathcal{N}_k|$ denotes the degree[3] (technically, the *in-degree*) of agent k, which is equal to the size of its neighborhood, and the symbol $\mathsf{deg}_{\mathsf{max}}$ denotes the maximum degree across the network:

$$\mathsf{deg}_{\mathsf{max}} \triangleq \max_{k\in\{1,2,\ldots,K\}} \mathsf{deg}_k. \tag{4.56}$$

Since the combination matrix must be left stochastic, the sum along each of its columns is equal to 1. As a result, each node k has a nonempty neighborhood, i.e., $\mathsf{deg}_k > 0$ for $k = 1, 2, \ldots, K$.

4.6.1 Left Stochastic Policies

The first two rows of Table 4.1 show two popular policies to construct a left stochastic matrix. The *uniform-averaging rule* is perhaps the simplest one. Each agent k scales the observations received from neighbor j (possibly including the case $j = k$) with a uniform weight. Since all the nonzero weights used by agent k must add up to 1, the uniform weight must be equal to $1/\mathsf{deg}_k$.

In the rule reported on the second row, named *relative-degree rule*, agent k sets, for all agents $j \in \mathcal{N}_k$,

$$a_{jk} = \frac{\mathsf{deg}_j}{\sum\limits_{m\in\mathcal{N}_k} \mathsf{deg}_m}, \tag{4.57}$$

that is, agent k scales the information received from agent j proportionally to the degree of agent j, where the proportionality factor (the sum in the denominator) serves to guarantee that the weights add up to 1. One difference between the uniform-averaging and the relative-degree rules is that in the latter case agent k should know the degree deg_j of each neighbor $j \in \mathcal{N}_k$, whereas in the former case it must know only its own degree deg_k.

[3]According to (4.1), the neighborhood of agent k includes agent k itself when there is a self-loop. In this case, the degree deg_k also counts agent k. Note that this definition of degree differs from other definitions used in the literature, where agent k is excluded from the neighborhood and consequently from the degree.

Table 4.1: Popular policies to construct the combination matrix $A = [a_{jk}]$. The second column lists the properties of the graph and indicates whether the matrix is left stochastic (LS) or doubly stochastic (DS).

Entries of the combination matrix A	Type of graph & matrix
1. Uniform-averaging rule $a_{jk} = \begin{cases} \dfrac{1}{\mathsf{deg}_k} & \text{if } j \in \mathcal{N}_k, \\ 0 & \text{otherwise.} \end{cases}$	directed, LS
2. Relative-degree rule $a_{jk} = \begin{cases} \dfrac{\mathsf{deg}_j}{\sum_{m \in \mathcal{N}_k} \mathsf{deg}_m} & \text{if } j \in \mathcal{N}_k, \\ 0 & \text{otherwise.} \end{cases}$	directed, LS
3. Laplacian rule $a_{jk} = \begin{cases} a & \text{if } j \in \mathcal{N}_k \backslash \{k\}, \\ 1 - a\,(\mathsf{deg}_k - 1) & \text{if } j = k, \\ 0 & \text{otherwise.} \end{cases}$	undirected self-loops for all k symmetric DS
4. Metropolis $a_{jk} = \begin{cases} \dfrac{1}{\max\{\mathsf{deg}_j, \mathsf{deg}_k\}} & \text{if } j \in \mathcal{N}_k \backslash \{k\}, \\ 1 - \sum_{m \in \mathcal{N}_k \backslash \{k\}} a_{mk} & \text{if } j = k, \\ 0 & \text{otherwise.} \end{cases}$	undirected self-loops for all k symmetric DS

4.6.2 Doubly Stochastic Policies

One common situation is when the support graph of A is *undirected* and each agent uses its own information, which means that *all* nodes in the graph have a self-loop. For this scenario we illustrate two popular policies, referred to as the Laplacian and Metropolis combination rules.

The *Laplacian rule*, which appears in the third row of Table 4.1, relies on the use of the so-called Laplacian matrix of the network graph, denoted by $L = [l_{jk}]$ and defined as follows [22, 56, 100, 151, 153]:

$$l_{jk} = \begin{cases} -1 & \text{if } j \in \mathcal{N}_k \backslash \{k\}, \\ \mathsf{deg}_k - 1 & \text{if } j = k, \\ 0 & \text{otherwise.} \end{cases} \tag{4.58}$$

The Laplacian rule constructs the combination matrix A from L by setting

$$A = I - a\,L \tag{4.59}$$

for some scalar a that must guarantee that all entries a_{jk} with $j \in \mathcal{N}_k$ are positive. It is straightforward to check that this condition imposes the following constraint on the scalar a:

$$0 < a < \frac{1}{\mathsf{deg}_{\mathsf{max}} - 1}. \tag{4.60}$$

We are assuming that $\mathsf{deg}_{\mathsf{max}} > 1$, since each agent has a self-loop, which means that the case $\mathsf{deg}_{\mathsf{max}} = 1$ would correspond to the trivial case where all agents are connected only to themselves.

It is readily verified that the matrix A in (4.59) is left stochastic. Moreover, by construction, the Laplacian matrix L is symmetric, which means that A in (4.59) is also symmetric. But since A is left stochastic, the symmetry ensures that A is *doubly stochastic*. Note, however, that a doubly stochastic matrix need not be symmetric in general. It is also important to note that undirected graphs do not imply that the combination matrix must be doubly stochastic. For example, it can be verified that if we apply the uniform-averaging rule to a general undirected graph we do not obtain a doubly stochastic matrix.

Let us briefly comment on the choice of the parameter a. We have shown in the previous sections conditions for the powers of A to be convergent. In particular, they converge when A is primitive. The choice of the scalar a determines the value of the second largest-magnitude eigenvalue and therefore the rate of convergence — see (4.8) and (4.9). It is shown in [32,

172] how a can be chosen to maximize the convergence rate when knowledge of the Laplacian matrix is available. An alternative popular choice is

$$a = \frac{1}{\mathsf{deg}_{\mathsf{max}}}, \tag{4.61}$$

which is also referred to as the *maximum-degree rule.*[4] In the following, when we refer to the Laplacian rule we implicitly imply that a is computed according to (4.61).

Note that with the Laplacian rule *all agents* in the network use *one and the same weight.* Moreover, in order to implement this rule each agent needs to know the maximum degree from across the network. A different combination policy is the *Metropolis rule*, which replaces the maximum degree from across the network with the maximum degree between agents j and k, and is accordingly also referred to as the *local-degree rule.* Specifically, beyond its own degree, agent k must only know the degrees of its neighbors $j \in \mathcal{N}_k \backslash \{k\}$. It is readily seen that the Metropolis rule also yields a symmetric and doubly stochastic combination matrix.

[4]Actually, the maximum-degree and the local-degree rules in [172] use $\mathsf{deg}_{\mathsf{max}} - 1$ in place of $\mathsf{deg}_{\mathsf{max}}$, but this choice would imply that the node(s) featuring maximum degree will not have a self-loop.

Chapter 5

Social Learning with Geometric Averaging

The derivations in Chapter 3 motivated the following social learning strategy with geometric averaging (see listing (3.16)):

$$\boldsymbol{\psi}_{k,t}(\theta) \propto \boldsymbol{\mu}_{k,t-1}(\theta)\ell_k(\boldsymbol{x}_{k,t}|\theta), \tag{5.1a}$$

$$\boldsymbol{\mu}_{k,t}(\theta) \propto \prod_{j\in\mathcal{N}_k} [\boldsymbol{\psi}_{j,t}(\theta)]^{a_{jk}}. \tag{5.1b}$$

An alternative representation is obtained by grouping the two steps, which yields

$$\boldsymbol{\mu}_{k,t}(\theta) \propto \prod_{j\in\mathcal{N}_k} [\boldsymbol{\mu}_{j,t-1}(\theta)\ell(\boldsymbol{x}_{j,t}|\theta)]^{a_{jk}}. \tag{5.2}$$

In this chapter we examine the long-term properties of $\boldsymbol{\mu}_{k,t}$ as $t \to \infty$ and identify the hypothesis $\vartheta^\star$ that is learned by the agents.

To avoid repetitions, we collect in the following assumption two common conditions that will be used to prove all the results in the remainder of this text.

Assumption 5.1 (Combination matrix and initial beliefs).

i) **Combination matrix.** The $K \times K$ combination matrix $A = [a_{jk}]$ is left stochastic:

$$\sum_{j=1}^{K} a_{jk} = 1, \qquad a_{jk} \geq 0. \tag{5.3}$$

ii) **Initial Beliefs.** For each agent $k = 1, 2, \ldots, K$, the initial belief vector has strictly positive entries:

$$\mu_{k,0}(\theta) > 0 \quad \forall\theta \in \Theta. \tag{5.4}$$

Condition i) was motivated in Chapter 3, when we showed that under the two optimal pooling rules that we derived (namely, the geometric and arithmetic averaging rules), each agent must employ convex combination weights; this fact translates into the left stochastic property of the combination matrix. Condition ii) rules out the singular case where the agents begin the learning process by ignoring some hypotheses.

5.1 Belief Convergence

The convergence questions will be addressed by considering the following model for the data distributions and likelihoods characterizing the individual agents.

Assumption 5.2 (Data distributions and likelihoods). Each agent $k = 1, 2, \ldots, K$ at time $t = 1, 2, \ldots$ receives a data sample $\boldsymbol{x}_{k,t}$. The collections of K samples across the agents, $\{\boldsymbol{x}_{1,t}, \boldsymbol{x}_{2,t}, \ldots, \boldsymbol{x}_{K,t}\}$, are assumed iid over time. The probability (density or mass) function of $\boldsymbol{x}_{k,t}$ is denoted by f_k. Note that dependence across the agents (i.e., over space) is possible since f_k is a *marginal* probability function pertaining to agent k. To perform social learning, agent k employs likelihood models $\{\ell_{k,\theta}\}_{\theta \in \Theta}$ of the same nature as f_k (namely, for all $\theta \in \Theta$, $\ell_{k,\theta}$ is a pdf if f_k is a pdf, and a pmf otherwise).[1] We assume that, for $k = 1, 2, \ldots, K$, and for all $\theta \in \Theta$,

$$D(f_k || \ell_{k,\theta}) < \infty. \tag{5.5}$$

For later use, it is important to note that, under Assumptions 5.1 and 5.2, the beliefs $\boldsymbol{\psi}_{k,t}(\theta)$ and $\boldsymbol{\mu}_{k,t}(\theta)$ resulting from (5.1a) and (5.1b) are almost-surely positive for all k, t, and θ. First, observe that the likelihoods cannot be zero, but for an ensemble of realizations occurring with probability zero under f_k, otherwise condition (5.5) would be violated. This implies that the denominator arising from the Bayesian update (i.e., the denominator hidden by the proportionality sign in (5.1a)) is nonzero almost surely. Moreover, positivity of the likelihoods also implies that, starting from a belief $\boldsymbol{\mu}_{k,t-1}(\theta)$ that is nonzero at any θ, the intermediate belief $\boldsymbol{\psi}_{k,t}(\theta)$ in (5.1a) is nonzero. Now, since the combination matrix is left stochastic because of point i) of Assumption 5.1, then $\mathcal{N}_k$ is nonempty (see Definition 4.10). Therefore, Eq. (5.1b) ensures that $\boldsymbol{\mu}_{k,t}(\theta) > 0$. Positivity of the beliefs $\boldsymbol{\psi}_{k,t}(\theta)$ and $\boldsymbol{\mu}_{k,t}(\theta)$ can be extended to all times by induction, after noticing that the

[1] As usual, we drop the argument x in $f_k(x)$ and $\ell_k(x|\theta)$ and write f_k and $\ell_{k,\theta}$, respectively, to denote the pertinent pdf or pmf. However, in the latter notation we need to add the subscript θ to emphasize the dependence on the particular θ.

initial beliefs $\mu_{k,0}(\theta)$ are positive in view of point ii) of Assumption 5.1. From now on, we will implicitly exploit the positivity of the beliefs when we evaluate expressions where it matters, e.g., when we compute ratios between beliefs or the logarithm of a belief.

The next theorem establishes the convergence of $\boldsymbol{\mu}_{k,t}$ as $t \to \infty$.

Theorem 5.1 (Belief convergence). Let Assumptions 5.1 and 5.2 be satisfied. Since all left stochastic matrices are Cesàro-summable (Theorem 4.4), there exists a limiting matrix $A^\bullet = [a^\bullet_{jk}]$ such that

$$\lim_{t\to\infty} \frac{1}{t} \sum_{\tau=1}^{t} A^\tau = A^\bullet. \tag{5.6}$$

For each agent $k = 1, 2, \dots, K$, consider the following *network average* of KL divergences:

$$\bar{D}_k(\theta) \triangleq \sum_{j=1}^{K} a^\bullet_{jk}\, D(f_j || \ell_{j,\theta}). \tag{5.7}$$

If $\bar{D}_k(\theta)$ admits a unique minimizer $\vartheta^\star_k$, then

$$\boldsymbol{\mu}_{k,t}(\vartheta^\star_k) \xrightarrow[t\to\infty]{\text{a.s.}} 1 \tag{5.8}$$

and the beliefs about all hypotheses $\theta \neq \vartheta^\star_k$ vanish at an exponential rate:

$$\frac{\log \boldsymbol{\mu}_{k,t}(\theta)}{t} \xrightarrow[t\to\infty]{\text{a.s.}} \bar{D}_k(\vartheta^\star_k) - \bar{D}_k(\theta) < 0 \quad \forall \theta \neq \vartheta^\star_k. \tag{5.9}$$

Proof. In view of (5.2), for any $\theta \neq \vartheta^\star_k$ we can write

$$\begin{aligned}
\log \frac{\boldsymbol{\mu}_{k,t}(\vartheta^\star_k)}{\boldsymbol{\mu}_{k,t}(\theta)} &= \sum_{j\in\mathcal{N}_k} a_{jk} \left[\log \frac{\boldsymbol{\mu}_{j,t-1}(\vartheta^\star_k)}{\boldsymbol{\mu}_{j,t-1}(\theta)} + \log \frac{\ell_j(\boldsymbol{x}_{j,t}|\vartheta^\star_k)}{\ell_j(\boldsymbol{x}_{j,t}|\theta)} \right] \\
&= \sum_{j=1}^{K} a_{jk} \left[\log \frac{\boldsymbol{\mu}_{j,t-1}(\vartheta^\star_k)}{\boldsymbol{\mu}_{j,t-1}(\theta)} + \log \frac{\ell_j(\boldsymbol{x}_{j,t}|\vartheta^\star_k)}{\ell_j(\boldsymbol{x}_{j,t}|\theta)} \right],
\end{aligned} \tag{5.10}$$

where the last equality follows from the definition of $\mathcal{N}_k$ introduced in (4.1). To prove the claim of the theorem, we call upon Lemma D.3. First, we observe that (5.10) can be cast in the vector form (D.57), namely, in the form

$$\boldsymbol{z}_t = A^\mathsf{T}(\boldsymbol{z}_{t-1} + \boldsymbol{y}_t) \tag{5.11}$$

by setting

$$\boldsymbol{y}_t = \left[\log \frac{\ell_1(\boldsymbol{x}_{1,t}|\vartheta^\star_k)}{\ell_1(\boldsymbol{x}_{1,t}|\theta)}, \log \frac{\ell_2(\boldsymbol{x}_{2,t}|\vartheta^\star_k)}{\ell_2(\boldsymbol{x}_{2,t}|\theta)}, \dots, \log \frac{\ell_K(\boldsymbol{x}_{K,t}|\vartheta^\star_k)}{\ell_K(\boldsymbol{x}_{K,t}|\theta)} \right], \tag{5.12a}$$

$$\boldsymbol{z}_t = \left[\log \frac{\boldsymbol{\mu}_{1,t}(\vartheta^\star_k)}{\boldsymbol{\mu}_{1,t}(\theta)}, \log \frac{\boldsymbol{\mu}_{2,t}(\vartheta^\star_k)}{\boldsymbol{\mu}_{2,t}(\theta)}, \dots, \log \frac{\boldsymbol{\mu}_{K,t}(\vartheta^\star_k)}{\boldsymbol{\mu}_{K,t}(\theta)} \right], \tag{5.12b}$$

where we recall that in our notation all vectors are column vectors. Lemma D.3 requires that A is left stochastic and that the sequence $\{\boldsymbol{y}_t\}$ is formed by iid vectors whose entries have finite mean. Now, under Assumption 5.1, A is left stochastic, whereas, under Assumption 5.2, the collections $\{\boldsymbol{x}_{1,t}, \boldsymbol{x}_{2,t}, \ldots, \boldsymbol{x}_{K,t}\}$ are iid over time, implying that the sequence $\{\boldsymbol{y}_t\}$ is formed by iid vectors. It remains to show that all entries of $\boldsymbol{y}_t$ have finite mean. To this end, consider the jth entry of $\boldsymbol{y}_t$,

$$\log \frac{\ell_j(\boldsymbol{x}_{j,t}|\vartheta_k^\star)}{\ell_j(\boldsymbol{x}_{j,t}|\theta)} = \log \frac{f_j(\boldsymbol{x}_{j,t})}{\ell_j(\boldsymbol{x}_{j,t}|\theta)} - \log \frac{f_j(\boldsymbol{x}_{j,t})}{\ell_j(\boldsymbol{x}_{j,t}|\vartheta_k^\star)}. \tag{5.13}$$

In view of Assumption 5.2, both terms on the RHS of (5.13) have finite mean, which implies that the jth entry of the vector $\boldsymbol{y}_t$ has finite mean. We conclude that the sequence $\{\boldsymbol{y}_t\}$ satisfies the conditions required to invoke Lemma D.3. In particular, the vector $\bar{y}$ used in Lemma D.3 coincides with $\mathbb{E}\boldsymbol{y}_t$. We can therefore apply the claim of Lemma D.3 to conclude that

$$\frac{1}{t}\,\boldsymbol{z}_t \xrightarrow[t\to\infty]{\text{a.s.}} (A^\bullet)^\top\, \mathbb{E}\boldsymbol{y}_t. \tag{5.14}$$

Using the definition of $\boldsymbol{z}_t$ from (5.12b) and taking the expectation of the individual entries of $\boldsymbol{y}_t$ in (5.12a), we can rewrite (5.14) in terms of the kth entry as follows:

$$\begin{aligned}\frac{1}{t}\log\frac{\boldsymbol{\mu}_{k,t}(\vartheta_k^\star)}{\boldsymbol{\mu}_{k,t}(\theta)} &\xrightarrow[t\to\infty]{\text{a.s.}} \sum_{j=1}^{K} a_{jk}^\bullet\, \mathbb{E}_{f_j} \log \frac{\ell_j(\boldsymbol{x}_{j,t}|\vartheta_k^\star)}{\ell_j(\boldsymbol{x}_{j,t}|\theta)}\\ &= \sum_{j=1}^{K} a_{jk}^\bullet \Big[D(f_j||\ell_{j,\theta}) - D(f_j||\ell_{j,\vartheta_k^\star})\Big]\\ &= \bar{D}_k(\theta) - \bar{D}_k(\vartheta_k^\star).\end{aligned} \tag{5.15}$$

Since $\vartheta_k^\star$ is the unique minimizer of $\bar{D}_k(\theta)$, then the RHS of (5.15) is positive for all $\theta \neq \vartheta_k^\star$, yielding

$$\log\frac{\boldsymbol{\mu}_{k,t}(\vartheta_k^\star)}{\boldsymbol{\mu}_{k,t}(\theta)} \xrightarrow[t\to\infty]{\text{a.s.}} \infty \quad \forall\theta \neq \vartheta_k^\star, \tag{5.16}$$

which further implies (recall that the beliefs are bounded)

$$\boldsymbol{\mu}_{k,t}(\theta) \xrightarrow[t\to\infty]{\text{a.s.}} 0 \quad \forall\theta \neq \vartheta_k^\star. \tag{5.17}$$

Since the entries $\boldsymbol{\mu}_{k,t}(\theta)$ add up to 1, Eq. (5.8) follows. Finally, using (5.8) in (5.15), Eq. (5.9) is proved.

■

Theorem 5.1 provides a complete characterization of the learning behavior under the social learning strategy in (5.2). Consider first the limiting matrix $A^\bullet$ defined by (5.6). Since the matrix power A^t is representative of t iterated exchanges of information between neighboring agents over the graph, the matrix entry $a_{jk}^\bullet$ represents an asymptotic weight that agent k will use to scale the information received by agent j. These asymptotic weights play a role in the construction of the network average of KL divergences $\bar{D}_k(\theta)$ in (5.7), which is defined for each agent k and can be different across the agents. When $\bar{D}_k(\theta)$ has a unique minimizer $\vartheta_k^\star$, Eq. (5.8) reveals that the belief vector $\boldsymbol{\mu}_{k,t}$ will asymptotically place unit mass on $\vartheta_k^\star$.

Notably, $\bar{D}_k(\theta)$ is determined by the interplay between attributes of the graph and attributes of the statistical models: The limiting matrix $A^\bullet$ summarizes the ultimate effect of the network topology and combination weights, whereas the KL divergences summarize the features of the statistical models that are relevant to the learning problem. Specifically, we see from (5.7) that the limiting matrix entry $a_{jk}^\bullet$ represents the weight assigned by agent k to the KL divergence $D(f_j||\ell_{j,\theta})$, which quantifies the difference between the actual model f_j that governs the data of agent j, and the postulated likelihood model $\ell_{j,\theta}$ that agent j uses to update its beliefs. As a result, the function $\bar{D}_k(\theta)$ represents a global (across the agents) measure of discrepancy between the true models $\{f_j\}_{j=1}^K$ and the local models $\{\ell_{j,\theta}\}_{j=1}^K$. In the case where the observations are independent across the agents, this measure admits a straightforward interpretation, as explained in the next example.

Example 5.1 (Observations independent across the agents). When the combination matrix is doubly stochastic and irreducible, we know from (4.18) that the Perron vector entries are uniform, yielding

$$\bar{D}_k(\theta) = \frac{1}{K}\sum_{j=1}^{K} D(f_j||\ell_{j,\theta}) \quad \text{for } k = 1, 2, \ldots, K. \tag{5.18}$$

If the observations are independent across the agents (i.e., over space), we can introduce the joint distribution f:

$$f(x_{1,t}, x_{2,t}, \ldots, x_{K,t}) = \prod_{k=1}^{K} f_k(x_{k,t}). \tag{5.19}$$

Introducing also the joint likelihood model ℓ_θ defined by

$$\ell(x_{1,t}, x_{2,t}, \ldots, x_{K,t}|\theta) = \prod_{k=1}^{K} \ell_k(x_{k,t}|\theta), \tag{5.20}$$

and observing that the KL divergence is additive for independent observations, we find that the average KL divergence in (5.18) is, but for the scaling factor $1/K$, the KL divergence between the joint models, i.e.,

$$\bar{D}_k(\theta) = \frac{1}{K} D(f||\ell_\theta). \tag{5.21}$$

Obviously, $\bar{D}_k(\theta)$ admits a minimizer since it is defined over a discrete finite set, Θ. The requirement that the minimizer is unique means that we rule out the possibility that there exist multiple hypotheses that provide the best explanation for the data. The following example illustrates a setup where the uniqueness of the minimizer is easily explained.

Example 5.2 (Unique minimizer of (5.7)**).** Consider a left stochastic irreducible combination matrix A. In this case, from Theorem 4.4 we know that the limiting matrix in (5.6) is given by $A^{\bullet} = v\mathbb{1}^{\top}$, where v is the Perron vector of A. As a result, from (5.7) we see that the network average of KL divergences is the *same* for all agents, and given by

$$\bar{D}_k(\theta) = \sum_{j=1}^{K} v_j D(f_j || \ell_{j,\theta}) \quad \text{for } k = 1, 2, \ldots, K. \tag{5.22}$$

Assume further that for each agent k we have $f_k(x) = \ell_k(x|\vartheta^o)$, for some $\vartheta^o \in \Theta$. That is, the true distribution $f_k(x)$ coincides with the local likelihood corresponding to a true hypothesis ϑ^o, which is common to all agents. We will refer to this situation as the *objective evidence* scenario in Section 5.3. In this case, the network average of KL divergences from (5.22) becomes

$$\bar{D}_k(\theta) = \sum_{j=1}^{K} v_j D(\ell_{j,\vartheta^o} || \ell_{j,\theta}) \quad \text{for } k = 1, 2, \ldots, K, \tag{5.23}$$

from which we see that $\bar{D}_k(\vartheta^o) = 0$ for $k = 1, 2, \ldots, K$. This implies (since the KL divergence is nonnegative) that ϑ^o is a minimizer of $\bar{D}_k(\theta)$. Observe also that we can have $\ell_{j,\theta} = \ell_{j,\vartheta^o}$ for some hypotheses $\theta \neq \vartheta^o$, which means that agent j is not able to distinguish θ from ϑ^o. If all agents were under this condition, then we would get $\bar{D}_k(\theta) = 0$. This possibility is ruled out by the assumption of a unique minimizer, which therefore translates into the following *global identifiability* condition: $\bar{D}_k(\theta) \neq 0$ for all $\theta \neq \vartheta^o$, i.e., for each $\theta \neq \vartheta^o$ there exists at least one agent that is able to distinguish θ from ϑ^o. This condition is discussed later — see Assumption 5.4.

The take-away messages from Theorem 5.1 are: *i)* the belief vector of agent k converges to a probability vector placing unit mass on a single hypothesis $\vartheta_k^\star$; and *ii)* this hypothesis is generally agent-dependent and is the minimizer of the weighted combination (5.7) of KL divergences between actual and postulated models. However, in its present form, the theorem does not give much insight into *what* the agents learn, leaving open a number of fundamental questions. For example, when is $\vartheta_k^\star$ the same for all agents? In other words, when do the agents reach *agreement* through social learning? Is the value $\vartheta_k^\star$ related to some *true hypothesis* contained in the observed data? Are there situations where multiple truths coexist? In the next sections we shed light on these and other important aspects for both cases of connected and weak graphs.

5.2 Learning over Connected Graphs

One critical element influencing the social learning behavior is the network topology. In this section we consider *connected* graphs, already introduced in Definition 4.5. The next theorem establishes the belief convergence under this setting, as a straightforward application of Theorem 5.1.

Theorem 5.2 (Network agreement over connected graphs). Let Assumptions 5.1 and 5.2 be satisfied. Assume that the network graph is connected, let v be the Perron vector associated with the combination matrix A, and consider the following *network average* of KL divergences:

$$D_{\mathsf{net}}(\theta) \triangleq \sum_{k=1}^{K} v_k D(f_k||\ell_{k,\theta}). \tag{5.24}$$

If $D_{\mathsf{net}}(\theta)$ admits a unique minimizer $\vartheta^\star$, then for $k = 1, 2, \ldots, K$,

$$\boldsymbol{\mu}_{k,t}(\vartheta^\star) \xrightarrow[t\to\infty]{\text{a.s.}} 1. \tag{5.25}$$

Proof. Since A is a left stochastic irreducible matrix, from (4.22) it follows that

$$A^\bullet = \lim_{t\to\infty} \frac{1}{t} \sum_{\tau=1}^{t} A^\tau = v\mathbb{1}^\top, \tag{5.26}$$

which means that the limiting matrix $A^\bullet$ in (5.6) has rank one with identical columns given by v. In this case, definition (5.7) reduces to (5.24) for all k, and then Eq. (5.25) follows from (5.8).

■

Theorem 5.2 reveals that *connected graphs enable agreement* among agents. To gain further insight into the mechanism that leads to agreement over graphs, it is useful to compare the network average $D_{\mathsf{net}}(\theta)$ in (5.24) against its general version $\bar{D}_k(\theta)$ in (5.7). To this end, let us rewrite these two network averages as follows:

$$\bar{D}_k(\theta) = \sum_{j=1}^{K} a^\bullet_{jk}\, D(f_j||\ell_{j,\theta}), \tag{5.27}$$

$$D_{\mathsf{net}}(\theta) = \sum_{j=1}^{K} v_j D(f_j||\ell_{j,\theta}). \tag{5.28}$$

We see that in (5.27) the KL divergence of the jth agent is scaled by a limiting weight, $a^\bullet_{jk}$, which *depends on the particular agent k under* *consideration.* In contrast, in (5.27) the KL divergence of the jth agent is scaled by the Perron vector entry v_j, which *does not depend on* k. This is because the time-average of the matrix powers converges to a matrix with all columns equal to the Perron vector v. As a result, while the network average $\bar{D}_k(\theta)$ determines the performance of agent k, the network average $D_{\mathsf{net}}(\theta)$ determines the performance of all agents. This explains why the agents behave equally and (when $D_{\mathsf{net}}(\theta)$ has a unique minimizer $\vartheta^\star$) are able to reach agreement, with the belief vector of *every* agent converging to a probability vector that places unit mass on *one and the same* hypothesis $\vartheta^\star$.

Example 5.3 (Agreement). Consider a network of $K = 12$ agents, connected according to the topology displayed in the top left panel of Figure 5.1. The network graph is undirected, and there are no self-loops. The combination matrix A is designed following the uniform-averaging rule (see Table 4.1), resulting in a left stochastic matrix. It can be verified that there exists a path between any two nodes in both directions, thus the graph is connected, i.e., the combination matrix is irreducible. We have evaluated the Perron vector associated with A, which is equal to

$$v = \frac{1}{30}\,[5,\ 2,\ 2,\ 2,\ 2,\ 2,\ 2,\ 2,\ 2,\ 2,\ 2,\ 5]\,. \tag{5.29}$$

Moreover, it can be verified that A has two eigenvalues on the unit circle: the eigenvalue 1 that must be present since A is left stochastic, and another eigenvalue equal to -1. As a result, A is *not* primitive.

Each agent $k = 1, 2, \ldots, 12$ observes streaming observations $\boldsymbol{x}_{k,1}, \boldsymbol{x}_{k,2}, \ldots$ distributed according to some true model $f_k(x)$. The agents are partitioned into the following clusters (displayed with different colors in Figure 5.1):

$$\begin{aligned} \mathcal{C}_1 &= \{1, 2, 3, 4\},\\ \mathcal{C}_2 &= \{5, 6, 7, 8\},\\ \mathcal{C}_3 &= \{9, 10, 11, 12\}, \end{aligned} \tag{5.30}$$

and the true models are assumed to be common to all agents belonging to the same cluster. That is, denoting by $g_c(x)$ the true model pertaining to cluster $\mathcal{C}_c$, with $c = 1, 2, 3$, we have $f_k(x) = g_c(x)$ for all $k \in \mathcal{C}_c$. The true model $g_c(x)$ is a unit-variance Gaussian pdf with mean ν_c, where

$$\nu_1 = 0.8, \quad \nu_2 = 1.6, \quad \nu_3 = 2.4. \tag{5.31}$$

We assume that all agents have common likelihoods, that is, $\ell_k(x|\theta) = \ell(x|\theta)$ for all k. Each likelihood $\ell(x|\theta)$, when regarded as a function of x, is a unit-variance Gaussian pdf with mean $\nu_\theta = \theta$, for $\theta \in \Theta = \{1, 2, 3\}$. The top right panel of Figure 5.1 shows the likelihoods (solid line) and the true models (dashed line).

The asymptotic beliefs resulting from the social learning process are characterized in Theorem 5.2, where we see that the agents will agree on the hypothesis $\vartheta^\star$ that minimizes

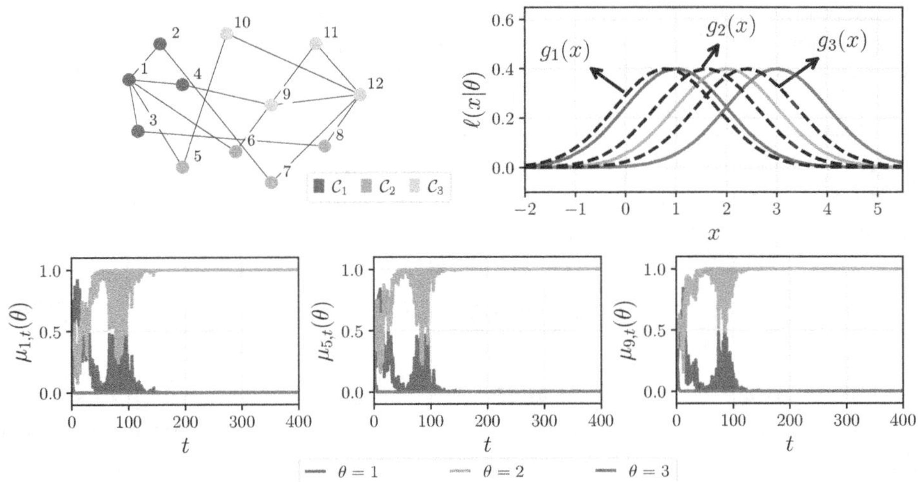

Figure 5.1: (*Top left*) Network topology showing the different clusters $\mathcal{C}_c$ corresponding to Example 5.3. The graph is undirected and there are no self-loops. (*Top right*) Likelihood models $\ell(x|\theta)$ (solid line) and true models $g_c(x)$ (dashed line). (*Bottom*) Belief evolution over 400 iterations for agents 1, 5, and 9. We see that, as t grows, the agents place their full belief mass on the unique minimizer $\vartheta^\star = 2$.

the network average of KL divergences $D_{\text{net}}(\theta)$ defined in (5.24). In this example, $D_{\text{net}}(\theta)$ is given by

$$D_{\text{net}}(\theta) = \sum_{k=1}^{12} v_k D(f_k||\ell_\theta) = \sum_{c=1}^{3} D(g_c||\ell_\theta) \times \sum_{k\in\mathcal{C}_c} v_k, \tag{5.32}$$

where the entries of the Perron vector are obtained from (5.29). We can compute the KL divergence between Gaussian distributions with the same variance using (2.45), which yields, for $\theta \in \Theta$,

$$\begin{aligned} D(g_1||\ell_\theta) &= \frac{1}{2}(0.8-\theta)^2, \\ D(g_2||\ell_\theta) &= \frac{1}{2}(1.6-\theta)^2, \\ D(g_3||\ell_\theta) &= \frac{1}{2}(2.4-\theta)^2, \end{aligned} \tag{5.33}$$

from which we see that (5.5) holds. Using (5.32), the network average of KL divergences can thus be written as

$$\begin{aligned} D_{\text{net}}(\theta) &= \frac{1}{2}(0.8-\theta)^2 \sum_{k\in\mathcal{C}_1} v_k + \frac{1}{2}(1.6-\theta)^2 \sum_{k\in\mathcal{C}_2} v_k + \frac{1}{2}(2.4-\theta)^2 \sum_{k\in\mathcal{C}_3} v_k \\ &= \frac{11}{60}(0.8-\theta)^2 + \frac{2}{15}(1.6-\theta)^2 + \frac{11}{60}(2.4-\theta)^2, \end{aligned} \tag{5.34}$$

with hypothesis-specific values

$$D_{\text{net}}(1) = 0.414, \quad D_{\text{net}}(2) = 0.314, \quad D_{\text{net}}(3) = 1.214. \tag{5.35}$$

The minimizer of $D_{\text{net}}(\theta)$ is therefore $\vartheta^\star = 2$. The bottom panels of Figure 5.1 show the evolution of the beliefs of agents $1, 5$, and 9 over 400 iterations. We observe that, although the agents belong to different clusters with different true models, they all agree asymptotically on the same hypothesis $\vartheta^\star = 2$.

We see from (5.24) and (5.25) that, over connected graphs, all agents end up solving the minimization problem

$$\vartheta^\star = \arg\min_{\theta\in\Theta} \sum_{k=1}^{K} v_k D(f_k || \ell_{k,\theta}). \tag{5.36}$$

In other words, all agents will agree on the hypothesis θ that minimizes a global (across the agents) measure of discrepancy between the true and likelihood models. If we compare this conclusion with the single-agent case studied in Lemma 2.3 and Example 2.4, we find that over there, the minimizer $\vartheta^\star$ had a useful interpretation as corresponding to the likelihood model that gives the best match with the true model f. The conclusion is not as straightforward in the multi-agent case since different agents can now have *different* true models f_k, and the question of what $\vartheta^\star$ means and how it relates to the true and likelihood models becomes more elaborate.

5.3 Objective Evidence

In this section we assume that the true generative model f_k agrees with one of the likelihood models at some *true hypothesis* (denoted by ϑ^o), i.e., $f_k = \ell_{k,\vartheta^o}$. In other words, the likelihood set for each agent includes the true generative model.

Assumption 5.3 (Objective evidence). Each agent $k = 1, 2, \ldots, K$ at time $t = 1, 2, \ldots$ receives a data sample $\boldsymbol{x}_{k,t}$. The collections of K samples across the agents, $\{\boldsymbol{x}_{1,t}, \boldsymbol{x}_{2,t}, \ldots, \boldsymbol{x}_{K,t}\}$, are assumed iid over time. To perform social learning, agent k employs likelihood models $\{\ell_{k,\theta}\}_{\theta\in\Theta}$, and each data sample $\boldsymbol{x}_{k,t}$ is distributed according to ℓ_{k,ϑ^o}, namely, the true underlying hypothesis is $\vartheta^o \in \Theta$. Moreover, we assume that, for $k = 1, 2, \ldots, K$ and for all θ and θ' belonging to Θ,

$$D(\ell_{k,\theta} || \ell_{k,\theta'}) < \infty. \tag{5.37}$$

Under this assumption, and as we are going to see, it is expected that a good social learning strategy should ultimately discover the true hypothesis by placing increasing mass on ϑ^o as more data are collected.

In order to distinguish the true hypothesis ϑ^o from another hypothesis θ, it is necessary that the data collected by the agents have different statistical properties under the two hypotheses. This is not always the case. For example, consider an agent observing a sinusoidal signal through a sensor that is able to detect only the amplitude of the signal, but not its phase. Now, if ϑ^o and θ correspond to two signals with the same amplitude but different phases, the data observed by the agent will have the same statistical properties under ϑ^o and θ, implying that ϑ^o is indistinguishable from θ. Formally, agent k will be unable to distinguish ϑ^o from θ when the likelihoods are the same under the two hypotheses, i.e., when

$$D(\ell_{k,\vartheta^o}||\ell_{k,\theta}) = 0. \tag{5.38}$$

When condition (5.38) is satisfied for at least one $\theta \neq \vartheta^o$, we say that the learning problem is *locally unidentifiable* for agent k. The qualification "locally" highlights the fact that agent k would be unable, if learning in isolation, to identify correctly ϑ^o. Note that local unidentifiability is typical in social learning, as individual agents have often a partial view regarding the phenomenon of interest, and their local data tend to be insufficient to identify correctly the true underlying hypothesis. This is one reason why the agents are motivated to cooperate.

However, local unidentifiability does not preclude the network from identifying the true model ϑ^o. This is because, through repeated social learning steps, the sharing of information will help *all* agents overcome their *individual* limitations and allow them to attain their learning goal. For this to happen, local identifiability is not necessary and the following global condition is in fact sufficient.

Assumption 5.4 (Global identifiability). For each hypothesis $\theta \neq \vartheta^o$, we assume that there exists at least one agent k (which can be different for different θ) such that

$$D(\ell_{k,\vartheta^o}||\ell_{k,\theta}) > 0. \tag{5.39}$$

In other words, global identifiability requires that, for each hypothesis $\theta \neq \vartheta^o$, there exists at least one agent that is able to distinguish it from ϑ^o. Note that this is a significantly weaker condition than requiring local identifiability for *all* agents. At one extreme, we may have a problem that is locally unidentifiable for all agents, but globally identifiable. Referring back to the sinusoidal signal example, consider now two agents using different types of sensors. Agent 1 is able to detect the amplitude of the signal but not the phase, whereas agent 2 is able to detect the phase but not the amplitude. If both amplitude and phase are relevant to reveal a hypothesis of interest, then none of the agents is in a position to identify the hypothesis. However, working together, they would be able to learn properly by combining their local information. The next result, which is a corollary of Theorem 5.2, ascertains that global identifiability is sufficient to guarantee truth learning under objective evidence.

Corollary 5.1 (Truth learning over connected graphs). Let Assumptions 5.1, 5.3, and 5.4 be satisfied. If the network graph is connected, then for $k = 1, 2, \ldots, K$,

$$\boldsymbol{\mu}_{k,t}(\vartheta^o) \xrightarrow[t\to\infty]{\text{a.s.}} 1. \tag{5.40}$$

Proof. The claim in (5.40) will be proved if we show that the true hypothesis ϑ^o coincides with the minimizer $\vartheta^\star$ defined in the statement of Theorem 5.2. To see that this is the case, note that under Assumption 5.3 we have $f_k = \ell_{k,\vartheta^o}$, which implies that the network average of KL divergences (5.24) becomes

$$D_{\text{net}}(\theta) = \sum_{k=1}^{K} v_k D(\ell_{k,\vartheta^o} || \ell_{k,\theta}). \tag{5.41}$$

Clearly, $D_{\text{net}}(\vartheta^o)$ is equal to 0. From Assumption 5.4, for each $\theta \neq \vartheta^o$, there exists at least one agent k for which $D(\ell_{k,\vartheta^o} || \ell_{k,\theta}) > 0$. From this assumption and the fact that $v_k > 0$ for $k = 1, 2, \ldots, K$, we have that

$$D_{\text{net}}(\theta) > 0, \quad \theta \neq \vartheta^o. \tag{5.42}$$

Hence, ϑ^o minimizes $D_{\text{net}}(\theta)$ and it therefore coincides with $\vartheta^\star$ from Theorem 5.2. ■

In summary, we see that Assumption 5.4 provides one important motivation for agents to cooperate in social learning. When the learning problem is *locally* unidentifiable, meaning that an individual agent can have one or more hypotheses $\theta \neq \vartheta^o$ that are indistinguishable from the true hypothesis (zero KL divergence), then this agent will not be able to learn well *individually*. In contrast, under the *global* identifiability condition (5.39), Corollary 5.1 reveals that each agent in the network will now be able to identify the true hypothesis by cooperating with its neighbors.

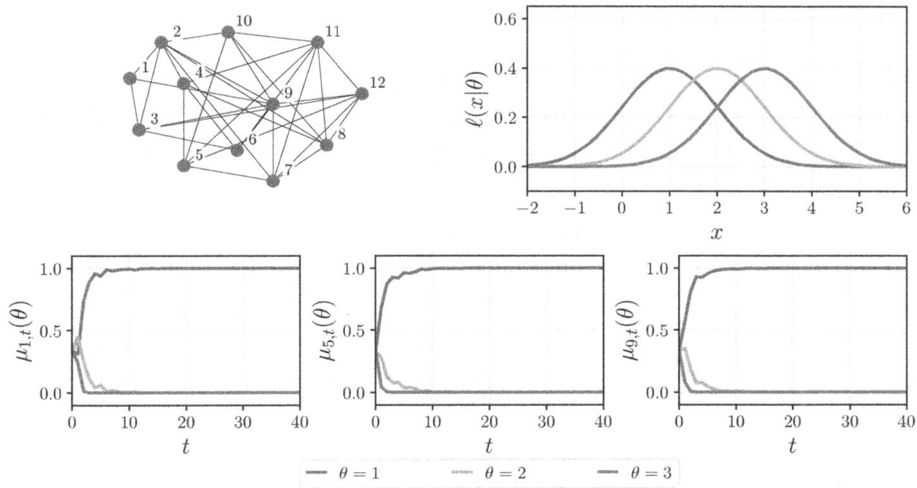

Figure 5.2: (*Top left*) Network topology used in Example 5.4. The graph is undirected and all agents are assumed to have a self-loop, not shown in the figure. (*Top right*) Likelihood models. (*Bottom*) Belief evolution over 40 iterations for agents 1, 5, and 9. We see that, as t grows, the agents place their full belief mass on the true hypothesis $\vartheta^o = 1$.

Example 5.4 (Truth learning). We consider the network topology shown in Figure 5.2. The graph is undirected and can be verified to be connected. Moreover, all agents have a self-loop, not shown in the figure. On top of this graph we build a combination matrix by using the Metropolis combination policy (see Table 4.1), which yields a doubly stochastic matrix. It follows from (4.18) that the Perron vector is uniform, which in this case yields $v = (1/12)\,\mathbb{1}$.

The network operates under the objective evidence model (Assumption 5.3). In other words, the streams of data $\boldsymbol{x}_{k,1}, \boldsymbol{x}_{k,2}, \ldots$ are drawn according to a true distribution $\ell_k(x|\vartheta^o)$ for each agent k. Specifically, the true underlying hypothesis is $\vartheta^o = 1$. The observations are statistically independent across the agents.

We assume that the agents have common likelihood models, i.e., $\ell_k(x|\theta) = \ell(x|\theta)$ for all k, and that $\ell(x|\theta)$ is a unit-variance Gaussian pdf with mean $\nu_\theta = \theta$, for $\theta \in \Theta = \{1, 2, 3\}$ — see the top right panel of Figure 5.2. We can verify that both (5.37) and Assumption 5.4 hold. The network average of KL divergences is given by

$$D_{\mathsf{net}}(\theta) = \frac{1}{12}\sum_{k=1}^{12} D(\ell_{k,\vartheta^o}||\ell_{k,\theta}) = D(\ell_{\vartheta^o}||\ell_\theta) = \frac{1}{2}(\vartheta^o - \theta)^2, \tag{5.43}$$

where the last equality follows from (2.45). In the bottom panels of Figure 5.2, we plot the belief evolution for agents 1, 5, and 9 over 40 iterations. We see that all agents agree asymptotically on the true hypothesis ϑ^o, as predicted by Corollary 5.1.

5.4 Subjective Evidence

There are many situations where it not possible to define a "true" hypothesis. For example, assume that two agents are forming their opinions about a particular candidate $\theta \in \{\text{candidate } 1, \text{candidate } 2\}$ in an election competition. Agent 1 belongs to a certain group that is biased toward candidate 1, and, hence, the evidence collected by agent 1 pushes the choice in favor of this candidate. The situation is reversed for agent 2. In this case we can talk of *subjective evidence*, and the fundamental question arises as to where the social learning strategy will converge. To start with, let us formalize the concept of subjective evidence in our framework.

Assumption 5.5 (Subjective evidence). Each agent $k = 1, 2, \ldots, K$ at time $t = 1, 2, \ldots$ receives a data sample $\boldsymbol{x}_{k,t}$. The collections of K samples across the agents, $\{\boldsymbol{x}_{1,t}, \boldsymbol{x}_{2,t}, \ldots, \boldsymbol{x}_{K,t}\}$, are assumed iid over time. To perform social learning, agent k employs likelihood models $\{\ell_{k,\theta}\}_{\theta\in\Theta}$, and each data sample $\boldsymbol{x}_{k,t}$ is distributed according to ℓ_{k,ϑ^o}, namely, the "locally true" underlying hypothesis at agent k is $\vartheta_k^o \in \Theta$. Moreover, we assume that, for $k = 1, 2, \ldots, K$ and for all θ and θ' belonging to Θ,

$$D(\ell_{k,\theta}||\ell_{k,\theta'}) < \infty. \tag{5.44}$$

According to Theorem 5.2, the key measure for determining on which opinion the agents will agree is the network average of KL divergences in (5.24) or, more specifically, its minimizer $\vartheta^\star$. Under the subjective evidence model defined by Assumption 5.5, this network average of KL divergences reduces to

$$D_{\mathsf{net}}(\theta) = \sum_{k=1}^{K} v_k D(\ell_{k,\vartheta_k^o}||\ell_{k,\theta}). \tag{5.45}$$

We will now examine how the local models and the network topology lead to the prevalence of some particular hypotheses.

Example 5.5 (How majority builds a common opinion). We consider the same network, combination matrix and likelihood models used in the previous example, but we now assume that the network operates under the subjective evidence model (Assumption 5.5). More specifically, the network is divided into two clusters,

$$\begin{aligned} \mathcal{C}_1 &= \{1, 2\}, \\ \mathcal{C}_2 &= \{3, 4, 5, 6, 7, 8, 9, 10, 11, 12\}, \end{aligned} \tag{5.46}$$

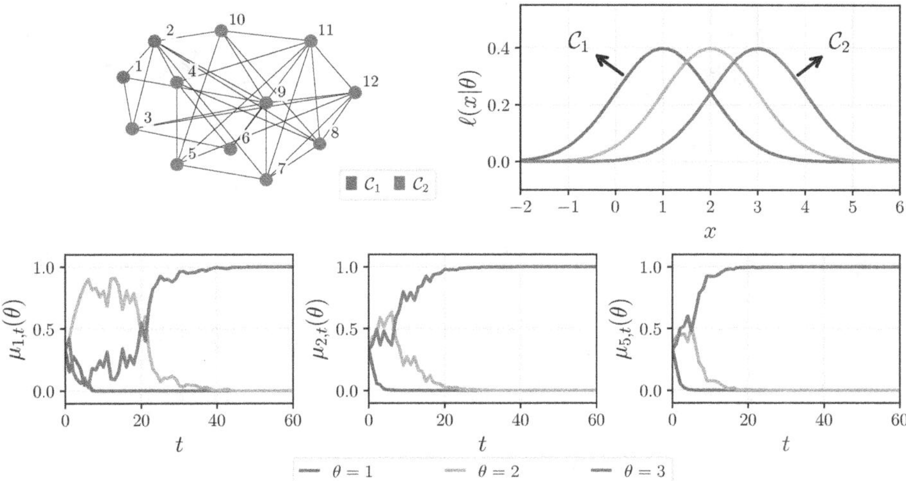

Figure 5.3: (*Top left*) Network topology showing the different clusters $\mathcal{C}_c$ corresponding to Example 5.5. The graph is undirected and all agents are assumed to have a self-loop, not shown in the figure. (*Top right*) Likelihood models. (*Bottom*) Belief evolution over 60 iterations for agents 1, 2, and 5. We see that, as t grows, the agents place their full belief mass on the unique minimizer $\vartheta^\star = 3$.

for which the true models are given by $f_k(x) = \ell(x|1)$ for $k \in \mathcal{C}_1$, and $f_k(x) = \ell(x|3)$ for $k \in \mathcal{C}_2$ — see the top right panel of Figure 5.3. It is readily verified that (5.44) holds. Moreover, in the simulations the observations have been generated as statistically independent across the agents.

Upon communicating during social learning, the agents will likely receive contrasting opinions, because the different clusters "promote" different hypotheses. However, since almost all agents belong to cluster $\mathcal{C}_2$, we expect that this conflict is resolved in favor of hypothesis 3. We now prove that this is actually the case in this example. To this end, let us write explicitly the network average of KL divergences

$$D_{\text{net}}(\theta) = \frac{2}{12} D(\ell_1 || \ell_\theta) + \frac{10}{12} D(\ell_3 || \ell_\theta) = \frac{2}{24}(1-\theta)^2 + \frac{10}{24}(3-\theta)^2, \tag{5.47}$$

from which we can compute the hypothesis-specific values

$$D_{\text{net}}(1) = \frac{5}{3}, \quad D_{\text{net}}(2) = \frac{1}{2}, \quad D_{\text{net}}(3) = \frac{1}{3}. \tag{5.48}$$

Therefore, the minimizer of the network average of KL divergences is $\vartheta^\star = 3$. In the bottom panels of Figure 5.3, we plot the belief evolution for agents 1, 2, and 5 over 60 iterations, which shows that the majority cluster $\mathcal{C}_2$ is able to steer the network's opinion toward hypothesis 3. Note that agents 1 and 2 belong to cluster $\mathcal{C}_1$, which promotes instead hypothesis 1.

Example 5.6 (How centrality builds a common opinion). In this example, we equalize the two clusters in Example 5.5 and set them according to

$$\begin{aligned} \mathcal{C}_1 &= \{1, 2, 3, 4, 5, 6\}, \\ \mathcal{C}_2 &= \{7, 8, 9, 10, 11, 12\}. \end{aligned} \tag{5.49}$$

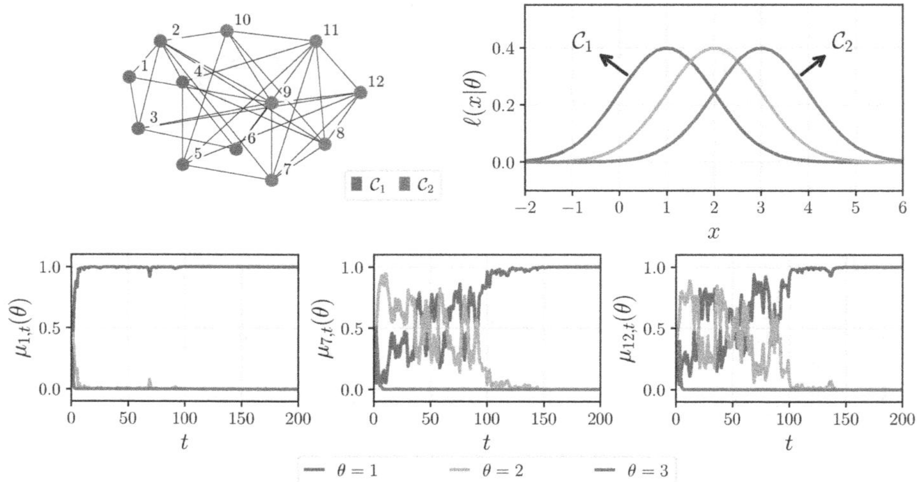

Figure 5.4: (*Top left*) Network topology showing the different clusters $\mathcal{C}_c$ corresponding to Example 5.6. The graph is undirected and all agents are assumed to have a self-loop, not shown in the figure. (*Top right*) Likelihood models. (*Bottom*) Belief evolution over 200 iterations for agents 1, 7, and 12. We see that, as t grows, the agents place their full belief mass on the unique minimizer $\vartheta^\star = 1$.

The true and likelihood models are the same as in Example 5.5, and they can be seen in the top right panel of Figure 5.4. Again, the true model of cluster $\mathcal{C}_1$ is $\ell(x|1)$, and the true model of cluster $\mathcal{C}_2$ is $\ell(x|3)$, indicating that conflicting evidence is observed by the agents.

Since the clusters have equal size, we cannot expect a majority rule to drive the agents' opinions. We now show how a different network attribute, namely, *centrality*, becomes important. From Theorem 5.2 we know that the network average $D_{\mathsf{net}}(\theta)$ determines the target hypothesis $\vartheta^\star$ the agents will agree on. Under the subjective evidence model, $D_{\mathsf{net}}(\theta)$ takes the specific form in (5.45). In the weighted combination of KL divergences appearing in (5.45), the impact of a particular agent k is enhanced by increasing the value of its own Perron vector entry v_k. Accordingly, v_k represents a measure of the *centrality* of agent k. This interpretation is actually not limited to social learning — see the explanation following Theorem 4.4.

Since different Perron vector entries reflect the different degree of influence of the agents, to highlight the role of agent centrality we would like to assign a nonuniform Perron vector. To this end, we now construct a left stochastic combination matrix (on top of the same network topology shown in top left panel of Figure 5.4) by using the procedure described in [170], which allows us to choose a predefined Perron vector. In this example, we choose in particular the following Perron vector:

$$v = \frac{1}{60}[8, 8, 8, 8, 8, 8, 2, 2, 2, 2, 2, 2]. \tag{5.50}$$

The combination matrix is built with the following rule:

$$a_{jk} = \begin{cases} v_j & \text{if } j \in \mathcal{N}_k \setminus \{k\}, \\ 1 - \sum\limits_{j \in \mathcal{N}_k \setminus \{k\}} v_j & \text{if } j = k, \\ 0 & \text{otherwise.} \end{cases} \tag{5.51}$$

We can verify that the resulting combination matrix A is left stochastic with the Perron vector specified in (5.50). Under this design, although the number of agents in each cluster is the same, their importance in the network is very distinct. From (5.50) we see that significantly larger centrality scores are given to agents belonging to cluster $\mathcal{C}_1$ in comparison with agents in cluster $\mathcal{C}_2$. This will impact the asymptotic beliefs in the network, as we see next.

First, we write down the expression for the network average of KL divergences,

$$\mathcal{D}_{\text{net}}(\theta) = \frac{8}{10} D(\ell_1 || \ell_\theta) + \frac{2}{10} D(\ell_3 || \ell_\theta) = \frac{8}{20}(1-\theta)^2 + \frac{2}{20}(3-\theta)^2, \tag{5.52}$$

from which we can compute the hypothesis-specific values

$$\mathcal{D}_{\text{net}}(1) = 0.4, \quad \mathcal{D}_{\text{net}}(2) = 0.5, \quad \mathcal{D}_{\text{net}}(3) = 1.6. \tag{5.53}$$

The minimizer of the network average of KL divergences is now $\vartheta^\star = 1$. In the bottom panels of Figure 5.4, we plot the beliefs of agents $1, 7$, and 12 over 200 iterations. All these agents tend to place their full belief mass on hypothesis 1. In other words, the cluster with the largest centrality scores, $\mathcal{C}_1$, is able to determine the network's opinion. Note that agents 7 and 12 belong to cluster $\mathcal{C}_2$, which promotes a hypothesis different from $\vartheta^\star$.

Example 5.7 (Truth is somewhere in between). In the last two examples we considered two distinct elements that determine the final agents' opinions, namely, cluster size and agent centrality. We now remove both these elements and examine how the opinion formation mechanism changes. We consider the same network topology used in the last two examples, with balanced clusters (see the top left panel of Figure 5.5), and with a Metropolis combination matrix (see Table 4.1), so that all agents share the same centrality score since the Perron vector has equal entries. The true and likelihood models are kept unchanged with respect to the last two examples. As was the case before, the observations are statistically independent across the agents.

We can evaluate the network average of KL divergences as

$$\mathcal{D}_{\text{net}}(\theta) = \frac{6}{12} D(\ell_1 || \ell_\theta) + \frac{6}{12} D(\ell_3 || \ell_\theta) = \frac{6}{24}(1-\theta)^2 + \frac{6}{24}(3-\theta)^2, \tag{5.54}$$

from which we can compute the hypothesis-specific values

$$\mathcal{D}_{\text{net}}(1) = 1, \quad \mathcal{D}_{\text{net}}(2) = 0.5, \quad \mathcal{D}_{\text{net}}(3) = 1. \tag{5.55}$$

Therefore, the minimizer of the network average of KL divergences is $\vartheta^\star = 2$. In the bottom panels of Figure 5.5, we plot the evolution of beliefs of agents $1, 5$, and 9 over 40 iterations. The curves show that, although the clusters $\mathcal{C}_1$ and $\mathcal{C}_2$ observe evidence supporting, respectively, hypotheses 1 and 3, neither cluster is able to exert a domineering

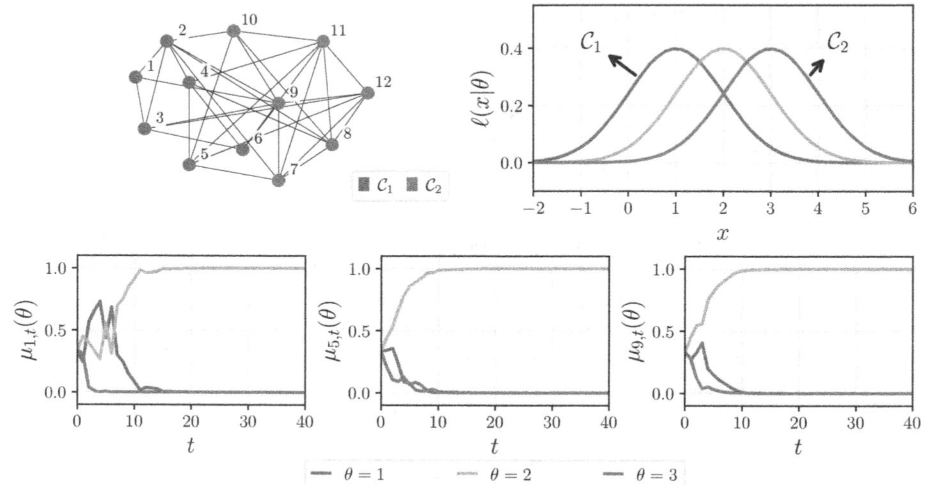

Figure 5.5: (*Top left*) Network topology showing the different clusters $\mathcal{C}_c$ corresponding to Example 5.7. The graph is undirected and all agents are assumed to have a self-loop, not shown in the figure. (*Top right*) Likelihood models. (*Bottom*) Belief evolution over 40 iterations for agents 1, 5, and 9. We see that, as t grows, the agents place their full belief mass on the unique minimizer $\vartheta^\star = 2$.

influence. Instead, the conflicting evidence drives the agents to place their full belief mass on the intermediate hypothesis $\vartheta^\star = 2$.

How can we explain this effect? One interpretation is that, in the presence of *conflicting evidence*, the agents opt for a conservative choice. Referring to real-life situations, we can think of one person betting on a soccer match between teams 1 and 2. Assume that discordant solicitations come from the environment, i.e., the person receives data suggesting to bet on the victory of team 1, as well as data suggesting to bet on the victory of team 2. If there is no sufficient evidence to let one suggestion prevail, then the most plausible choice would be to bet on a draw!

5.5 Fake Evidence

There is another specialization of the general model in Assumption 5.2 that is useful in social learning applications. It is the case where some agents observe data generated according to the true hypothesis ϑ^o, while the other agents observe data following "fake" distributions.

Assumption 5.6 (Fake evidence). Each agent $k = 1, 2, \ldots, K$ at time $t = 1, 2, \ldots$ receives a data sample $\boldsymbol{x}_{k,t}$. The collections of K samples across the agents,

$\{\boldsymbol{x}_{1,t}, \boldsymbol{x}_{2,t}, \ldots, \boldsymbol{x}_{K,t}\}$, are assumed iid over time. The probability (density or mass) function of $\boldsymbol{x}_{k,t}$ is denoted by f_k. To perform social learning, agent k employs likelihood models $\{\ell_{k,\theta}\}_{\theta\in\Theta}$ of the same nature as f_k (namely, for all $\theta \in \Theta$, $\ell_{k,\theta}$ is a pdf if f_k is a pdf, and a pmf otherwise).
There exists a true hypothesis $\vartheta^o \in \Theta$ and the agents are divided into two categories, *truthful* and *untruthful*. The data samples of the truthful agents are distributed according to the likelihoods corresponding to a common true hypothesis ϑ^o, i.e., $f_k = \ell_{k,\vartheta^o}$ when agent k is truthful. When agent k is untruthful, its data samples are instead drawn from some arbitrary f_k. We assume that, for $k = 1, 2, \ldots, K$ and for all $\theta \in \Theta$,

$$D(f_k||\ell_{k,\theta}) < \infty. \tag{5.56}$$

The fundamental question arising from the model in Assumption 5.6 is whether the untruthful agents can bias the choices of the truthful agents and preclude them from learning the true hypothesis ϑ^o.

Example 5.8 (One fake agent). Consider the network topology displayed in the top left panel of Figure 5.6. The graph is undirected and all agents are assumed to have a self-loop, not shown in the figure. This graph can be verified to be strong. On top of it, we construct a Metropolis combination matrix — see Table 4.1. We focus on the fake evidence case (Assumption 5.6), where the network is "contaminated" by the presence of one untruthful agent, namely agent 12. The likelihoods follow the Gaussian models used in the last examples.

The true model is the same across all *truthful* agents and corresponds to a true likelihood $\ell(x|\vartheta^o)$, i.e., $f_k(x) = \ell(x|\vartheta^o)$ for $k = 1, 2, \ldots, 11$. In contrast, the observations of the untruthful agent are drawn from a true model $f_{12}(x)$, which is set as a unit-variance Gaussian pdf with mean $\nu_{12} = 20$. In this scenario, we can verify that (5.56) holds. Moreover, in the simulations the observations have been generated as statistically independent across the agents. The true model of the untruthful agent and the likelihoods are displayed in the top right panel of Figure 5.6. We consider the true underlying hypothesis to be $\vartheta^o = 1$.

The network average of KL divergences is given by

$$D_{\mathsf{net}}(\theta) = \frac{11}{12}D(\ell_{\vartheta^o}||\ell_\theta) + \frac{1}{12}D(f_{12}||\ell_\theta) = \frac{11}{24}(1-\theta)^2 + \frac{1}{24}(20-\theta)^2, \tag{5.57}$$

from which we can compute the hypothesis-specific values

$$D_{\mathsf{net}}(1) = 15.04, \quad D_{\mathsf{net}}(2) = 13.96, \quad D_{\mathsf{net}}(3) = 13.88. \tag{5.58}$$

The network average of KL divergences is thus minimized at $\vartheta^\star = 3$. In the bottom panels of Figure 5.6, we plot the evolution of beliefs of agents 1, 5, and 12 over 80 iterations. We see that the presence of the untruthful agent is sufficient to lead the network astray, by forcing all agents to place their full belief mass on the wrong hypothesis $\vartheta^\star = 3$.

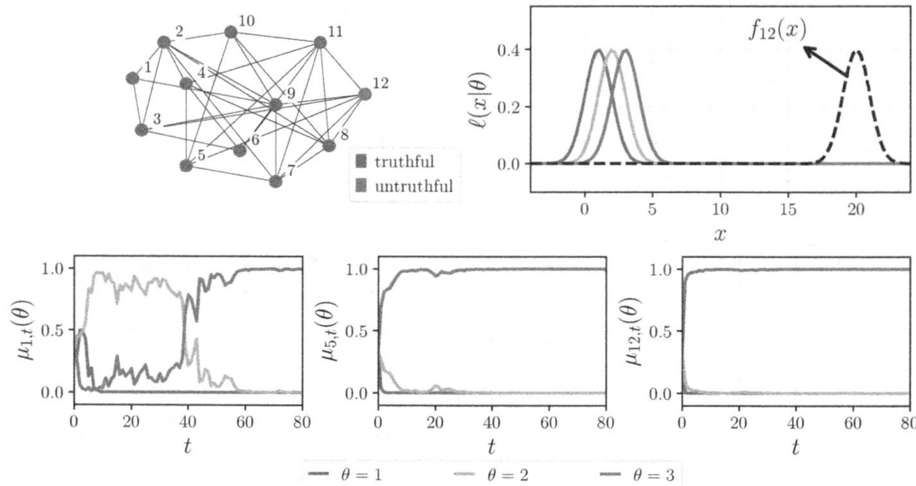

Figure 5.6: (*Top left*) Network topology showing truthful and untruthful agents corresponding to Example 5.8. The graph is undirected and all agents are assumed to have a self-loop, not shown in the figure. (*Top right*) Likelihood models $\ell(x|\theta)$ (solid line) and true model $f_{12}(x)$ of the untruthful agent (dashed line). The true model of the truthful agents is $\ell(x|1)$ (blue line). (*Bottom*) Belief evolution over 80 iterations for agents 1, 5, and 12. We see that, as t grows, the agents place their full belief mass on the unique minimizer $\vartheta^\star = 3$.

5.6 Learning over Weak Graphs

The discussion in the earlier sections focused on examining belief propagation over connected graphs. We now examine what happens over weak graphs, which were introduced in Section 4.5. We recall that, over a weak graph, the agents are partitioned into two groups, $\mathcal{S}$ and $\mathcal{R}$, containing *sending* and *receiving* networks, respectively.

Theorem 5.3 (Mind control over weak graphs). Let Assumptions 5.1 and 5.2 be satisfied. Assume that the network graph is weak. According to Theorem 4.5, we have

$$A^\bullet = \lim_{t\to\infty} \frac{1}{t} \sum_{\tau=1}^{t} A^\tau = \left[\begin{array}{c|c} V & W \\ \hline 0 & 0 \end{array}\right], \tag{5.59}$$

where the matrices V and W are defined by (4.28) and (4.29), respectively. For each $s = 1, 2, \ldots, S$, if agent k belongs to the sth sending network, its asymptotic beliefs can be derived directly from Theorem 5.2. This is because the neighborhood $\mathcal{N}_k$ contains only agents from the sth sending network. In contrast, the agents in the receiving networks exhibit the following distinct behavior. Using (5.59), for each agent $k \in \mathcal{R}$ we can rewrite the network average of KL divergences in (5.7) as

$$\bar{D}_k(\theta) = \sum_{j \in \mathcal{S}} w_{jk} D(f_j || \ell_{j,\theta}). \tag{5.60}$$

If $\bar{D}_k(\theta)$ admits a unique minimizer $\vartheta_k^\star$, then for all $k \in \mathcal{R}$,

$$\boldsymbol{\mu}_{k,t}(\vartheta_k^\star) \xrightarrow[t \to \infty]{\text{a.s.}} 1. \tag{5.61}$$

Proof. The result follows from Theorem 5.1 once we apply Theorem 4.6 and consequently replace the general matrix $A^\bullet$ in (5.6) with the particular matrix in (5.59). ■

Equation (5.60) contains the essential elements to understand the learning mechanism over weak graphs. First, the behavior of agents belonging to $\mathcal{R}$ is determined solely by KL divergences relative to agents belonging to $\mathcal{S}$. This is a remarkable conclusion that leads to a phenomenon we refer to as *mind control* [118, 147, 148], since the learning behavior of the agents in the receiving networks is completely controlled by the agents in the sending networks.

Second, the dependence of the weights w_{jk} in (5.60) on the agent index k reveals that the learning behavior can also be distinct across the agents in the receiving networks, leading to a phenomenon we refer to as *discord* [118, 147, 148]. This behavior is in sharp contrast with what happens over connected graphs, where we have seen in (5.25) that all agents reach *agreement* on a common hypothesis $\vartheta^\star$.

Example 5.9 (Truth learning under objective evidence). Consider 12 agents partitioned into the following clusters:

$$\begin{aligned} \mathcal{C}_1 &= \{1, 2, 3, 4\}, \\ \mathcal{C}_2 &= \{5, 6, 7, 8\}, \\ \mathcal{C}_3 &= \{9, 10, 11, 12\}. \end{aligned} \tag{5.62}$$

The agents are connected according to the weak graph shown in the top left panel of Figure 5.7, which is made of two sending networks and one receiving network. The clusters $\mathcal{C}_1$ and $\mathcal{C}_2$ correspond to the two sending networks, whereas cluster $\mathcal{C}_3$ correspond to the receiving network. That is, we have $\mathcal{S} = \mathcal{C}_1 \cup \mathcal{C}_2$ and $\mathcal{R} = \mathcal{C}_3$. All agents are assumed to have a self-loop (not shown in the figure) and, according to the weak-graph model, the edges from the two sending networks to the receiving network are directed. All other edges are chosen as undirected. Moreover, the combination matrix constructed with the uniform-averaging rule — see Table 4.1.

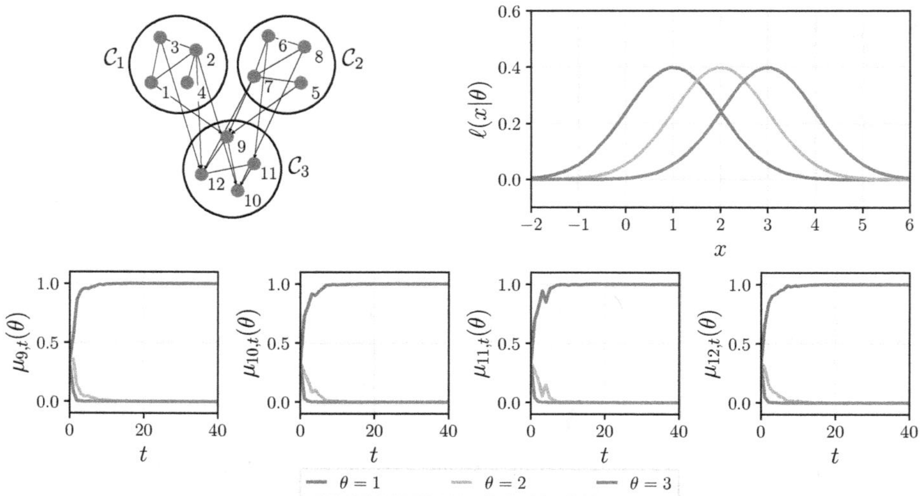

Figure 5.7: (*Top left*) Network topology showing the sending networks, i.e., clusters $\mathcal{C}_1$ and $\mathcal{C}_2$, and the receiving network, i.e., cluster $\mathcal{C}_3$, used in Example 5.9. Undirected edges are represented without arrows, and all agents have a self-loop, not shown in the figure. (*Top right*) Likelihood models. (*Bottom*) Belief evolution for the agents in the receiving network over 40 iterations. We see that, as t grows, the agents place their full belief mass on the common true hypothesis $\vartheta^o = 1$.

We assume that all agents operate under the objective evidence model with true hypothesis $\vartheta^o = 1$ and that they use the same Gaussian likelihoods adopted in the last examples (see the top right panel of Figure 5.7). Moreover, in the simulations the observations are drawn as statistically independent across the agents. From (5.60), the network average of KL divergences for agent $k \in \mathcal{R}$ is given by

$$\bar{D}_k(\theta) = \frac{1}{2}\sum_{j\in\mathcal{S}} w_{jk}(\vartheta^o - \theta)^2 = \frac{1}{2}(1-\theta)^2, \tag{5.63}$$

where we used the fact that $\sum_{j\in\mathcal{S}} w_{jk} = 1$. Thus, $\bar{D}_k(\theta)$ is clearly minimized at $\vartheta_k^\star = \vartheta^o = 1$ for any agent $k \in \mathcal{R}$. The bottom panels of Figure 5.7 show the evolution of beliefs over time for all agents in the receiving network. We see that, in this case, the agents in the receiving network asymptotically place their full belief mass on the true hypothesis ϑ^o in accordance with Theorem 5.3.

Example 5.9 shows a situation where all agents in a weak graph are able to learn the truth. As a matter of fact, it is possible to give a complete characterization of truth learning for the case of objective evidence. First, we must distinguish between the behavior of sending and receiving networks.

Consider first the sending networks, and observe that the agents in each sending network receive information only from agents in the same sending network. This implies that, when we run the social learning algorithm in listing (3.16), the beliefs of the agents in the sth sending network are actually produced by the same social learning algorithm run with a combination matrix equal to the submatrix A_s. Since A_s is irreducible, from Theorem 5.2 we conclude that the agents in each sending network will learn the truth if the problem is *globally identifiable within that network*, i.e., if the *individual sending network* satisfies Assumption 5.4.

On the other hand, for agents in the receiving networks, we can rewrite (5.60) under objective evidence, yielding

$$\bar{D}_k(\theta) = \sum_{j \in \mathcal{S}} w_{jk}\, D(\ell_{j,\vartheta^o} || \ell_{j,\theta}). \tag{5.64}$$

We see that $\bar{D}_k(\vartheta^o) = 0$ and, hence, ϑ^o is a minimizer for $\bar{D}_k(\theta)$ because the KL divergence is nonnegative. This minimizer is unique when $\bar{D}_k(\theta) > 0$ for all $\theta \neq \vartheta^o$. In view of (5.64), this condition is met when, for each $\theta \neq \vartheta^o$, there exists at least one agent $j \in \mathcal{S}$ satisfying

$$w_{jk}\, D(\ell_{j,\vartheta^o} || \ell_{j,\theta}) > 0, \tag{5.65}$$

which means that agent j is able to distinguish θ from ϑ^o (i.e., that $D(\ell_{j,\vartheta^o} || \ell_{j,\theta}) > 0$) and is connected to agent k through some path (i.e., $w_{jk} > 0$). Note that, according to this definition, the problem might be unidentifiable for some sending network s, but identifiable for the *ensemble* of sending networks that are connected to agent k. In this case, agent k will learn the truth, even if agents belonging to the sth sending network will not. This happens because the agents belonging to s receive information only from agents within *their own* sending network, while agent k benefits from information received from other sending networks.

Example 5.10 (Mind control). Consider the same weak graph, combination matrix, and likelihoods used in Example 5.9. Recall that the agents were organized into three clusters according to (5.62), with the clusters $\mathcal{C}_1$ and $\mathcal{C}_2$ representing the sending networks, and cluster $\mathcal{C}_3$ representing the receiving network.

Concerning the true distributions, we assume that the true models $\{f_k(x)\}$ vary across the clusters, while the agents within the same cluster share the same true model. Accordingly, denoting by $g_c(x)$ the true model pertaining to cluster $\mathcal{C}_c$, with $c = 1, 2, 3$, we have $f_k(x) = g_c(x)$ for all $k \in \mathcal{C}_c$. The true model $g_c(x)$ is a unit-variance Gaussian pdf with mean ν_c, where

$$\nu_1 = 0.8, \quad \nu_2 = 1.2, \quad \nu_3 = 3.2. \tag{5.66}$$

Moreover, in the simulations the observations are drawn as statistically independent across the agents. The true and likelihood models can be seen in the top right panel of

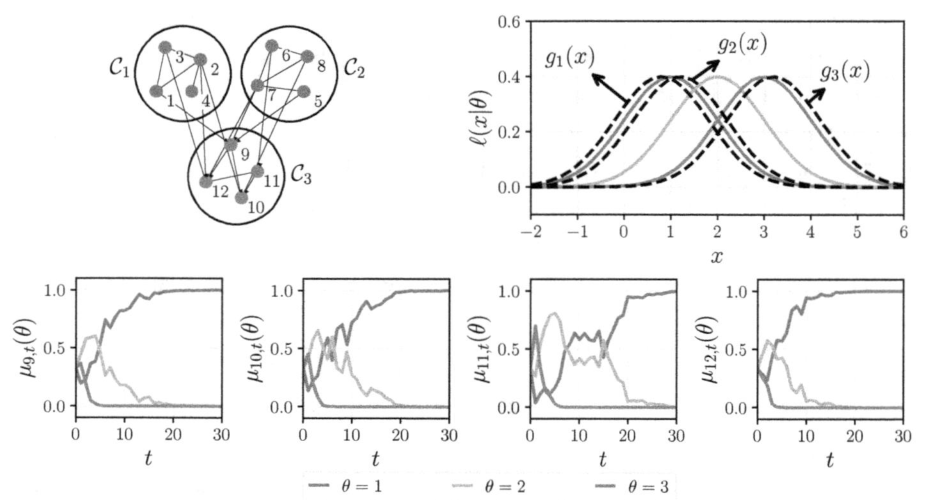

Figure 5.8: (*Top left*) Network topology showing the sending networks, i.e., clusters $\mathcal{C}_1$ and $\mathcal{C}_2$, and the receiving network, i.e., cluster $\mathcal{C}_3$, used in Example 5.10. Undirected edges are represented without arrows, and all agents have a self-loop, not shown in the figure. (*Top right*) Likelihood models $\ell(x|\theta)$ (solid line) and true models $g_c(x)$ (dashed line). (*Bottom*) Belief evolution for the agents in the receiving network over 30 iterations. We see that, as t grows, these agents place all their belief mass on the common hypothesis $\vartheta_k^\star = 1$.

Figure 5.8. The models pertaining to the agents in the sending networks lie closer to $\ell(x|1)$, therefore providing evidence supporting hypothesis 1. In contrast, the agents in the receiving network observe data streams whose distribution is closer to $\ell(x|3)$, thus supporting hypothesis 3.

The network average of KL divergences is given by

$$\begin{aligned}\bar{D}_k(\theta) &= \frac{1}{2}\sum_{j\in\mathcal{C}_1} w_{jk}(\nu_1-\theta)^2 + \frac{1}{2}\sum_{j\in\mathcal{C}_2} w_{jk}(\nu_2-\theta)^2 \\ &= \frac{1}{2}(0.8-\theta)^2\sum_{j\in\mathcal{C}_1} w_{jk} + \frac{1}{2}(1.2-\theta)^2\sum_{j\in\mathcal{C}_2} w_{jk}.\end{aligned} \tag{5.67}$$

Using the fact that

$$\sum_{j\in\mathcal{S}} w_{jk} = \sum_{j\in\mathcal{C}_1} w_{jk} + \sum_{j\in\mathcal{C}_2} w_{jk} = 1, \tag{5.68}$$

we can evaluate the hypothesis-specific values of $\bar{D}_k(\theta)$ as

$$\bar{D}_k(1) = \frac{(0.2)^2}{2}\sum_{j\in\mathcal{C}_1} w_{jk} + \frac{(0.2)^2}{2}\sum_{j\in\mathcal{C}_2} w_{jk} = 0.02, \tag{5.69}$$

$$\bar{D}_k(2) = \frac{(1.2)^2}{2}\sum_{j\in\mathcal{C}_1} w_{jk} + \frac{(0.8)^2}{2}\sum_{j\in\mathcal{C}_2} w_{jk} = 0.32 + 0.4\sum_{j\in\mathcal{C}_1} w_{jk}, \tag{5.70}$$

$$\bar{D}_k(3) = \frac{(2.2)^2}{2}\sum_{j\in\mathcal{C}_1} w_{jk} + \frac{(1.8)^2}{2}\sum_{j\in\mathcal{C}_2} w_{jk} = 1.62 + 0.8\sum_{j\in\mathcal{C}_1} w_{jk}. \tag{5.71}$$

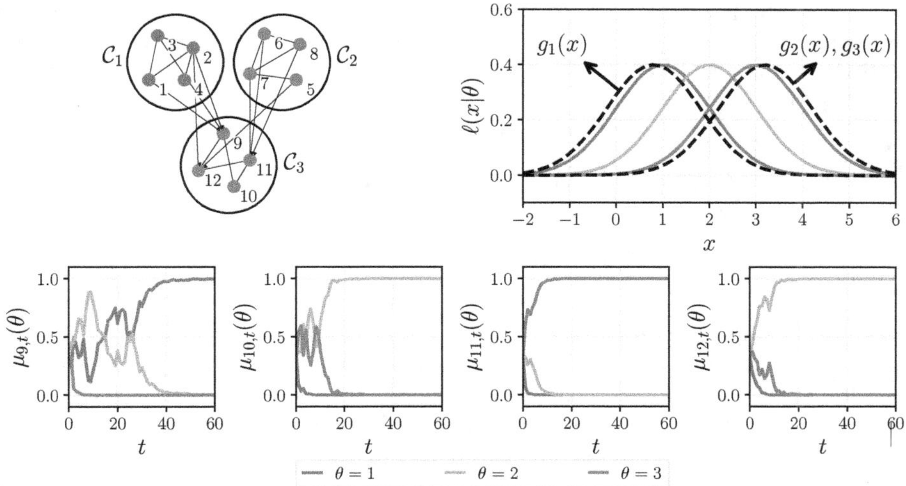

Figure 5.9: (*Top left*) Network topology showing the sending networks, i.e., clusters $\mathcal{C}_1$ and $\mathcal{C}_2$, and the receiving network, i.e., cluster $\mathcal{C}_3$, used in Example 5.11. Undirected edges are represented without arrows, and all agents have a self-loop, not shown in the figure. (*Top right*) Likelihood models $\ell(x|\theta)$ (solid line) and true models $g_c(x)$ (dashed line). (*Bottom*) Belief evolution for the agents in the receiving network over 60 iterations. We see that, as t grows, discord across the agents emerges, since their beliefs are concentrated on hypotheses $\vartheta_k^\star$ that depend on the particular agent k. Specifically, we have: $\vartheta_9^\star = 1$, $\vartheta_{10}^\star = 2$, $\vartheta_{11}^\star = 3$, and $\vartheta_{12}^\star = 2$.

Since the weights w_{jk} are nonnegative, from (5.71) we conclude that the minimizer of the network average of KL divergences is $\vartheta_k^\star = 1$ for any agent $k \in \mathcal{R}$. In the bottom panels of Figure 5.8, we see that, despite observing private data generated according to hypothesis 3, the agents in the receiving network asymptotically ignore this local information and place their full belief mass on the hypothesis supported by the sending networks, i.e., $\vartheta_k^\star = 1$.

Example 5.11 (Discord). In this example we start from the setting used in Example 5.10, and modify the network topology and the true distributions as follows. Concerning the topology, we consider a weak graph with the same sending and receiving networks used in Example 5.10, however with different connectivity between these networks, resulting in the graph shown in Figure 5.9.

Concerning the true distributions, as done in Example 5.10 we assume that $f_k(x) = g_c(x)$ for all $k \in \mathcal{C}_c$, with $c = 1, 2, 3$. The true model $g_c(x)$ is a unit-variance Gaussian pdf with mean ν_c, where

$$\nu_1 = 0.8, \qquad \nu_2 = 3.2, \qquad \nu_3 = 3.2. \tag{5.72}$$

Moreover, in the simulations the observations are generated as statistically independent across the agents. The true and likelihood models can be seen in the top right panel of Figure 5.9. The true model pertaining to the first sending network (cluster $\mathcal{C}_1$) is closer to $\ell(x|1)$, whereas the true model of cluster $\mathcal{C}_2$ is closer to $\ell(x|3)$. This means that, differently from what happened in Example 5.10, the two sending networks now provide conflicting information to the agents in the receiving network. Moreover, these agents share the same true model as cluster $\mathcal{C}_2$, which would in principle suggest further support for hypothesis 3.

The network average of KL divergences is computed for any agent $k \in \mathcal{R}$ as

$$\begin{aligned}\bar{D}_k(\theta) &= \frac{1}{2}\sum_{j\in\mathcal{C}_1} w_{jk}(\nu_1-\theta)^2 + \frac{1}{2}\sum_{j\in\mathcal{C}_2} w_{jk}(\nu_2-\theta)^2 \\ &= \frac{1}{2}(0.8-\theta)^2\sum_{j\in\mathcal{C}_1} w_{jk} + \frac{1}{2}(3.2-\theta)^2\sum_{j\in\mathcal{C}_2} w_{jk}.\end{aligned} \tag{5.73}$$

The minimization of (5.73) is not so easily found as was the case for Examples 5.9 and 5.10. As a matter of fact, the solution in this example is agent-dependent and varies according to the connectivity of each agent in the receiving network with respect to the sending networks $\mathcal{C}_1$ and $\mathcal{C}_2$. Specifically, the effect of this connectivity is represented in (5.73) by the cumulative weights $\sum_{j\in\mathcal{C}_1} w_{jk}$ and $\sum_{j\in\mathcal{C}_2} w_{jk}$, which are reported in Table 5.1.

Table 5.1: Cumulative weights incorporating the effect from each of the two sending networks $\mathcal{C}_1$ and $\mathcal{C}_2$ to each agent k in the receiving network.

Agent k	$\sum_{j\in\mathcal{C}_1} w_{jk}$	$\sum_{j\in\mathcal{C}_2} w_{jk}$
9	0.8	0.2
10	0.5	0.5
11	0.2	0.8
12	0.5	0.5

The cumulative weights from Table 5.1 quantify the influence of each sending network ($\mathcal{C}_1$ and $\mathcal{C}_2$) on each agent k in the receiving network. For example, we see that agent 9 is mostly influenced by the sending network $\mathcal{C}_1$, while agent 11 is mostly influenced by the sending network $\mathcal{C}_2$. Agents 10 and 12 are affected equally by the two sending networks. Inserting into (5.73) the values reported in Table 5.1, we obtain the following minimizers for $\bar{D}_k(\theta)$:

$$\vartheta_9^\star = 1, \quad \vartheta_{10}^\star = 2, \quad \vartheta_{11}^\star = 3, \quad \vartheta_{12}^\star = 2, \tag{5.74}$$

which reveal two remarkable effects. Agent 9 sees the sending network $\mathcal{C}_1$ as the most influential, and is accordingly steered toward the hypothesis promoted by $\mathcal{C}_1$. The situation is reversed for agent 11, which is in fact more influenced by the sending network $\mathcal{C}_2$.

A second phenomenon is observed for agents 10 and 12, for which no domineering sending network emerges. In this case, the agents opt for hypothesis 2, according to the *truth-is-somewhere-in-between* effect observed in Example 5.7.

Chapter 6

Error Probability Performance

The main focus of the previous chapter was to study the convergence of the belief vectors under the social learning strategy with geometric averaging summarized in listing (3.16). In particular, Theorem 5.2 revealed that, over connected graphs, all agents asymptotically place the full belief mass on some target hypothesis $\vartheta^\star$ that optimizes a global measure of matching between the data and the likelihood models. For example, under the objective evidence model in Section 5.3, all agents tend to promote with full confidence the true underlying hypothesis ϑ^o.

These results focus only on what happens as $t \to \infty$. It is equally important to examine the performance of social learning for finite t, which is the focus of the current chapter. To do so, one useful index of performance is the *error probability* of each agent k at each time instant t. This measure is formally defined in the next section as the probability that the belief vector $\boldsymbol{\mu}_{k,t}$ is not maximized at the target hypothesis $\vartheta^\star$. We already know from Theorem 5.2 that the probability of error converges to 0 for all agents. However, the result does not provide information about how fast the probability will approach zero.

Unfortunately, for general data distributions and likelihood models, a closed-form characterization for the error probability is a formidable task. For this reason, we will focus instead on the asymptotic analysis (for large t) of the error probability. In particular, in Theorem 6.2 we will show an asymptotic normality result that can be used to approximate the error probability through closed-form expressions involving the Gaussian distribution. Then, in Theorem 6.3, we will perform a large deviation analysis to calculate the error exponents that reveal how fast the error probability converges to 0 as $t \to \infty$.

It is interesting to remark that Theorems 5.2, 6.2, and 6.3 form a standard path in asymptotic statistics [159, 166]. This observation is not surprising once we recognize that, after unfolding the recursion from (5.10) (and ignoring a transient term that depends on the initial beliefs), the logarithmic belief ratios will be sums of independent random variables, namely, of logarithmic likelihood ratios scaled by coefficients arising from powers of the combination matrix. And for sums of independent variables, one can typically carry out the following three-step asymptotic analysis. First, one appeals to the law of large numbers to characterize the convergence of the sum (divided by t) toward some deterministic value, as we did in the proof of Theorem 5.2. Second, one can characterize the asymptotic distribution of the sum through central limit theorems leading to Gaussian approximations, as we will do in Theorem 6.2. As a third step, one traditionally appeals to the theory of large deviations to characterize the probability of deviating from the prescribed limiting value, and this type of analysis is carried out in Theorem 6.3.

For the performance analysis in this chapter, we continue to work under Assumptions 5.1 and 5.2, and focus on primitive graphs and a unique target hypothesis $\vartheta^\star$, as stated in the next assumption.

Assumption 6.1 (Primitive graphs and unique minimizer $\vartheta^\star$). We assume that the network graph is primitive and focus on the network average of KL divergences encountered in Theorem 5.2,

$$D_{\mathsf{net}}(\theta) = \sum_{k=1}^{K} v_k D(f_k || \ell_{k,\theta}), \tag{6.1}$$

where v is the Perron vector associated with the left stochastic combination matrix A. As done before, we assume that $D_{\mathsf{net}}(\theta)$ has a unique minimizer

$$\vartheta^\star = \arg\min_{\theta\in\Theta} D_{\mathsf{net}}(\theta), \tag{6.2}$$

which in the sequel will be referred to as the target hypothesis.

6.1 Useful Statistical Descriptors

Before carrying out our analysis, it is convenient to introduce several quantities of interest. For ease of reference, the major symbols used in our analysis are listed in Table 6.1.

Table 6.1: Notation relevant to the performance analysis of social learning.

$f_k(x)$	True distribution governing the data of agent k
$\ell_k(x\|\theta)$	Likelihood model of agent k
$\boldsymbol{\mu}_{k,t}(\theta)$	belief assigned to hypothesis θ by agent k at time t
$\boldsymbol{\mu}_{k,t}$	$H \times 1$ vector stacking the entries $\boldsymbol{\mu}_{k,t}(\theta)$
$v = [v_k]$	Perron vector
$D_{\mathsf{net}}(\theta)$	Network average of KL divergences, $\sum_{k=1}^{K} v_k D(f_k\|\|\ell_{k,\theta})$
$\vartheta^\star$	Target hypothesis that minimizes $D_{\mathsf{net}}(\theta)$
$\boldsymbol{\lambda}_{k,t}(\theta)$	Log likelihood ratio, $\log \dfrac{\ell_k(\boldsymbol{x}_{k,t}\|\vartheta^\star)}{\ell_k(\boldsymbol{x}_{k,t}\|\theta)}$, $\theta \neq \vartheta^\star$
$\boldsymbol{\lambda}_{k,t}$	$(H-1) \times 1$ vector stacking the entries $\boldsymbol{\lambda}_{k,t}(\theta)$, $\theta \neq \vartheta^\star$
$\bar{\lambda}_k$	Expected value of $\boldsymbol{\lambda}_{k,t}$
Σ_k	$(H-1) \times (H-1)$ covariance matrix of $\boldsymbol{\lambda}_{k,t}$
$\Lambda_k(s;\theta)$	Logarithmic moment generating function (LMGF) of $\boldsymbol{\lambda}_{k,t}(\theta)$
$\boldsymbol{\lambda}_{\mathsf{net},t}(\theta)$	Network average of log likelihood ratios, $\sum_{k=1}^{K} v_k\, \boldsymbol{\lambda}_{k,t}(\theta)$, $\theta \neq \vartheta^\star$
$\boldsymbol{\lambda}_{\mathsf{net},t}$	$(H-1) \times 1$ vector stacking the entries $\boldsymbol{\lambda}_{\mathsf{net},t}(\theta)$
$\bar{\lambda}_{\mathsf{net}}$	Expected value of $\boldsymbol{\lambda}_{\mathsf{net},t}$
Σ_{net}	$(H-1) \times (H-1)$ covariance matrix of $\boldsymbol{\lambda}_{\mathsf{net},t}$
$\Lambda_{\mathsf{net}}(s;\theta)$	Logarithmic moment generating function of $\boldsymbol{\lambda}_{\mathsf{net},t}(\theta)$
$\boldsymbol{\beta}_{k,t}(\theta)$	Log belief ratio, $\log \dfrac{\boldsymbol{\mu}_{k,t}(\vartheta^\star)}{\boldsymbol{\mu}_{k,t}(\theta)}$, $\theta \neq \vartheta^\star$
$\boldsymbol{\beta}_{k,t}$	$(H-1) \times 1$ vector stacking the entries $\boldsymbol{\beta}_{k,t}(\theta)$
$\bar{\boldsymbol{\beta}}_{k,t}$	Time-scaled version of $\boldsymbol{\beta}_{k,t}$, namely, $\bar{\boldsymbol{\beta}}_{k,t} \triangleq \dfrac{\boldsymbol{\beta}_{k,t}}{t}$
$p_{k,t}$	Instantaneous error probability of agent k at time t

6.1.1 Log Likelihood Ratios

First, we introduce the *log likelihood ratio*[1]

$$\boldsymbol{\lambda}_{k,t}(\theta) \triangleq \log \frac{\ell_k(\boldsymbol{x}_{k,t}|\vartheta^\star)}{\ell_k(\boldsymbol{x}_{k,t}|\theta)}, \quad \theta \neq \vartheta^\star, \tag{6.3}$$

and its expectation

$$\bar{\lambda}_k(\theta) \triangleq \mathbb{E}\boldsymbol{\lambda}_{k,t}(\theta) = D(f_k||\ell_{k,\theta}) - D(f_k||\ell_{k,\vartheta^\star}). \tag{6.4}$$

Note that, under Assumption 5.2, the log likelihood ratios are almost-surely well defined, since, in view of (5.5), the numerator and denominator in (6.3) are equal to 0 with zero probability. Note also that $\bar{\lambda}_k(\theta)$ does not depend on t since, in view of Assumption 5.2, the expectation in (6.4) is computed assuming that $\boldsymbol{x}_{k,t}$ is distributed according to some true underlying stationary model $f_k(x)$, i.e., we continue to assume invariant distribution over time. When we omit the argument θ and write $\boldsymbol{\lambda}_{k,t}$, we will be referring to the $(H-1)\times 1$ *vector* of log likelihood ratios

$$\boldsymbol{\lambda}_{k,t} = [\boldsymbol{\lambda}_{k,t}(1), \boldsymbol{\lambda}_{k,t}(2), \dots, \boldsymbol{\lambda}_{k,t}(H-1)], \tag{6.5}$$

where, without loss of generality, we consider that the set of hypotheses is $\Theta = \{1, 2, \dots, H\}$ and that the hypotheses have been ordered in such a way that $\vartheta^\star = H$. To avoid confusion, we recall that in our notation all vectors are column vectors. Likewise, we introduce the $(H-1)\times 1$ vector

$$\bar{\lambda}_k = \mathbb{E}\boldsymbol{\lambda}_{k,t} \tag{6.6}$$

that collects the expected values $\bar{\lambda}_k(\theta)$ for $\theta \neq \vartheta^\star$.

We continue by defining the *network average* of log likelihood ratios, for all $\theta \neq \vartheta^\star$,

$$\boldsymbol{\lambda}_{\mathsf{net},t}(\theta) \triangleq \sum_{k=1}^{K} v_k\, \boldsymbol{\lambda}_{k,t}(\theta) \tag{6.7}$$

or, in vector form,

$$\boldsymbol{\lambda}_{\mathsf{net},t} = \sum_{k=1}^{K} v_k\, \boldsymbol{\lambda}_{k,t}. \tag{6.8}$$

The weight assigned to the log likelihood ratio of the kth agent is given by the kth entry, v_k, of the Perron vector that is associated with irreducible

[1] In order to avoid confusion, we remark that in [25] the symbol $\boldsymbol{\lambda}_{k,t}$ was used to denote log belief ratios instead of log likelihood ratios. In this book we adopt a more suggestive notation: We use the symbol λ (lambda) to denote log *likelihood* ratios, and the symbol β (beta) to denote log *belief* ratios — see the forthcoming section.

matrices, i.e., with connected graphs — see Theorem 4.1. It is also useful to introduce the expected vector

$$\bar{\lambda}_{\mathsf{net}} \triangleq \mathbb{E}\boldsymbol{\lambda}_{\mathsf{net},t} = \sum_{k=1}^{K} v_k \bar{\lambda}_k, \tag{6.9}$$

whose θth entry, in view of (6.1) and (6.4), is given by

$$\bar{\lambda}_{\mathsf{net}}(\theta) = \mathbb{E}\boldsymbol{\lambda}_{\mathsf{net},t}(\theta) = \sum_{k=1}^{K} v_k \bar{\lambda}_k(\theta) = D_{\mathsf{net}}(\theta) - D_{\mathsf{net}}(\vartheta^\star) > 0, \tag{6.10}$$

where positivity results from the uniqueness of $\vartheta^\star$ in (6.2).

The average variable $\boldsymbol{\lambda}_{\mathsf{net},t}$ plays a fundamental role in the description of the social learning performance. In fact, we will discover in this chapter that different statistical descriptors of $\boldsymbol{\lambda}_{\mathsf{net},t}$ (mean, covariance matrix, generating functions) characterize at different levels of refinement the asymptotic properties of a fundamental decision statistic used to evaluate the performance, namely, the log belief ratios introduced in the next section.

6.1.2 Log Belief Ratios

In order to characterize the learning performance, it is convenient to work in terms of the logarithmic ratio between the belief about $\vartheta^\star$ and the belief about $\theta \neq \vartheta^\star$. Therefore, with reference to the beliefs of agent k at time t, we introduce the *log belief ratio*

$$\boldsymbol{\beta}_{k,t}(\theta) \triangleq \log \frac{\boldsymbol{\mu}_{k,t}(\vartheta^\star)}{\boldsymbol{\mu}_{k,t}(\theta)}, \qquad \theta \neq \vartheta^\star. \tag{6.11}$$

Observe that the ratio is well defined since, as already remarked, under conditions (5.4) and (5.5), the beliefs $\boldsymbol{\mu}_{k,t}(\theta)$ remain almost-surely nonzero for any θ during the algorithm evolution. As we did for the log likelihood ratio, it is also useful to introduce the $(H-1)\times 1$ vector of log belief ratios

$$\boldsymbol{\beta}_{k,t} = \left[\boldsymbol{\beta}_{k,t}(1), \boldsymbol{\beta}_{k,t}(2), \ldots, \boldsymbol{\beta}_{k,t}(H-1)\right]. \tag{6.12}$$

Note that the log belief ratio vector $\boldsymbol{\beta}_{k,t}$ has $H-1$ entries, whereas the belief vector $\boldsymbol{\mu}_{k,t}$ has H entries. However, we must recall that $\boldsymbol{\mu}_{k,t}$ has only $H-1$ degrees of freedom, since it is a probability vector, which implies that once $H-1$ entries are given, the remaining entry is obtained from the condition $\sum_{\theta\in\Theta} \boldsymbol{\mu}_{k,t}(\theta) = 1$. The next theorem shows that the H-dimensional vector $\boldsymbol{\mu}_{k,t}$ can be fully reconstructed given knowledge of the $(H-1)$-dimensional vector $\boldsymbol{\beta}_{k,t}$. In the theorem, we use normal font for $\mu_{k,t}$ and $\beta_{k,t}$ to emphasize that the result pertains to the functional dependence between beliefs and log belief ratios, with the particular statistical distributions being immaterial here.

Theorem 6.1 (Sufficiency of log belief ratios). Let $0 < \mu_{k,t}(\theta) < 1$ for all $\theta \in \Theta$. The belief vector $\mu_{k,t}$ is a deterministic function of the log belief ratio vector $\beta_{k,t}$. Specifically, we have that

$$\mu_{k,t}(\theta) = \begin{cases} \dfrac{e^{-\beta_{k,t}(\theta)}}{1 + \sum\limits_{\theta' \neq \vartheta^\star} e^{-\beta_{k,t}(\theta')}} & \text{if } \theta \neq \vartheta^\star, \\ \\ \dfrac{1}{1 + \sum\limits_{\theta' \neq \vartheta^\star} e^{-\beta_{k,t}(\theta')}} & \text{if } \theta = \vartheta^\star. \end{cases} \tag{6.13}$$

Proof. Consider $\theta \neq \vartheta^\star$. From (6.11) we have

$$\mu_{k,t}(\theta) = \mu_{k,t}(\vartheta^\star) e^{-\beta_{k,t}(\theta)}. \tag{6.14}$$

Since the belief vector is a probability vector, and, hence, its entries must add up to 1, we must have

$$\mu_{k,t}(\vartheta^\star) + \sum_{\theta' \neq \vartheta^\star} \mu_{k,t}(\theta') = 1. \tag{6.15}$$

Using (6.14) in the summation appearing in (6.15), we conclude that

$$\mu_{k,t}(\vartheta^\star) + \sum_{\theta' \neq \vartheta^\star} \mu_{k,t}(\vartheta^\star) e^{-\beta_{k,t}(\theta')} = 1, \tag{6.16}$$

which is equivalent to

$$\mu_{k,t}(\vartheta^\star) = \frac{1}{1 + \sum\limits_{\theta' \neq \vartheta^\star} e^{-\beta_{k,t}(\theta')}}, \tag{6.17}$$

and (6.13) is proved for the case $\theta = \vartheta^\star$. The expression in (6.13) for $\theta \neq \vartheta^\star$ follows by substituting (6.17) into (6.14).

■

6.1.3 Error Probabilities

One natural way for the agents to make a decision is to select the hypothesis or hypotheses that maximize the belief. Under this rule, the occurrence of a wrong decision by agent k at time t corresponds to the occurrence of the event

$$\mathcal{E}_{k,t} \triangleq \left\{ \vartheta^\star \neq \arg\max_{\theta \in \Theta} \boldsymbol{\mu}_{k,t}(\theta) \right\}. \tag{6.18}$$

Therefore, the *instantaneous* error probability of agent k at time t can be defined as

$$p_{k,t} \triangleq \mathbb{P}\left[\mathcal{E}_{k,t}\right] = \mathbb{P}\left[\vartheta^\star \neq \arg\max_{\theta\in\Theta} \boldsymbol{\mu}_{k,t}(\theta)\right]. \tag{6.19}$$

It is useful to rewrite the error probability as a function of the log belief ratios. To this end, observe that the event within brackets in (6.19) corresponds to stating that the belief is not maximized at $\vartheta^\star$, which in turn corresponds to affirming that the log belief ratios in (6.11) are less than or equal to 0 for at least one $\theta \neq \vartheta^\star$. That is, the occurrence of an error corresponds to the event

$$\mathcal{E}_{k,t} = \left\{\exists\theta \neq \vartheta^\star \text{ such that } \boldsymbol{\beta}_{k,t}(\theta) \leq 0\right\}, \tag{6.20}$$

which can be rewritten as the union of events where any log belief ratio is less than or equal to 0, i.e.,

$$\mathcal{E}_{k,t} = \bigcup_{\theta\neq\vartheta^\star} \left\{\boldsymbol{\beta}_{k,t}(\theta) \leq 0\right\}. \tag{6.21}$$

We can thus write the probability of error as

$$p_{k,t} = \mathbb{P}\left[\bigcup_{\theta\neq\vartheta^\star} \left\{\boldsymbol{\beta}_{k,t}(\theta) \leq 0\right\}\right]. \tag{6.22}$$

6.2 Normal Approximation for Large t

In this section we prove that the random vector $\boldsymbol{\beta}_{k,t}$ (properly shifted and scaled) is asymptotically normal as $t \to \infty$. To this end, we will assume finiteness of second-order moments for the log likelihood ratios $\boldsymbol{\lambda}_{k,t}(\theta)$. In order to state the asymptotic normality result, it is useful to introduce some additional quantities, which appear listed in Table 6.1. First, we define the $(H-1)\times(H-1)$ covariance matrix of the vector of log likelihood ratios at every agent k:

$$\Sigma_k \triangleq \mathbb{E}\left[\left(\boldsymbol{\lambda}_{k,t} - \bar{\lambda}_k\right)\left(\boldsymbol{\lambda}_{k,t} - \bar{\lambda}_k\right)^{\mathsf{T}}\right]. \tag{6.23}$$

Likewise, we introduce the covariance matrix of the network average vector $\boldsymbol{\lambda}_{\mathsf{net},t}$ defined by (6.8):

$$\Sigma_{\mathsf{net}} \triangleq \mathbb{E}\left[\left(\boldsymbol{\lambda}_{\mathsf{net},t} - \bar{\lambda}_{\mathsf{net}}\right)\left(\boldsymbol{\lambda}_{\mathsf{net},t} - \bar{\lambda}_{\mathsf{net}}\right)^{\mathsf{T}}\right]. \tag{6.24}$$

Finally, we introduce a symbol for the log belief ratio divided by t:

$$\bar{\boldsymbol{\beta}}_{k,t} \triangleq \frac{\boldsymbol{\beta}_{k,t}}{t}. \tag{6.25}$$

Theorem 6.2 (Asymptotic normality under geometric averaging). Let Assumptions 5.1, 5.2, and 6.1 be satisfied, and let $\mathscr{G}(0, \Sigma)$ denote a random vector having a zero-mean multivariate Gaussian distribution with covariance matrix Σ. If the covariance matrices Σ_k have finite entries, then for $k = 1, 2, \dots, K$,

$$\sqrt{t}\left(\bar{\boldsymbol{\beta}}_{k,t} - \bar{\lambda}_{\mathsf{net}}\right) \xrightarrow[t\to\infty]{\mathrm{d}} \mathscr{G}\left(0, \Sigma_{\mathsf{net}}\right). \tag{6.26}$$

Proof. Exploiting (5.2), (6.11), and the definition of $\mathcal{N}_k$ from (4.1), we obtain the recursion, for $\theta \neq \vartheta^\star$,

$$\boldsymbol{\beta}_{k,t}(\theta) = \sum_{j=1}^{K} a_{jk}\left[\boldsymbol{\beta}_{j,t-1}(\theta) + \boldsymbol{\lambda}_{j,t}(\theta)\right], \tag{6.27}$$

which can be unfolded to arrive at the equality

$$\boldsymbol{\beta}_{k,t}(\theta) = \sum_{j=1}^{K}[A^t]_{jk}\boldsymbol{\beta}_{j,0}(\theta) + \sum_{\tau=1}^{t}\sum_{j=1}^{K}[A^\tau]_{jk}\,\boldsymbol{\lambda}_{j,t-\tau+1}(\theta), \tag{6.28}$$

where we recall that $[A^t]_{jk}$ denotes the (j,k) entry of the matrix power A^t. This relation can be rewritten in the following vector form by using the log likelihood and log belief *vectors* defined in (6.5) and (6.12), respectively:

$$\begin{aligned}\boldsymbol{\beta}_{k,t} &= \sum_{j=1}^{K}[A^t]_{jk}\boldsymbol{\beta}_{j,0} + \sum_{\tau=1}^{t}\sum_{j=1}^{K}[A^\tau]_{jk}\,\boldsymbol{\lambda}_{j,t-\tau+1}\\ &\stackrel{\mathrm{d}}{=} \sum_{j=1}^{K}[A^t]_{jk}\boldsymbol{\beta}_{j,0} + \sum_{\tau=1}^{t}\sum_{j=1}^{K}[A^\tau]_{jk}\,\boldsymbol{\lambda}_{j,\tau},\end{aligned} \tag{6.29}$$

where the symbol $\stackrel{\mathrm{d}}{=}$ denotes equality in distribution, which holds because the data are iid over time. Using the definitions of $\bar{\lambda}_{\mathsf{net}}$ and $\bar{\boldsymbol{\beta}}_{k,t}$ provided in (6.9) and (6.25), respectively, in view of (6.29) we can write

$$\begin{aligned}
&\sqrt{t}\left(\bar{\beta}_{k,t}-\bar{\lambda}_{\mathsf{net}}\right)\\
&\stackrel{\mathrm{d}}{=}\frac{1}{\sqrt{t}}\sum_{j=1}^{K}[A^t]_{jk}\beta_{j,0}+\sqrt{t}\left(\frac{1}{t}\sum_{\tau=1}^{t}\sum_{j=1}^{K}[A^\tau]_{jk}\,\boldsymbol{\lambda}_{j,\tau}-\sum_{j=1}^{K}v_j\bar{\lambda}_j\right)\\
&=\frac{1}{\sqrt{t}}\sum_{j=1}^{K}[A^t]_{jk}\beta_{j,0}+\sqrt{t}\left(\frac{1}{t}\sum_{\tau=1}^{t}\sum_{j=1}^{K}[A^\tau]_{jk}\,\boldsymbol{\lambda}_{j,\tau}-\frac{1}{t}\sum_{\tau=1}^{t}\sum_{j=1}^{K}v_j\bar{\lambda}_j\right)\\
&=\frac{1}{\sqrt{t}}\sum_{j=1}^{K}[A^t]_{jk}\beta_{j,0}+\frac{1}{\sqrt{t}}\sum_{\tau=1}^{t}\sum_{j=1}^{K}\left([A^\tau]_{jk}-v_j\right)\bar{\lambda}_j\\
&+\frac{1}{\sqrt{t}}\sum_{\tau=1}^{t}\sum_{j=1}^{K}[A^\tau]_{jk}\left(\boldsymbol{\lambda}_{j,\tau}-\bar{\lambda}_j\right).
\end{aligned}\tag{6.30}$$

Since $0\le[A^t]_{jk}\le 1$, the first term on the RHS vanishes as $t\to\infty$. In addition, since the matrix A is assumed to be primitive, we can use the bound in (4.25) to conclude that the second term on the RHS also vanishes as $t\to\infty$. Accordingly, in view of Slutsky's theorem (applied to vectors — see (D.39)) the claim of the theorem will be proved if we show that the third term converges in distribution (see Definition D.4) to a Gaussian random vector with mean zero and covariance matrix Σ_{net}. To this end, we call upon Theorem D.9, applied to the sequence

$$\boldsymbol{y}_t=\sum_{j=1}^{K}[A^t]_{jk}\left(\boldsymbol{\lambda}_{j,t}-\bar{\lambda}_j\right).\tag{6.31}$$

We now verify that this sequence satisfies conditions (D.52), (D.53), and (D.54). It is immediately seen that condition (D.52) is satisfied since $\mathbb{E}\boldsymbol{\lambda}_{j,t}=\bar{\lambda}_j$, implying that $\mathbb{E}\boldsymbol{y}_t=0$. Consider next condition (D.53). We will show that it is satisfied with limiting covariance matrix equal to Σ_{net}, i.e., we will establish that

$$\lim_{t\to\infty}\frac{1}{t}\sum_{\tau=1}^{t}\mathbb{E}\left[\boldsymbol{y}_\tau\boldsymbol{y}_\tau^{\mathsf{T}}\right]=\Sigma_{\mathsf{net}}.\tag{6.32}$$

Applying the definition of Σ_{net} from (6.24), Eq. (6.32) becomes

$$\begin{aligned}
\lim_{t\to\infty}\frac{1}{t}\sum_{\tau=1}^{t}\mathbb{E}\left[\boldsymbol{y}_\tau\boldsymbol{y}_\tau^{\mathsf{T}}\right]&=\mathbb{E}\left[\left(\boldsymbol{\lambda}_{\mathsf{net},t}-\bar{\lambda}_{\mathsf{net}}\right)\left(\boldsymbol{\lambda}_{\mathsf{net},t}-\bar{\lambda}_{\mathsf{net}}\right)^{\mathsf{T}}\right]\\
&=\mathbb{E}\left[\left(\boldsymbol{\lambda}_{\mathsf{net},1}-\bar{\lambda}_{\mathsf{net}}\right)\left(\boldsymbol{\lambda}_{\mathsf{net},1}-\bar{\lambda}_{\mathsf{net}}\right)^{\mathsf{T}}\right],
\end{aligned}\tag{6.33}$$

where in the last step we replaced $\boldsymbol{\lambda}_{\mathsf{net},t}$ with $\boldsymbol{\lambda}_{\mathsf{net},1}$ because the vectors $\boldsymbol{\lambda}_{\mathsf{net},t}$ are identically distributed over time.

Now, we recall that the *Cesàro limit* of the sequence is equal to the limit of the sequence (when the latter exists).[2] Therefore, to prove (6.33) it will be sufficient to

[2] Given a real-valued sequence $\{z_\tau\}$, its Cesàro limit is defined as the limit of the sequence of arithmetic means $\bar{z}_t=(1/t)\sum_{\tau=1}^{t}z_\tau$. Note that the Cesàro limit might exist even when

establish the following result:

$$\lim_{\tau\to\infty} \mathbb{E}\left[\boldsymbol{y}_\tau \boldsymbol{y}_\tau^\mathsf{T}\right] = \mathbb{E}\left[\left(\boldsymbol{\lambda}_{\mathsf{net},1} - \bar{\lambda}_{\mathsf{net}}\right)\left(\boldsymbol{\lambda}_{\mathsf{net},1} - \bar{\lambda}_{\mathsf{net}}\right)^\mathsf{T}\right]. \tag{6.35}$$

By substituting (6.31) into the LHS of (6.35), and (6.7) into the RHS, Eq. (6.35) can be equivalently rewritten as

$$\begin{aligned} &\lim_{\tau\to\infty} \mathbb{E}\left[\sum_{j=1}^{K}\sum_{j'=1}^{K} [A^\tau]_{jk}[A^\tau]_{j'k}\left(\boldsymbol{\lambda}_{j,\tau} - \bar{\lambda}_j\right)\left(\boldsymbol{\lambda}_{j',\tau} - \bar{\lambda}_{j'}\right)^\mathsf{T}\right] \\ &= \lim_{\tau\to\infty} \mathbb{E}\left[\sum_{j=1}^{K}\sum_{j'=1}^{K} [A^\tau]_{jk}[A^\tau]_{j'k}\left(\boldsymbol{\lambda}_{j,1} - \bar{\lambda}_j\right)\left(\boldsymbol{\lambda}_{j',1} - \bar{\lambda}_{j'}\right)^\mathsf{T}\right] \\ &= \mathbb{E}\left[\sum_{j=1}^{K}\sum_{j'=1}^{K} v_j v_{j'}\left(\boldsymbol{\lambda}_{j,1} - \bar{\lambda}_j\right)\left(\boldsymbol{\lambda}_{j',1} - \bar{\lambda}_{j'}\right)^\mathsf{T}\right], \end{aligned} \tag{6.36}$$

where, in the intermediate step, we replaced $\boldsymbol{\lambda}_{j,\tau}$ and $\boldsymbol{\lambda}_{j',\tau}$ with $\boldsymbol{\lambda}_{j,1}$ and $\boldsymbol{\lambda}_{j',1}$ due to the identical distribution over time. Proving (6.36) is equivalent to proving that, for all $\theta, \theta' \in \Theta$,

$$\begin{aligned} &\lim_{\tau\to\infty} \mathbb{E}\left[\sum_{j=1}^{K}\sum_{j'=1}^{K} [A^\tau]_{jk}[A^\tau]_{j'k}\left(\boldsymbol{\lambda}_{j,1}(\theta) - \bar{\lambda}_j(\theta)\right)\left(\boldsymbol{\lambda}_{j',1}(\theta') - \bar{\lambda}_{j'}(\theta')\right)\right] \\ &= \mathbb{E}\left[\sum_{j=1}^{K}\sum_{j'=1}^{K} v_j v_{j'}\left(\boldsymbol{\lambda}_{j,1}(\theta) - \bar{\lambda}_j(\theta)\right)\left(\boldsymbol{\lambda}_{j',1}(\theta') - \bar{\lambda}_{j'}(\theta')\right)\right]. \end{aligned} \tag{6.37}$$

Let us verify that (6.37) holds. For this purpose, observe first that

$$\lim_{\tau\to\infty} [A^\tau]_{jk} = v_j \tag{6.38}$$

in view of (4.23), which implies

$$\begin{aligned} &\sum_{j=1}^{K}\sum_{j'=1}^{K} [A^\tau]_{jk}[A^\tau]_{j'k}\left(\boldsymbol{\lambda}_{j,1}(\theta) - \bar{\lambda}_j(\theta)\right)\left(\boldsymbol{\lambda}_{j',1}(\theta') - \bar{\lambda}_{j'}(\theta')\right) \\ &\xrightarrow[\tau\to\infty]{\text{a.s.}} \sum_{j=1}^{K}\sum_{j'=1}^{K} v_j v_{j'}\left(\boldsymbol{\lambda}_{j,1}(\theta) - \bar{\lambda}_j(\theta)\right)\left(\boldsymbol{\lambda}_{j',1}(\theta') - \bar{\lambda}_{j'}(\theta')\right). \end{aligned} \tag{6.39}$$

Therefore, Eq. (6.37) would be proved if we could interchange the limit and the expectation. In view of the dominated convergence theorem (Theorem D.6), this operation is legitimate if the τ-dependent random variables on the LHS of (6.39) are upper bounded by a τ-independent random variable that has finite mean. We now show that this is

---

the sequence $\{z_\tau\}$ does not admit a limit. However, when $\{z_\tau\}$ admits a limit, the following implication is known to hold [52, Thm. 4.2.3]:

$$\lim_{\tau\to\infty} z_\tau = z \quad \Longrightarrow \quad \lim_{t\to\infty} \bar{z}_t = z. \tag{6.34}$$

actually the case. By applying the triangle inequality and noting that $0 \leq [A^\tau]_{jk} \leq 1$ for all τ, j, and k, we can write

$$\left| \sum_{j=1}^{K} \sum_{j'=1}^{K} [A^\tau]_{jk} [A^\tau]_{j'k} \left(\boldsymbol{\lambda}_{j,1}(\theta) - \bar{\lambda}_j(\theta) \right) \left(\boldsymbol{\lambda}_{j',1}(\theta') - \bar{\lambda}_{j'}(\theta') \right) \right|$$
$$\leq \sum_{j=1}^{K} \sum_{j'=1}^{K} \left| \boldsymbol{\lambda}_{j,1}(\theta) - \bar{\lambda}_j(\theta) \right| \times \left| \boldsymbol{\lambda}_{j',1}(\theta') - \bar{\lambda}_{j'}(\theta') \right| \triangleq \boldsymbol{z}^\star. \tag{6.40}$$

Since the log likelihood ratios have finite second moment by assumption, and since we have the inequality

$$\left| \boldsymbol{\lambda}_{j,1}(\theta) - \bar{\lambda}_j(\theta) \right| \times \left| \boldsymbol{\lambda}_{j',1}(\theta') - \bar{\lambda}_{j'}(\theta') \right|$$
$$\leq \frac{1}{2} \left[\left(\boldsymbol{\lambda}_{j,1}(\theta) - \bar{\lambda}_j(\theta) \right)^2 + \left(\boldsymbol{\lambda}_{j',1}(\theta') - \bar{\lambda}_{j'}(\theta') \right)^2 \right], \tag{6.41}$$

the random variable $\boldsymbol{z}^\star$ defined in (6.40) has finite mean, as desired. We can accordingly call upon the dominated convergence theorem to establish that (6.37) holds. This concludes the verification of condition (D.53).

It remains to show that the Lindeberg condition (D.54) is satisfied, namely, that

$$\frac{1}{t} \sum_{\tau=1}^{t} \mathbb{E}\left[\|\boldsymbol{y}_\tau\|^2 \, \mathbb{I}\left[\|\boldsymbol{y}_\tau\|^2 > \varepsilon\, t \right] \right] = 0. \tag{6.42}$$

To this end, we note that, since

$$\sum_{j=1}^{K} [A^\tau]_{jk} = 1 \tag{6.43}$$

and $[A^\tau]_{jk} \geq 0$, we can apply Jensen's inequality (see Theorem C.5 and in particular (C.10)) to the squared norm of (6.31) to get

$$\|\boldsymbol{y}_\tau\|^2 \leq \sum_{j=1}^{K} [A^\tau]_{jk} \|\boldsymbol{\lambda}_{j,\tau} - \bar{\lambda}_j\|^2 \leq \sum_{j=1}^{K} \|\boldsymbol{\lambda}_{j,\tau} - \bar{\lambda}_j\|^2 \triangleq \boldsymbol{z}^\star_\tau. \tag{6.44}$$

The condition $\|\boldsymbol{y}_\tau\|^2 \leq \boldsymbol{z}^\star_\tau$ further implies

$$\|\boldsymbol{y}_\tau\|^2 \, \mathbb{I}\left[\|\boldsymbol{y}_\tau\|^2 > \varepsilon\, t \right] \leq \boldsymbol{z}^\star_\tau \, \mathbb{I}\left[\boldsymbol{z}^\star_\tau > \varepsilon\, t \right]. \tag{6.45}$$

Note that the random variables $\boldsymbol{z}^\star_\tau$ defined in (6.44) are identically distributed. Therefore, in view of (6.45) we have

$$\frac{1}{t} \sum_{\tau=1}^{t} \mathbb{E}\left[\|\boldsymbol{y}_\tau\|^2 \, \mathbb{I}\left[\|\boldsymbol{y}_\tau\|^2 > \varepsilon\, t \right] \right] \leq \frac{1}{t} \sum_{\tau=1}^{t} \mathbb{E}\left[\boldsymbol{z}^\star_\tau \, \mathbb{I}\left[\boldsymbol{z}^\star_\tau > \varepsilon\, t \right] \right]$$
$$= \mathbb{E}\left[\boldsymbol{z}^\star_1 \, \mathbb{I}\left[\boldsymbol{z}^\star_1 > \varepsilon\, t \right] \right]. \tag{6.46}$$

Since we can write

$$\boldsymbol{z}^\star_1 \, \mathbb{I}\left[\boldsymbol{z}^\star_1 > \varepsilon\, t \right] \leq \boldsymbol{z}^\star_1 \tag{6.47}$$

and since $\boldsymbol{z}_1^\star$ has finite mean (because the log likelihood ratios have finite variances), we can apply the dominated convergence theorem (Theorem D.6) to $\boldsymbol{z}_1^\star \mathbb{I}[\boldsymbol{z}_1^\star > \varepsilon t]$ to conclude that the RHS of (6.46) vanishes as $t \to \infty$. This also implies that the LHS vanishes, which means that the Lindeberg condition holds. Then the proof is complete by applying Theorem D.9 to the last term on the RHS of (6.30), with the choice of $\boldsymbol{y}_t$ in (6.31).

■

Example 6.1 (Gaussian approximation). We consider a network of $K = 10$ agents that communicate according to the topology in Figure 6.1. The graph is undirected, and all agents are assumed to have a self-loop, not shown in the figure. The graph can be verified to be strong. On top of it, a Metropolis combination matrix (see Table 4.1) is constructed, which results in a doubly stochastic matrix, therefore yielding a uniform Perron vector $v = [v_k]$, with $v_k = 1/K$ for $k = 1, 2, \ldots, K$. The agents wish to solve

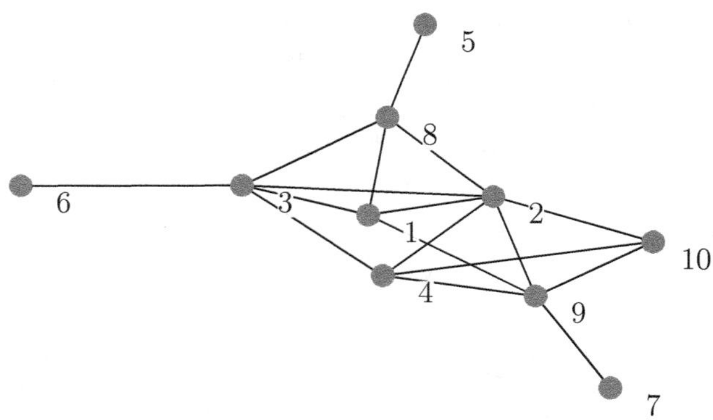

Figure 6.1: Network topology used in Example 6.1. The graph is undirected and all agents are assumed to have a self-loop (not shown in the figure).

a social learning problem with three hypotheses, i.e., $\theta \in \{1, 2, 3\}$. The observations $\boldsymbol{x}_{k,t} \in \{0, 1\}$, for each agent k and time t, are all distributed as balanced Bernoulli random variables (i.e., with $\mathbb{P}[\boldsymbol{x}_{k,t} = 0] = 0.5$), and are independent across k and t. We assume identical Bernoulli likelihood models across the agents, namely,

$$\ell_k(x|\theta) = q_\theta \, \mathbb{I}[x = 0] + (1 - q_\theta) \, \mathbb{I}[x = 1], \tag{6.48}$$

where the hypothesis-dependent probabilities q_θ are

$$q_1 = 0.52, \qquad q_2 = 0.48, \qquad q_3 = 0.5. \tag{6.49}$$

According to this setup, we are considering the objective evidence model described in Section 5.3, since the observations are distributed according to a true underlying hypothesis, in this case hypothesis $\vartheta^o = 3$. Using (6.48), the log likelihood ratio between the true hypothesis ϑ^o and a hypothesis $\theta \in \{1,2\}$ is computed as

$$\begin{aligned}\boldsymbol{\lambda}_{k,t}(\theta) &= \log\frac{\ell_k(\boldsymbol{x}_{k,t}|\vartheta^o)}{\ell_k(\boldsymbol{x}_{k,t}|\theta)} = \log\frac{\ell_k(\boldsymbol{x}_{k,t}|3)}{\ell_k(\boldsymbol{x}_{k,t}|\theta)}\\ &= \mathbb{I}[\boldsymbol{x}_{k,t}=0]\log\frac{0.5}{q_\theta} + \mathbb{I}[\boldsymbol{x}_{k,t}=1]\log\frac{0.5}{1-q_\theta}\\ &= -\log 2 - \mathbb{I}[\boldsymbol{x}_{k,t}=0]\log q_\theta - \mathbb{I}[\boldsymbol{x}_{k,t}=1]\log(1-q_\theta).\end{aligned} \tag{6.50}$$

From (6.50) we can compute the mean of $\boldsymbol{\lambda}_{k,t}(\theta)$ as

$$\bar{\lambda}_k(\theta) = \mathbb{E}\boldsymbol{\lambda}_{k,t}(\theta) = -\log 2 - \frac{1}{2}\log q_\theta - \frac{1}{2}\log(1-q_\theta). \tag{6.51}$$

By combining (6.50) and (6.51), and performing straightforward algebraic manipulations, we can write

$$\begin{aligned}\boldsymbol{\lambda}_{k,t}(\theta) - \bar{\lambda}_k(\theta) &= \frac{1}{2}\mathbb{I}[\boldsymbol{x}_{k,t}=0]\log\frac{1-q_\theta}{q_\theta} + \frac{1}{2}\mathbb{I}[\boldsymbol{x}_{k,t}=1]\log\frac{q_\theta}{1-q_\theta}\\ &= \frac{1}{2}(-1)^{1-\boldsymbol{x}_{k,t}}\log\frac{q_\theta}{1-q_\theta}.\end{aligned} \tag{6.52}$$

We observe that the random variables $\boldsymbol{\lambda}_{k,t}(1)-\bar{\lambda}_k(1)$ and $\boldsymbol{\lambda}_{k,t}(2)-\bar{\lambda}_k(2)$ are proportional, i.e., they are deterministically related. Accordingly, their covariance matrix must be singular. In fact, from (6.52) we can compute the covariance matrix

$$\Sigma_k = \frac{1}{4}\times\begin{bmatrix}\left(\log\frac{q_1}{1-q_1}\right)^2 & \log\frac{q_1}{1-q_1}\log\frac{q_2}{1-q_2}\\ \log\frac{q_1}{1-q_1}\log\frac{q_2}{1-q_2} & \left(\log\frac{q_2}{1-q_2}\right)^2\end{bmatrix}, \tag{6.53}$$

whose determinant is seen to be 0. Since in this example the observations are identically distributed across the agents, and since the Metropolis matrix is doubly stochastic (hence, the Perron vector has all entries equal to $1/K$), the network covariance matrix Σ_{net} from (6.24) is equal to

$$\Sigma_{\text{net}} = \frac{1}{4K}\times\begin{bmatrix}\left(\log\frac{q_1}{1-q_1}\right)^2 & \log\frac{q_1}{1-q_1}\log\frac{q_2}{1-q_2}\\ \log\frac{q_1}{1-q_1}\log\frac{q_2}{1-q_2} & \left(\log\frac{q_2}{1-q_2}\right)^2\end{bmatrix}. \tag{6.54}$$

Note that, thanks to the factor K appearing in (6.54), the variances (i.e., the diagonal entries of Σ_{net}) decrease as the number of agents increases. This is one example that shows the benefits of cooperation, since a reduced variance is representative of a higher learning accuracy. We will examine more closely the benefits of cooperation in Section 6.3.1, in terms of another performance indicator, namely, the large deviation exponents that will be seen to govern the decay to 0 of the error probability.

In view of the aforementioned proportionality (i.e., perfect correlation) between the random variables $\boldsymbol{\lambda}_{k,t}(1)-\bar{\lambda}_k(1)$ and $\boldsymbol{\lambda}_{k,t}(2)-\bar{\lambda}_k(2)$, it is redundant to examine the joint evolution of the log belief ratios $\bar{\boldsymbol{\beta}}_{k,t}(1)$ and $\bar{\boldsymbol{\beta}}_{k,t}(2)$. We focus instead on their individual evolution. More specifically, in each panel of Figure 6.2, we display a histogram computed from 5000 independent realizations of the shifted and scaled variable $\sqrt{t}\left(\bar{\boldsymbol{\beta}}_{k,t}(\theta) - \bar{\lambda}_{\text{net}}(\theta)\right)$, for $k=2$ and $\theta=1$. Different panels refer to different

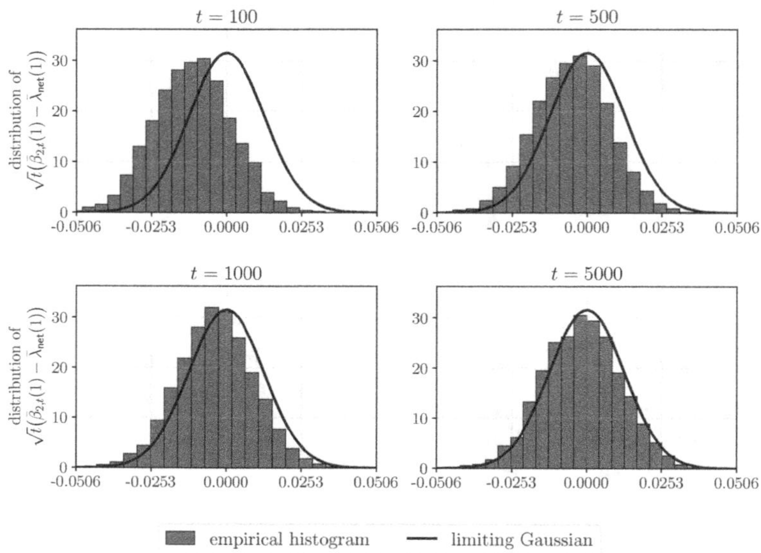

Figure 6.2: Histograms computed from 5000 independent realizations of the shifted and scaled variable $\sqrt{t}\left(\bar{\beta}_{k,t}(\theta) - \bar{\lambda}_{\mathsf{net}}(\theta)\right)$, for $k = 2$ and $\theta = 1$, in the setting of Example 6.1. Different panels refer to different values of t. The histograms are compared against a zero-mean Gaussian distribution with variance $\Sigma_{\mathsf{net}}(1,1)$ (black curves), where the covariance matrix Σ_{net} is reported in (6.54).

values of t. In view of Theorem 6.2, this shifted and scaled variable must follow, for sufficiently large t, a zero-mean Gaussian distribution with variance $\Sigma_{\mathsf{net}}(1,1)$, where $\Sigma_{\mathsf{net}}(\theta,\theta')$ denotes the (θ,θ') entry of the covariance matrix Σ_{net} in (6.54). The pdf of this limiting Gaussian distribution is represented by the black curves in Figure 6.2. Examining the four panels of the figure (which correspond to different values of t), we see that the empirical and limiting distributions become in fact similar as t increases.[3]

6.3 Large Deviations for Large t

In this section we resort to the theory of large deviations introduced in Appendix E, to obtain the following type of asymptotic characterization for the error probability [59, 60]:

$$p_{k,t} = \exp\left\{-t\left[\Psi + o(1)\right]\right\} \tag{6.55}$$

for a certain value Ψ that is called the *error exponent*. The symbol $o(1)$ denotes here a quantity that approaches zero as $t \to \infty$ — see Table 1.1.

[3]We remark that convergence in distribution refers to the convergence of cumulative distribution functions and not of probability density functions. Therefore, Figure 6.2 should not be interpreted in the sense of showing convergence of pdfs.

We conclude from (6.55) that the leading exponential order (as $t \to \infty$) is given by the term $-t\,\Psi$. Equation (6.55) can be equivalently rewritten as

$$\lim_{t\to\infty} \frac{1}{t} \log p_{k,t} = -\Psi. \tag{6.56}$$

In place of (6.55) or (6.56), a compact and common notation to indicate equality to the leading exponential order is [52]

$$p_{k,t} \doteq e^{-\Psi\, t}. \tag{6.57}$$

The error exponent Ψ is a compact statistical descriptor of the social learning performance; it can be used to compare different systems or to optimize different parameters (e.g., the network graph, the likelihood models) to achieve the maximum decay rate for the error probability. For example, it makes sense to compare two different networks implementing a social learning algorithm in terms of their exponents; the network featuring the largest exponent will be considered superior since its probability vanishes faster. We will see relevant examples of this type of comparison in Chapter 13.

The theory of large deviations has been exploited in [9] for binary hypothesis testing, and in [106] for social learning with geometric averaging, under the objective evidence model. The next theorem considers the more general setting in Assumption 5.2.

Before stating the theorem, it is necessary to introduce the logarithmic moment generating function (LMGF), a.k.a. cumulant generating function, of the log likelihood ratios (see Appendix E.1.2):

$$\Lambda_k(s;\theta) \triangleq \log \mathbb{E} \exp\Big\{ s\, \boldsymbol{\lambda}_{k,t}(\theta) \Big\}, \tag{6.58}$$

where $s \in \mathbb{R}$ and the expectation is computed under the true model $f_k(x)$, which does not change over time, and this explains why $\Lambda_k(s;\theta)$ does not depend on t. It is also useful to introduce the LMGF of the network average of log likelihood ratios $\boldsymbol{\lambda}_{\mathsf{net},t}(\theta)$ defined by (6.7):

$$\Lambda_{\mathsf{net}}(s;\theta) \triangleq \log \mathbb{E} \exp\Big\{ s\, \boldsymbol{\lambda}_{\mathsf{net},t}(\theta) \Big\} \tag{6.59}$$

and its Fenchel-Legendre transform (see Appendix E.1.1)

$$\Lambda^*_{\mathsf{net}}(y;\theta) = \sup_{s\in\mathbb{R}} \Big(sy - \Lambda_{\mathsf{net}}(s;\theta) \Big), \quad y \in \mathbb{R}. \tag{6.60}$$

Theorem 6.3 (Error exponents under geometric averaging). Let Assumptions 5.1, 5.2, and 6.1 be satisfied. If, for $k = 1, 2, \ldots, K$ and for all $\theta \neq \vartheta^\star$,

$$\Lambda_k(s;\theta) < \infty \quad \forall s \in \mathbb{R}, \tag{6.61}$$

then

$$\mathbb{P}\left[\beta_{k,t}(\theta) \leq 0\right] \doteq e^{-\Psi(\theta)\,t}, \tag{6.62}$$

where

$$\Psi(\theta) \triangleq \Lambda^*_{\mathsf{net}}(0;\theta) = -\inf_{s\in\mathbb{R}} \Lambda_{\mathsf{net}}(s;\theta) > 0. \tag{6.63}$$

Moreover, the error probability for each agent k is dominated by the worst-case (i.e., the smallest) exponent:

$$p_{k,t} \doteq e^{-\Psi\,t}, \quad \Psi = \min_{\theta\neq\vartheta^\star} \Psi(\theta). \tag{6.64}$$

Proof. To prove the theorem we will study the large deviations of the time-scaled log belief ratio $\bar{\beta}_{k,t}(\theta)$ — see (6.25). The proof of the theorem involves: *i)* calling upon the Gärtner-Ellis theorem (Theorem E.2) to provide the exponential characterization of the log beliefs for the individual hypotheses $\theta \neq \vartheta^\star$, namely, Eq. (6.62); and *ii)* using classic probabilistic bounds to obtain, from the individual error exponents, the exponent of the overall error probability $p_{k,t}$, namely, Eq. (6.64).

We start with step *i)*. Let

$$\Lambda_{1/t}(s) \triangleq \log \mathbb{E} \exp\left\{s\,\bar{\beta}_{k,t}(\theta)\right\} \tag{6.65}$$

denote the LMGF of the time-scaled log belief ratio $\bar{\beta}_{k,t}(\theta)$. For simplicity, we omitted the dependence of $\Lambda_{1/t}(s)$ on k and θ. Consider now the Gärtner-Ellis theorem with the asymptotic parameter ε chosen as $\varepsilon = 1/t$, with $t \to \infty$. Examining the claim of Theorem E.2, and in particular condition (E.159), we see that if we establish that

$$\lim_{t\to\infty} \frac{1}{t}\Lambda_{1/t}(s\,t) = \Lambda_{\mathsf{net}}(s;\theta), \tag{6.66}$$

then we can conclude that (6.62) holds with exponent $\Psi(\theta)$ given by (6.63). Let us accordingly prove that (6.66) holds. In view of (6.28), the LMGF $\Lambda_{1/t}(s)$ can be computed as

$$\Lambda_{1/t}(s) = \log \mathbb{E} \exp\left\{\frac{s}{t}\,\beta_{k,t}(\theta)\right\} = \frac{s}{t}\sum_{j=1}^{K}[A^t]_{jk}\beta_{j,0}(\theta) + \sum_{\tau=1}^{t}\widehat{\Lambda}_\tau\left(\frac{s}{t}\right). \tag{6.67}$$

In the last step we exploited the fact that the random variables $\boldsymbol{\lambda}_{j,t-\tau+1}(\theta)$ are independent over time (we recall that the LMGF of the sum of independent random variables is equal to the sum of the LMGFs of the random variables) and introduced the function

$$\begin{aligned}\widehat{\Lambda}_\tau(s) &\triangleq \log \mathbb{E} \exp\left\{s\sum_{j=1}^{K}[A^\tau]_{jk}\,\boldsymbol{\lambda}_{j,t-\tau+1}(\theta)\right\}\\ &= \log \mathbb{E} \exp\left\{s\sum_{j=1}^{K}[A^\tau]_{jk}\,\boldsymbol{\lambda}_{j,1}(\theta)\right\},\end{aligned} \tag{6.68}$$

where the equality follows from the identical distribution over time. From (6.67) we can write

$$\frac{1}{t}\Lambda_{1/t}(st) - \Lambda_{\mathsf{net}}(s;\theta) = \frac{s}{t}\sum_{j=1}^{K}[A^t]_{jk}\beta_{j,0}(\theta) + \frac{1}{t}\sum_{\tau=1}^{t}\left(\widehat{\Lambda}_\tau(s) - \Lambda_{\mathsf{net}}(s;\theta)\right). \tag{6.69}$$

To prove (6.66), we show that both terms on the RHS of (6.69) vanish as $t \to \infty$. Since $0 \le [A^t]_{jk} \le 1$, the first term vanishes as $t \to \infty$. Regarding the second term, in view of (4.23) we have the following convergence:

$$\exp\left\{ s\sum_{j=1}^{K}[A^\tau]_{jk}\,\boldsymbol{\lambda}_{j,1}(\theta)\right\} \xrightarrow[\tau\to\infty]{\text{a.s.}} \exp\left\{ s\sum_{j=1}^{K} v_j \boldsymbol{\lambda}_{j,1}(\theta)\right\}. \tag{6.70}$$

Moreover, using (6.43) and applying Jensen's inequality (see Theorem C.5 and in particular (C.10)) to the exponential function, we can write

$$\exp\left\{ s\sum_{j=1}^{K}[A^\tau]_{jk}\,\boldsymbol{\lambda}_{j,1}(\theta)\right\} \le \sum_{j=1}^{K}[A^\tau]_{jk}\exp\left\{s\,\boldsymbol{\lambda}_{j,1}(\theta)\right\} \le \sum_{j=1}^{K}\exp\left\{s\,\boldsymbol{\lambda}_{j,1}(\theta)\right\}. \tag{6.71}$$

Note that the RHS of (6.71) has finite mean in view of (6.61). Therefore, Eq. (6.71) guarantees that the random variable

$$\exp\left\{ s\sum_{j=1}^{K}[A^\tau]_{jk}\,\boldsymbol{\lambda}_{j,1}(\theta)\right\} \tag{6.72}$$

is upper bounded by a random variable (independent of τ) with finite mean. This allows us to call upon the dominated convergence theorem (Theorem D.6) and conclude from (6.70) that

$$\lim_{\tau\to\infty}\mathbb{E}\exp\left\{ s\sum_{j=1}^{K}[A^\tau]_{jk}\,\boldsymbol{\lambda}_{j,1}(\theta)\right\} = \mathbb{E}\exp\left\{ s\sum_{j=1}^{K} v_j\boldsymbol{\lambda}_{j,1}(\theta)\right\}, \tag{6.73}$$

which, taking the logarithm and using (6.7), (6.59), and (6.68), is equivalent to

$$\lim_{\tau\to\infty}\widehat{\Lambda}_\tau(s) = \Lambda_{\mathsf{net}}(s;\theta). \tag{6.74}$$

Equation (6.74) implies that the second term on the RHS of (6.69) vanishes — see footnote 2 in this chapter. This concludes the proof of (6.66).

It is now legitimate to call upon Theorem E.2 (with the choice $\varepsilon = 1/t$), which establishes that the following large deviation principle (see Definition E.2) holds for all sets $\mathcal{S}$ (the infimum over an empty set is taken as ∞):

$$\begin{aligned}-\inf_{y\in\text{int}(\mathcal{S})}\Lambda^*_{\mathsf{net}}(y;\theta) &\le \liminf_{t\to\infty}\frac{1}{t}\log\mathbb{P}\left[\bar{\boldsymbol{\beta}}_{k,t}(\theta)\in\mathcal{S}\right]\\ &\le \limsup_{t\to\infty}\frac{1}{t}\log\mathbb{P}\left[\bar{\boldsymbol{\beta}}_{k,t}(\theta)\in\mathcal{S}\right] \le -\inf_{y\in\text{cl}(\mathcal{S})}\Lambda^*_{\mathsf{net}}(y;\theta),\end{aligned} \tag{6.75}$$

where $\text{int}(\mathcal{S})$ and $\text{cl}(\mathcal{S})$ denote the interior and the closure of $\mathcal{S}$, respectively, and where $\Lambda^*_{\mathsf{net}}(y;\theta)$ is the Fenchel-Legendre transform of $\Lambda_{\mathsf{net}}(s;\theta)$ — see (6.60). The function

$\Lambda^*_{\mathsf{net}}(y;\theta)$ is also referred to, in the theory of large deviations, as the *rate function* — see Appendix F. Note that $\Lambda_{\mathsf{net}}(s;\theta)$ is finite for all $s \in \mathbb{R}$ because so are by assumption the individual LMGFs $\Lambda_k(s)$ — see footnote 6 in Appendix F. Accordingly, the function $\Lambda_{\mathsf{net}}(s;\theta)$ and its Fenchel-Legendre transform $\Lambda^*_{\mathsf{net}}(y;\theta)$ possess all the regularity properties listed in Lemma E.1. Consider in particular the choice $\mathcal{S} = (-\infty, 0]$, and observe that $\bar{\lambda}_{\mathsf{net}}(\theta) > 0$ due to Assumption 6.1. By exploiting the aforementioned regularity properties, we can compute the infimum and supremum appearing in (6.75) as (see, also Figures E.1 and E.2 for typical shapes of the rate function)

$$\inf_{y\in\mathrm{int}(\mathcal{S})} \Lambda^*_{\mathsf{net}}(y;\theta) = \inf_{y\in\mathrm{cl}(\mathcal{S})} \Lambda^*_{\mathsf{net}}(y;\theta) = \Lambda^*_{\mathsf{net}}(0;\theta), \tag{6.76}$$

which means that $\mathcal{S} = (-\infty, 0]$ is a continuity set of the function $\Lambda^*_{\mathsf{net}}(y;\theta)$ or an Λ^*_{net}-continuity set — see (E.155). Substituting (6.76) into (6.75), we obtain

$$\lim_{t\to\infty} \frac{1}{t} \log \mathbb{P}\left[\bar{\boldsymbol{\beta}}_{k,t}(\theta) \le 0\right] = -\Lambda^*_{\mathsf{net}}(0;\theta), \tag{6.77}$$

where, in view of (6.60), the rate function evaluated at $y = 0$ can be computed as

$$\Lambda^*_{\mathsf{net}}(0;\theta) = \sup_{s\in\mathbb{R}} \Big(-\Lambda_{\mathsf{net}}(s;\theta) \Big) = -\inf_{s\in\mathbb{R}} \Lambda_{\mathsf{net}}(s;\theta) > 0. \tag{6.78}$$

The inequality in (6.78) holds since, in view of Lemma E.1, the rate function $\Lambda^*_{\mathsf{net}}(y;\theta)$ is nonnegative and is equal to 0 only when y is equal to the mean of the random variable whose LMGF is $\Lambda_{\mathsf{net}}(s;\theta)$. This random variable is $\boldsymbol{\lambda}_{\mathsf{net},t}(\theta)$ and its mean is $\bar{\lambda}_{\mathsf{net}}(\theta)$. Since we have $0 \ne \bar{\lambda}_{\mathsf{net}}(\theta)$, we conclude that $\Lambda^*_{\mathsf{net}}(0;\theta) > 0$. Grouping (6.77), (6.78), and the definition of $\Psi(\theta)$ in (6.63) (and further observing that $\mathbb{P}[\boldsymbol{\beta}_{k,t}(\theta) \le 0] = \mathbb{P}[\bar{\boldsymbol{\beta}}_{k,t}(\theta) \le 0]$ because $\bar{\boldsymbol{\beta}}_{k,t}(\theta) = \boldsymbol{\beta}_{k,t}(\theta)/t$), the proof of (6.62) is complete.

It remains to prove that the exponential characterization (6.62) for the probability $\mathbb{P}[\boldsymbol{\beta}_{k,t}(\theta) \le 0]$ implies the exponential characterization (6.64) for the *overall* error probability $p_{k,t}$. To this end, observe that in view of (6.22), $p_{k,t}$ can be bounded as follows (with the lower bound holding for all $\theta \ne \vartheta^\star$):

$$\mathbb{P}\left[\boldsymbol{\beta}_{k,t}(\theta) \le 0\right] \le p_{k,t} \le \sum_{\theta\ne\vartheta^\star} \mathbb{P}\left[\boldsymbol{\beta}_{k,t}(\theta) \le 0\right], \tag{6.79}$$

where the upper bound is the union bound [65].

Using the lower bound in (6.79) along with (6.77) we can write

$$\liminf_{t\to\infty} \frac{1}{t} \log p_{k,t} \ge \max_{\theta\ne\vartheta^\star} \Big(-\Psi(\theta) \Big) = -\min_{\theta\ne\vartheta^\star} \Psi(\theta) = -\Psi, \tag{6.80}$$

where Ψ is defined in (6.64).

Let us now focus on the upper bound in (6.79). By definition, for all $\theta \ne \vartheta^\star$ we have that $\Psi \le \Psi(\theta)$. Accordingly, the convergence in (6.77) implies that, given an arbitrary $\varepsilon > 0$, for sufficiently large t we can write

$$\mathbb{P}\left[\boldsymbol{\beta}_{k,t}(\theta) \le 0\right] \le e^{-(\Psi-\varepsilon)t}. \tag{6.81}$$

Using (6.81) in the RHS of (6.79) yields

$$\frac{1}{t} \log p_{k,t} \le \frac{1}{t} \log(H-1) - \Psi + \varepsilon, \tag{6.82}$$

Due to the arbitrariness of ε, we have

$$\limsup_{t\to\infty} \frac{1}{t} \log p_{k,t} \leq -\Psi. \tag{6.83}$$

Grouping (6.80) and (6.83), we obtain the desired claim.

■

6.3.1 Benefits of Cooperation

One useful insight that can be gained from Theorem 6.3 relates to the benefits of cooperation. In Chapter 5 we have seen that cooperation is rewarding since it allows to overcome the limited view that agents experience when the learning problems are not locally identifiable. Using the results from Theorem 6.3, it is possible to reveal another benefit of cooperation, namely, that *cooperation can improve learning accuracy.* We illustrate this aspect through an example.

Example 6.2 (Cooperation improves learning accuracy). Consider K agents connected according to a primitive graph associated with a doubly stochastic combination matrix, yielding a Perron vector with uniform entries $v_k = 1/K$ for $k = 1, 2, \ldots, K$. The observations are statistically independent across the agents. Moreover, the likelihood models $\ell_{k,\theta}$ and and the true distributions f_k are equal across the agents. These models guarantee that each agent could learn the target hypothesis $\vartheta^\star$ individually. Therefore, in this case cooperation is not useful to resolve local unidentifiability issues. However, we will now show that cooperation boosts the learning performance. In particular, the performance will be measured in terms of the error exponent Ψ in (6.64).

To evaluate Ψ, we need to evaluate first the hypothesis-dependent error exponents $\Psi(\theta)$ in (6.63), where the LMGF $\Lambda_{\mathsf{net}}(s;\theta)$ was defined in (6.59). To start with, recall from (6.58) that $\Lambda_k(s;\theta)$ denotes the LMGF of $\boldsymbol{\lambda}_{k,t}(\theta)$, and observe that $\Lambda_{\mathsf{net}}(s;\theta)$ is given by

$$\Lambda_{\mathsf{net}}(s;\theta) = \sum_{k=1}^{K} \Lambda_k(v_k s;\theta) = \sum_{k=1}^{K} \Lambda_k(s/K;\theta), \tag{6.84}$$

where in the first equality we used the fact that the data are independent across the agents (and, hence, the LMGF of $\boldsymbol{\lambda}_{\mathsf{net},t}(\theta)$ in (6.7) is given by the sum of the LMGFs of the variables $v_k\,\boldsymbol{\lambda}_{k,t}(\theta)$), whereas in the second equality we replaced each Perron vector entry v_k by $1/K$ since the combination matrix is doubly stochastic. Moreover, since the random variables $\boldsymbol{\lambda}_{k,t}(\theta)$ are identically distributed across the agents, from (6.84) we can also write

$$\Lambda_{\mathsf{net}}(s;\theta) = K\Lambda_k(s/K;\theta), \tag{6.85}$$

where the particular choice of $k = 1, 2, \ldots, K$ is immaterial.

Using (6.63), we can compute $\Psi(\theta)$ for the network of K agents. We can also specialize (6.63) to the case of an *individual* agent working in isolation. The corresponding exponent will be denoted by $\Psi_{\mathsf{ind}}(\theta)$. In summary, from (6.63) we obtain

$$\Psi(\theta) = -\inf_{s\in\mathbb{R}} \Lambda_{\mathsf{net}}(s;\theta), \qquad \Psi_{\mathsf{ind}}(\theta) = -\inf_{s\in\mathbb{R}} \Lambda_{\mathsf{ind}}(s;\theta). \tag{6.86}$$

Exploiting (6.85) and (6.86), we obtain

$$\Psi(\theta) = -\inf_{s\in\mathbb{R}} \Lambda_{\mathsf{net}}(s;\theta) = -K \inf_{s\in\mathbb{R}} \Lambda_{\mathsf{ind}}(s/K;\theta) = -K \inf_{s\in\mathbb{R}} \Lambda_{\mathsf{ind}}(s;\theta) = K\Psi_{\mathsf{ind}}(\theta). \quad (6.87)$$

Referring to the worst-case exponent in (6.64), we finally obtain

$$\Psi = K\Psi_{\mathsf{ind}}. \quad (6.88)$$

We thus find that the network error exponent is K times larger than the error exponent of a standalone agent; this implies that the error probability vanishes exponentially faster (by a factor K) in the social learning case. Intuitively, a network of K agents observes K times as much data as a single agent at each time instant. Cooperation among the agents allows to exploit this increased knowledge and yields the aforementioned improvement in the learning performance. Note also that a similar effect was already observed in terms of the K-fold variance reduction in Example 6.1.

Moreover, in Chapter 13 we will also discuss the connections between the performance of the individual agents in the network and the performance of an ideal centralized system that has access to all observations. We will see that independence across the agents and doubly stochastic matrices lead to an asymptotic equivalence of each agent with the centralized system. However, if we remove these conditions the agents can incur a performance loss with respect to the centralized system, and we will discuss an alternative social learning strategy to address this issue.

Example 6.3 (Error exponents). Consider the same setup used in Example 6.1. We want to specialize to this example the large deviation characterization provided by Theorem 6.3. To this end, we proceed to evaluate the error exponents $\Psi(\theta)$ in (6.63). They can be computed by exploiting the characterization available for the rate function of Bernoulli variables from Example E.3, as we now illustrate.

From (6.50) we know that $\boldsymbol{\lambda}_{k,t}(\theta)$ is equal to $-\log 2 - \log q_\theta$ if $\boldsymbol{x}_{k,t} = 0$ and to $-\log 2 - \log(1-q_\theta)$ if $\boldsymbol{x}_{k,t} = 1$, which is equivalent to the representation

$$\boldsymbol{\lambda}_{k,t}(\theta) = \log\frac{0.5}{q_\theta} + \log\frac{q_\theta}{1-q_\theta}\boldsymbol{x}_{k,t} = a_\theta + b_\theta\,\boldsymbol{x}_{k,t}, \quad (6.89)$$

where

$$a_\theta \triangleq \log\frac{0.5}{q_\theta}, \qquad b_\theta \triangleq \log\frac{q_\theta}{1-q_\theta}. \quad (6.90)$$

Equation (6.89) reveals that $\boldsymbol{\lambda}_{k,t}(\theta)$ is a shifted and scaled version of the Bernoulli variable $\boldsymbol{x}_{k,t}$ (which, in this example, has equiprobable outcomes 0 and 1 under the true underlying model). From the definition of an LMGF, it is readily verified that if a random variable $\boldsymbol{x}$ has LMGF $\Lambda(s)$, then a shifted and scaled variable $a+b\boldsymbol{x}$ has LMGF equal to

$$as + \Lambda(bs). \quad (6.91)$$

Applying this property to (6.89), and denoting by $\Lambda_{\mathsf{Ber}}(s)$ the LMGF of the Bernoulli variable $\boldsymbol{x}_{k,t}$ (i.e., the LMGF from (E.45) with the choice $p = 1/2$), we get

$$\Lambda_k(s;\theta) = a_\theta s + \Lambda_{\mathsf{Ber}}(b_\theta s). \quad (6.92)$$

Since the observations are iid across the agents, for the evaluation of the network LMGF $\Lambda_{\mathsf{net}}(s;\theta)$ we can appeal to (6.85), obtaining

$$\Lambda_{\mathsf{net}}(s;\theta) = a_\theta s + K\Lambda_{\mathsf{Ber}}(b_\theta s/K). \quad (6.93)$$

From the definition of the Fenchel-Legendre transform, it is straightforward to verify that if a function $g(s)$ has Fenchel-Legendre transform $g^*(y)$, the following three properties hold, for any choice of the constants $a \in \mathbb{R}$, $b \neq 0$, and $c > 0$:

$$as + g(s) \to g^*(y-a), \quad g(bs) \to g^*(y/b), \quad cg(s/c) \to cg^*(y), \tag{6.94}$$

where the arrow indicates application of the Fenchel-Legendre transform. Using these three properties in (6.93), we obtain

$$\Lambda^*_{\text{net}}(y;\theta) = K\Lambda^*_{\text{Ber}}\left(\frac{y-a_\theta}{b_\theta}\right). \tag{6.95}$$

Replacing Λ^*_{Ber} with the expression for the rate function of a Bernoulli random variable with equiprobable outcomes (i.e., Eq. (E.58) with probability $p = 1/2$), we obtain, for the case $b_\theta > 0$,

$$\Lambda^*_{\text{net}}(y;\theta) = \begin{cases} KD_b\left(\dfrac{y-a_\theta}{b_\theta}\middle\|\dfrac{1}{2}\right) & \text{if } a_\theta \le y \le a_\theta + b_\theta, \\ \infty & \text{otherwise,} \end{cases} \tag{6.96}$$

and for the case $b_\theta < 0$,

$$\Lambda^*_{\text{net}}(y;\theta) = \begin{cases} KD_b\left(\dfrac{y-a_\theta}{b_\theta}\middle\|\dfrac{1}{2}\right) & \text{if } a_\theta + b_\theta \le y \le a_\theta, \\ \infty & \text{otherwise.} \end{cases} \tag{6.97}$$

In the last two equations, the notation $D_b(r'\|r'')$ is a shortcut for the KL divergence (see Definition B.4) between the two binary pmfs $[r', 1-r']$ and $[r'', 1-r'']$, namely,

$$D_b(r'\|r'') \triangleq r' \log\frac{r'}{r''} + (1-r')\log\frac{1-r'}{1-r''}. \tag{6.98}$$

Using the definitions of a_θ and b_θ from (6.90) in (6.96) and (6.97), we get

$$\Lambda^*_{\text{net}}(y;\theta) = \begin{cases} KD_b\left(\dfrac{y-\log\frac{0.5}{q_\theta}}{\log\frac{q_\theta}{1-q_\theta}}\middle\|\dfrac{1}{2}\right) & \text{if } y_{\text{min}} \le y \le y_{\text{max}}, \\ \infty & \text{otherwise,} \end{cases} \tag{6.99}$$

where

$$y_{\text{min}} = \log\frac{0.5}{\max(q_\theta, 1-q_\theta)}, \qquad y_{\text{max}} = \log\frac{0.5}{\min(q_\theta, 1-q_\theta)}. \tag{6.100}$$

We can now compute the error exponent $\Psi(\theta)$ by evaluating the rate function $\Lambda^*_{\text{net}}(y;\theta)$ at $y = 0$, yielding

$$\Psi(\theta) = KD_b\left(\frac{\log\frac{q_\theta}{0.5}}{\log\frac{q_\theta}{1-q_\theta}}\middle\|\frac{1}{2}\right). \tag{6.101}$$

With the choices in (6.49), from (6.101) we obtain the numerical values

$$\Psi(1) = 2\times 10^{-3}, \qquad \Psi(2) = 2\times 10^{-3} \tag{6.102}$$

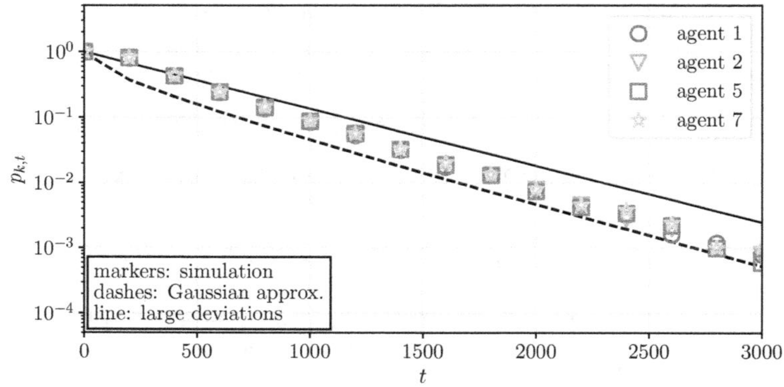

Figure 6.3: Error probability $p_{k,t}$ as a function of t, for $k = 1, 2, 5, 7$, in the setting of Example 6.3. Markers refer to the empirical error probability estimated from 5000 Monte Carlo runs. The dashed line refers to the theoretical error probability in (6.22) computed using the Gaussian approximation in (6.104). The solid line refers to the function $e^{-\Psi t}$, with error exponent Ψ predicted by the large deviation analysis in Theorem 6.3.

and the error exponent Ψ in (6.64) is thus

$$\Psi = \min_{\theta \in \{1,2\}} \Psi(\theta) = 2 \times 10^{-3}. \tag{6.103}$$

In Figure 6.3 we display the error probability $p_{k,t}$ defined in (6.19), as a function of t, for the agents listed in the legend. The markers in the figure represent probabilities estimated empirically from 5000 Monte Carlo runs. The dashed line represents the error probability curve computed by using the following Gaussian approximation for the scaled log belief ratio:

$$\bar{\beta}_{k,t}(\theta) \approx \mathscr{G}\left(\bar{\lambda}_{\mathsf{net}}, \frac{1}{t}\Sigma_{\mathsf{net}}(\theta,\theta)\right), \tag{6.104}$$

which is obtained from Theorem 6.2. The solid line represents the function $e^{-\Psi t}$, where Ψ is the error exponent provided by (6.103).

Figure 6.3 shows that, in the considered example, the Gaussian approximation can be used to estimate the error probability with good accuracy in a certain range, say, for $1500 < t < 3000$. However, we know from Appendix E (see Example E.5) that the Gaussian approximation does not offer theoretical convergence guarantees on the tails, i.e., as the error probability decreases. On the other hand, the error exponent Ψ, while not being useful to approximate the error probability curves, is guaranteed to provide a faithful prediction of their *exponential rate of decay*, that is, of their *slope* (in the considered logarithmic scale for the vertical axis).

Example 6.4 (Chernoff information). Consider the situation where: *i)* the observations are statistically independent across the agents; *ii)* they follow the objective evidence model (see Section 5.3) with a common underlying hypothesis ϑ^o; and *iii)* the combination matrix is doubly stochastic.

Since the observations are independent across the agents and the matrix is doubly stochastic (hence, its Perron vector has equal entries), to compute the LMGF of the network variable $\boldsymbol{\lambda}_{\mathsf{net},t}(\theta)$ we can use (6.84) to obtain

$$\Lambda_{\mathsf{net}}(s;\theta) = \sum_{k=1}^{K} \Lambda_k(s/K;\theta). \tag{6.105}$$

In view of Theorem 6.3, the error exponent of the social learning strategy is given by

$$\Lambda^*_{\mathsf{net}}(0;\theta) = -\inf_{s\in\mathbb{R}} \sum_{k=1}^{K} \Lambda_k(s/K;\theta) = -\inf_{s\in\mathbb{R}} \sum_{k=1}^{K} \Lambda_k(s;\theta). \tag{6.106}$$

In view of the independence across the agents, and using (6.3) and (6.58), the last sum in (6.106) can also be represented as

$$\log \mathbb{E}\left[\prod_{k=1}^{K} \exp\left\{s\boldsymbol{\lambda}_{k,t}(\theta)\right\}\right] = \log \mathbb{E}\left[\left(\frac{\prod_{k=1}^{K} \ell_k(\boldsymbol{x}_{k,t}|\vartheta^o)}{\prod_{k=1}^{K} \ell_k(\boldsymbol{x}_{k,t}|\theta)}\right)^s\right]. \tag{6.107}$$

Now, given a random vector $\boldsymbol{z}$ and two pdfs or pmfs $f(z)$ and $g(z)$, the quantity

$$C(f,g) \triangleq -\inf_{s\in\mathbb{R}} \log \mathbb{E}\left[\left(\frac{f(\boldsymbol{z})}{g(\boldsymbol{z})}\right)^s\right] \tag{6.108}$$

is referred to as the *Chernoff information* between f and g [44, 59, 60]. In view of (6.107), the error exponent $\Lambda^*_{\mathsf{net}}(0;\theta)$ is the Chernoff information between the two pdfs or pmfs

$$\prod_{k=1}^{K} \ell_k(x_{k,t}|\vartheta^o), \qquad \prod_{k=1}^{K} \ell_k(x_{k,t}|\theta), \tag{6.109}$$

defined over the aggregate of observations across the agents, namely, $[\boldsymbol{x}_{1,t}, \boldsymbol{x}_{2,t}, \ldots, \boldsymbol{x}_{K,t}]$. The Chernoff information is one fundamental quantity to characterize the performance of optimal Bayesian hypothesis testing, originally used for the binary case [44], and later for the multi-hypothesis case [107]. We will comment on these aspects more closely when dealing with the comparison between social learning and Bayesian learning in Section 13.1.

Chapter 7

Social Learning with Arithmetic Averaging

In this chapter we consider the social learning algorithm with arithmetic averaging, reported in listing (3.24) and replicated here for ease of reference:

$$\boldsymbol{\psi}_{k,t}(\theta) \propto \boldsymbol{\mu}_{k,t-1}(\theta)\ell_k(\boldsymbol{x}_{k,t}|\theta), \tag{7.1a}$$

$$\boldsymbol{\mu}_{k,t}(\theta) = \sum_{j\in\mathcal{N}_k} a_{jk}\boldsymbol{\psi}_{j,t}(\theta). \tag{7.1b}$$

Pooling the beliefs by means of an arithmetic average is perhaps the simplest and most direct solution. Despite this simplicity, however, establishing the convergence of the beliefs under arithmetic averaging is significantly more challenging than it is under geometric averaging. This is because, under geometric averaging, it is possible to reduce the analysis to the study of log belief ratios that can be expressed in terms of sums of independent log likelihood ratios. Due to this property, one can then appeal to the strong law of large numbers to obtain convergence results under great generality — see Theorem 5.1. As a matter of fact, in Chapter 5 we were able to examine a number of useful cases, including continuous and discrete distributions, connected and weak graphs, objective and subjective evidence, and the presence of fake agents. In comparison, the available results on convergence of the beliefs under arithmetic averaging are more limited, mostly focusing on data belonging to discrete finite sets (in the forthcoming treatment we remove this restriction), connected graphs, and the objective evidence model.

7.1 Modeling Assumptions

The convergence results in this chapter are stated under the objective evidence model (Assumption 5.3), i.e., the observations $\boldsymbol{x}_{k,t}$ are distributed according to some true likelihood $\ell_k(x|\vartheta^o)$, where $\vartheta^o \in \Theta$.

Recall that the index k appearing in the observation $\boldsymbol{x}_{k,t}$ and the likelihoods $\ell_k(x|\vartheta^o)$ indicates that the network agents are allowed to be heterogeneous. This heterogeneity affects the inferential capabilities of the individual agents. For example, as was seen in Section 5.3, given a certain hypothesis θ, agent k might not be able to distinguish θ from the true hypothesis ϑ^o. This happens when agent k has the same model for θ and ϑ^o, i.e., when $D(\ell_{k,\vartheta^o}||\ell_{k,\theta}) = 0$. In this case, we say that θ is *indistinguishable* from ϑ^o by agent k. More generally, we define the set of *indistinguishable hypotheses* (which we will refer to, for brevity, as *indistinguishable set*) for each agent k as

$$\mathcal{I}_k \triangleq \Big\{\theta \in \Theta \backslash \{\vartheta^o\} \text{ such that } D(\ell_{k,\vartheta^o}||\ell_{k,\theta}) = 0\Big\}. \tag{7.2}$$

We also define the set of *distinguishable hypotheses* (referred to as *distinguishable set*) for agent k as

$$\mathcal{D}_k \triangleq \Theta \backslash \Big(\mathcal{I}_k \cup \{\vartheta^o\}\Big). \tag{7.3}$$

As was done in Section 5.3, to enable all agents to learn the truth we require the *global* identifiability condition formulated in Assumption 5.4. That is, we assume that for each $\theta \neq \vartheta^o$, there exists at least one agent that is able to distinguish θ from ϑ^o.

Before establishing the convergence result, we introduce the following assumption, which excludes the case where the true likelihood $\ell_k(x|\vartheta^o)$ can be constructed as a convex combination of the *distinguishable* likelihoods $\ell_k(x|\theta)$ for $\theta \in \mathcal{D}_k$.

Assumption 7.1 (Convex independent likelihoods). For each agent k whose distinguishable set $\mathcal{D}_k$ is nonempty, the true likelihood ℓ_{k,ϑ^o} is not a convex combination of the likelihoods $\{\ell_{k,\theta}\}_{\theta \in \mathcal{D}_k}$ of the distinguishable hypotheses. This means that for all convex combination weights $\{q(\theta)\}_{\theta \in \mathcal{D}_k}$ (i.e., nonnegative weights such that $\sum_{\theta \in \mathcal{D}_k} q(\theta) = 1$), we have

$$\ell_{k,\vartheta^o} \neq \sum_{\theta \in \mathcal{D}_k} q(\theta)\ell_{k,\theta}. \tag{7.4}$$

Assumption 7.1 is a sufficient condition that is useful to prove our results. It is typically satisfied when the agents employ parametric families of likelihoods, where different hypotheses are identified by different values of the parameters. For example, in a Gaussian, exponential, or binomial family it is not possible to represent one likelihood as the convex combination of other likelihoods within the same family.

Thus, when the likelihoods belong to some structured parametric family, Assumption 7.1 is typically satisfied. We now show that it can be satisfied also in the somehow opposite case where the likelihoods are chosen in an *unstructured* manner. Specifically, let $\mathcal{X}_k$ be a discrete finite set, and recall that a probability mass function on $\mathcal{X}_k$ is a point lying in the probability simplex $\Delta_{|\mathcal{X}_k|}$. Assume that the likelihoods are picked uniformly at random from the probability simplex. Then, when the cardinality of $\mathcal{X}_k$ is larger than the cardinality of the distinguishable set $\mathcal{D}_k$, the probability of picking a set of likelihoods that violate Assumption 7.1 is zero. This is because: *i)* the dimension of the probability simplex is $d' = |\mathcal{X}_k| - 1$, whereas the dimension of the convex hull generated by $|\mathcal{D}_k|$ likelihoods is at most $d'' = |\mathcal{D}_k| - 1 < d'$; and *ii)* if we pick some points uniformly at random from a continuous space of dimension d', the probability that they lie in a given space of dimension $d'' < d'$ is zero.

The next example shows one case where Assumption 7.1 is violated.

Example 7.1 (Discrete observation space with two elements). Consider a discrete observation space $\mathcal{X}_k = \{a, b\}$. For each $\theta \in \Theta$, the pmf

$$\ell_{k,\theta} = [\ell_k(a|\theta),\, \ell_k(b|\theta)] \tag{7.5}$$

can be represented by a point $(p_1, p_2) \in \mathbb{R}^2$ — see Figure 7.1. More specifically, $\ell_{k,\theta}$ lies in Δ_2, the probability simplex in $\mathbb{R}^2$, which is a line segment. Consider now three hypotheses, $\theta_1, \theta_2, \theta_3$, for which agent k has three distinct likelihoods. It follows that one of them must necessarily lie in between the other two likelihoods, as shown in Figure 7.1. Referring to this figure, when $\vartheta^o = \theta_3$ condition (7.4) would be violated.

7.2 Belief Convergence

For observations modeled as discrete random variables with finite support, truth learning under arithmetic averaging is established in [131, Thm. 5], without requiring Assumption 7.1. By introducing Assumption 7.1, in the next theorem we are able to cover also the cases of discrete random

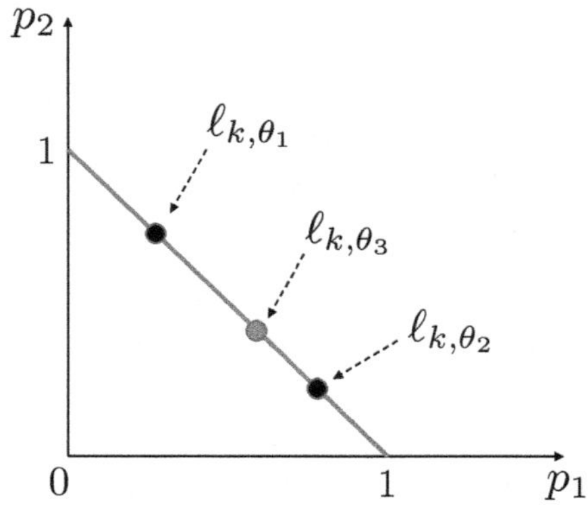

Figure 7.1: Illustration for Example 7.1.

variables with infinite support and of continuous random variables. For discrete random variables with finite support, the proof of the theorem below constitutes an alternative (perhaps simpler) proof with respect to the one offered in [131], albeit at the expense of introducing an additional assumption.

Theorem 7.1 (Belief convergence). Let Assumptions 5.1, 5.3, 5.4, and 7.1 be satisfied. If the network graph is connected, then for $k = 1, 2, \ldots, K$,

$$\boldsymbol{\mu}_{k,t}(\vartheta^o) \xrightarrow[t\to\infty]{\text{a.s.}} 1. \tag{7.6}$$

Before presenting the proof of Theorem 7.1, we introduce some relevant intermediate results. We start by defining the following quantity for any convex combination vector $q \in \Delta_H$:

$$d_k(q) \triangleq \mathbb{E} \log \frac{\ell_k(\boldsymbol{x}_{k,t}|\vartheta^o)}{\sum\limits_{\theta\in\Theta} q(\theta)\ell_k(\boldsymbol{x}_{k,t}|\theta)}, \tag{7.7}$$

which is a KL divergence because *i)* the denominator is a pdf (or pmf) as it is a convex combination of pdfs (or pmfs); and *ii)* under Assumption 5.3, the expectation is computed considering the true likelihood $\ell_k(x|\vartheta^o)$. Furthermore, given the underlying probability space $(\Omega, \mathscr{F}, \mathbb{P})$, we introduce the filtration (see Definition D.5) generated by the belief vectors of all agents, namely, the sequence of sub-σ-fields

$$\mathcal{F}_t \triangleq \sigma\left(\{\boldsymbol{\mu}_{k,0}\}_{k=1}^K, \{\boldsymbol{\mu}_{k,1}\}_{k=1}^K, \ldots, \{\boldsymbol{\mu}_{k,t}\}_{k=1}^K\right), \quad t = 0, 1, \ldots \tag{7.8}$$

Note that $\mathcal{F}_0 = \sigma\left(\{\boldsymbol{\mu}_{k,0}\}_{k=1}^K\right) = \{\emptyset, \Omega\}$ is the trivial σ-field, since we are modeling the initial beliefs as deterministic.

Before proceeding with the analysis, it is important to remark that, as was the case for geometric averaging, even under arithmetic averaging the beliefs $\boldsymbol{\psi}_{k,t}(\theta)$ and $\boldsymbol{\mu}_{k,t}(\theta)$ remain nonzero almost surely if Assumptions 5.1 and 5.2 are satisfied. In fact, we have already observed (see the discussion before Theorem 5.1) that under Assumption 5.2 the likelihoods are almost-surely positive, implying that *i)* the denominator in the Bayesian update step (7.1a) is almost-surely positive; and *ii)* starting from a belief $\boldsymbol{\mu}_{k,t-1}(\theta)$ that is nonzero at any θ, the intermediate belief $\boldsymbol{\psi}_{k,t}(\theta)$ in (7.1a) is nonzero. Now, since in view of point i) of Assumption 5.1 the combination matrix is left stochastic, for each agent k there exists at least one agent j such that $a_{jk} > 0$ (see Definition 4.10). Thus, Eq. (7.1b) implies that $\boldsymbol{\mu}_{k,t}(\theta) > 0$. Moreover, since from point ii) in Assumption 5.1 the initial beliefs $\mu_{k,0}(\theta)$ are nonzero, positivity of the beliefs $\boldsymbol{\psi}_{k,t}(\theta)$ and $\boldsymbol{\mu}_{k,t}(\theta)$ extends to all t by induction. The aforementioned properties will be useful in the following development, where we will work with log beliefs such as $\log \boldsymbol{\mu}_{k,t}(\vartheta^o)$. Since the beliefs are nonzero almost surely, the random variable $\log \boldsymbol{\mu}_{k,t}(\vartheta^o)$ is well-posed. Moreover, since belief vectors are probability vectors, we also have $\boldsymbol{\mu}_{k,t}(\vartheta^o) < 1$, which implies that $\log \boldsymbol{\mu}_{k,t}(\vartheta^o)$ is a negative random variable (recall that when we say that a random variable is negative we mean that it is smaller than 0 almost surely).

Lemma 7.1 (Useful submartingale). Let Assumptions 5.1 and 5.3 be satisfied. Assume that the network graph is connected, let v be the Perron vector associated with the combination matrix A, and define the random variables, for $t = 0, 1, \dots,$

$$\boldsymbol{m}_t \triangleq \sum_{k=1}^K v_k \log \boldsymbol{\mu}_{k,t}(\vartheta^o). \tag{7.9}$$

Then the following properties hold:

i) For $t = 1, 2, \dots,$

$$\mathbb{E}\left[\boldsymbol{m}_t | \mathcal{F}_{t-1}\right] \geq \boldsymbol{m}_{t-1} + \sum_{k=1}^K v_k d_k(\boldsymbol{\mu}_{k,t-1}). \tag{7.10}$$

ii) The sequence $\{\boldsymbol{m}_t\}_{t=0}^\infty$ is a **negative** submartingale (see Definition D.6) with respect to the filtration $\{\mathcal{F}_t\}_{t=0}^\infty$ in (7.8), and there exists a random variable $\boldsymbol{m}_\infty$ such that

$$\boldsymbol{m}_t \xrightarrow[t\to\infty]{\text{a.s.}} \boldsymbol{m}_\infty. \tag{7.11}$$

iii) The sequence of expected values $\mathbb{E}\boldsymbol{m}_t$ has a finite limit.

Proof. Taking the logarithm of (7.1b), we can write

$$
\begin{aligned}
\log \boldsymbol{\mu}_{k,t}(\vartheta^o) &= \log\left(\sum_{j\in\mathcal{N}_k} a_{jk}\boldsymbol{\psi}_{j,t}(\vartheta^o)\right)\\
&\stackrel{\text{(a)}}{=} \log\left(\sum_{j\in\mathcal{N}_k} a_{jk}\frac{\boldsymbol{\mu}_{j,t-1}(\vartheta^o)\ell_j(\boldsymbol{x}_{j,t}|\vartheta^o)}{\sum\limits_{\theta\in\Theta}\boldsymbol{\mu}_{j,t-1}(\theta)\ell_j(\boldsymbol{x}_{j,t}|\theta)}\right)\\
&\stackrel{\text{(b)}}{\geq} \sum_{j\in\mathcal{N}_k} a_{jk}\log\left(\frac{\boldsymbol{\mu}_{j,t-1}(\vartheta^o)\ell_j(\boldsymbol{x}_{j,t}|\vartheta^o)}{\sum\limits_{\theta\in\Theta}\boldsymbol{\mu}_{j,t-1}(\theta)\ell_j(\boldsymbol{x}_{j,t}|\theta)}\right)\\
&= \sum_{j\in\mathcal{N}_k} a_{jk}\log\boldsymbol{\mu}_{j,t-1}(\vartheta^o) + \sum_{j\in\mathcal{N}_k} a_{jk}\log\left(\frac{\ell_j(\boldsymbol{x}_{j,t}|\vartheta^o)}{\sum\limits_{\theta\in\Theta}\boldsymbol{\mu}_{j,t-1}(\theta)\ell_j(\boldsymbol{x}_{j,t}|\theta)}\right), \qquad (7.12)
\end{aligned}
$$

where in (a) we used (7.1a) and in (b) we used Jensen's inequality (see Theorem C.5 and in particular (C.10)) in view of the concavity of the logarithm. Taking the expectation of the LHS and RHS of (7.12) conditioned on $\mathcal{F}_{t-1}$ yields, for $t = 1, 2, \ldots,$

$$
\begin{aligned}
\mathbb{E}\left[\log\boldsymbol{\mu}_{k,t}(\vartheta^o)|\mathcal{F}_{t-1}\right] &\geq \sum_{j\in\mathcal{N}_k} a_{jk}\log\boldsymbol{\mu}_{j,t-1}(\vartheta^o)\\
&+\sum_{j\in\mathcal{N}_k} a_{jk}\mathbb{E}\left[\log\left(\frac{\ell_j(\boldsymbol{x}_{j,t}|\vartheta^o)}{\sum\limits_{\theta\in\Theta}\boldsymbol{\mu}_{j,t-1}(\theta)\ell_j(\boldsymbol{x}_{j,t}|\theta)}\right)\Bigg|\mathcal{F}_{t-1}\right]. \qquad (7.13)
\end{aligned}
$$

Assumption 5.3 implies that the observations at time t are independent of the past observations, and, hence, of the previous-lag belief vector $\boldsymbol{\mu}_{j,t-1}$. Moreover, once we condition on the filtration $\mathcal{F}_{t-1}$, the random vector $\boldsymbol{\mu}_{j,t-1}$ is frozen. As a result, the expectation on the second term on the RHS of (7.13) corresponds to a KL divergence between the true likelihood ℓ_{j,ϑ^o} and a mixture of likelihoods $\sum_{\theta\in\Theta}\boldsymbol{\mu}_{j,t-1}(\theta)\ell_{j,\theta}$. In other words, using definition (7.7), the second term on the RHS of (7.13) can be represented as

$$
\sum_{j\in\mathcal{N}_k} a_{jk}d_j(\boldsymbol{\mu}_{j,t-1}). \qquad (7.14)
$$

In view of the definition of $\mathcal{N}_k$ from (4.1) and using (7.14), we can rewrite (7.13) as

$$
\mathbb{E}\left[\log\boldsymbol{\mu}_{k,t}(\vartheta^o)|\mathcal{F}_{t-1}\right] \geq \sum_{j=1}^{K} a_{jk}\log\boldsymbol{\mu}_{j,t-1}(\vartheta^o) + \sum_{j=1}^{K} a_{jk}d_j(\boldsymbol{\mu}_{j,t-1}). \qquad (7.15)
$$

Since the combination matrix A is left stochastic (see part i) of Assumption 5.1) and the network graph is assumed to be connected, it follows from Definition 4.6 and Lemma 4.3 that A is an irreducible matrix with spectral radius $\rho(A) = 1$. From the Perron-Frobenius theorem (Theorem 4.1) it follows that we can define the Perron vector v, which, we recall, has positive entries and satisfies the relation

$$
Av = v. \qquad (7.16)
$$

Expanding this equation in terms of the individual entries of Av and v, we get

$$\sum_{k=1}^{K} a_{jk} v_k = v_j, \qquad j = 1, 2, \ldots, K. \tag{7.17}$$

Multiplying both sides of (7.15) by v_k, summing over k, and using (7.17) yields (7.10), which proves part i) of the lemma.

To prove part ii), observe that the nonnegativity of the KL divergence implies $d_k(\boldsymbol{\mu}_{k,t-1}) \geq 0$; it then follows from part i) that

$$\mathbb{E}\left[\boldsymbol{m}_t | \mathcal{F}_{t-1}\right] \geq \boldsymbol{m}_{t-1}. \tag{7.18}$$

Note that $\boldsymbol{m}_t$ is a negative random variable since the entries of the Perron vector are positive and all the beliefs are almost surely strictly less than 1 — see the discussion before the statement of the theorem. Taking the expectation of both sides of (7.18), we can write

$$0 > \mathbb{E}\boldsymbol{m}_t \geq \mathbb{E}\boldsymbol{m}_{t-1} \geq \cdots \geq m_0, \tag{7.19}$$

which implies that $\boldsymbol{m}_t$ has finite mean for $t = 0, 1, \ldots$ (note that m_0 is finite since the initial beliefs are nonzero in view of point ii) in Assumption 5.1). Therefore, in view of (7.18), we conclude that the sequence $\{\boldsymbol{m}_t\}_{t=0}^{\infty}$ is a negative submartingale (see Definition D.6). Then part ii) follows from the martingale convergence theorem — see in particular Corollary D.1. Finally, part iii) follows from (7.19), which implies that the sequence of expectations is a convergent sequence (since it is nondecreasing and bounded from above).

■

Lemma 7.2 (All agents discard the distinguishable hypotheses). Let Assumptions 5.1, 5.3, and 7.1 be satisfied, and assume that the network graph is connected, with a combination matrix A having Perron vector v. Then, for $k = 1, 2, \ldots, K$ and for all $\theta \in \mathcal{D}_k$,

$$\boldsymbol{\mu}_{k,t}(\theta) \xrightarrow[t\to\infty]{\text{P}} 0. \tag{7.20}$$

Proof. Under the considered assumptions, we can use the results from Lemma 7.1. Taking the expectation in (7.10), we get

$$\mathbb{E}\boldsymbol{m}_t \geq \mathbb{E}\boldsymbol{m}_{t-1} + \sum_{k=1}^{K} v_k \mathbb{E} d_k(\boldsymbol{\mu}_{k,t-1}). \tag{7.21}$$

Using (7.21) along with the fact that the KL divergence is nonnegative, we see that

$$0 \leq \sum_{k=1}^{K} v_k \mathbb{E} d_k(\boldsymbol{\mu}_{k,t-1}) \leq \mathbb{E}\boldsymbol{m}_t - \mathbb{E}\boldsymbol{m}_{t-1}, \tag{7.22}$$

which, in view of part ii) of Lemma 7.1, implies that the RHS of (7.22) converges to 0. Therefore, we can apply the squeeze (or sandwich) theorem [144, Thm. 3.19] to (7.22), obtaining

$$\lim_{t\to\infty}\sum_{k=1}^{K} v_k \mathbb{E} d_k(\boldsymbol{\mu}_{k,t-1}) = 0. \tag{7.23}$$

Since $v_k > 0$ for all k and $d_k(\boldsymbol{\mu}_{k,t-1})$ is a nonnegative random variable, it follows that

$$\lim_{t\to\infty} \mathbb{E} d_k(\boldsymbol{\mu}_{k,t-1}) = 0, \tag{7.24}$$

which means that $d_k(\boldsymbol{\mu}_{k,t-1})$ converges to 0 in the 1st mean, i.e., in the L_1 norm — see Definition D.3. In view of (D.17), this implies that $d_k(\boldsymbol{\mu}_{k,t-1})$ converges to 0 in probability, namely,

$$d_k(\boldsymbol{\mu}_{k,t-1}) \xrightarrow[t\to\infty]{\mathrm{P}} 0 \tag{7.25}$$

for $k = 1, 2, \ldots, K$. Using Pinsker's inequality (Theorem C.7) we can lower bound the KL divergence $d_k(\boldsymbol{\mu}_{k,t-1})$ and write

$$d_k(\boldsymbol{\mu}_{k,t-1}) \geq \frac{1}{2} D_{\mathsf{TV}}^2 \left(\ell_{k,\vartheta^o}\,,\, \sum_{\theta\in\Theta} \boldsymbol{\mu}_{k,t-1}(\theta)\ell_{k,\theta} \right), \tag{7.26}$$

where the symbol D_{TV} denotes the total variation distance, whose expression is provided in Definition C.1.

Consider now an agent k for which $|\mathcal{D}_k| > 0$. Letting

$$\boldsymbol{q}(\theta) = \frac{\boldsymbol{\mu}_{k,t-1}(\theta)}{\sum\limits_{\theta'\in\mathcal{D}_k} \boldsymbol{\mu}_{k,t-1}(\theta')}, \quad \theta \in \mathcal{D}_k, \tag{7.27}$$

we can write

$$\begin{aligned}
&\ell_k(\boldsymbol{x}_{k,t}|\vartheta^o) - \sum_{\theta\in\Theta} \boldsymbol{\mu}_{k,i-1}(\theta)\ell_k(\boldsymbol{x}_{k,t}|\theta)\\
&= \left(1 - \sum_{\theta\in\mathcal{I}_k\cup\{\vartheta^o\}} \boldsymbol{\mu}_{k,i-1}(\theta)\right)\ell_k(\boldsymbol{x}_{k,t}|\vartheta^o) - \sum_{\theta\in\mathcal{D}_k} \boldsymbol{\mu}_{k,i-1}(\theta)\ell_k(\boldsymbol{x}_{k,t}|\theta)\\
&= \left(\ell_k(\boldsymbol{x}_{k,t}|\vartheta^o) - \sum_{\theta\in\mathcal{D}_k} \boldsymbol{q}(\theta)\ell_k(\boldsymbol{x}_{k,t}|\theta)\right) \sum_{\theta'\in\mathcal{D}_k} \boldsymbol{\mu}_{k,i-1}(\theta'),
\end{aligned} \tag{7.28}$$

which, in view of the formulas for the total variation distance in Definition C.1, implies

$$\begin{aligned}
&D_{\mathsf{TV}}\left(\ell_{k,\vartheta^o}\,,\, \sum_{\theta\in\Theta} \boldsymbol{\mu}_{k,t-1}(\theta)\ell_{k,\theta} \right)\\
&= \left| \sum_{\theta\in\mathcal{D}_k} \boldsymbol{\mu}_{k,t-1}(\theta) \right| \times D_{\mathsf{TV}}\left(\ell_{k,\vartheta^o}\,,\, \sum_{\theta\in\mathcal{D}_k} \boldsymbol{q}(\theta)\ell_{k,\theta} \right).
\end{aligned} \tag{7.29}$$

We now show that the total variation distance appearing on the RHS is lower bounded by a strictly positive value d_{min}. Let w be a vector belonging to the probability simplex

$\Delta_{|\mathcal{D}_k|}$. Denote the entries of w by $w(\theta)$, for $\theta \in \mathcal{D}_k$, and consider the total variation distance

$$D_{\mathsf{TV}}\left(\ell_{k,\vartheta^o}\,,\ \sum_{\theta\in\mathcal{D}_k} w(\theta)\ell_{k,\theta}\right) = g(w) \tag{7.30}$$

regarded as a function of w. It is readily verified that $g(w)$ is continuous with respect to w. We want to characterize the infimum of $g(w)$ over $\Delta_{\mathcal{D}_k}$. Since the probability simplex is a compact set (i.e., it is closed and bounded), from the extreme value theorem [144], the infimum of $g(w)$ over $\Delta_{\mathcal{D}_k}$ is in fact a minimum that is attained at some point(s) of the set. Denoting by d_{min} this minimum, we must have $g(w) = d_{\mathsf{min}}$ for some $w \in \mathcal{D}_k$. Since the total variation distance is nonnegative, $d_{\mathsf{min}} \geq 0$. Now, if $d_{\mathsf{min}} = 0$, then the total variation in (7.30) would be equal to 0 for some $w \in \mathcal{D}_k$. This would mean ℓ_{k,ϑ^o} could be written as a convex combination of the likelihoods $\{\ell_{k,\theta}\}_{\theta\in\mathcal{D}_k}$, violating Assumption 7.1. We conclude that $d_{\mathsf{min}} > 0$.

In summary, we have shown that $g(w) \geq d_{\mathsf{min}} > 0$ for all $w \in \mathcal{D}_k$. As a result, the total variation distance appearing on the RHS of (7.29) can be lower bounded as follows:

$$D_{\mathsf{TV}}\left(\ell_{k,\vartheta^o}\,,\ \sum_{\theta\in\mathcal{D}_k} \boldsymbol{q}(\theta)\ell_{k,\theta}\right) \geq d_{\mathsf{min}} > 0. \tag{7.31}$$

Combining (7.26), (7.29), and (7.31), we obtain

$$d_k(\boldsymbol{\mu}_{k,t-1}) \geq \frac{d_{\mathsf{min}}^2}{2}\left|\sum_{\theta\in\mathcal{D}_k} \boldsymbol{\mu}_{k,t-1}(\theta)\right|^2. \tag{7.32}$$

Since d_{min} is positive, we conclude from (7.25) that, for $k = 1, 2, \ldots, K$ and for all $\theta \in \mathcal{D}_k$,

$$\boldsymbol{\mu}_{k,t}(\theta) \xrightarrow[t\to\infty]{\mathrm{p}} 0. \tag{7.33}$$

■

Lemma 7.3 (All agents learn the truth in probability). Let Assumptions 5.1, 5.3, 5.4, and 7.1 be satisfied. If the network graph is connected, then for $k = 1, 2, \ldots, K$,

$$\boldsymbol{\mu}_{k,t}(\vartheta^o) \xrightarrow[t\to\infty]{\mathrm{p}} 1, \tag{7.34}$$

where we remark that the convergence holds in probability (while in Theorem 7.1 we strengthen this result by proving almost-sure convergence).

Proof. We start by showing that, for an agent k and a hypothesis θ, we have

$$\boldsymbol{\mu}_{k,t}(\theta) \xrightarrow[t\to\infty]{\mathrm{p}} 0, \tag{7.35}$$

then the same result holds for *all* other agents in the network.

Using (7.1b), under condition (7.35) we can write

$$\boldsymbol{\mu}_{k,t}(\theta) = \sum_{j\in\mathcal{N}_k} a_{jk}\boldsymbol{\psi}_{j,t}(\theta) \xrightarrow[t\to\infty]{\mathrm{p}} 0. \tag{7.36}$$

Now, let $0 < \varepsilon < 1$. Since $a_{jk} > 0$ for $j \in \mathcal{N}_k$, and since $\boldsymbol{\psi}_{j,t}(\theta)$ is nonnegative, then for all $j \in \mathcal{N}_k$ the following implication holds:

$$a_{jk}\boldsymbol{\psi}_{j,t}(\theta) > \varepsilon \implies \sum_{j \in \mathcal{N}_k} a_{jk}\boldsymbol{\psi}_{j,t}(\theta) > \varepsilon. \tag{7.37}$$

This further implies that

$$\mathbb{P}\left[a_{jk}\boldsymbol{\psi}_{j,t}(\theta) > \varepsilon\right] \leq \mathbb{P}\left[\sum_{j \in \mathcal{N}_k} a_{jk}\boldsymbol{\psi}_{j,t}(\theta) > \varepsilon\right] \tag{7.38}$$

for all $j \in \mathcal{N}_k$, and using (7.36) in (7.38) we conclude that

$$\boldsymbol{\psi}_{j,t}(\theta) \xrightarrow[t\to\infty]{\text{p}} 0. \tag{7.39}$$

Now we would like to show that the convergence result in (7.39) holds for $\boldsymbol{\mu}_{j,t}(\theta)$ as well. Actually, we would not need to prove this result when $\theta \in \mathcal{D}_j$, since in this case we already know from Lemma 7.2 that $\boldsymbol{\mu}_{j,t}(\theta)$ converges to 0 in probability. However, the following derivation holds for any θ. In view of (7.1a), the belief $\boldsymbol{\mu}_{j,t-1}(\theta)$ can be represented as

$$\begin{aligned}\boldsymbol{\mu}_{j,t-1}(\theta) &= \boldsymbol{\psi}_{j,t}(\theta) \sum_{\theta' \in \Theta} \boldsymbol{\mu}_{j,t-1}(\theta') \frac{\ell_j(\boldsymbol{x}_{j,t}|\theta')}{\ell_j(\boldsymbol{x}_{j,t}|\theta)} \\ &\leq \boldsymbol{\psi}_{j,t}(\theta) \sum_{\theta' \in \Theta} \frac{\ell_j(\boldsymbol{x}_{j,t}|\theta')}{\ell_j(\boldsymbol{x}_{j,t}|\theta)}.\end{aligned} \tag{7.40}$$

Observe that the random variable defined by the sum in (7.40) has constant distribution over time, and that $\boldsymbol{\psi}_{j,t}(\theta)$ vanishes in probability in view of (7.39). Therefore, we can apply Slutsky's theorem (in particular, Eq. (D.38) in Theorem D.4) to the RHS of (7.40), concluding that $\boldsymbol{\mu}_{j,t}(\theta) \xrightarrow[t\to\infty]{\text{p}} 0$ for all $j \in \mathcal{N}_k$.

In summary, we have shown that the following implication holds

$$\boldsymbol{\mu}_{k,t}(\theta) \xrightarrow[t\to\infty]{\text{p}} 0 \implies \boldsymbol{\mu}_{j,t}(\theta) \xrightarrow[t\to\infty]{\text{p}} 0 \quad \forall j \in \mathcal{N}_k. \tag{7.41}$$

Consider now the neighbors $j' \in \mathcal{N}_j$ of an agent $j \in \mathcal{N}_k$. Repeating the same steps used to prove the implication in (7.41), we get

$$\boldsymbol{\mu}_{j,t}(\theta) \xrightarrow[t\to\infty]{\text{p}} 0 \implies \boldsymbol{\mu}_{j',t}(\theta) \xrightarrow[t\to\infty]{\text{p}} 0 \quad \forall j' \in \mathcal{N}_j. \tag{7.42}$$

Since the network graph is assumed to be connected, we can repeat this process so as to reach all agents in the network, finally establishing that

$$\text{if } \boldsymbol{\mu}_{k,t}(\theta) \xrightarrow[t\to\infty]{\text{p}} 0 \text{ for an agent } k, \text{ then } \boldsymbol{\mu}_{j,t}(\theta) \xrightarrow[t\to\infty]{\text{p}} 0 \text{ for } j = 1, 2, \ldots, K. \tag{7.43}$$

Let us now consider an agent k for which $\mathcal{D}_k$ is nonempty. Such an agent must necessarily exist in view of Assumption 5.4. From Lemma 7.2 we know that

$$\boldsymbol{\mu}_{k,t}(\theta) \xrightarrow[t\to\infty]{\text{p}} 0 \quad \forall \theta \in \mathcal{D}_k. \tag{7.44}$$

In view of (7.43), this implies that $\boldsymbol{\mu}_{j,t}(\theta) \xrightarrow[t\to\infty]{\mathrm{p}} 0$ for $j = 1, 2, \ldots, K$ and for all $\theta \in \mathcal{D}_k$. Repeating the above argument for all agents with nonempty distinguishable set $\mathcal{D}_k$, we have that

$$\boldsymbol{\mu}_{j,t}(\theta) \xrightarrow[t\to\infty]{\mathrm{p}} 0 \tag{7.45}$$

for $j = 1, 2, \ldots, K$ and for all $\theta \in \bigcup_{k=1}^{K} \mathcal{D}_k$. Since Assumption 5.4 imposes that all hypotheses $\theta \neq \vartheta^o$ are distinguishable for at least one agent, it follows that $\bigcup_{k=1}^{K} \mathcal{D}_k = \Theta \backslash \{\vartheta^o\}$, which, in view of (7.45), means that, for all agents, the beliefs about the false hypotheses vanish in probability. Therefore, all agents learn the truth in probability, and the proof is complete.

■

Proof of Theorem 7.1. Lemma 7.3 ensures that the whole network learns the truth in probability. Therefore, we have that

$$\sum_{k=1}^{K} v_k \log \boldsymbol{\mu}_{k,t}(\vartheta^o) \xrightarrow[t\to\infty]{\mathrm{p}} 0. \tag{7.46}$$

Using part ii) of Lemma 7.1, and since almost-sure convergence implies convergence in probability, we have that

$$\sum_{k=1}^{K} v_k \log \boldsymbol{\mu}_{k,t}(\vartheta^o) \xrightarrow[t\to\infty]{\mathrm{a.s.}} 0. \tag{7.47}$$

Since $v_k > 0$ and $\log \boldsymbol{\mu}_{k,t}(\vartheta^o) < 0$, it follows that

$$\log \boldsymbol{\mu}_{k,t}(\vartheta^o) \xrightarrow[t\to\infty]{\mathrm{a.s.}} 0, \tag{7.48}$$

which is equivalent to

$$\boldsymbol{\mu}_{k,t}(\vartheta^o) \xrightarrow[t\to\infty]{\mathrm{a.s.}} 1, \tag{7.49}$$

and the proof of Theorem 7.1 is complete.

■

To illustrate the result of Theorem 7.1, we introduce the following example.

Example 7.2 (Truth learning under arithmetic averaging). We consider the same setting used in Example 5.4, which is now briefly summarized. The network graph is reported in Figure 7.2 (it is undirected and all agents are assumed to have a self-loop, not shown in the figure). On top of this graph, a combination matrix is built by using the Metropolis combination rule — see Table 4.1.

The agents have common likelihood models, i.e., $\ell_k(x|\theta) = \ell(x|\theta)$ for all k, and $\ell(x|\theta)$ is a unit-variance Gaussian pdf with mean $\nu_\theta = \theta$, for $\theta \in \Theta = \{1, 2, 3\}$ — see the top right panel of Figure 7.2. The network operates under the objective evidence model (Assumption 5.3), with the true underlying hypothesis being $\vartheta^o = 1$.

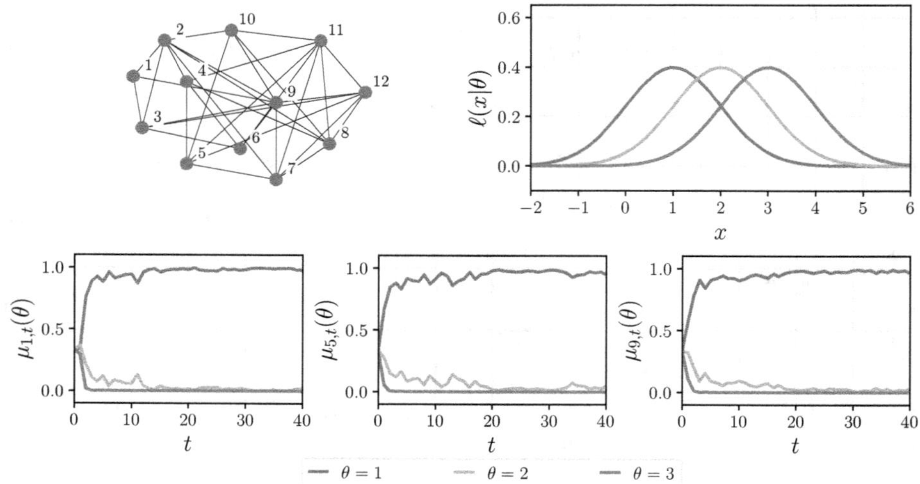

Figure 7.2: (*Top left*) Network topology used in Example 7.2. The graph is undirected and all agents are assumed to have a self-loop, not shown in the figure. (*Top right*) Likelihood models. (*Bottom*) Belief evolution over 40 iterations for agents 1, 5, and 9. We see that, as t grows, the agents place their full belief mass on the true hypothesis $\vartheta^o = 1$.

While in Example 5.4 the agents used geometric averaging, in the present example they implement the social learning strategy with arithmetic averaging seen in (7.1a) and (7.1b).

In the bottom panels of Figure 7.2, we plot the belief evolution for agents 1, 5, and 9 over 40 iterations. We see that all agents agree asymptotically on the true hypothesis ϑ^o, as predicted by Theorem 7.1.

Chapter 8

Adaptive Social Learning

We have seen in Chapter 5 that, as the amount of streaming data grows, the belief vector converges to an ideal belief vector that places unit mass on the true hypothesis (or on the hypothesis corresponding to the minimizer of the network average of KL divergences). In other words, if the amount of streaming data is sufficiently large, maximum credibility is assigned to the target hypothesis whereas no credibility is assigned to other hypotheses. Remarkably, the learning performance continues to improve steadily as more evidence is collected, and we know from (5.9) that the convergence to the ideal belief is exponentially fast. Such continuous improvement has a subtle and often overlooked effect of making the agents *stubborn* and unable to react quickly enough to drifts in the underlying operational conditions (such as a changing target hypothesis). This is a serious limitation, both from design and behavioral perspectives. From the design viewpoint, there are several applications where adaptation is a critical requirement for the deployment of learning systems in highly dynamic and uncertain environments. From the behavioral viewpoint, we would like the social learning models to capture the cognitive abilities of groups of animals or humans, who tend to adapt well to changing conditions.

8.1 Stubbornness of Agents

Let us illustrate the slow reaction time to drifts in the environment by applying the social learning algorithm (3.16) to a problem involving weather forecasting, with three possible hypotheses: *sunny*, *cloudy*, and *rainy*.

Consider 10 agents linked by a connected graph and assume that the streaming observations collected by the agents drive them to believe that

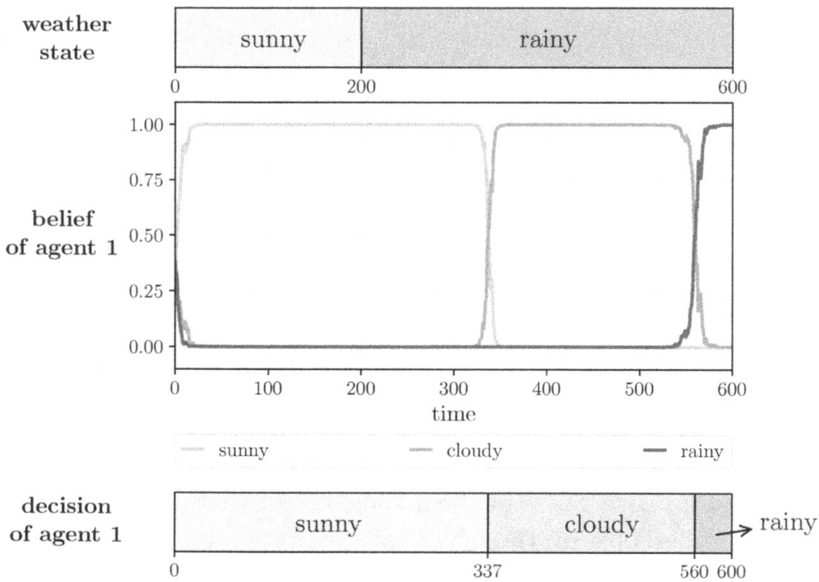

Figure 8.1: *Traditional* social learning strategy — see listing (3.16). (*Top*) Evolution of the weather state. The state drifts at time 200 from "sunny" to "rainy". (*Center*) Belief evolution for agent 1. (*Bottom*) The instantaneous decision of agent 1, taken by choosing the hypothesis that maximizes the belief at the current instant. We see that traditional social learning is not able to adapt to the new state of nature: It takes until about time 580 to correctly identify the "rainy" state (blue), after having assigned for a long intermediate period maximal belief to a wrong state, namely, the "cloudy" state (green).

"tomorrow will be sunny." In this way, their belief vectors will converge to place maximal mass on the state corresponding to sunny weather. After some time, the observations available for the decision evolve in response to changes in meteorological conditions, with the most recent evidence suggesting that "tomorrow will be rainy." In this case, the agents will unfortunately show some significant inertia to changing their beliefs to place maximal mass on the state corresponding to rainy weather.

This effect is illustrated in Figure 8.1, where we display the time evolution of the true[1] state (top), the beliefs of agent 1 (center), and its decisions (bottom). Similar behavior can be observed for the other agents. In line with the evidence suggested by the initial set of data, the belief

[1]To avoid confusion, note that in a weather forecasting problem, data are collected at a certain time instant to predict the weather state relative to a future time instant, e.g., one day ahead. Accordingly, when we say that the true state changes from sunny to rainy at time t, we do not mean that it is actually starting to rain at time t. We mean instead that the data collected at time t are compatible with the statistical model corresponding to rainy weather, rather than sunny weather.

mass assigned to the hypothesis "sunny" (see the yellow curve) becomes close to 1 after a few iterations, i.e., the network arrives quickly at the correct determination about the state of nature.

The state of nature changes to rainy at instant $t = 200$, but a long time passes before the agent perceives the drift. The belief starts to change only at $t \approx 350$, when, however, the agent *still does not detect the true state.* Indeed, it first transitions to believing that it is cloudy (green curve) before switching to believing that it is rainy (blue curve) many iterations later, at $t \approx 580$. This example shows that, under the traditional social learning strategy (3.16), the agents are not able to react sufficiently fast and to adapt their beliefs to track drifts in the environment. While they are able to learn very well *until* the change, they show a delayed reaction *after* the change, needing many iterations to overcome their stubbornness and opt for the correct hypothesis.

8.2 Adaptive Update

In order to instill adaptation into the social learning algorithm, we must make it more reactive to the incoming data and less dependent on the past beliefs. Referring back to the general scheme for non-Bayesian social learning in Figure 3.3, the step where the algorithm blends past and new information is the update step. We recall that the goal of this step, for each agent k at time t, is to modify the past belief vector $\mu_{k,t-1}$ into an *intermediate* belief vector $\psi_{k,t}$ by incorporating the likelihood $\ell_k(x_{k,t}|\theta)$ of the new data sample $x_{k,t}$. For this task, the traditional social learning algorithm (3.16) relies on a *Bayesian update* with prior $\mu_{k,t-1}$ and likelihood $\ell_k(x_{k,t}|\theta)$, namely,

$$\psi_{k,t}(\theta) = \mu_{k,t}^{\mathsf{Bu}}(\theta) \triangleq \frac{\mu_{k,t-1}(\theta)\ell_k(x_{k,t}|\theta)}{\sum_{\theta' \in \Theta} \mu_{k,t-1}(\theta')\ell_k(x_{k,t}|\theta')}. \tag{8.1}$$

We illustrated in Figure 8.1 that this learning approach infuses some stubbornness into the behavior of the agents. Therefore, to construct an *adaptive* social learning algorithm, we now focus on modifying the update rule (8.1) by adjusting the computation of the intermediate belief vector $\psi_{k,t}$.

To this end, we will make use of another belief vector similar to the one introduced in (2.67), namely,

$$\mu_{k,t}^{\mathsf{lik}}(\theta) \triangleq \frac{\ell_k(x_{k,t}|\theta)}{\sum\limits_{\theta' \in \Theta} \ell_k(x_{k,t}|\theta')}. \tag{8.2}$$

We referred to this belief as the "likelihood" posterior since it basically turns the likelihood into a *belief* by suitable normalization. Note that the belief vector $\mu_{k,t}^{\mathsf{lik}}$ corresponds to a Bayesian update similar to (8.1), albeit obtained with a *uniform prior* that gives equal preference to all hypotheses (i.e., by replacing $\mu_{k,t-1}(\theta)$ in (8.1) with $1/H$). Comparing (8.2) with (8.1), we see that (8.2) *ignores the past belief and relies solely on the new data.* This property of $\mu_{k,t}^{\mathsf{lik}}$ will be exploited in the next sections to construct a social learning strategy that is more reactive to new data.

In this construction, we will follow the approach used in Section 2.3, where we showed that the Bayesian posterior is the minimizer of cost functions based on information-theoretic measures. The new update rule will rely on suitable modifications of these cost functions, which infuse the learning algorithm with an *adaptation* capability. In particular, we will propose two approaches and interpretations, which will lead to the same adaptive rule.

8.2.1 Adaptive Update: First Approach

In traditional social learning, the intermediate belief vector $\psi_{k,t}$ is obtained through a Bayesian update, i.e., it is computed using (8.1). As a result, $\psi_{k,t}$ is obviously the solution to the optimization problem

$$\psi_{k,t} = \mu_{k,t}^{\mathsf{Bu}} = \arg\min_{p \in \Delta_H} D\left(p || \mu_{k,t}^{\mathsf{Bu}}\right). \tag{8.3}$$

One way to induce faster reaction to new data is to modify this construction by combining two different KL divergences: One divergence is based on the Bayesian update $\mu_{k,t}^{\mathsf{Bu}}$ (which accounts for *past* information through the past belief vector $\mu_{k,t-1}$, and for *new* data through the likelihood) and another divergence is based on the "likelihood" posterior $\mu_{k,t}^{\mathsf{lik}}$ (which, as already mentioned, employs *only new* data). Specifically, we now seek to construct $\psi_{k,t}$ by considering instead

$$\psi_{k,t} \triangleq \arg\min_{p \in \Delta_H} \left\{(1-\delta) D\left(p || \mu_{k,t}^{\mathsf{Bu}}\right) + \delta D\left(p || \mu_{k,t}^{\mathsf{lik}}\right)\right\}, \tag{8.4}$$

where $0 < \delta < 1$ is a weight used to tune the degree of adaptation of the resulting update rule. We see from (8.4) that, when $\delta \to 0$, we recover the

Bayesian update (8.3). In comparison, as δ moves away from zero, the role of $D(p||\mu^{\mathsf{lik}}_{k,t})$ is magnified. This is one way to promote adaptation by giving more relevance to new evidence and depressing the convictions arising from the past. The extreme case $\delta = 1$ would correspond to $\psi_{k,t} = \mu^{\mathsf{lik}}_{k,t}$, i.e., to a social learning algorithm that throws away the past information at each time instant. Such algorithm would push adaptation to the limit, in the sense that the beliefs at time t would depend only on the data observed at time t, without exploiting more fully the information collected over previous time instants. In the next section we will comment more closely on the choice of δ and its impact on the trade-off between learning performance and adaptation capacity.

It is possible to solve (8.4) and obtain a closed-form expression for the intermediate belief vector. Indeed, through straightforward manipulations we can write

$$
\begin{aligned}
&(1-\delta)D(p||\mu^{\mathsf{Bu}}_{k,t}) + \delta D(p||\mu^{\mathsf{lik}}_{k,t}) \\
&= (1-\delta)\sum_{\theta\in\Theta} p(\theta)\log\frac{p(\theta)}{\mu_{k,t-1}\ell(x_{k,t}|\theta)} \\
&+ \delta\sum_{\theta\in\Theta} p(\theta)\log\frac{p(\theta)}{\ell(x_{k,t}|\theta)} + \text{const.} \\
&= \sum_{\theta\in\Theta} p(\theta)\log\left(\frac{p(\theta)}{\mu_{k,t-1}\ell(x_{k,t}|\theta)}\right)^{1-\delta} \\
&+ \sum_{\theta\in\Theta} p(\theta)\log\left(\frac{p(\theta)}{\ell(x_{k,t}|\theta)}\right)^{\delta} + \text{const.} \\
&= \sum_{\theta\in\Theta} p(\theta)\log\frac{p(\theta)}{\mu^{1-\delta}_{k,t-1}\ell(x_{k,t}|\theta)} + \text{const.} \\
&= \underbrace{\sum_{\theta\in\Theta} p(\theta)\log\frac{p(\theta)}{\dfrac{\mu^{1-\delta}_{k,t-1}(\theta)\ell(x_{k,t}|\theta)}{\sum\limits_{\theta'\in\Theta}\mu^{1-\delta}_{k,t-1}(\theta')\ell(x_{k,t}|\theta')}}}_{\text{KL divergence}} + \text{const.}
\end{aligned}
\tag{8.5}
$$

The constant terms collect quantities that do not depend on p. According to (8.5), we can nullify the final KL divergence and, hence, minimize the cost function in (8.4), with the unique choice

$$
\psi_{k,t}(\theta) = \frac{\mu^{1-\delta}_{k,t-1}(\theta)\ell(x_{k,t}|\theta)}{\sum_{\theta'\in\Theta}\mu^{1-\delta}_{k,t-1}(\theta')\ell(x_{k,t}|\theta')}. \tag{8.6}
$$

Compared with (8.1), we now see that the past belief $\mu_{k,t-1}(\theta)$ is raised to the power $1-\delta$.

8.2.2 Adaptive Update: Second Approach

We can motivate the same construction (8.6) by following an alternative argument. Referring back to (2.72), we know that the Bayesian update (8.1) for the intermediate belief is also the result of solving the following optimization problem

$$\psi_{k,t} = \mu_{k,t}^{\mathsf{Bu}} = \arg\min_{p\in\Delta_H} \left\{ H(p,\mu_{k,t-1}) + D(p||\mu_{k,t}^{\mathsf{lik}}) \right\} \tag{8.7}$$

formulated in terms of: *i)* the cross-entropy $H(p,\mu_{k,t-1})$ between the candidate belief vector p and the past belief vector $\mu_{k,t-1}$; and *ii)* the KL divergence $D(p||\mu_{k,t}^{\mathsf{lik}})$ between p and the "likelihood" posterior $\mu_{k,t}^{\mathsf{lik}}$ in (8.2). Again, in order to endow this construction with an adaptation ability, we incorporate weighting and modify (8.7) into

$$\psi_{k,t} = \arg\min_{p\in\Delta_H} \left\{ (1-\delta)H(p,\mu_{k,t-1}) + D(p||\mu_{k,t}^{\mathsf{lik}}) \right\}, \tag{8.8}$$

with $0<\delta<1$. As was the case before, this choice for the weighting allows us to recover the traditional Bayesian update (8.1) when $\delta \to 0$, and the limiting solution $\psi_{k,t} = \mu_{k,t}^{\mathsf{lik}}$ when $\delta = 1$. As δ moves away from zero, the cross-entropy term that incorporates the past information (through the past belief vector $\mu_{k,t-1}$) is given progressively less importance. In this way, we enhance the role of the new information, which is incorporated into the KL divergence involving the belief $\mu_{k,t}^{\mathsf{lik}}$ (which depends solely on the new data). We can also solve (8.8) in closed form by means of the following manipulations:

$$\begin{aligned}
&(1-\delta)H(p,\mu_{k,t-1}) + D(p||\mu_{k,t}^{\mathsf{lik}}) \\
&\quad= (1-\delta)\sum_{\theta\in\Theta} p(\theta)\log\frac{1}{\mu_{k,t-1}(\theta)} + \sum_{\theta\in\Theta} p(\theta)\log\frac{p(\theta)}{\mu_{k,t}^{\mathsf{lik}}(\theta)} \\
&\quad= (1-\delta)\sum_{\theta\in\Theta} p(\theta)\log\frac{1}{\mu_{k,t-1}(\theta)} + \sum_{\theta\in\Theta} p(\theta)\log\frac{p(\theta)}{\ell(x_{k,t}|\theta)} + \text{const.} \\
&\quad= \sum_{\theta\in\Theta} p(\theta)\log\frac{1}{\mu_{k,t-1}^{1-\delta}(\theta)} + \sum_{\theta\in\Theta} p(\theta)\log\frac{p(\theta)}{\ell(x_{k,t}|\theta)} + \text{const.} \\
&\quad= \sum_{\theta\in\Theta} p(\theta)\log\frac{p(\theta)}{\mu_{k,t-1}^{1-\delta}(\theta)\ell(x_{k,t}|\theta)} + \text{const.}
\end{aligned}$$

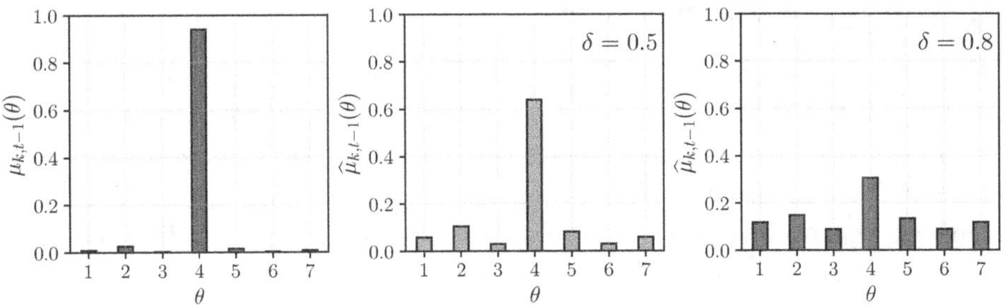

Figure 8.2: An example illustrating why (8.10) is a flattened version of $\mu_{k,t-1}$.

$$= \underbrace{\sum_{\theta\in\Theta} p(\theta)\log\frac{p(\theta)}{\frac{\mu_{k,t-1}^{1-\delta}(\theta)\ell(x_{k,t}|\theta)}{\sum_{\theta'\in\Theta}\mu_{k,t-1}^{1-\delta}(\theta')\ell(x_{k,t}|\theta')}}}_{\text{KL divergence}} + \text{const.} \tag{8.9}$$

It follows that the optimal solution to (8.8) coincides with (8.6).

8.2.3 Interpretation as a Bayesian Update

It is possible to show that (8.6) corresponds to a *Bayesian* update applied to a modified prior. To this end, we decompose (8.6) into two steps. First, agent k uses the past belief $\mu_{k,t-1}(\theta)$ to construct a new belief as follows:

$$\widehat{\mu}_{k,t-1}(\theta) = \frac{\mu_{k,t-1}^{1-\delta}(\theta)}{\sum\limits_{\theta'\in\Theta}\mu_{k,t-1}^{1-\delta}(\theta')}, \tag{8.10}$$

where the normalization is meant to ensure that $\widehat{\mu}_{k,t-1}$ is a probability vector. Second, agent k applies Bayes' rule by taking as the prior the *modified* belief vector $\widehat{\mu}_{k,t-1}$, yielding

$$\psi_{k,t}(\theta) = \frac{\widehat{\mu}_{k,t-1}(\theta)\ell_k(x_{k,t}|\theta)}{\sum\limits_{\theta'\in\Theta}\widehat{\mu}_{k,t-1}(\theta')\ell_k(x_{k,t}|\theta')}. \tag{8.11}$$

These two steps combined are equivalent to (8.6). The exponentiation and normalization in (8.10) has the physical meaning of *flattening* the belief vector, i.e., of making it more uniform across θ, as shown in Figure 8.2 for two values of δ. In this way, if an agent had a particularly peaked belief around a certain hypothesis, perhaps due to a bias accumulated over time, flattening the belief helps give more credit to new data.

8.2.4 Adaptive Social Learning

Referring back to Figure 3.3, we can now use the adaptive rule (8.6) in the *general update* block, in place of the traditional Bayesian update that was used before. For the combination rule we focus on geometric averaging. The resulting social learning algorithm is detailed in listing (8.13). By grouping the update and combination steps, the overall belief evolution can be represented in the following compact form:

$$\mu_{k,t}(\theta) \propto \prod_{j\in\mathcal{N}_k} \left[\mu_{j,t-1}^{1-\delta}(\theta)\ell(x_{j,t}|\theta)\right]^{a_{jk}}. \tag{8.12}$$

This recursion replaces (5.2) and is referred to as the *adaptive social learning* (ASL) strategy. It was originally proposed in [25], where it is also possible to find an alternative form for the update step that will be discussed later in Section 8.5.

Adaptive social learning (ASL)

start from the prior belief vectors $\mu_{k,0}$ for $k = 1, 2, \ldots, K$
choose an adaptation parameter δ, with $0 < \delta < 1$
for $t = 1, 2, \ldots$
  for $k = 1, 2, \ldots, K$
    agent k observes $x_{k,t}$
    for $\theta = 1, 2, \ldots, H$

$$\psi_{k,t}(\theta) = \frac{\mu_{k,t-1}^{1-\delta}(\theta)\ell_k(x_{k,t}|\theta)}{\sum\limits_{\theta'\in\Theta} \mu_{k,t-1}^{1-\delta}(\theta')\ell_k(x_{k,t}|\theta')} \qquad \text{(self-learning)}$$

    end
  end

  for $k = 1, 2, \ldots, K$
    for $\theta = 1, 2, \ldots, H$

$$\mu_{k,t}(\theta) = \frac{\prod_{j\in\mathcal{N}_k}[\psi_{j,t}(\theta)]^{a_{jk}}}{\sum\limits_{\theta'\in\Theta}\prod_{j\in\mathcal{N}_k}[\psi_{j,t}(\theta')]^{a_{jk}}} \qquad \text{(cooperation)}$$

    end
  end
end

(8.13)

8.3 Learning versus Adaptation

In the social learning framework discussed in Chapter 5, a stationary setting is assumed, where the streaming observations collected by the agents are generated from fixed models $\{f_k\}$, and all relevant system attributes (e.g., network topology, likelihood models) do not change over time. The goal of the social learning strategy was to maximize the belief about the target model $\vartheta^\star$ that provides the best explanation for the data. In comparison, under an *adaptive* setting where the system attributes can change over time, the learning algorithm now needs to satisfy at least two requirements. While effective learning (i.e., convergence) must be guaranteed under stationary conditions, it is also critical to guarantee adaptation (i.e., tracking) under drifting conditions, such as drifting of $\vartheta^\star$. Therefore, an *adaptive* social learning algorithm should allow agents to react more readily to these drifts and start learning under the new conditions, within a tolerable reaction time.

The trade-off between learning and adaptation translates into a trade-off between steady-state performance (how *well* an algorithm learns) and convergence rate (how *fast* it learns during its transient phase). As is typical of adaptive strategies, an algorithm with faster convergence properties is able to track better albeit at the expense of worse learning performance. A systematic analysis of the learning/adaptation trade-off requires us to define more formally the concepts of learning and adaptation, and to develop a proper technical framework in order to quantify this trade-off.

Learning. In our context, "learning" means "guessing the right model" after sufficient time. As we have explained in the previous chapters, the right model is formally identified by a target hypothesis $\vartheta^\star$ that minimizes a suitable cost function providing a degree of fitting between the data (i.e., the underlying true generative models) and the likelihood models employed by the agents. The learning performance of a social learning algorithm is assessed by means of a *steady-state* analysis, where the statistical conditions are assumed to remain stationary and the amount of data is sufficiently large to neglect transient effects related to the initial state. The analysis then focuses on evaluating the probability that an agent guesses the target hypothesis $\vartheta^\star$.

In Chapter 5 we showed that traditional social learning with geometric averaging (see listing (3.16)) enables all agents to learn the desired target with *vanishing* error probability. We will see that this is not the case for the ASL strategy. A residual error probability remains over time, and it depends on δ. The analysis in Chapter 9 will quantify the size of this error.

Adaptation. When the generative models or other system attributes change during the learning process, the social learning algorithm will need to react to these drifts so as to guarantee proper learning and tracking under the new conditions. The adaptation ability of the algorithm will be measured by how long it takes to reach the steady-state regime corresponding to these new conditions. This stage of the algorithm is usually referred to as the *transient phase*, and the time to reach the steady-state regime is called the *adaptation time*. A detailed analysis of the transient phase of the ASL strategy will be carried out in Chapter 10.

8.4 Adaptive Setting

In Chapters 5 and 6 we characterized the learning behavior of *nonadaptive* social learning with geometric averaging. Here and in the next two chapters, we will be dealing with the learning behavior of the ASL strategy. The derivations in the aforementioned chapters (see for example Theorem 5.1) exploited the recursive form in (5.11), whose converging behavior was established by applying the strong law of large numbers. Unfortunately, for the ASL strategy in (8.12), the introduction of the *adaptation parameter* δ changes completely the picture. In fact, we will see that the beliefs $\boldsymbol{\mu}_{k,t}(\theta)$ will no longer converge to deterministic values (such as 1 or 0) as $t \to \infty$. They will converge instead to *random* variables (we will see later that such randomness plays a critical role in enabling adaptation and tracking). The characterization of these random variables and the associated error probability will be demanding. We start by defining the observational model used to study adaptive social learning.

Definition 8.1 (Observational model for adaptive social learning). Assume that the ASL algorithm in listing (8.13) has been running until a certain time t_0, after which the system conditions (e.g., the true model) change.[2] Then, adaptation is quantified by characterizing the transient phase, starting at $t_0 + 1$, that the system undergoes before reaching the steady state. To examine the steady-state behavior (as $t \to \infty$), we assume that from $t_0 + 1$ onward the system remains stationary. Specifically, each agent $k = 1, 2, \dots, K$ at time $t = t_0 + 1, t_0 + 2, \dots$ receives a data sample $\boldsymbol{x}_{k,t}$. The collections of K samples across the agents, $\{\boldsymbol{x}_{1,t}, \boldsymbol{x}_{2,t}, \dots, \boldsymbol{x}_{K,t}\}$, are assumed iid over time. The probability (density or mass) function of $\boldsymbol{x}_{k,t}$ is denoted by f_k. To perform social learning, agent k employs likelihood models $\{\ell_{k,\theta}\}_{\theta \in \Theta}$ of the same nature as f_k (namely, for all $\theta \in \Theta$, $\ell_{k,\theta}$ is a pdf if f_k is a pdf, and a pmf otherwise).

Due to the structure of the recursion in (8.12), to examine it from $t_0 + 1$ onward we do not need all the past belief vectors, but only the belief vectors $\{\mu_{k,t_0}\}_{k=1}^{K}$. Moreover, by examining (8.12), and from the same argument used in the proof of Theorem 5.1, it is straightforward to see that if the algorithm is initialized with a belief vector placing nonzero mass on all $\theta \in \Theta$, the belief will remain nonzero at any θ during the algorithm evolution (with probability 1). This implies that the condition of positive initial beliefs that we have been using so far can be translated into the assumption of positive beliefs at t_0. For convenience of notation and without loss of generality, in the following analysis we set $t_0 = 0$.

8.4.1 Steady-State Error Probabilities

In Chapter 5 we were able to establish exact convergence, as $t \to \infty$, of the traditional social learning strategy (5.2) to the target hypothesis $\vartheta^\star$. Now, because of the adaptation requirement, even when $t \to \infty$ there will be a nontrivial probability of arriving at an erroneous decision. Accordingly, we need to introduce an error probability that will be useful: *i)* in the steady-state analysis, to quantify the learning performance; and *ii)* in the transient analysis, to measure the adaptation time.[3]

We thus introduce the *steady-state* error probability

$$p_k(\delta) \triangleq \lim_{t \to \infty} p_{k,t}, \tag{8.14}$$

where we made explicit the dependence on δ of the limiting probability since in the following we will examine its behavior as $\delta \to 0$. There are two fundamental questions related to the concept of steady-state error probability. The first question regards its *existence*, which in principle is not guaranteed. Theorem 9.1 will provide an affirmative answer to this question by characterizing the steady-state behavior of the log belief ratios. The second question regards the *evaluation* of $p_k(\delta)$. An exact evaluation is generally a formidable task. Therefore, to tackle this critical problem, in Chapter 9 we will perform an asymptotic analysis in the regime of small δ.

---

[2]Actually, for our analysis to hold, it is not required that a change occurs at $t_0 + 1$. In other words, t_0 can be any arbitrary time instant. However, in adaptive social learning we are mainly interested in examining what happens after a change.

[3]Other metrics to quantify learning and adaptation are possible. For example, in the theory of quickest detection, usual metrics are the rate of false alarms (which in our setting can be connected to the average number of samples between mistakenly chosen hypotheses in steady state), and the expected time to detect a change (which in our setting can be connected to the adaptation time) [14, 141, 163].

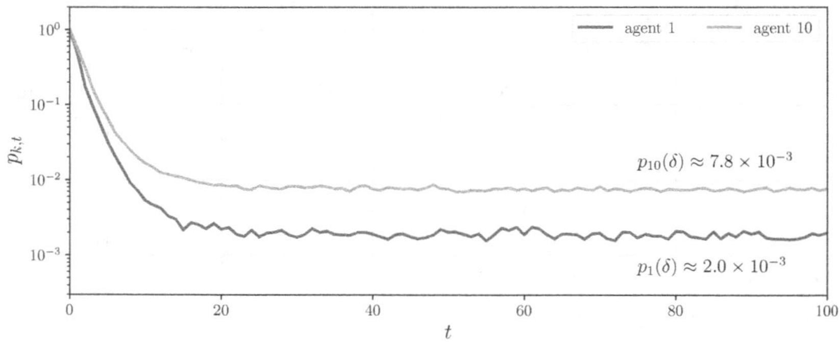

Figure 8.3: Illustrative example showing the evolution of the error probability of two agents in a network running the ASL algorithm.

In Figure 8.3 we show an example of evolution for the error probability $p_{k,t}$ (estimated empirically via Monte Carlo simulation) of two agents in a network implementing the ASL strategy (8.12). We see that the instantaneous error probability $p_{k,t}$ converges toward a steady-state *nonzero* value $p_k(\delta)$ as t increases. It is useful to remark that this behavior is different from that of traditional social learning studied in Chapter 5 where, *under stationary conditions*, the error probability of each agent was shown to vanish as time elapses. This is one instance of the learning/adaptation trade-off: Nonadaptive strategies can increase their accuracy indefinitely under stationary conditions. However, astronomically low values of the error probabilities lead to a detrimental inertia in responding to nonstationary conditions.

8.5 Variation on ASL

Another adaptive rule can be obtained in lieu of (8.6) by revisiting the information-theoretic approach used in Section 8.2.1. The modification consists of replacing the *Bayesian* update $\mu_{k,t}^{\mathsf{Bu}}$ that was used in the optimization problem (8.4), with the *previous-lag* belief $\mu_{k,t-1}$, yielding

$$\psi_{k,t} = \arg\min_{p \in \Delta_H} \left\{ (1-\delta) D(p||\mu_{k,t-1}) + \delta D(p||\mu_{k,t}^{\mathsf{lik}}) \right\}, \tag{8.15}$$

where, as usual, $0 < \delta < 1$. Note that, while for $\delta \to 0$ the adaptive rule arising from (8.4) led back to the traditional Bayesian update, the alternative rule (8.15) will instead ignore the new data and stick to the old belief vector $\mu_{k,t-1}$.

We can solve (8.15) to get a closed-form expression for the intermediate belief vector. To this end, we manipulate the cost function in (8.15) as follows:

$$
\begin{aligned}
&(1-\delta)D(p||\mu_{k,t-1}) + \delta D(p||\mu_{k,t}^{\mathsf{lik}}) \\
&= (1-\delta)\sum_{\theta\in\Theta} p(\theta)\log\frac{p(\theta)}{\mu_{k,t-1}(\theta)} + \delta\sum_{\theta\in\Theta} p(\theta)\log\frac{p(\theta)}{\ell(x_{k,t}|\theta)} + \text{const.} \\
&= \sum_{\theta\in\Theta} p(\theta)\log\left(\frac{p(\theta)}{\mu_{k,t-1}(\theta)}\right)^{1-\delta} + \sum_{\theta\in\Theta} p(\theta)\log\left(\frac{p(\theta)}{\ell(x_{k,t}|\theta)}\right)^{\delta} + \text{const.} \\
&= \sum_{\theta\in\Theta} p(\theta)\log\frac{p(\theta)}{\mu_{k,t-1}^{1-\delta}(\theta)\ell^{\delta}(x_{k,t}|\theta)} + \text{const.} \\
&= \underbrace{\sum_{\theta\in\Theta} p(\theta)\log\frac{p(\theta)}{\dfrac{\mu_{k,t-1}^{1-\delta}(\theta)\ell^{\delta}(x_{k,t}|\theta)}{\sum_{\theta'\in\Theta}\mu_{k,t-1}^{1-\delta}(\theta')\ell^{\delta}(x_{k,t}|\theta')}}}_{\text{KL divergence}} + \text{const.} \qquad (8.16)
\end{aligned}
$$

and we conclude that the cost function is minimized by the choice

$$\psi_{k,t}(\theta) \propto \mu_{k,t-1}^{1-\delta}(\theta)\ell^{\delta}(x_{k,t}|\theta). \qquad (8.17)$$

Compare now (8.17) with (8.6). Both rules discount the past belief $\mu_{k,t-1}(\theta)$ by raising it to the power $1-\delta$ (recall that $0<\delta<1$). The fundamental difference is that in (8.17) the likelihood $\ell(x_{k,t}|\theta)$ is also discounted, since it is raised to the power δ. Note that, while (as was explained in Section 8.2.3) Eq. (8.6) can be interpreted as a *Bayesian* update with modified prior, this is no longer true for (8.17). This is because the integral with respect to x of the likelihood exponentiated to δ is not equal to 1. We will examine more closely these aspects later in Example 9.2, where we will discover that the ASL update rule (8.6) and its variation (8.17) exhibit an interesting commonality as well as an important distinguishing feature. They will be shown to be equivalent in terms of decisions (i.e., the hypothesis maximizing the beliefs will ultimately be the same under both strategies), but the credibility assigned to the hypotheses (i.e., the values of the belief-vector entries) can be very different under the two strategies.

Interpretation as a diffusion strategy. Consider now the adaptive rule (8.17), followed by the geometric-averaging rule

$$\mu_{k,t}(\theta) \propto \prod_{j\in\mathcal{N}_k} [\psi_{j,t}(\theta)]^{a_{jk}}. \qquad (8.18)$$

It is useful to rewrite (8.17) and (8.18) in terms of the log belief ratios in (6.11) and the log likelihood ratios in (6.3), yielding

$$\log \frac{\psi_{k,t}(\vartheta^\star)}{\psi_{k,t}(\theta)} = (1-\delta)\beta_{k,t-1}(\theta) + \delta\, \lambda_{k,t}(\theta), \tag{8.19a}$$

$$\beta_{k,t}(\theta) = \sum_{j=1}^{K} a_{jk} \log \frac{\psi_{j,t}(\vartheta^\star)}{\psi_{j,t}(\theta)}. \tag{8.19b}$$

The iterative algorithm described by (8.19a) and (8.19b) is in the form of a standard *diffusion* algorithm (of the adapt-then-combine type) with constant step-size δ [151, 152, 155]. In particular, the RHS of (8.19a) can be rewritten as

$$\beta_{k,t-1}(\theta) - \delta \underbrace{\left(\beta_{k,t-1} - \lambda_{k,t}(\theta)\right)}_{\text{stochastic gradient}}, \tag{8.20}$$

which is an iteration of a stochastic gradient descent algorithm with step-size δ and instantaneous risk function at agent k given by

$$J_k(\beta) = \frac{1}{2}\,\mathbb{E}\left[\left(\beta - \boldsymbol{\lambda}_{k,t}(\theta)\right)^2\right], \tag{8.21}$$

where β is the (scalar) optimization variable and $\boldsymbol{\lambda}_{k,t}(\theta)$ is the random input variable. It is worth mentioning that, in the context of estimation and detection, the *single-agent* version of (8.19a) and (8.19b) is also known as exponentially-weighted-moving-average (EWMA) control chart or geometric-moving-average (GMA) control chart [143]. This terminology arises because, by recursive application of the weight $(1-\delta)$, the same data is assigned an overall weight that changes over time, specifically decaying with the exponential (or geometric) law $(1-\delta)^t$.

Chapter 9

Learning Accuracy under ASL

The adaptation properties of the ASL strategy (8.13) are enabled by a learning mechanism that is fundamentally different from that of traditional social learning. To see why, let us assume that the system conditions remain stable for a sufficiently long time interval. With traditional social learning, the belief assigned to the target hypothesis $\vartheta^\star$ will converge to 1 as $t \to \infty$. As we have already observed, such assignment of full credibility has the downside that, in the face of drifting conditions, the algorithm becomes stubborn and remains stuck in its past determination for a long time before moving on to track the changes.

In contrast, we will see that for the ASL strategy the beliefs will *not* converge as time elapses: They will fluctuate indefinitely, exhibiting a random behavior including in steady state. This everlasting randomness is critical to ensure that the algorithm will adapt quickly to a change in the environment. This is because the random fluctuations keep the algorithm more dynamic, preventing it from being "trapped" into a conviction arising from past data. This behavior is commonly encountered in the theory of stochastic optimization when one employs stochastic gradient algorithms with constant step-size. In this context, when the optimizers drift over time, random fluctuations help the algorithm move away from a current stationary point and start tracking the new optimizer [151, 154, 155]. On the technical side, however, the random character preserved by the belief even when $t \to \infty$, makes the steady-state analysis significantly more challenging.

In order to carry out a meaningful steady-state analysis, the fundamental preliminary step becomes to establish whether the random belief fluctuations allow the belief vectors to reach a steady state as $t \to \infty$, in the sense that they converge to some limiting random vectors with a fixed distribution. We will ascertain that this is the case. Once this fact is established, the learning performance will then be assessed by examining the *statistical* behavior of the beliefs in steady state. We will provide an accurate characterization of such statistical behavior in the regime of small adaptation parameters, i.e., by performing an asymptotic analysis as $\delta \to 0$. Under this regime, we will show that the steady-state belief vector places unit mass on $\vartheta^\star$ with probability converging to 1 as $\delta \to 0$. The properties of this convergence will be characterized in detail through an asymptotic normality result and a large deviation analysis.

In the next chapter, we will furthermore characterize the transient performance by obtaining closed-form relations that reveal how the adaptation time grows with smaller δ. When all is said and done, the analysis will reveal that the well-known learning/adaptation trade-off from classic learning theory [155] continues to exist under social learning: Smaller (resp., larger) values of δ imply higher (resp., lower) learning accuracy and slower (resp., faster) adaptation or response time.

Before proceeding with the analysis, we remark that the relevant descriptors (e.g., log likelihood ratios, log belief ratios) and the pertinent notation have been introduced in Chapter 6. In particular, the average variable $\boldsymbol{\lambda}_{\mathsf{net},t}$ in (6.8), which played a critical role in the performance of traditional social learning, will be seen to play an equally important role to characterize the steady-state ($t \to \infty$) performance of the ASL strategy in the regime of small adaptation parameters ($\delta \to 0$).

9.1 Steady-State Analysis

We start by examining the evolution of the log belief ratios. Exploiting (8.12), (6.11), and (6.3), we can readily establish the following recursion, for all $\theta \neq \vartheta^\star$ (compare with (6.27)):

$$\boldsymbol{\beta}_{k,t}(\theta) = \sum_{j \in \mathcal{N}_k} a_{jk} \Big[(1-\delta)\boldsymbol{\beta}_{j,t-1}(\theta) + \boldsymbol{\lambda}_{j,t}(\theta)\Big]. \qquad (9.1)$$

This recursion can be unfolded to obtain

$$\beta_{k,t}(\theta) = \underbrace{(1-\delta)^t \sum_{j=1}^{K} [A^t]_{jk} \beta_{j,0}(\theta)}_{\text{transient term}} + \sum_{\tau=1}^{t} \sum_{j=1}^{K} (1-\delta)^{\tau-1} [A^\tau]_{jk} \, \boldsymbol{\lambda}_{j,t-\tau+1}(\theta), \tag{9.2}$$

where we recall that $A = [a_{jk}]$ denotes the left stochastic combination matrix.

The goal of the steady-state analysis is to examine the learning behavior of the algorithm for large t. To this end, in the next theorem we start by establishing that the log belief ratio vector $\boldsymbol{\beta}_{k,t}$ converges *in distribution*, as $t \to \infty$, to a certain steady-state *random* vector $\boldsymbol{\beta}_k$. This means that the probability distribution of $\boldsymbol{\beta}_{k,t}$ converges to the probability distribution of $\boldsymbol{\beta}_k$ — see Definition D.4. As a notational remark, we will be denoting steady-state variables (i.e., limiting variables obtained as $t \to \infty$) by omitting the subscript t from the corresponding non-asymptotic notation.

Theorem 9.1 (Steady-state log belief ratios). Let Assumptions 5.1, 5.2, and 6.1 be satisfied.[1] Then, for $k = 1, 2, \ldots, K$, the vector of log belief ratios $\boldsymbol{\beta}_{k,t}$ converges in distribution as $t \to \infty$ to a random vector $\boldsymbol{\beta}_k$:

$$\boldsymbol{\beta}_{k,t} \xrightarrow[t\to\infty]{\mathrm{d}} \boldsymbol{\beta}_k. \tag{9.3}$$

Furthermore, the entries of $\boldsymbol{\beta}_k$ are given by the random variables

$$\boldsymbol{\beta}_k(\theta) \triangleq \sum_{j=1}^{K} \sum_{\tau=1}^{\infty} (1-\delta)^{\tau-1} [A^\tau]_{jk} \, \boldsymbol{\lambda}_{j,\tau}(\theta), \qquad \theta \neq \vartheta^\star, \tag{9.4}$$

where each of the inner series is (almost surely) absolutely convergent.

Proof. We are interested in characterizing, for each agent k, the asymptotic behavior of the random vector $\boldsymbol{\beta}_{k,t}$ as $t \to \infty$. In particular, we want to establish that it converges in distribution. In view of (9.2), the log belief vector $\boldsymbol{\beta}_{k,t}$ can be written as

$$\boldsymbol{\beta}_{k,t} = (1-\delta)^t \sum_{j=1}^{K} [A^t]_{jk} \beta_{j,0}(\theta) + \widehat{\boldsymbol{\beta}}_{k,t}, \tag{9.5}$$

[1]We remark that, in the proof of this theorem, we only use the definition of $\vartheta^\star$ from Assumption 6.1, but we do not need the fact that the graph is primitive. For this reason, the result of the theorem still holds if we replace the primitive graph with a connected graph in Assumption 6.1.

where, for $\theta \neq \vartheta^\star$, the entries of the vector $\widehat{\boldsymbol{\beta}}_{k,t}$ are defined as

$$\widehat{\boldsymbol{\beta}}_{k,t}(\theta) \triangleq \sum_{j=1}^{K}\sum_{\tau=1}^{t}(1-\delta)^{\tau-1}[A^{\mathsf{T}}]_{jk}\,\boldsymbol{\lambda}_{j,t-\tau+1}(\theta). \tag{9.6}$$

Since the first term on the RHS of (9.5) converges (deterministically) to 0 as $t \to \infty$, in view of the vector version of Slutsky's theorem (see (D.39)), to establish (9.3) it is sufficient to prove that

$$\widehat{\boldsymbol{\beta}}_{k,t} \xrightarrow[t\to\infty]{\mathrm{d}} \boldsymbol{\beta}_k. \tag{9.7}$$

We will now establish that (9.7) holds.

Let us start by observing from (9.6) that $\widehat{\boldsymbol{\beta}}_{k,t}$ can be represented as

$$\widehat{\boldsymbol{\beta}}_{k,t} = g_{k,t,\delta}\Big(\{\boldsymbol{\lambda}_{j,1}\}_{j=1}^{K}, \{\boldsymbol{\lambda}_{j,2}\}_{j=1}^{K}, \ldots, \{\boldsymbol{\lambda}_{j,t}\}_{j=1}^{K}\Big) \tag{9.8}$$

to highlight that the random vector $\boldsymbol{\beta}_{k,t}$ is a certain function $g_{k,t,\delta}$ of the log likelihood ratios $\{\boldsymbol{\lambda}_{j,1}\}_{j=1}^{K}, \{\boldsymbol{\lambda}_{j,2}\}_{j=1}^{K}, \ldots, \{\boldsymbol{\lambda}_{j,t}\}_{j=1}^{K}$, collected from all agents up to time t. Consider now the vector $\boldsymbol{\beta}_{k,t}^{\leftarrow}$, whose θth entry is given by

$$\boldsymbol{\beta}_{k,t}^{\leftarrow}(\theta) = \sum_{j=1}^{K}\sum_{\tau=1}^{t}(1-\delta)^{\tau-1}[A^{\mathsf{T}}]_{jk}\,\boldsymbol{\lambda}_{j,\tau}(\theta), \tag{9.9}$$

which corresponds to (9.6) with the log likelihood ratios $\boldsymbol{\lambda}_{j,t-\tau+1}(\theta)$ taken in reverse order. In view of (9.8), this means that we can write

$$\boldsymbol{\beta}_{k,t}^{\leftarrow} = g_{k,t,\delta}\Big(\{\boldsymbol{\lambda}_{j,t}\}_{j=1}^{K}, \{\boldsymbol{\lambda}_{j,t-1}\}_{j=1}^{K}, \ldots, \{\boldsymbol{\lambda}_{j,1}\}_{j=1}^{K}\Big). \tag{9.10}$$

However, since the data are iid over time, the reverse ordering does not alter the *distribution* of the resulting random vector, which means that

$$\widehat{\boldsymbol{\beta}}_{k,t} \stackrel{\mathrm{d}}{=} \boldsymbol{\beta}_{k,t}^{\leftarrow}, \tag{9.11}$$

where $\stackrel{\mathrm{d}}{=}$ denotes equality in distribution. Accordingly, since $\widehat{\boldsymbol{\beta}}_{k,t}$ and $\boldsymbol{\beta}_{k,t}^{\leftarrow}$ share the same distribution for all t, to establish (9.7) it suffices to establish that

$$\boldsymbol{\beta}_{k,t}^{\leftarrow} \xrightarrow[t\to\infty]{\mathrm{d}} \boldsymbol{\beta}_k. \tag{9.12}$$

In view of Lemma F.3, each of the K inner partial sums in (9.9) converges *almost surely* to the random variable defined by the series

$$\sum_{\tau=1}^{\infty}(1-\delta)^{\tau-1}[A^{\mathsf{T}}]_{jk}\,\boldsymbol{\lambda}_{j,\tau}(\theta), \tag{9.13}$$

which, in particular, according to Lemma F.3 is almost surely an *absolutely convergent* series. Note that the assumptions of Lemma F.3 are met because the random variables $\boldsymbol{\lambda}_{j,\tau}(\theta)$ have finite first moment in view of (5.5), and the weights $[A^{\mathsf{T}}]_{jk}$ are nonnegative and bounded by 1. Summing (9.13) over $j = 1, 2, \ldots K$, we have in fact shown that

$$\boldsymbol{\beta}_{k,t}^{\leftarrow} \xrightarrow[t\to\infty]{\mathrm{a.s.}} \boldsymbol{\beta}_k \tag{9.14}$$

where the θth entry of the limiting random vector $\boldsymbol{\beta}_k$ is given by

$$\boldsymbol{\beta}_k(\theta) = \sum_{j=1}^{K} \sum_{\tau=1}^{\infty} (1-\delta)^{\tau-1} [A^{\tau}]_{jk} \, \boldsymbol{\lambda}_{j,\tau}(\theta). \tag{9.15}$$

Since almost-sure convergence implies convergence in distribution, in view of (9.11) we obtain (9.12), which completes the proof.

■

As a corollary of Theorem 9.1, we characterize the steady-state belief vector.

Corollary 9.1 (Steady-state belief vector). Under the same assumptions used in Theorem 9.1, for $k = 1, 2, \dots, K$ the belief vector $\boldsymbol{\mu}_{k,t}$ from (8.12) converges in distribution as $t \to \infty$ to a steady-state belief vector $\boldsymbol{\mu}_k$:

$$\boldsymbol{\mu}_{k,t} \xrightarrow[t\to\infty]{\mathrm{d}} \boldsymbol{\mu}_k, \tag{9.16}$$

where the entries of $\boldsymbol{\mu}_k$ are defined as follows:

$$\boldsymbol{\mu}_k(\theta) = \begin{cases} \dfrac{e^{-\boldsymbol{\beta}_k(\theta)}}{1 + \sum\limits_{\theta' \neq \vartheta^\star} e^{-\boldsymbol{\beta}_k(\theta')}} & \text{if } \theta \neq \vartheta^\star, \\ \\ \dfrac{1}{1 + \sum\limits_{\theta' \neq \vartheta^\star} e^{-\boldsymbol{\beta}_k(\theta')}} & \text{if } \theta = \vartheta^\star. \end{cases} \tag{9.17}$$

Proof. By applying the continuous mapping theorem (Theorem D.3) to the belief vector defined by (6.13), we conclude that the convergence in (9.3) implies the convergence of $\boldsymbol{\mu}_{k,t}$ to the expressions in (9.17).

■

It is useful to provide some comments on Theorem 9.1. First, we have that the random series in (9.4) is (almost surely) absolutely convergent, which means that the steady-state random vector $\boldsymbol{\beta}_k$ can be meaningfully defined. For this convergence to hold, in Theorem 9.1 we did not need existence of second or higher-order moments of the log likelihood ratios $\boldsymbol{\lambda}_{k,t}(\theta)$. It was enough to assume finite mean, a condition guaranteed by (5.5) applied to (6.4).

Second, consider the random sums (9.6) and (9.9), namely,

$$\widehat{\boldsymbol{\beta}}_{k,t}(\theta) = \sum_{j=1}^{K} \sum_{\tau=1}^{t} (1-\delta)^{\tau-1} [A^{\mathsf{T}}]_{jk} \, \boldsymbol{\lambda}_{j,t-\tau+1}(\theta) \tag{9.18}$$

and

$$\boldsymbol{\beta}_{k,t}^{\leftarrow}(\theta) = \sum_{j=1}^{K} \sum_{\tau=1}^{t} (1-\delta)^{\tau-1} [A^{\mathsf{T}}]_{jk} \, \boldsymbol{\lambda}_{j,\tau}(\theta). \tag{9.19}$$

It is important to notice that (9.4) does not correspond to letting $t \to \infty$ in (9.18). Indeed, the series in (9.4) is obtained from (9.18) by first taking the summands indexed by $t - \tau + 1$ in *reverse* order and then letting $t \to \infty$. In other words, the series in (9.4) is obtained by considering the limiting value of $\boldsymbol{\beta}_{k,t}^{\leftarrow}(\theta)$.

To gain further insight, in the left panel of Figure 9.1, we display one realization of the random sums in (9.18) and (9.19), for $\theta = 2$. The random sum $\widehat{\boldsymbol{\beta}}_{k,t}(\theta)$, displayed with solid line, exhibits persistent random fluctuations as time elapses. In contrast, the random sum $\boldsymbol{\beta}_{k,t}^{\leftarrow}(\theta)$, displayed with dashed line, converges as time elapses; actually, it converges to the value $\boldsymbol{\beta}_{k}(\theta)$ defined by (9.4). The right panel shows a different realization of the two random sums. We see that the limiting value $\boldsymbol{\beta}_{k}(\theta)$ is different in the two panels, which emphasizes that this limiting value is random.

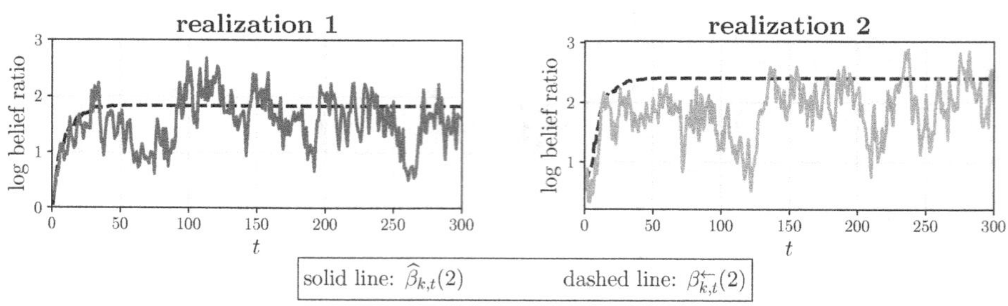

Figure 9.1: Illustrative curves showing a comparison of the random sequences $\boldsymbol{\beta}_{k,t}(\theta)$ and $\boldsymbol{\beta}_{k,t}^{\leftarrow}(\theta)$.

The profoundly different behavior of $\widehat{\boldsymbol{\beta}}_{k,t}(\theta)$ and $\boldsymbol{\beta}_{k,t}^{\leftarrow}(\theta)$ arises from the different ordering of the summands in (9.18) and (9.19). In particular, in (9.19) the most recent term, corresponding to $\boldsymbol{\lambda}_{j,t}(\theta)$, is scaled by the *smallest* weight $(1-\delta)^{t-1}$. As $t \to \infty$, this weight vanishes, and the series converges (almost surely). In contrast, in (9.18) the term $\boldsymbol{\lambda}_{j,t}(\theta)$ is scaled by the *highest* weight $(1-\delta)^0 = 1$, which does not vanish as $t \to \infty$, thus keeping fluctuations alive. These persistent random fluctuations imply that the agents will never converge to accept one hypothesis with full certainty. Making the agents more "doubtful" renders them more reactive to changes, enabling the adaptation mechanisms discussed in detail in the forthcoming analysis.

Even though the sums in (9.18) and (9.19) exhibit a markedly different behavior in terms of their time evolution (i.e., on the sample paths), Theorem 9.1 ensures that their *probability distributions* converge as $t \to \infty$ to the same distribution, namely, to the distribution of the limiting variable $\boldsymbol{\beta}_k(\theta)$. This equivalence can be explained as follows. Consider one of the panels in Figure 9.1, and focus on a sufficiently large t (say, $t = 300$). We see that the corresponding values $\widehat{\boldsymbol{\beta}}_{k,300}(\theta)$ and $\boldsymbol{\beta}^{\leftarrow}_{k,300}(\theta)$ are different from each other. However, if we now repeat the experiment in Figure 9.1 several times, the realizations of $\widehat{\boldsymbol{\beta}}_{k,300}(\theta)$ across different experiments will be distributed similarly to the realizations of $\boldsymbol{\beta}^{\leftarrow}_{k,300}(\theta)$.

The existence of a steady-state vector $\boldsymbol{\beta}_k$ to which $\boldsymbol{\beta}_{k,t}$ converges in distribution, makes the definition of a steady-state error probability meaningful. That is, along with the *instantaneous* error probability introduced in (6.22),

$$p_{k,t} = \mathbb{P}\left[\bigcup_{\theta \neq \vartheta^\star} \left\{\boldsymbol{\beta}_{k,t}(\theta) \leq 0\right\}\right], \tag{9.20}$$

we introduce the *steady-state* error probability (making explicit the dependence on δ is important for the following treatment)

$$p_k(\delta) \triangleq \mathbb{P}\left[\bigcup_{\theta \neq \vartheta^\star} \left\{\boldsymbol{\beta}_{k}(\theta) \leq 0\right\}\right]. \tag{9.21}$$

Now, we learned from Theorem 9.1 that the probability distribution of $\boldsymbol{\beta}_{k,t}$ converges to the probability distribution of the *steady-state* vector $\boldsymbol{\beta}_k$ and, hence, from (9.20),[2]

$$\lim_{t\to\infty} p_{k,t} = p_k(\delta). \tag{9.25}$$

[2]Actually, there is one subtlety that must be considered to infer (9.25) from the convergence in distribution of $\boldsymbol{\beta}_{k,t}$ to $\boldsymbol{\beta}_k$. Indeed, let $\mathcal{S} = \left\{z \in \mathbb{R}^{H-1} : z_1 > 0, z_2 > 0, \ldots, z_{H-1} > 0\right\}$ and observe that

$$1 - p_{k,t} = \mathbb{P}\left[\boldsymbol{\beta}_{k,t} \in \mathcal{S}\right]. \tag{9.22}$$

Then, Eq. (9.25) is equivalent to

$$\lim_{t\to\infty} \mathbb{P}\left[\boldsymbol{\beta}_{k,t} \in \mathcal{S}\right] = \mathbb{P}\left[\boldsymbol{\beta}_{k} \in \mathcal{S}\right]. \tag{9.23}$$

According to (D.15), Eq. (9.23) follows from the convergence in distribution (9.3), provided that $\mathbb{P}\left[\boldsymbol{\beta}_k \in \partial\mathcal{S}\right] = 0$, i.e., provided that the distribution of the limiting random vector $\boldsymbol{\beta}_k$ assigns zero probability to the boundary $\partial\mathcal{S}$ of the set $\mathcal{S}$. However, from Lemma F.2 we know that the

Before concluding this section, we remark that Theorem 9.1 constitutes only a first step toward the evaluation of the ASL performance, since it establishes only the existence of a steady-state error probability without providing any explicit form for it. Obtaining an analytical formula for the steady-state error probability is in general a formidable task. In the next sections we tackle this challenging problem by focusing on an asymptotic characterization of $\boldsymbol{\beta}_k$ in the regime of small δ.

9.2 Small-δ Regime

We will provide three types of asymptotic results, namely, a weak law of small adaptation parameters, an asymptotic normality result, and a large deviation analysis. This type of characterization was applied to *binary* adaptive detection in [119, 120, 123]. In this text we focus instead on *adaptive social learning*. The results that we are going to present were originally proved in [25] with reference to the objective evidence model from Section 5.3 and (regarding the asymptotic normality and the large deviations) under the assumption of statistically independent observations across the agents. Here we generalize these results by considering arbitrary true models $f_k(x)$ (see Definition 8.1) and by removing the independence assumption.

Weak law of small δ (Theorem 9.2). We will show that, for small δ, the *scaled* steady-state vector $\delta \times \boldsymbol{\beta}_k$ is concentrated on a deterministic quantity, namely, the mean $\bar{\lambda}_{\mathsf{net}}$ of the vector $\boldsymbol{\lambda}_{\mathsf{net},t}$ defined by (6.8). This concentration property will guarantee that, with high probability as $\delta \to 0$, the target hypothesis $\vartheta^\star$ is chosen by each agent. Moreover, we show that the steady-state belief about the target hypothesis converges to 1 as $\delta \to 0$. This result will require only finiteness of the first moments of the log likelihood ratios $\boldsymbol{\lambda}_{k,t}(\theta)$.

Asymptotic normality (Theorem 9.3). We will ascertain that the

---

series

$$\boldsymbol{\beta}_k(\theta) = \sum_{\tau=1}^{\infty} \sum_{j=1}^{K} (1-\delta)^{\tau-1} [A^\tau]_{jk} \, \boldsymbol{\lambda}_{j,\tau}(\theta) \tag{9.24}$$

is a continuous random variable if the random variables $\sum_{j=1}^{K} (1-\delta)^{\tau-1}[A^\tau]_{jk} \, \boldsymbol{\lambda}_{j,\tau}(\theta)$ are *not* deterministic from a certain τ onward. Since the case where these variables become deterministic appears to be pathological, we can safely assume that $\mathbb{P}[\boldsymbol{\beta}_k(\theta) = 0] = 0$ for all $\theta \neq \vartheta^\star$, which further implies $\mathbb{P}[\boldsymbol{\beta}_k \in \partial\mathcal{S}] = 0$.

steady-state log belief ratios (properly shifted and scaled) are asymptotically normal for small δ. From this property we will also construct a Gaussian approximation for the error probability $p_k(\delta)$ of each individual agent. For these results we assume finiteness of the variance of the log likelihood ratio $\boldsymbol{\lambda}_{k,t}(\theta)$. We remark that earlier results of asymptotic normality for adaptive distributed detection were established under the stronger requirement of finiteness of the third-order moment [119].

Large deviations (Theorem 9.4). We will characterize the exponential rate of decay of the error probability $p_k(\delta)$ as $\delta \to 0$. This result will require the existence of the moment generating function of the log likelihood ratios $\boldsymbol{\lambda}_{k,t}(\theta)$.

Notably, the above three results reflect perfectly a traditional path in asymptotic statistics [159, 166]. It is also interesting to note that the requirements in terms of finiteness of moments are the same that we encounter in the classic theorems, that is, first moments for the weak law of large numbers [159, 166], second moments for the central limit theorem [159, 166], and moment generating function for large deviations [59, 60]. However, in order to avoid misunderstanding, it is necessary to clarify one fundamental difference between our small-δ analysis and classic results. Let us refer, for example, to the asymptotic normality result. In the traditional setting considered in statistics, one examines the Gaussian behavior exhibited by sums of random variables when the number of terms of the sum goes to infinity. In contrast, the result in Theorem 9.3 does *not* affirm that the sums involved in (9.2) converge to a Gaussian random variable as $t \to \infty$. As a matter of fact, we have shown in Theorem 9.1 that these sums converge to some random variable $\boldsymbol{\beta}_k(\theta)$, but this variable is *not Gaussian*, in general. Theorem 9.3 deals instead with the behavior of the limiting variable $\boldsymbol{\beta}_k(\theta)$ as δ goes to 0. The same distinction applies to the other two asymptotic results, namely, the weak law of small adaptation parameters and the large deviation analysis. For this reason, as already explained in [123], the correct way to deal with the asymptotic regime of small δ in the adaptation context involves the following two steps:

- First, it is necessary to introduce a proper steady-state vector (i.e., the vector $\boldsymbol{\beta}_k$ in Theorem 9.1), which already embodies the effect of combining an infinite number of summands.

- Then, one needs to characterize the asymptotic behavior of $\boldsymbol{\beta}_k$ as δ goes to 0.

It is worth noting that, in the adaptation literature, the critical role of the first step is usually not emphasized. This is because the adaptation literature mostly focuses on regression/estimation problems, where one usually quantifies the performance by evaluating convergence of *moments* [151, 154]. In contrast, when dealing with social learning, we need to characterize the statistical descriptors, i.e., the log belief ratios $\boldsymbol{\beta}_{k,t}(\theta)$, or the beliefs $\boldsymbol{\mu}_{k,t}(\theta)$, in order to quantify the performance, e.g., through the probability of choosing the correct hypothesis. In order to evaluate probabilities at the steady state, it is critical to obtain first a representation of the steady-state random variables, which is what we did in Theorem 9.1.

In preparation for the technical analysis, it is convenient to introduce the following scaled version of the limiting random vector $\boldsymbol{\beta}_k$:

$$\boldsymbol{b}_k \triangleq \delta \times \boldsymbol{\beta}_k. \tag{9.26}$$

We remark that the error probability in (9.25) can be equivalently rewritten in terms of the scaled vector $\boldsymbol{b}_k$:

$$p_k(\delta) = \mathbb{P}\left[\bigcup_{\theta \neq \vartheta^\star} \left\{\boldsymbol{b}_k(\theta) \leq 0\right\}\right]. \tag{9.27}$$

Table 9.1: Notation relevant to the ASL performance analysis.

$\boldsymbol{\beta}_k$	Steady-state ($t \to \infty$) log belief ratio vector of agent k in (9.4)
$\boldsymbol{b}_k$	Scaled version of $\boldsymbol{\beta}_k$, namely, $\boldsymbol{b}_k = \delta \times \boldsymbol{\beta}_k$

9.3 Consistency of Adaptive Social Learning

In this section we focus on characterizing the learning behavior of the ASL strategy as $\delta \to 0$. The main result enabling this characterization is the weak law of small adaptation parameters. This law establishes that, as $\delta \to 0$, for each agent k the scaled steady-state log belief ratio vector $\boldsymbol{b}_k$ is concentrated on *the same* deterministic quantity $\bar{\lambda}_{\mathsf{net}}$, namely, the expected value of the network average log likelihood ratio vector.

Theorem 9.2 (Weak law of small adaptation parameters). Let Assumptions 5.1, 5.2, and 6.1 be satisfied. Then, for $k = 1, 2, \ldots, K$,

$$\boldsymbol{b}_k \xrightarrow[\delta\to 0]{\mathrm{p}} \bar{\lambda}_{\mathsf{net}}. \tag{9.28}$$

Proof. Consider the θth entry of the scaled steady-state belief vector,

$$\boldsymbol{b}_k(\theta) = \delta \boldsymbol{\beta}_k(\theta) = \delta \sum_{j=1}^{K} \sum_{\tau=1}^{\infty} (1-\delta)^{\tau-1} [A^\tau]_{jk}\, \boldsymbol{\lambda}_{j,\tau}(\theta). \tag{9.29}$$

We now want to apply Lemma F.5 to each one of the K inner series in (9.29). Consider the jth series, and apply Lemma F.5 with the choices

$$\alpha_\tau = [A^\tau]_{jk}, \qquad \boldsymbol{y}_\tau = \boldsymbol{\lambda}_{j,\tau}(\theta), \qquad \boldsymbol{z}(\delta) = \delta \sum_{\tau=1}^{\infty} (1-\delta)^{\tau-1} \alpha_\tau\, \boldsymbol{y}_\tau. \tag{9.30}$$

In view of (4.25), α_τ meets condition (F.4) with $\alpha = v_j$, and by definition

$$\mathbb{E}\boldsymbol{y}_\tau = \bar{\lambda}_j(\theta). \tag{9.31}$$

Therefore, from Lemma F.5 we conclude that

$$\boldsymbol{z}(\delta) \xrightarrow[\delta\to 0]{\mathrm{p}} v_j \bar{\lambda}_j(\theta). \tag{9.32}$$

Since the sum of random variables converging in probability converges in probability to the sum of the limiting variables (see property P1 in Lemma D.1), we conclude that

$$\boldsymbol{b}_k(\theta) \xrightarrow[\delta\to 0]{\mathrm{p}} \sum_{j=1}^{K} v_j \bar{\lambda}_j(\theta) = \bar{\lambda}_{\mathsf{net}}(\theta). \tag{9.33}$$

Moreover, since the convergence in probability of random vectors is equivalent to the convergence in probability of their entries (see property P2 in Lemma D.1), Eq. (9.33) is equivalent to (9.28), and the proof is complete.

■

From Theorem 9.2 we can immediately establish a first form of consistency of the ASL strategy, namely, that the probability of error $p_k(\delta)$ vanishes as $\delta \to 0$.

Corollary 9.2 (ASL consistency). Under the same assumptions used in Theorem 9.2, for $k = 1, 2, \ldots, K$,

$$\lim_{\delta\to 0} p_k(\delta) = 0. \tag{9.34}$$

Proof. In view of Assumption 6.1, the network average of KL divergences $D_{\mathsf{net}}(\theta)$ admits a unique minimizer $\vartheta^\star$, yielding $\bar{\lambda}_{\mathsf{net}}(\theta) > 0$ for all $\theta \neq \vartheta^\star$ — see (6.10). Then, Eq. (9.28) implies that, for all $\theta \neq \vartheta^\star$,

$$\lim_{\delta \to 0} \mathbb{P}\left[\boldsymbol{b}_k(\theta) \leq 0\right] = 0, \tag{9.35}$$

which, using the union bound in (9.27), gives (9.34). ■

Corollary 9.2 expresses consistency in terms of the probability of making a wrong choice, i.e., it reveals that such probability vanishes as $\delta \to 0$. We now present another corollary of Theorem 9.2, which strengthens the concept of consistency by showing that, as $\delta \to 0$, the belief vector displays the desired behavior of placing unit mass on the target hypothesis $\vartheta^\star$.

Corollary 9.3 (Belief behavior under ASL). Under the same assumptions used in Theorem 9.2, for $k = 1, 2, \ldots, K$,

$$\boldsymbol{\mu}_k(\vartheta^\star) \xrightarrow[\delta \to 0]{\mathrm{p}} 1. \tag{9.36}$$

Proof. Substituting definition (9.26) into the second relation in (9.17), we obtain

$$\boldsymbol{\mu}_k(\vartheta^\star) = \frac{1}{1 + \displaystyle\sum_{\theta \neq \vartheta^\star} \exp\left\{-\frac{\boldsymbol{b}_k(\theta)}{\delta}\right\}}. \tag{9.37}$$

On the other hand, from (9.28) and the positivity condition (6.10) we have that, for $\theta \neq \vartheta^\star$,

$$\boldsymbol{b}_k(\theta) \xrightarrow[\delta \to 0]{\mathrm{p}} \bar{\lambda}_{\mathsf{net}}(\theta) > 0. \tag{9.38}$$

From (9.38) we conclude that all arguments of the exponential functions in (9.37) diverge to $-\infty$ in probability as $\delta \to 0$ [159]; this fact implies (9.36). ■

Example 9.1 (ASL consistency for vanishing δ). We consider $K = 10$ agents connected according to the strong graph displayed in the left panel of Figure 9.2 (the graph is undirected, and we assume that all agents have a self-loop, not shown in the figure). The combination matrix is designed using the uniform-averaging rule, resulting in a left stochastic matrix — see Table 4.1.

The network is tasked with the following learning problem. Recall that a Laplace random variable with mean $\bar{x}$ and scale parameter σ has the following pdf:

$$\frac{1}{2\sigma} \exp\left\{-\frac{|x - \bar{x}|}{\sigma}\right\}. \tag{9.39}$$

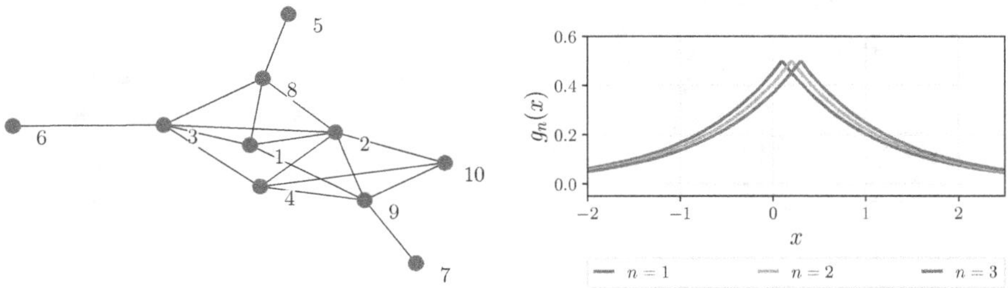

Figure 9.2: (*Left*) Network topology used in Example 9.1. The graph is undirected and all agents are assumed to have a self-loop, not shown in the figure. (*Right*) Family of Laplace densities used in the example.

In our example we consider a family of Laplace pdfs, seen in the right panel of Figure 9.2, in the form

$$g_n(x) = \frac{1}{2} e^{-|x-0.1\,n|}, \qquad n = 1, 2, 3, \tag{9.40}$$

that is, with unit scale parameter and mean equal to $0.1\,n$. The likelihood models adopted by the agents are chosen from among these Laplace densities, in the way specified in Table 9.2. For example, from the first row, each of the agents $k \in \{1, 2, 3\}$ uses the likelihood models

$$\ell_k(x|1) = g_1(x), \quad \ell_k(x|2) = g_1(x), \quad \ell_k(x|3) = g_3(x). \tag{9.41}$$

Table 9.2: Identifiability setup for the learning problem in Example 9.1.

Agent k	**Likelihood model: $\ell_k(x\|\theta)$**		
	$\theta = 1$	$\theta = 2$	$\theta = 3$
$1 - 3$	$g_1(x)$	$g_3(x)$	$g_3(x)$
$4 - 6$	$g_1(x)$	$g_1(x)$	$g_3(x)$
$7 - 10$	$g_3(x)$	$g_2(x)$	$g_3(x)$

To make the setting more interesting, we assume that the inference problem is *locally unidentifiable for all agents*. For example, we see from Table 9.2 that agent 3 is not able to distinguish $\theta = 1$ from $\theta = 2$, since the model corresponding to these two hypotheses coincide, namely, $\ell_3(x|1) = \ell_3(x|2) = g_1(x)$.

Regarding the generation of the data $\{\boldsymbol{x}_{k,t}\}$, they are iid across the agents and over time, and we focus on the objective evidence model described in Section 5.3, where there exists a common true hypothesis ϑ^o. In this example we set $\vartheta^o = 3$. According to Table 9.2, the data of agents $1 - 6$ obey model $g_3(x)$, whereas for agents $7 - 10$ the true model is $g_1(x)$. As we already know from the previous chapters, in this case the target hypothesis that minimizes the network average of KL divergences $D_{\mathsf{net}}(\theta)$ is the true hypothesis, namely, $\vartheta^\star = \vartheta^o$.

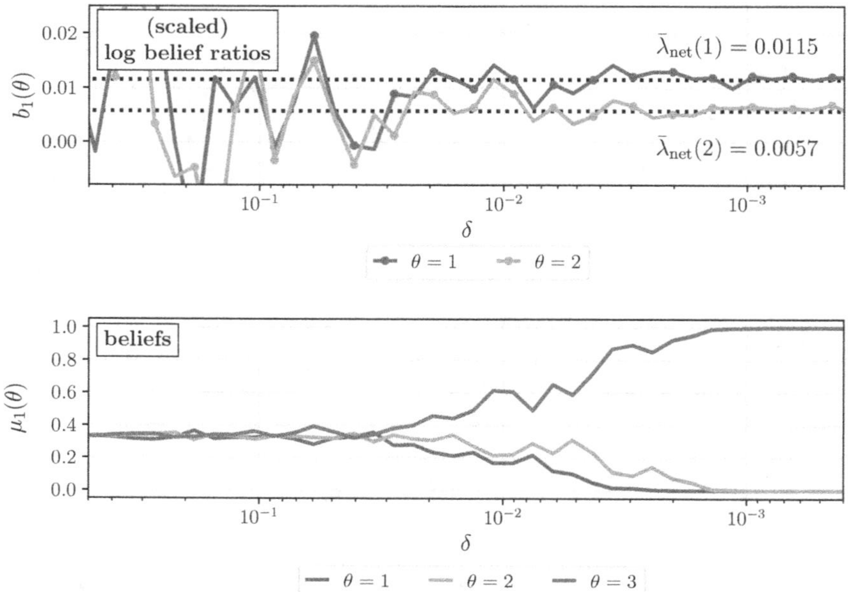

Figure 9.3: Consistency of the ASL strategy — see Example 9.1. (*Top*) According to the weak law of small adaptation parameters (Theorem 9.2), as $\delta \to 0$ the entries of the (scaled, steady-state) log belief ratio vector for agent 1 are concentrated on the corresponding entries of the deterministic vector $\bar{\lambda}_{\text{net}}$. (*Bottom*) According to Corollary 9.3, the steady-state belief vector for agent 1 tends to place unit mass on the target hypothesis $\vartheta^\star = 3$ as $\delta \to 0$.

In order to examine the steady-state behavior *empirically*, we let the ASL algorithm run for a sufficiently long period of time. To be conservative, in view of the prescriptions that we will obtain later from Chapter 10, the duration of this period is chosen to be at least one order of magnitude larger than the inverse of the adaptation parameter, $1/\delta$.

In our simulation, we consider the evolution of the ASL algorithm over $T = 10000$ time samples, after which we assume that the algorithm has reached the steady state, namely, in terms of log belief ratios we assume that

$$\delta \times \boldsymbol{\beta}_{k,T} \approx \delta \times \boldsymbol{\beta}_k = \boldsymbol{b}_k. \tag{9.42}$$

From Theorem 9.2 we know that, as δ approaches zero, the vectors $\boldsymbol{b}_k$ for all agents tend to be concentrated on $\bar{\lambda}_{\text{net}}$. This effect is shown in the top panel of Figure 9.3, where, for each value of δ (50 values for δ, uniformly spaced in the log domain, are chosen from the interval $[0.0001, 1)$), we run an independent experiment and report the corresponding values of the scaled log belief ratios $\delta \times \boldsymbol{\beta}_{1,T}(\theta) \approx \boldsymbol{b}_1(\theta)$, for hypotheses $\theta = 1$ and $\theta = 2$. We see the weak law of small adaptation parameter arising, since the limiting log belief ratios tend to be concentrated on $\bar{\lambda}_{\text{net}}(\theta)$. Moreover, in the bottom panel of Figure 9.3 we display the corresponding behavior for the beliefs, revealing that, in accordance with Corollary 9.3, as $\delta \to 0$, the belief about the target hypothesis tends to 1.

Example 9.2 (Belief behavior under the alternative update rule in (8.17)**).** In this example we examine the performance of the alternative update rule introduced in Section 8.5. To this end, let us combine the two steps (8.19a) and (8.19b) into a single step (we use the superscript "diff" because the strategy in Section 8.5 was shown to be equivalent to a diffusion strategy):

$$\boldsymbol{\beta}_{k,t}^{\mathsf{diff}}(\theta) = \sum_{j\in\mathcal{N}_k} a_{jk} \left\{(1-\delta)\boldsymbol{\beta}_{j,t-1}^{\mathsf{diff}}(\theta) + \delta\,\boldsymbol{\lambda}_{j,t}(\theta)\right\}, \tag{9.43}$$

where, in comparison with (9.1), we now have an additional factor δ multiplying $\boldsymbol{\lambda}_{j,t}(\theta)$. Unfolding the recursion in (9.43) we arrive at

$$\boldsymbol{\beta}_{k,t}^{\mathsf{diff}}(\theta) = (1-\delta)^t \sum_{j=1}^{K}[A^t]_{jk}\boldsymbol{\beta}_{j,0}^{\mathsf{diff}}(\theta) + \delta\sum_{\tau=1}^{t}\sum_{j=1}^{K}(1-\delta)^{\tau-1}[A^\tau]_{jk}\boldsymbol{\lambda}_{j,t-\tau+1}(\theta). \tag{9.44}$$

In (9.2) we arrived at a similar expression for the ASL strategy, namely,

$$\boldsymbol{\beta}_{k,t}(\theta) = (1-\delta)^t \sum_{j=1}^{K}[A^t]_{jk}\boldsymbol{\beta}_{j,0}(\theta) + \sum_{\tau=1}^{t}\sum_{j=1}^{K}(1-\delta)^{\tau-1}[A^\tau]_{jk}\,\boldsymbol{\lambda}_{j,t-\tau+1}(\theta). \tag{9.45}$$

We see that in both (9.44) and (9.45) there is a first term that dies out exponentially with time, and which is due to the initial state. Therefore, the relevant terms that determine the evolution of the algorithms over time are given by the trailing summations appearing in (9.44) and (9.45). Comparing these terms, we see that they differ only by a scaling factor δ. Recalling that $\boldsymbol{\beta}_{k,t}^{\mathsf{diff}}(\theta)$ and $\boldsymbol{\beta}_{k,t}(\theta)$ represent log belief ratios, as far as we have to maximize these ratios over $\theta \neq \vartheta^\star$, the scaling factor is immaterial. We conclude that the two strategies are equivalent in terms of selection of the maximum-credibility opinion! However, this does not mean that the *beliefs* of the two strategies would take on the same values and, as we will now show, in terms of belief formation there is a more sensible difference between the two strategies.

To illustrate this phenomenon, we start by expressing the belief of the diffusion strategy in terms of the log belief ratios $\boldsymbol{\beta}_{k,t}^{\mathsf{diff}}(\theta)$ in (9.43) (see Theorem 6.1) obtaining in particular

$$\boldsymbol{\mu}_{k,t}^{\mathsf{diff}}(\vartheta^\star) = \frac{1}{1+\sum_{\theta\neq\vartheta^\star} e^{-\boldsymbol{\beta}_{k,t}^{\mathsf{diff}}(\theta)}}. \tag{9.46}$$

Then, we focus on the steady state. Regarding the ASL strategy, from Theorem 9.1 we know that

$$\boldsymbol{\beta}_{k,t} \xrightarrow[t\to\infty]{\mathrm{d}} \boldsymbol{\beta}_k. \tag{9.47}$$

However, since the transient terms in (9.44) and (9.45) can be ignored thanks to Slutsky's theorem (Theorem D.4), by considering that the second term in (9.44) is equal to the second term in (9.45) multiplied by δ, we arrive at

$$\boldsymbol{\beta}_{k,t}^{\mathsf{diff}} \xrightarrow[t\to\infty]{\mathrm{d}} \delta\times\boldsymbol{\beta}_k = \boldsymbol{b}_k, \tag{9.48}$$

where in the equality we applied the definition of $\boldsymbol{b}_k$ from Table 9.1. Using (9.48) in (9.46), we conclude from the continuous mapping theorem (Theorem D.3) that the beliefs in (9.46) converge in distribution, as $t \to \infty$, to a *steady-state* belief vector $\boldsymbol{\mu}_k^{\mathsf{diff}}$ whose $\vartheta^\star$th entry is given by (compare with (9.37)):

$$\boldsymbol{\mu}_k^{\mathsf{diff}}(\vartheta^\star) = \frac{1}{1 + \sum_{\theta \neq \vartheta^\star} e^{-\boldsymbol{b}_k(\theta)}}. \tag{9.49}$$

Applying Theorem 9.2, from (9.49) and the continuous mapping theorem (now with reference to the convergence as $\delta \to 0$), we obtain

$$\boldsymbol{\mu}_k^{\mathsf{diff}}(\vartheta^\star) \xrightarrow[\delta \to 0]{\mathrm{p}} \frac{1}{1 + \sum_{\theta \neq \vartheta^\star} e^{-\bar{\lambda}_{\mathsf{net}}(\theta)}}. \tag{9.50}$$

Therefore, for the diffusion strategy the belief vector tends, as $\delta \to 0$, to a deterministic vector that does *not* place unit mass on the target hypothesis $\vartheta^\star$, even if it is always maximized at $\vartheta^\star$. In other words, when we move toward the nonadaptive solution (since as $\delta \to 0$ we are going to give equal credit to *all* data from the initial time instant up to the present one), we do not recover the behavior of traditional social learning.

In summary, we conclude that the two considered adaptive strategies, using the updates (8.6) and (8.17), respectively, are equivalent in terms of selection of the maximum-credibility hypothesis, but they differ in terms of belief formation. In particular, as $\delta \to 0$, with the ASL update rule (8.6) the agents tend to place all the belief mass on the target hypothesis $\vartheta^\star$, i.e., their beliefs about $\vartheta^\star$ converge to 1. In comparison, with the diffusion update rule (8.17) the agents' beliefs converges to a deterministic belief vector whose maximum entry is located at $\vartheta^\star$, but is not equal to 1. This difference might matter, e.g., from a *behavioral* perspective, namely, to understand which update strategy reflects better the way of reasoning that an individual agent uses in social learning environments.

Theorem 9.2 establishes the convergence of the error probability to 0 as $\delta \to 0$. However, it does not reveal *how* this probability vanishes. In the next two sections, we characterize the behavior of the error probability in greater detail. First, in Section 9.4 we establish that the vector of log belief ratios (properly shifted and scaled) follows a Gaussian distribution for small δ. Then, in Section 9.5 we show that the error probability of each individual agent decays exponentially with the inverse of the adaptation parameter, $1/\delta$. We also provide a detailed characterization of the error exponent. These results are useful because they reveal how the accuracy of the algorithm varies with the adaptation parameter δ, providing manageable formulas for performance evaluation and highlighting the fundamental scaling laws of adaptive social learning.

9.4 Normal Approximation for Small δ

In this section we show that the random vector $\boldsymbol{b}_k$, when properly shifted and scaled, is asymptotically normal as $\delta \to 0$. To this end, we will assume finiteness of second-order moments for the log likelihood ratios $\boldsymbol{\lambda}_{k,t}(\theta)$. We recall that, according to Table 6.1, the covariance matrix of $\boldsymbol{\lambda}_{k,t}$ is denoted by Σ_k, whereas the covariance matrix of the average variable $\boldsymbol{\lambda}_{\mathsf{net},t}$ is denoted by Σ_{net}. In addition, we introduce the following notation for the first two moments of the scaled steady-state log belief ratios $\boldsymbol{b}_k$:

$$\bar{b}_k \triangleq \mathbb{E}\boldsymbol{b}_k, \qquad \Sigma_{b_k} \triangleq \mathbb{E}\left[\left(\boldsymbol{b}_k - \bar{b}_k\right)\left(\boldsymbol{b}_k - \bar{b}_k\right)^{\mathsf{T}}\right]. \tag{9.51}$$

Using Lemmas F.4 and F.6, it is possible to express the mean and covariance matrix of the random vector $\boldsymbol{b}_k$ as[3]

$$\bar{b}_k = \bar{\lambda}_{\mathsf{net}} + O(\delta), \quad \Sigma_{b_k} = \frac{\delta}{2}\Sigma_{\mathsf{net}} + O(\delta^2), \tag{9.52}$$

where the notation $O(\delta)$ represents a quantity such that the ratio $O(\delta)/\delta$ remains bounded as $\delta \to 0$ — see Table 1.1. We see from (9.52) that, as $\delta \to 0$, there are leading terms that do not depend on the agent index k. The impact of the agents is implicitly included in the higher-order corrections, i.e., in the $O(\cdot)$ terms. Moreover, Eq. (9.52) reveals that, for small δ, the first two moments of the scaled steady-state log belief ratios are determined by the first two moments of the *network average* of log likelihood ratios. In particular, for small δ, the first relation in (9.52) reveals that $\bar{b}_k$ approximates $\bar{\lambda}_{\mathsf{net}}$, whereas the second relation reveals that Σ_{b_k} approximates $\delta\,\Sigma_{\mathsf{net}}/2$. We are now ready to state our asymptotic normality theorem.

Theorem 9.3 (Asymptotic normality under ASL). Let Assumptions 5.1, 5.2, and 6.1 be satisfied, and let $\mathscr{G}(0,\Sigma)$ denote a random vector having a zero-mean multivariate Gaussian distribution with covariance matrix Σ. If the covariance matrices Σ_k have finite entries, then for $k = 1, 2, \ldots, K$,

$$\frac{\boldsymbol{b}_k - \bar{\lambda}_{\mathsf{net}}}{\sqrt{\delta}} \xrightarrow[\delta\to 0]{\mathrm{d}} \mathscr{G}\left(0, \frac{1}{2}\,\Sigma_{\mathsf{net}}\right). \tag{9.53}$$

Proof. In the proof, it is convenient to use the notation

$$\boldsymbol{g} = [\boldsymbol{g}(1), \boldsymbol{g}(2), \ldots, \boldsymbol{g}(H-1)] \tag{9.54}$$

[3] Technically, Lemma F.6 deals with variances and not covariances. However, the result for covariances is obtained following the same arguments used to prove Lemma F.6.

for a zero-mean Gaussian random vector with covariance matrix equal to $\Sigma_{\mathsf{net}}/2$. According to this notation, claim (9.53) is reformulated as

$$\frac{\boldsymbol{b}_k - \bar{\lambda}_{\mathsf{net}}}{\sqrt{\delta}} \xrightarrow[\delta\to 0]{\mathrm{d}} \boldsymbol{g}. \tag{9.55}$$

When dealing with convergence in distribution of random vectors, the standard path is to reduce the vector problem to a scalar problem through the so-called Cramér-Wold device — see Theorem D.2. Using the Cramér-Wold device, the claim in (9.55) will be proved if we show that, for any sequence of real numbers $c(1), c(2), \ldots, c(H-1)$,

$$\sum_{\theta\neq\vartheta^\star} c(\theta)\,\frac{\boldsymbol{b}_k(\theta) - \bar{\lambda}_{\mathsf{net}}(\theta)}{\sqrt{\delta}} \xrightarrow[\delta\to 0]{\mathrm{d}} \sum_{\theta\neq\vartheta^\star} c(\theta)\boldsymbol{g}(\theta). \tag{9.56}$$

Let us now examine the LHS of (9.56). Recalling the definition of $\boldsymbol{b}_k$ from Table 9.1 and using (9.15), we get

$$\sum_{\theta\neq\vartheta^\star} c(\theta)\boldsymbol{b}_k(\theta) = \sum_{j=1}^{K}\delta\sum_{\tau=1}^{\infty}(1-\delta)^{\tau-1}[A^\tau]_{jk}\sum_{\theta\neq\vartheta^\star} c(\theta)\boldsymbol{\lambda}_{j,\tau}(\theta). \tag{9.57}$$

To establish (9.56), we will call upon Lemma F.7. It is convenient to introduce an ad-hoc notation that matches the notation used in Appendix F. Let us set, for $j = 1, 2, \ldots, K$ and $\tau \in \mathbb{N}$,

$$\boldsymbol{y}_{j,\tau} = \sum_{\theta\neq\vartheta^\star} c(\theta)\boldsymbol{\lambda}_{j,\tau}(\theta), \qquad \boldsymbol{y}_\tau = [\boldsymbol{y}_{1,\tau}, \boldsymbol{y}_{2,\tau}, \ldots, \boldsymbol{y}_{K,\tau}], \tag{9.58}$$

$$\alpha_{j,\tau} = [A^\tau]_{jk}, \qquad \alpha_\tau = [\alpha_{1,\tau}, \alpha_{2,\tau}, \ldots, \alpha_{K,\tau}], \qquad \alpha = v, \tag{9.59}$$

$$\boldsymbol{z}_t(\delta) = \delta\sum_{\tau=1}^{t}(1-\delta)^{\tau-1}\alpha_\tau^\mathsf{T}\boldsymbol{y}_\tau, \qquad \boldsymbol{z}(\delta) = \delta\sum_{\tau=1}^{\infty}(1-\delta)^{\tau-1}\alpha_\tau^\mathsf{T}\boldsymbol{y}_\tau, \tag{9.60}$$

$$\boldsymbol{y}_{\mathsf{ave},\tau} = v^\mathsf{T}\boldsymbol{y}_\tau = \sum_{\theta\neq\vartheta^\star} c(\theta)\boldsymbol{\lambda}_{\mathsf{net},\tau}(\theta), \tag{9.61}$$

$$\bar{y}_{\mathsf{ave}} = \mathbb{E}\left[v^\mathsf{T}\boldsymbol{y}_\tau\right] = \sum_{\theta\neq\vartheta^\star} c(\theta)\bar{\lambda}_{\mathsf{net}}(\theta), \tag{9.62}$$

$$\sigma^2_{\mathsf{ave}} = \mathsf{VAR}\left[v^\mathsf{T}\boldsymbol{y}_\tau\right] = \sum_{\theta\neq\vartheta^\star}\sum_{\theta'\neq\vartheta^\star} c(\theta)c(\theta')\Sigma_{\mathsf{net}}(\theta,\theta'), \tag{9.63}$$

where $\Sigma_{\mathsf{net}}(\theta,\theta')$ is the (θ,θ') entry of Σ_{net}. We see that the random variables $\boldsymbol{z}_t(\delta)$ in (9.60) match the structure of the random sums used in Definition F.2. In particular, condition (F.37) is verified in view of (4.25), with the sequence of vectors α_τ converging to the Perron vector v. Moreover, $\boldsymbol{y}_\tau$ has finite second moment since it is a linear combination of random vectors with finite second moments. It is therefore legitimate to invoke Lemma F.7 to infer the following convergence in distribution:

$$\frac{\boldsymbol{z}(\delta) - \bar{y}_{\mathsf{ave}}}{\sqrt{\delta}} \xrightarrow[\delta\to 0]{\mathrm{d}} \mathscr{G}\left(0, \frac{1}{2}\sigma^2_{\mathsf{ave}}\right). \tag{9.64}$$

On the other hand, exploiting Eqs. (9.57)–(9.63), from straightforward algebraic manipulations one can verify the identity:

$$\frac{\boldsymbol{z}(\delta) - \bar{y}_{\mathsf{ave}}}{\sqrt{\delta}} = \sum_{\theta\neq\vartheta^\star} c(\theta)\,\frac{\boldsymbol{b}_k(\theta) - \bar{\lambda}_{\mathsf{net}}(\theta)}{\sqrt{\delta}}, \tag{9.65}$$

which, in view of (9.64), implies

$$\sum_{\theta \neq \vartheta^\star} c(\theta) \frac{\boldsymbol{b}_k(\theta) - \bar{\lambda}_{\mathsf{net}}(\theta)}{\sqrt{\delta}} \xrightarrow[\delta \to 0]{\mathrm{d}} \mathscr{G}\left(0, \frac{1}{2}\sigma^2_{\mathsf{ave}}\right). \tag{9.66}$$

We see that (9.66) would correspond to (9.56) if the linear combination on the RHS of (9.56) (which is a zero-mean Gaussian variable since it is a linear combination of zero-mean Gaussian variables) has variance $\sigma^2_{\mathsf{ave}}/2$. This turns out to be the case, since we have

$$\mathsf{VAR}\left[\sum_{\theta \neq \vartheta^\star} c(\theta)\boldsymbol{g}(\theta)\right] = \frac{1}{2}\sum_{\theta \neq \vartheta^\star}\sum_{\theta' \neq \vartheta^\star} c(\theta)c(\theta')\Sigma_{\mathsf{net}}(\theta, \theta') = \frac{1}{2}\sigma^2_{\mathsf{ave}} \tag{9.67}$$

and the proof is complete.

■

Example 9.3 (Gaussian approximation). With reference to the same setting used in Example 9.1, we consider $T = 10000$ time samples, where again all agents are collecting data under a true hypothesis $\vartheta^o = 3$. We assume that the ASL algorithm has reached the steady state at T, allowing us to write

$$\delta \times \boldsymbol{\beta}_{k,T} \approx \delta \times \boldsymbol{\beta}_k = \boldsymbol{b}_k. \tag{9.68}$$

In each panel of Figure 9.4, we display 200 independent realizations of the shifted and scaled vector

$$\frac{\delta \times \boldsymbol{\beta}_{k,T} - \bar{\lambda}_{\mathsf{net}}}{\sqrt{\delta}} \approx \frac{\boldsymbol{b}_k - \bar{\lambda}_{\mathsf{net}}}{\sqrt{\delta}}, \tag{9.69}$$

for $k = 6$. From Theorem 9.3 it follows that, in steady state, this shifted and scaled vector must follow, for sufficiently small δ, a zero-mean bivariate Gaussian distribution with covariance matrix $\Sigma_{\mathsf{net}}/2$. The red dashed lines in the figure represent two confidence regions [90] relative to the bivariate Gaussian density with covariance matrix $\Sigma_{\mathsf{net}}/2$. Specifically, the smaller and larger ellipses correspond to confidence levels (i.e., integrals of the bivariate density over the considered elliptical regions) equal to 0.68 and 0.95, respectively. Examining the four panels of Figure 9.4, which display different values of δ, we see that the empirical and limiting distributions tend to overlap.

The values of $\bar{\lambda}_{\mathsf{net}}$ and Σ_{net} necessary to obtain Figure 9.4 have been computed analytically. The mean $\bar{\lambda}_k$ can be evaluated analytically by using the characterization for the distribution of $\boldsymbol{\lambda}_{k,t}$ provided later in Example 9.6. Specifically, since the mean is given by the first derivative of the log moment generating function (LMGF) evaluated at zero, to compute the mean we exploited the explicit expressions obtained in (9.140) and (9.141). The covariance matrix Σ_k has been evaluated by computing the expected values in (6.23) through numerical integration, using the pertinent Laplace distributions. Once $\bar{\lambda}_k$ and Σ_k are computed for all agents, the desired network quantities, $\bar{\lambda}_{\mathsf{net}}$ and Σ_{net}, are computed through (6.10) and (6.24), respectively (further exploiting, for the covariance, the independence across the agents).

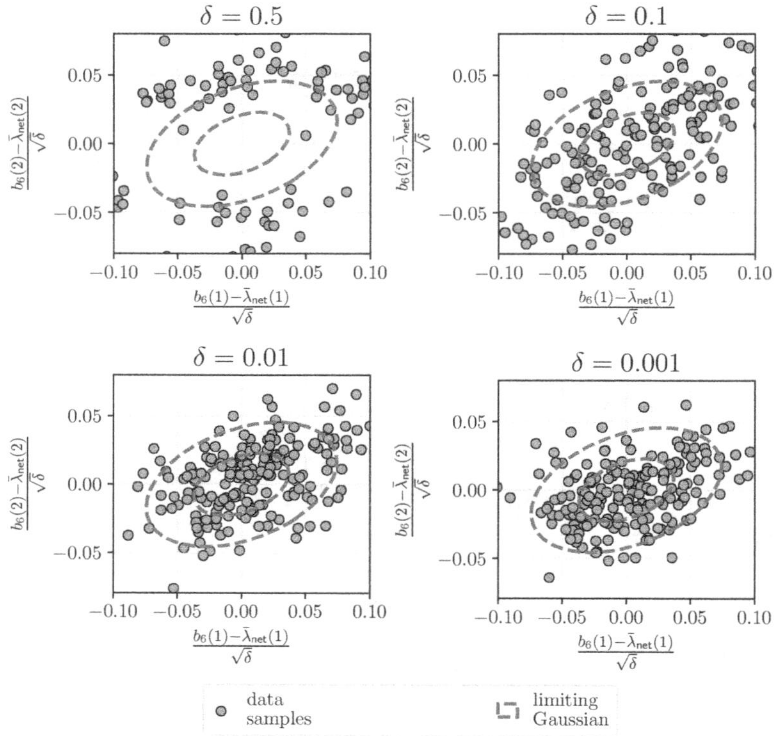

Figure 9.4: Asymptotic normality under ASL, Example 9.3. The green circles represent 200 independent realizations of the shifted and scaled vector $(\delta \times \boldsymbol{\beta}_{k,T} - \bar{\lambda}_{\text{net}})/\sqrt{\delta} \approx (\boldsymbol{b}_k - \bar{\lambda}_{\text{net}})/\sqrt{\delta}$, for $k = 6$ and $T = 10000$. The red dashed lines represent two confidence ellipses relative to the bivariate Gaussian density with covariance matrix $\Sigma_{\text{net}}/2$. Specifically, the smaller and larger ellipses correspond to confidence levels 0.68 and 0.95, respectively.

From Theorem 9.3 we can construct the following approximation for small δ:

$$\boldsymbol{b}_k \approx \mathscr{G}\left(\bar{\lambda}_{\text{net}}, \frac{\delta}{2}\Sigma_{\text{net}}\right), \tag{9.70}$$

which does *not* depend on the agent index k. However, we see from (9.52) that the moments $\bar{\lambda}_{\text{net}}$ and $(\delta/2)\Sigma_{\text{net}}$ represent approximations, for small δ, of the *actual* moments of $\boldsymbol{b}_k$. As a result, we can capture possible differences across the agents by replacing $\bar{\lambda}_{\text{net}}$ and $(\delta/2)\Sigma_{\text{net}}$ in (9.70) with their exact counterparts $\bar{b}_k$ and Σ_{b_k}, yielding the *agent-dependent* approximation

$$\boldsymbol{b}_k \approx \mathscr{G}\left(\bar{b}_k, \Sigma_{b_k}\right). \tag{9.71}$$

Using Lemmas F.4 and F.6, the quantities $\bar{b}_k$ and Σ_{b_k} can be evaluated from the series in (9.4). For the mean, from Lemma F.4 we have

$$\bar{b}_k = \mathbb{E}\boldsymbol{b}_k = \delta \times \mathbb{E}\boldsymbol{\beta}_k = \delta \sum_{\tau=1}^{\infty}\sum_{j=1}^{K}(1-\delta)^{\tau-1}[A^{\tau}]_{jk}\,\bar{\lambda}_j, \tag{9.72}$$

whereas Lemma F.6 (applied to covariances in place of variances, see footnote 3) allows us to write the covariance matrix as

$$\Sigma_{b_k} = \delta^2 \sum_{\tau=1}^{\infty}(1-\delta)^{2(\tau-1)} \times \mathbb{E}\left[\left(\sum_{j=1}^{K}[A^{\tau}]_{jk}\left(\boldsymbol{\lambda}_{j,\tau}-\bar{\lambda}_j\right)\right)\left(\sum_{j'=1}^{K}[A^{\tau}]_{j'k}\left(\boldsymbol{\lambda}_{j',\tau}-\bar{\lambda}_{j'}\right)\right)^{\mathsf{T}}\right]. \tag{9.73}$$

We see that Eq. (9.72) requires only knowledge of the mean of the log likelihood ratios. In comparison, to evaluate analytically (9.73) one needs also knowledge of the dependence across the agents. A simplified expression holds when the observations are independent across the agents, in which case Eq. (9.74) reduces to

$$\Sigma_{b_k} = \delta^2 \sum_{\tau=1}^{\infty}\sum_{j=1}^{K}(1-\delta)^{2(\tau-1)}\left([A^{\tau}]_{jk}\right)^2 \Sigma_j. \tag{9.74}$$

In practice, the above computations are performed by truncating the series appearing in (9.72), (9.73), and (9.74).

The next example focuses on the evaluation of the error probability by means of approximations (9.70) and (9.71).

Example 9.4 (Error probabilities). We focus on the evaluation of the error probabilities with reference to the setting used in the previous example. The results are shown in Figure 9.5. Consider first the curves displaying the *empirical* error probabilities, which are evaluated via Monte Carlo simulation. We see an interesting phenomenon emerging. The curves corresponding to distinct agents, displayed as functions of the *inverse* of the adaptation parameter, $1/\delta$, stay nearly parallel (in a logarithmic scale). This highlights at least two facts. First, as $\delta \to 0$, the error probabilities decay exponentially with $1/\delta$, approximately with the same slope in logarithmic scale. Second, distinct agents have distinct error probabilities. Examining the network topology in Figure 9.2, we observe that the ordering of the probability curves reflects the properties of the network graph. For example, agent 5, which has fewer connections, features a higher error probability. In contrast, agent 1, which has more connections, features a lower error probability.

Let us now focus on evaluating the error probabilities by using the Gaussian approximations (9.70) and (9.71). The mean $\bar{\lambda}_k$ and the covariance matrix Σ_k can be obtained as explained in the previous example. From $\bar{\lambda}_k$ and Σ_k we compute the network

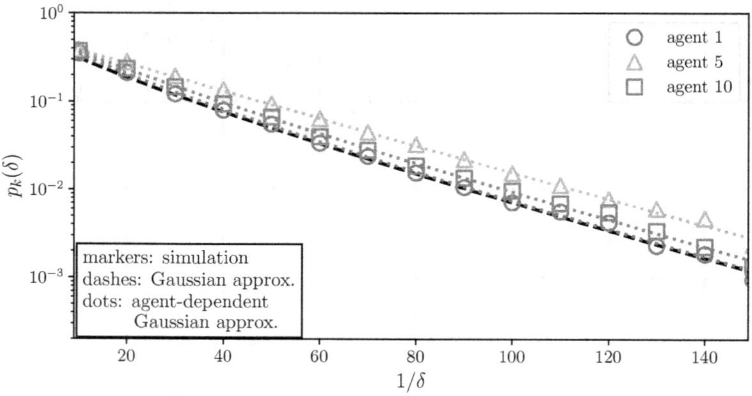

Figure 9.5: Steady-state error probability $p_k(\delta)$ as a function of $1/\delta$, for $k = 1, 5, 10$, in the setting of Example 9.4. Markers refer to the empirical error probability estimated from 20000 Monte Carlo runs. The dashed line refers to the theoretical error probability in (9.21) computed using the Gaussian approximation in (9.70). Dotted lines refer to the theoretical error probability in (9.21) computed, for agents 1, 5, and 10, using the agent-dependent Gaussian approximation in (9.71).

quantities $\bar{\lambda}_{\mathsf{net}}$ and Σ_{net} necessary to evaluate (9.70). The moments necessary to evaluate (9.71), $\bar{b}_k$ and Σ_{b_k}, have been computed by using truncated versions of the series in (9.72) and (9.74), respectively (in particular, we resort to (9.74) because the model adopted in Example 9.1 considers independence across the agents).

We see from Figure 9.5 that the error probabilities computed using approximation (9.70) do not fit well the empirical error probabilities. On the other hand, once we observed that the performance varies across the agents, we should have expected that approximation (9.70) would not perform well, because it does not depend on the particular agent. In comparison, Figure 9.5 shows that the agent-dependent approximation in (9.71) captures well the differences across the agents.

9.5 Large Deviations for Small δ

In Section 6.3 we exploited *large deviations* [59, 60] to characterize the decay of the error probability $p_{k,t}$ in traditional social learning as $t \to \infty$. However, we have learned from Section 9.1 that in adaptive social learning the error probability does not vanish anymore as $t \to \infty$; it converges instead to a *steady-state* value $p_k(\delta)$. Moreover, Corollary 9.2 guarantees that $p_k(\delta)$ vanishes as the adaptation parameter δ approaches 0. Accordingly, in this section we use the theory of large deviations to characterize the decay of the *steady-state* error probability as $\delta \to 0$. More formally, the large deviation analysis will furnish the following type of representation [59, 60]:

$$p_k(\delta) = \exp\left\{-\frac{1}{\delta}\Big[\Phi + o(1)\Big]\right\} \tag{9.75}$$

for a certain *error exponent* Φ. We denote by $o(1)$ a quantity that approaches zero as $\delta \to 0$ — see Table 1.1. We conclude from (9.75) that the leading exponential order (as $\delta \to 0$) is given by the term $-\Phi/\delta$. Taking logarithms, Eq. (9.75) can be equivalently written as

$$\lim_{\delta \to 0} \delta \log p_k(\delta) = -\Phi. \tag{9.76}$$

In place of (9.75) or (9.76), we also use the following more compact notation to indicate equality to the leading exponential order [52]:

$$p_k(\delta) \doteq e^{-\Phi/\delta}. \tag{9.77}$$

As was the case for nonadaptive social learning in Chapter 6, also for adaptive social learning the error exponent is a compact performance descriptor, which is useful to compare different systems or to optimize different parameters (e.g., the network graph, the likelihood models) [95].

The next theorem provides the large deviation characterization of the ASL strategy. It is useful to recall that, according to Table 6.1, the LMGF of $\boldsymbol{\lambda}_{k,t}$ is denoted by $\Lambda_k(s;\theta)$, whereas the LMGF of the average variable $\boldsymbol{\lambda}_{\mathsf{net},t}$ is denoted by $\Lambda_{\mathsf{net}}(s;\theta)$.

Theorem 9.4 (Error exponents under ASL). Let Assumptions 5.1, 5.2, and 6.1 be satisfied. Assume that, for $k = 1, 2, \ldots, K$ and for all $\theta \neq \vartheta^\star$,

$$\Lambda_k(s;\theta) < \infty \quad \forall s \in \mathbb{R}, \tag{9.78}$$

and introduce the function

$$\phi(s;\theta) = \int_0^s \frac{\Lambda_{\mathsf{net}}(\varsigma;\theta)}{\varsigma} d\varsigma, \tag{9.79}$$

along with its Fenchel-Legendre transform (see Appendix E.1.1)

$$\phi^*(y;\theta) = \sup_{s \in \mathbb{R}} \Big(sy - \phi(s;\theta)\Big). \tag{9.80}$$

Then

$$\mathbb{P}\left[\boldsymbol{b}_k(\theta) \leq 0\right] \doteq e^{-\Phi(\theta)/\delta}, \quad \Phi(\theta) \triangleq \phi^*(0;\theta) = -\inf_{s \in \mathbb{R}} \phi(s;\theta) > 0. \tag{9.81}$$

Moreover, the error probability for each agent is dominated by the worst-case (i.e., the smallest) exponent:

$$p_k(\delta) \doteq e^{-\Phi/\delta}, \quad \Phi = \min_{\theta \neq \vartheta^\star} \Phi(\theta). \tag{9.82}$$

Proof. We start by establishing the large deviation characterization of the probability $\mathbb{P}[\boldsymbol{b}_k(\theta) \le 0]$ provided by (9.81). To this end, we will call upon Lemma F.9. It is convenient to introduce an ad-hoc notation that matches the notation used in Appendix F. Let us set, for $j = 1, 2, \dots, K$ and $\tau \in \mathbb{N}$,

$$\boldsymbol{y}_{j,\tau} = \boldsymbol{\lambda}_{j,\tau}(\theta), \qquad \boldsymbol{y}_\tau = [\boldsymbol{y}_{1,\tau}, \boldsymbol{y}_{2,\tau}, \dots, \boldsymbol{y}_{K,\tau}], \tag{9.83}$$

$$\alpha_{j,\tau} = [A^\tau]_{jk}, \qquad \alpha_\tau = [\alpha_{1,\tau}, \alpha_{2,\tau}, \dots, \alpha_{K,\tau}], \qquad \alpha = v, \tag{9.84}$$

$$\boldsymbol{z}_t(\delta) = \delta \sum_{\tau=1}^{t} (1-\delta)^{\tau-1} \alpha_\tau^\mathsf{T} \boldsymbol{y}_\tau, \qquad \boldsymbol{z}(\delta) = \delta \sum_{\tau=1}^{\infty} (1-\delta)^{\tau-1} \alpha_\tau^\mathsf{T} \boldsymbol{y}_\tau, \tag{9.85}$$

$$\boldsymbol{y}_{\mathsf{ave},\tau} = v^\mathsf{T} \boldsymbol{y}_\tau = \boldsymbol{\lambda}_{\mathsf{net},\tau}(\theta), \qquad \Lambda_{\mathsf{ave}}(s) = \log \mathbb{E} \exp\left\{ s\, \boldsymbol{y}_{\mathsf{ave},\tau} \right\}, \tag{9.86}$$

$$\Lambda_{\boldsymbol{z}_t}(s) = \log \mathbb{E} \exp\left\{ s\, \boldsymbol{z}_t(\delta) \right\}, \qquad \Lambda_\delta(s) = \log \mathbb{E} \exp\left\{ s\, \boldsymbol{z}(\delta) \right\}. \tag{9.87}$$

It is possible to verify that the random variables $\boldsymbol{y}_{j,\tau}$ in (9.83) satisfy the conditions required by Lemma F.9. Applying Lemma F.9 to the random series $\boldsymbol{z}(\delta)$ in (9.85), we obtain

$$\lim_{\delta \to 0} \delta \Lambda_\delta(s/\delta) = \int_0^s \frac{\Lambda_{\mathsf{ave}}(\varsigma)}{\varsigma} d\varsigma. \tag{9.88}$$

Exploiting definitions (9.83)–(9.87), it is possible to verify the identities

$$\boldsymbol{b}_k = \boldsymbol{z}(\delta), \qquad \Lambda_{\mathsf{net}}(\varsigma; \theta) = \Lambda_{\mathsf{ave}}(\varsigma), \tag{9.89}$$

which means that Eq. (9.88) is equivalent to

$$\lim_{\delta \to 0} \delta \Lambda_{\boldsymbol{b}_k}(s/\delta) = \int_0^s \frac{\Lambda_{\mathsf{net}}(\varsigma; \theta)}{\varsigma} d\varsigma = \phi(s; \theta), \tag{9.90}$$

where $\Lambda_{\boldsymbol{b}_k}$ denotes the LMGF of $\boldsymbol{b}_k$ and in the last equality we used (9.79).

The convergence in (9.90) allows us to call upon the Gärtner-Ellis theorem (Theorem E.2), implying that the following large deviation principle (see Definition E.2) holds for all sets $\mathcal{S}$ (the infimum over an empty set is taken as ∞):

$$\begin{aligned} -\inf_{y \in \mathrm{int}(\mathcal{S})} \phi^*(y; \theta) &\le \liminf_{\delta \to 0} \delta \log \mathbb{P}[\boldsymbol{b}_k(\theta) \in \mathcal{S}] \\ &\le \limsup_{\delta \to 0} \delta \log \mathbb{P}[\boldsymbol{b}_k(\theta) \in \mathcal{S}] \le -\inf_{y \in \mathrm{cl}(\mathcal{S})} \phi^*(y; \theta), \end{aligned} \tag{9.91}$$

where $\phi^*(y; \theta)$ is the Fenchel-Legendre transform of $\phi(s; \theta)$ — see (9.80). We recall that the function $\phi^*(y; \theta)$ is also referred to, in the theory of large deviations, as the *rate function* — see Appendix F. Note that $\phi(s; \theta)$ in (9.90) is the integral transformation used in Lemma E.2, applied to the LMGF $\Lambda_{\mathsf{net}}(s; \theta)$ of the random variable $\boldsymbol{\lambda}_{\mathsf{net},t}(\theta)$. Note also that $\Lambda_{\mathsf{net}}(s; \theta)$ is finite for all $s \in \mathbb{R}$ because so are by assumption the individual LMGFs $\Lambda_k(s; \theta)$ — see footnote 6 in Appendix F. Accordingly, the function $\phi(s; \theta)$ and its Fenchel-Legendre transform $\phi^*(y;\theta)$ possess all the regularity properties listed in Lemma E.2. Consider the choice $\mathcal{S} = (-\infty, 0]$, and observe that $\bar{\lambda}_{\text{net}}(\theta) > 0$ due to Assumption 6.1. Exploiting the aforementioned regularity properties, we conclude that the infima appearing in (9.91) are given by (see also Figure E.3 for a typical shape of the rate function)

$$\inf_{y \in \text{int}(\mathcal{S})} \phi^*(y;\theta) = \inf_{y \in \text{cl}(\mathcal{S})} \phi^*(y;\theta) = \phi^*(0;\theta), \tag{9.92}$$

i.e., $\mathcal{S} = (-\infty, 0]$ is a continuity set of the function $\phi^*(y;\theta)$ or a $\phi^\star$-continuity set — see (E.155). Using (9.92) in (9.91), we get

$$\lim_{\delta \to 0} \delta \log \mathbb{P}\left[\boldsymbol{b}_k(\theta) \leq 0\right] = -\phi^*(0;\theta). \tag{9.93}$$

Substituting the explicit definition of the rate function from (9.80), we have

$$\phi^*(0;\theta) = \sup_{s \in \mathbb{R}} \Big(-\phi(s;\theta) \Big) = -\inf_{s \in \mathbb{R}} \phi(s;\theta) > 0, \tag{9.94}$$

where the inequality holds because, in view of Lemma E.2, the rate function $\phi^*(y;\theta)$ is nonnegative and is equal to 0 only when y is equal to the mean of the random variable whose LMGF is $\Lambda_{\text{net}}(s;\theta)$. This random variable is $\boldsymbol{\lambda}_{\text{net},t}(\theta)$ and its mean is $\bar{\lambda}_{\text{net}}(\theta)$. Since we have $0 \neq \bar{\lambda}_{\text{net}}(\theta)$, we conclude that $\phi^*(0;\theta) > 0$. Combining (9.93), (9.94), and the definition of $\Phi(\theta)$ in (9.81), we have in fact established (9.81).

Let us move on to establishing (9.82). In light of (6.22), the error probability of *not* choosing $\vartheta^\star$ can be bounded as follows (with the lower bound holding for all $\theta \neq \vartheta^\star$):

$$\mathbb{P}\left[\boldsymbol{\beta}_{k,t}(\theta) \leq 0\right] \leq p_{k,t} \leq \sum_{\theta \neq \vartheta^\star} \mathbb{P}\left[\boldsymbol{\beta}_{k,t}(\theta) \leq 0\right], \tag{9.95}$$

where the upper bound is the union bound. In steady state, Eq. (9.95) implies

$$\mathbb{P}\left[\boldsymbol{\beta}_k(\theta) \leq 0\right] \leq p_k(\delta) \leq \sum_{\theta \neq \vartheta^\star} \mathbb{P}\left[\boldsymbol{\beta}_k(\theta) \leq 0\right] \tag{9.96}$$

or, equivalently, in terms of the vector of *scaled* log belief ratios $\boldsymbol{b}_k$,

$$\mathbb{P}\left[\boldsymbol{b}_k(\theta) \leq 0\right] \leq p_k(\delta) \leq \sum_{\theta \neq \vartheta^\star} \mathbb{P}\left[\boldsymbol{b}_k(\theta) \leq 0\right]. \tag{9.97}$$

Using the lower bound in (9.97), from (9.93) and the definition of Φ appearing in (9.82), we readily conclude that

$$\liminf_{\delta \to 0} \delta \log p_k(\delta) \geq \max_{\theta \neq \vartheta^\star} \Big(-\Phi(\theta) \Big) = -\min_{\theta \neq \vartheta^\star} \Phi(\theta) = -\Phi. \tag{9.98}$$

Let us now focus on the upper bound in (9.97). By definition, for all $\theta \neq \vartheta^\star$ we have that $\Phi \leq \Phi(\theta)$. Accordingly, the convergence in (9.93) implies that, given an arbitrary $\varepsilon > 0$, for sufficiently small δ we can write

$$\mathbb{P}\left[\boldsymbol{b}_k(\theta) \leq 0\right] \leq e^{-(\Phi - \varepsilon)/\delta}. \tag{9.99}$$

Using (9.99) in the RHS of (9.97), we obtain

$$p_k(\delta) \leq \sum_{\theta \neq \vartheta^\star} e^{-(\Phi - \varepsilon)/\delta} = (H-1)e^{-(\Phi - \varepsilon)/\delta}, \tag{9.100}$$

or

$$\delta \log p_k(\delta) \le \delta \log(H-1) - \Phi + \varepsilon. \tag{9.101}$$

Due to the arbitrariness of ε, we have

$$\limsup_{\delta\to 0} \delta \log p_k(\delta) \le -\Phi. \tag{9.102}$$

Grouping (9.98) and (9.102), we obtain the desired claim.

■

The main message conveyed by Theorem 9.4 is that the steady-state error probability of each individual agent converges to 0 as $\delta \to 0$, exponentially fast as a function of $1/\delta$. This exponential law provides a *universal* law for adaptive social learning, which is in line with the universal scaling law for adaptive distributed detection — see [123]. The exponent Φ governing the exponential decay depends on the statistical properties of $\boldsymbol{\lambda}_{\mathsf{net},t}(\theta)$, the network average of log likelihood ratios defined by (6.7).

9.5.1 Finiteness of Error Exponents

The next corollary gives some useful information about the error exponents.

Corollary 9.4 (Useful properties of the error exponents). Let the same assumptions used in Theorem 9.4 be satisfied, and let, for $\theta \neq \vartheta^\star$,

$$\lambda_{\mathsf{inf}}(\theta) \triangleq \inf\left(\mathrm{supp}_{\lambda_{\mathsf{net}}(\theta)}\right), \tag{9.103}$$

where $\mathrm{supp}_{\lambda_{\mathsf{net}}(\theta)}$ denotes the support of the distribution of $\boldsymbol{\lambda}_{\mathsf{net},t}(\theta)$ (see Definition E.1). If $\lambda_{\mathsf{inf}}(\theta) \ge 0$, the error exponent $\Phi(\theta)$ is infinite.
Instead, if $\lambda_{\mathsf{inf}}(\theta) < 0$, then the error exponent is finite and can be computed as

$$\Phi(\theta) = -\inf_{s\in\mathbb{R}} \phi(s;\theta) = -\phi(s_\theta^\star;\theta), \tag{9.104}$$

where $s_\theta^\star < 0$ is the unique nonzero solution to the equation

$$\Lambda_{\mathsf{net}}(s_\theta^\star;\theta) = 0. \tag{9.105}$$

Moreover, in this case the exponent can be upper bounded by

$$\Phi(\theta) < |s_\theta^\star|\, \bar{\lambda}_{\mathsf{net}}(\theta). \tag{9.106}$$

Proof. We have shown in the proof of Theorem 9.4 that the error exponent $\Phi(\theta)$ is given by the Fenchel-Legendre transform

$$\phi^*(y;\theta) = \sup_{s\in\mathbb{R}}\Big(sy - \phi(s;\theta)\Big) \tag{9.107}$$

evaluated at $y = 0$. The characterization of the rate function provided in Lemma E.2 reveals that, if $\lambda_{\mathsf{inf}}(\theta) \geq 0$, then $\phi^*(0;\theta) = \infty$, and the first claim of the lemma is proved.

Let us consider next the case $\lambda_{\mathsf{inf}}(\theta) < 0$. From (9.107) we can write

$$\phi^*(0;\theta) = \sup_{s\in\mathbb{R}} \Big(-\phi(s;\theta)\Big) = -\inf_{s\in\mathbb{R}} \phi(s;\theta). \tag{9.108}$$

Property Q0 from Lemma E.2 guarantees that $\phi(s;\theta)$ (as a function of s) is strictly convex and infinitely differentiable. In particular, from (E.105) we know that the first derivative $\phi'(s;\theta)$ satisfies

$$\lim_{s\to 0}\phi'(s;\theta) = \lim_{s\to 0}\frac{\Lambda_{\mathsf{net}}(s;\theta)}{s} = \Lambda'_{\mathsf{net}}(0;\theta) = \bar{\lambda}_{\mathsf{net}}(\theta) > 0, \tag{9.109}$$

where the last equality holds because the first derivative of the LMGF evaluated at $s = 0$ is equal to the mean (see (E.31)), and the inequality follows from (6.10). Moreover, as shown in the proof of Lemma E.2 (see (E.122)), we have

$$\lim_{s\to-\infty} \phi'(s;\theta) = \lambda_{\mathsf{inf}}(\theta) < 0. \tag{9.110}$$

Grouping (9.110) and (9.109), we conclude that $\phi'(s;\theta)$ (which is strictly increasing because $\phi(s;\theta)$ is strictly convex) increases monotonically from negative to positive values as s spans the interval $(-\infty, 0)$. As a result, there exists a unique value $s^\star_\theta < 0$ such that

$$\phi'(s^\star_\theta;\theta) = 0. \tag{9.111}$$

Moreover, exploiting the integral form of $\phi(s;\theta)$ in (9.79), we observe that

$$\phi'(s^\star_\theta;\theta) = 0 \iff \frac{\Lambda_{\mathsf{net}}(s^\star_\theta;\theta)}{s^\star_\theta} = 0 \iff \Lambda_{\mathsf{net}}(s^\star_\theta;\theta) = 0. \tag{9.112}$$

. The value $s^\star_\theta$ is the unique value where the first derivative of the strictly convex function $\phi(s;\theta)$ is equal to 0. Therefore, $s^\star_\theta$ is the unique minimizer of $\phi(s;\theta)$, implying that

$$\Phi(\theta) = \phi^*(0;\theta) = -\inf_{s\in\mathbb{R}} \phi(s;\theta) = -\phi(s^\star_\theta;\theta), \tag{9.113}$$

and (9.104) is proved. It remains to show that (9.106) holds. From the convexity properties of the LMGF (see Appendix E.1.2) we know that $\Lambda_{\mathsf{net}}(s;\theta)$ is strictly convex for all $s \in \mathbb{R}$. Therefore, exploiting Lemma A.1, specifically (A.3a), for all $s \neq 0$ we can write

$$\Lambda_{\mathsf{net}}(s;\theta) > s\Lambda'_{\mathsf{net}}(0;\theta) = s\bar{\lambda}_{\mathsf{net}}(\theta). \tag{9.114}$$

In particular, for $s < 0$ we will have the reverse inequality

$$\frac{\Lambda_{\mathsf{net}}(s;\theta)}{s} < \bar{\lambda}_{\mathsf{net}}(\theta). \tag{9.115}$$

Recalling that $s^\star_\theta < 0$ and using (9.115) in (9.79), we obtain

$$\Phi(\theta) = -\phi(s^\star_\theta;\theta) = -\int_0^{s^\star_\theta} \frac{\Lambda_{\mathsf{net}}(\varsigma;\theta)}{\varsigma}d\varsigma = \int_{-|s^\star_\theta|}^{0} \frac{\Lambda_{\mathsf{net}}(\varsigma;\theta)}{\varsigma}d\varsigma < |s^\star_\theta|\,\bar{\lambda}_{\mathsf{net}}(\theta), \tag{9.116}$$

and the proof is complete.

■

We see from Corollary 9.4 that when $\boldsymbol{\lambda}_{\text{net},t}(\theta) \geq 0$, the error exponent is infinite. This means that the convergence to 0 of the error probability is *super-exponential*, i.e., more favorable. However, this case is seldom verified, for the following reasons.

We focus for simplicity on the case where the observations are discrete random variables. First, let $\theta \neq \vartheta^\star$ and consider an agent k for which $\ell_{k,\theta} \neq \ell_{k,\vartheta^\star}$. Such an agent must exist because, in view of Assumption 6.1, the network divergence $D_{\text{net}}(\theta)$ has a unique minimizer $\vartheta^\star$. Letting

$$\mathcal{X}_= \triangleq \Big\{x : \ell_k(x|\theta) = \ell_k(x|\vartheta^\star)\Big\}, \qquad \mathcal{X}_{\neq} \triangleq \Big\{x : \ell_k(x|\theta) \neq \ell_k(x|\vartheta^\star)\Big\}, \tag{9.117}$$

we can write

$$\sum_{x\in\mathcal{X}_=} \ell_k(x|\theta) + \sum_{x\in\mathcal{X}_{\neq}} \ell_k(x|\theta) = 1 = \sum_{x\in\mathcal{X}_=} \ell_k(x|\theta) + \sum_{x\in\mathcal{X}_{\neq}} \ell_k(x|\vartheta^\star), \tag{9.118}$$

which implies

$$\sum_{x\in\mathcal{X}_{\neq}} \Big(\ell_k(x|\vartheta^\star) - \ell_k(x|\theta)\Big) = 0. \tag{9.119}$$

As a result, the log likelihood ratio $\log(\,\ell_k(x|\vartheta^\star)/\ell_k(x|\theta)\,)$ must take on positive and negative values. Therefore, excluding ad-hoc (and unrealistic) interactions between the likelihoods and the true joint distribution of the agents' observations, the network average of log likelihood ratios $\boldsymbol{\lambda}_{\text{net},t}(\theta)$ takes on positive and negative values with nonzero probability, implying that the point 0 is greater than the infimum of the support of the distribution of $\boldsymbol{\lambda}_{\text{net},t}(\theta)$.

9.5.2 Benefits of Cooperation

In Section 6.3.1 we discussed the benefits of cooperation for traditional social learning (with geometric averaging). The next example shows that cooperation is also rewarding in *adaptive* social learning.

Example 9.5 (Cooperation improves learning accuracy). We borrow the setup from Example 6.2, which is summarized here. We consider a network of K agents. The combination matrix is doubly stochastic and primitive, yielding a uniform Perron vector, i.e., $v_k = 1/K$ for $k = 1, 2, \ldots, K$. The observations are assumed independent across the agents. The likelihoods and the true distributions are equal across the agents, and would allow each agent to learn the target hypothesis $\vartheta^\star$ individually. Nevertheless, the agents cooperate over the network by implementing the ASL strategy. We now show that cooperation can boost the learning performance, which will be measured in terms of the error exponents in (9.81).

To compute these exponents, we need to evaluate first the LMGF $\Lambda_{\mathsf{net}}(s;\theta)$ appearing in (9.79). This LMGF was computed in (6.84); we repeat here the derivation for convenience of presentation. According to definition (6.59), we have

$$\Lambda_{\mathsf{net}}(s;\theta) \triangleq \log \mathbb{E} \exp\left\{s\, \boldsymbol{\lambda}_{\mathsf{net},t}(\theta)\right\}, \tag{9.120}$$

namely, $\Lambda_{\mathsf{net}}(s;\theta)$ is the LMGF of the network average of log likelihood ratios introduced in (6.7), which, in the considered case where $v_k = 1/K$ for all k, is equal to

$$\boldsymbol{\lambda}_{\mathsf{net},t}(\theta) = \frac{1}{K}\sum_{k=1}^{K} \boldsymbol{\lambda}_{k,t}(\theta). \tag{9.121}$$

We see that $\boldsymbol{\lambda}_{\mathsf{net},t}(\theta)$ is a linear combination of the log likelihood ratios. Since the LMGF of the sum of independent random variables is the sum of the LMGFs of the individual variables, the LMGF $\Lambda_{\mathsf{net}}(s;\theta)$ is given by

$$\Lambda_{\mathsf{net}}(s;\theta) = \sum_{k=1}^{K} \Lambda_k(s/K;\theta) = K\Lambda_k(s/K;\theta), \tag{9.122}$$

where

$$\Lambda_k(s;\theta) = \log \mathbb{E} \exp\left\{s\, \boldsymbol{\lambda}_{k,t}(\theta)\right\} \tag{9.123}$$

denotes the LMGF of the log likelihood ratio $\boldsymbol{\lambda}_{k,t}(\theta)$. We remark that $\Lambda_k(s;\theta)$ is one and the same for all k because the likelihoods and the data distributions are identical across the agents.

Using (9.122), the integral in (9.79) can be computed as

$$\phi(s;\theta) = \int_0^s \frac{\Lambda_{\mathsf{net}}(\varsigma;\theta)}{\varsigma} d\varsigma = K\int_0^s \frac{\Lambda_k(\varsigma/K;\theta)}{\varsigma} d\varsigma. \tag{9.124}$$

As a particular case, we obtain from (9.124) the integral corresponding to the case $K = 1$, i.e., to an *individual* agent working in isolation, namely,

$$\phi_{\mathsf{ind}}(s;\theta) = \int_0^s \frac{\Lambda_k(\varsigma;\theta)}{\varsigma} d\varsigma. \tag{9.125}$$

Returning to the general case in (9.124) and performing the change of variable $\varsigma' = \varsigma/K$, we get

$$\phi(s;\theta) = K \underbrace{\int_0^{s/K} \frac{\Lambda_k(\varsigma';\theta)}{\varsigma'} d\varsigma'}_{=\phi_{\mathsf{ind}}(s/K;\theta)} \tag{9.126}$$

or

$$\phi(s;\theta) = K\phi_{\mathsf{ind}}(s/K;\theta). \tag{9.127}$$

According to (9.81), the error exponents for the case of K agents and for the case of a single agent are, respectively,

$$\Phi(\theta) = -\inf_{s\in\mathbb{R}} \phi(s;\theta), \qquad \Phi_{\mathsf{ind}}(\theta) = -\inf_{s\in\mathbb{R}} \phi_{\mathsf{ind}}(s;\theta). \tag{9.128}$$

Exploiting (9.127) and (9.128), we obtain

$$\Phi(\theta) = -\inf_{s\in\mathbb{R}} \phi(s;\theta) = -K\inf_{s\in\mathbb{R}} \phi_{\mathsf{ind}}(s/K;\theta) = -K\inf_{s\in\mathbb{R}} \phi_{\mathsf{ind}}(s;\theta) = K\Phi_{\mathsf{ind}}(\theta). \tag{9.129}$$

Referring to the worst-case exponent in (9.82), we finally obtain

$$\Phi = K\Phi_{\text{ind}}, \tag{9.130}$$

which reveals that, for the ASL strategy, the *network* error exponent Φ is K times larger than the *individual* error exponent pertaining to a standalone agent. The same K-fold increase was observed in (6.88). However, recall that (6.88) referred to traditional social learning with geometric averaging. Specifically, the error exponents in (6.88) quantify the decay rate, as $t \to \infty$, of the error probabilities. In traditional social learning, these probabilities vanish as $t \to \infty$. In comparison, they do not vanish in the adaptive strategy, but they converge to steady-state probabilities, which vanish as the adaptation parameter δ approaches 0. The exponents in (9.130) quantify the decay rate of the steady-state probabilities as $\delta \to 0$.

Therefore, the comparison between (6.88) and (9.130) leads to the following remarkable conclusion: The learning mechanisms of traditional and adaptive social learning are different, resulting in two different types of error exponents to quantify the performance; nevertheless, under both scenarios, cooperation is rewarding, resulting in a K-fold increase of the error exponents with respect to a standalone agent.

Example 9.6 (Error exponents). We now revisit Example 9.1 in terms of error exponents. To this end, we need to compute first the LMGF of the log likelihood ratio $\boldsymbol{\lambda}_{k,t}(\theta)$ in (6.3). Since the likelihoods belong to the Laplace family described by (9.39), after some straightforward algebra the log likelihood ratio is found to be

$$\boldsymbol{\lambda}_{k,t}(\theta) = |\boldsymbol{x}_{k,t} - \bar{x}_k(\theta)| - |\boldsymbol{x}_{k,t} - \bar{x}_k(\vartheta^o)|, \tag{9.131}$$

where $\bar{x}_k(\theta)$ denotes the expectation of $\boldsymbol{x}_{k,t}$, computed under likelihood $\ell_k(x|\theta)$. For example, using Table 9.2, we see that

$$\bar{x}_1(1) = 0.1, \quad \bar{x}_4(3) = 0.3, \quad \bar{x}_7(2) = 0.2. \tag{9.132}$$

Next, we introduce the auxiliary quantity, for $\theta \neq \vartheta^o$,

$$e_{k,\theta} \triangleq \bar{x}_k(\theta) - \bar{x}_k(\vartheta^o), \tag{9.133}$$

as well as the centered variable

$$\widetilde{\boldsymbol{x}}_{k,t} = \boldsymbol{x}_{k,t} - \bar{x}_k(\vartheta^o). \tag{9.134}$$

Recalling that $\boldsymbol{x}_{k,t}$ is distributed according to the true underlying pdf $\ell_k(x|\vartheta^o)$, the pdf of the centered variable $\widetilde{\boldsymbol{x}}_{k,t}$ is given by $\ell_k(x + \bar{x}_k(\vartheta^o)|\vartheta^o)$, which is a Laplace pdf with zero mean and unit scale parameter, namely,

$$g_0(x) = \frac{1}{2}e^{-|x|}. \tag{9.135}$$

Using (9.133) and (9.134) in (9.131), we obtain

$$\boldsymbol{\lambda}_{k,t}(\theta) = |\widetilde{\boldsymbol{x}}_{k,t} - e_{k,\theta}| - |\widetilde{\boldsymbol{x}}_{k,t}|. \tag{9.136}$$

Consider first the case $e_{k,\theta} > 0$. The random variable $\boldsymbol{\lambda}_{k,t}(\theta)$ can be represented as

$$\boldsymbol{\lambda}_{k,t}(\theta) = \begin{cases} e_{k,\theta} & \text{if } \widetilde{\boldsymbol{x}}_{k,t} < 0, \\ e_{k,\theta} - 2\,\widetilde{\boldsymbol{x}}_{k,t} & \text{if } \widetilde{\boldsymbol{x}}_{k,t} \in [0, e_{k,\theta}], \\ -e_{k,\theta} & \text{if } \widetilde{\boldsymbol{x}}_{k,t} > e_{k,\theta}. \end{cases} \tag{9.137}$$

In order to evaluate the error exponents, it is necessary to compute the LMGF of $\boldsymbol{\lambda}_{k,t}(\theta)$. From (6.58) we know that this LMGF is defined as

$$\Lambda_k(s;\theta) = \log \mathbb{E} \exp\left\{ s\, \boldsymbol{\lambda}_{k,t}(\theta) \right\}. \tag{9.138}$$

To compute the expectation in (9.138), i.e., the moment generating function of $\boldsymbol{\lambda}_{k,t}$, we can exploit (9.137) and (9.135) and write

$$\begin{aligned}
&\mathbb{E} e^{s\, \boldsymbol{\lambda}_{k,t}(\theta)} \\
&= \int_{-\infty}^{0} e^{s\, e_{k,\theta}} g_0(x) dx + \int_{0}^{e_{k,\theta}} e^{s\,(e_{k,\theta}-2x)} g_0(x) dx + \int_{e_{k,\theta}}^{\infty} e^{-s\, e_{k,\theta}} g_0(x) dx \\
&= \frac{e^{s\, e_{k,\theta}}}{2} \int_{-\infty}^{0} e^{x} dx + \frac{e^{s\, e_{k,\theta}}}{2} \int_{0}^{e_{k,\theta}} e^{-(2s+1)x} dx + \frac{e^{-s\, e_{k,\theta}}}{2} \int_{e_{k,\theta}}^{\infty} e^{-x} dx \\
&= \frac{e^{s\, e_{k,\theta}}}{2} + \frac{e^{s\, e_{k,\theta}}}{2} \frac{1 - e^{-(2s+1)\, e_{k,\theta}}}{2s+1} + \frac{e^{-s\, e_{k,\theta}}\, e^{-e_{k,\theta}}}{2} \\
&= \frac{e^{s\, e_{k,\theta}}}{2} + \frac{e^{s\, e_{k,\theta}} - e^{-(s+1)\, e_{k,\theta}}}{2\,(2s+1)} + \frac{e^{-(s+1)\, e_{k,\theta}}}{2} \\
&= \frac{(s+1) e^{s\, e_{k,\theta}} + s e^{-(s+1)\, e_{k,\theta}}}{2s+1}
\end{aligned} \tag{9.139}$$

to arrive at

$$\Lambda_k(s;\theta) = \log\left(\frac{(s+1) e^{s\, e_{k,\theta}} + s e^{-(s+1)\, e_{k,\theta}}}{2s+1} \right). \tag{9.140}$$

Following similar steps for the case $e_{k,\theta} < 0$, we would find the following expression for the LMGF:

$$\Lambda_k(s;\theta) = \log\left(\frac{s e^{(s-1)\, e_{k,\theta}} + (s-1) e^{-s\, e_{k,\theta}}}{2s-1} \right). \tag{9.141}$$

Now, to compute the error exponents $\Phi(\theta)$ from (9.81), we need to compute the LMGF Λ_{net} appearing in (9.79). Recalling that Λ_{net} is the LMGF of the *network* variable $\boldsymbol{\lambda}_{\mathsf{net},t}(\theta) = \sum_{k=1}^{K} v_k \boldsymbol{\lambda}_{k,t}(\theta)$, we can write (as we also showed before in (6.84))

$$\Lambda_{\mathsf{net}}(s;\theta) = \sum_{k=1}^{K} \Lambda_k(v_k s;\theta), \tag{9.142}$$

which follows from the fact that the LMGF of the sum of independent random variables is the sum of the LMGFs of the individual variables. The error exponent $\Phi(\theta)$ is finally evaluated by: *i)* substituting (9.141) into (9.142); *ii)* evaluating numerically the integral in (9.79) to compute $\phi(0,\theta)$; and *iii)* computing $\Phi(\theta)$ from (9.81). Since the true state is $\vartheta^o = 3$, we need to evaluate $\Phi(\theta)$ for $\theta = 1$ and $\theta = 2$. Performing the aforementioned calculations, we obtain $\Phi(1) = 0.04778$ and $\Phi(2) = 0.03589$, which means that the dominant exponent is given by

$$\Phi = \min_{\theta \in \{1,2\}} \Phi(\theta) = 0.03589. \tag{9.143}$$

Now we illustrate the details of the numerical experiments. We let all agents execute the ASL algorithm for $T = 2000$ iterations and for 15 values of δ uniformly spaced in the interval $[1/150, 1/10]$. We run 20000 Monte Carlo experiments and compute

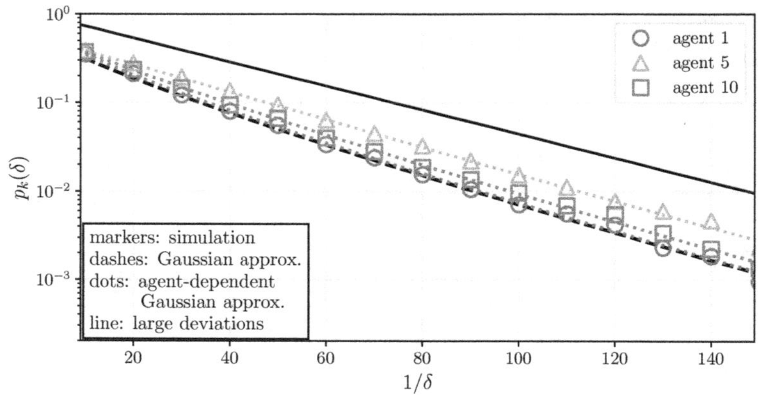

Figure 9.6: Steady-state error probability $p_k(\delta)$ as a function of $1/\delta$, for $k = 1, 5, 10$, in the setting of Example 9.6. Markers refer to the empirical error probability estimated from 20000 Monte Carlo runs. The dashed line refers to the theoretical error probability in (9.21) computed using the Gaussian approximation in (9.70). Dotted lines refer to the theoretical error probability in (9.21) computed, for agents 1, 5, and 10, using the agent-dependent Gaussian approximation in (9.71). The solid line refers to the function $e^{-\Phi/\delta}$, with error exponent Φ predicted by the large deviation analysis in Theorem 9.4.

the steady-state empirical probability of error for each agent and each value of δ. In Figure 9.6 the empirical probability curves of agents 1, 5, and 10 are compared against the theoretical error probability in (9.21) computed using the Gaussian approximations in (9.70) and (9.71). To highlight the exponential decay rate with error exponent Φ predicted by Theorem 9.4, in the figure we also plot the function $e^{-\Phi/\delta}$. We recall that this function should not be intended as an approximation for the error probabilities, but must be used only to capture the leading order exponential decay rate.

9.6 Main Performance Characteristics

The analysis carried out in this chapter revealed the following fundamental features.

Consistent social learning. Thanks to the weak law of small adaptation parameters proved in Theorem 9.2, we showed in Corollary 9.2 that with the ASL strategy each agent learns consistently, i.e., with vanishing probability of error as $\delta \to 0$. Moreover, we showed in Corollary 9.3 that a stronger notion of consistency applies, that is, as $\delta \to 0$ the belief of each agent about the target hypothesis $\vartheta^\star$ tends to 1.

Gaussian approximation. Theorem 9.3 showed that the vector $\boldsymbol{b}_k$, when properly shifted and scaled, is asymptotically normal. The theorem was exploited to derive the two Gaussian approximations in (9.70) and (9.71). In particular, the second approximation is able to capture differences arising across the error probabilities of the agents.

Large deviations. Theorem 9.4 revealed that the error probabilities of all agents decay exponentially fast, as $\delta \to 0$, with the inverse of the adaptation parameter, $1/\delta$. The exponent ruling this decay is the same for all agents.

Equivalence among agents? We saw in Figure 9.6 that the error probability curves of distinct agents stay nearly parallel (as functions of $1/\delta$, in the logarithmic-scale representation), which confirms that they are equivalent *at the leading order in the exponent.* On the other hand, we also saw that the performance of distinct agents *is not equalized as* δ *goes to* 0.

Observe that this is not in conflict with the theory of large deviations. Indeed, the equality to the leading exponential order in (9.77) does not imply in any way that we can approximate the probability of error as $e^{-\Phi/\delta}$, i.e., $p_k(\delta) \not\approx e^{-\Phi/\delta}$. This is because the large deviation analysis neglects subexponential corrections embodied in the $o(1)$ term. For example, consider two error probabilities, say $p_1(\delta) = e^{-\Phi/\delta}$ and $p_2(\delta) = 100\, e^{-\Phi/\delta}$. Since we can write

$$p_2(\delta) = 100\, e^{-\Phi/\delta} = e^{-\Phi/\delta + \log 100} = \exp\left\{ -\frac{1}{\delta}\Big[\Phi - \underbrace{\delta \log 100}_{o(1)}\Big]\right\}, \quad (9.144)$$

we see that the leading exponent of $p_2(\delta)$ is Φ. This is obviously the same as in $p_1(\delta)$. However, despite featuring the same error exponent as $p_1(\delta)$, probability $p_2(\delta)$ is two orders of magnitude larger than $p_1(\delta)$. In our setting, higher-order corrections in the error probabilities can reflect differences across the agents, arising due to various factors, for example, due to the difference between "central" agents with a high number of neighbors, as opposed to "peripheral" agents with few neighbors. For one instance of this behavior, refer back to the difference between the error probabilities of agents 1 and 5 in Figure 9.6, and to Figure 9.2 to see that agent 5 is more peripheral than agent 1. This richer behavior is not captured by the large deviation analysis, which is able to estimate only the error exponent.

Interestingly, the Gaussian approximation (9.71) is able to capture the discrepancies among the agents' error probabilities. However, we know that this approximation is not guaranteed to track the exact error probabilities as $\delta \to 0$. In order to find an approximation that captures the behavior of distinct agents *and* is exact for vanishing δ, a refined large deviation framework exists, usually referred to as "exact asymptotics" [8, 59, 60], which has been applied to binary adaptive detection in [120, 123].

Chapter 10

Adaptation under ASL

In this chapter we study another important aspect of adaptive social learning, namely, the transient behavior during the early stages of adaptation. To begin with, in Section 10.1 we provide a qualitative overview to illustrate the main rationale and goal of the transient analysis. Then, in Section 10.2 we quantify the adaptation capacity of the ASL strategy by characterizing the time necessary for the instantaneous error probability to get close to the steady-state value. Combining this characterization with the results available from Chapter 9, we arrive at a revealing description of the trade-off between learning and adaptation.

10.1 Qualitative Description of the Transient Phase

It is useful to provide a qualitative overview of the transient behavior of adaptive social learning in comparison with the traditional social learning strategy examined in Chapter 5. To this end, we consider the following illustrative example. We have a single agent (and, therefore, in this example we remove the subscript k from the notation) interested in solving a binary hypothesis problem with $\Theta = \{1, 2\}$. The likelihood models $\ell(x|1)$ and $\ell(x|2)$ employed by the agent are exact, namely, the data can originate from $\ell(x|1)$ or $\ell(x|2)$, depending on whether the true hypothesis is $\vartheta^o = 1$ or $\vartheta^o = 2$. To simplify the presentation, we assume symmetric KL divergences, i.e., we assume the validity of the following equality:[1]

$$\mathbb{E}_{\ell_1} \log \frac{\ell(\boldsymbol{x}|1)}{\ell(\boldsymbol{x}|2)} = \mathbb{E}_{\ell_2} \log \frac{\ell(\boldsymbol{x}|2)}{\ell(\boldsymbol{x}|1)} \triangleq \bar{\lambda} > 0, \tag{10.1}$$

where, as usual, the notation $\mathbb{E}_{\ell_\theta}$ denotes expectation under $\ell(x|\theta)$. We assume that at time $t = 1$ the true underlying hypothesis is $\vartheta^o = 1$, and the situation remains stationary until a certain time T_1, after which data start being generated according to $\vartheta^o = 2$. The purpose of the transient analysis is to examine how the algorithm is able to react to drifts. In this example, the drift is represented by the change in ϑ^o.

In order to study how the learning process progresses over time, it is sufficient to consider the time evolution of the log belief ratio

$$\check{\boldsymbol{\beta}}_t \triangleq \log \frac{\boldsymbol{\mu}_t(1)}{\boldsymbol{\mu}_t(2)}. \tag{10.2}$$

Note that in (10.2) we use symbol $\check{\boldsymbol{\beta}}_t$ to denote the log belief ratio, in place of the symbol $\boldsymbol{\beta}_t$ that was used before. This choice is meant to avoid confusion. Indeed, in our treatment the symbol $\boldsymbol{\beta}_t$ is always defined with the *true hypothesis* appearing in the numerator. Since in the following analysis the true hypothesis will change during the observation interval, using such a convention would require exchanging the numerator and denominator, thus adding unnecessary complexity.

In contrast, in the log belief ratio $\check{\boldsymbol{\beta}}_t$, we have hypothesis 1 in the numerator and hypothesis 2 in the denominator, irrespective of which hypothesis is true. This also means that, to classify correctly the hypotheses, we would like to have positive values of $\check{\boldsymbol{\beta}}_t$ when $\vartheta^o = 1$ and negative values when $\vartheta^o = 2$.

Applying the sequential Bayesian update strategy (2.21) to the considered single-agent binary setting, we obtain the following recursion (we use the superscripts na and ad to distinguish the nonadaptive and adaptive strategies, respectively):

$$\check{\boldsymbol{\beta}}_t^{\mathsf{na}} = \check{\boldsymbol{\beta}}_{t-1}^{\mathsf{na}} + \log \frac{\ell(\boldsymbol{x}_t|1)}{\ell(\boldsymbol{x}_t|2)}. \tag{10.3}$$

[1] We remark that the qualitative argument in this section does not rely on condition (10.1), which is made here only to simplify the example. It is interesting to note (and straightforward to verify) that condition (10.1) is always satisfied for all shift-in-mean problems with symmetric noise pdf, namely, when $\boldsymbol{x} = \boldsymbol{w} + m_1$ under $\ell(x|1)$ and $\boldsymbol{x} = \boldsymbol{w} + m_2$ under $\ell(x|2)$, where $\boldsymbol{w}$ is a zero-mean continuous random vector in $\mathbb{R}^d$ with an even pdf $f(w) = f(-w)$ (whose support is equal to $\mathbb{R}^d$ to guarantee that $\boldsymbol{x}$ has the same support under the two hypotheses, otherwise the detection problem becomes trivial).

Applying instead the adaptive update step of the ASL strategy (3.16) to the same single-agent binary case, we obtain the recursion

$$\check{\boldsymbol{\beta}}_t^{\mathsf{ad}} = (1-\delta)\check{\boldsymbol{\beta}}_{t-1}^{\mathsf{ad}} + \log\frac{\ell(\boldsymbol{x}_t|1)}{\ell(\boldsymbol{x}_t|2)}. \tag{10.4}$$

We assume flat priors for both (10.3) and (10.4), which implies $\check{\boldsymbol{\beta}}_0^{\mathsf{na}} = \check{\boldsymbol{\beta}}_0^{\mathsf{ad}} = 0$. We now examine separately the nonadaptive and adaptive strategies.

10.1.1 Nonadaptive Strategy

In order to appreciate the main trade-offs involved in the transient behavior, let us focus on the time evolution of the *expected* log belief ratio. Iterating (10.3) up to time T_1 and taking expectations we get

$$\mathbb{E}\check{\boldsymbol{\beta}}_{\mathsf{T}_1}^{\mathsf{na}} = \bar{\lambda}\,\mathsf{T}_1, \tag{10.5}$$

where $\bar{\lambda}$ is the symmetric KL divergence introduced in (10.1). Equation (10.5) shows that the expected value of the log belief ratio grows linearly with the duration T_1 of the stationarity interval. This linear growth is a reflection of the increasing knowledge acquired by the agent as it aggregates new information embodied in the log likelihood ratios. In the asymptotic regime, this knowledge becomes a certainty. In fact, we already know from Chapter 2 that $\check{\boldsymbol{\beta}}_{\mathsf{T}_1}^{\mathsf{na}} \to \infty$ almost surely as $\mathsf{T}_1 \to \infty$, which implies that if hypothesis 1 remains in force indefinitely, the belief of the agent about this hypothesis converges to 1. Unfortunately, this increasing confidence comes at the cost of an "elephant" memory that makes the algorithm slow in adaptation, as we now show.

To this end, let us examine the behavior for $t > \mathsf{T}_1$, recalling that from time $\mathsf{T}_1 + 1$ onward the true hypothesis switches to $\vartheta^o = 2$. We have that

$$\check{\boldsymbol{\beta}}_t^{\mathsf{na}} = \check{\boldsymbol{\beta}}_{\mathsf{T}_1}^{\mathsf{na}} + \sum_{\tau=\mathsf{T}_1+1}^{t} \log\frac{\ell(\boldsymbol{x}_\tau|1)}{\ell(\boldsymbol{x}_\tau|2)}, \qquad t > \mathsf{T}_1. \tag{10.6}$$

Then, from (10.5) and (10.6) we have that

$$\mathbb{E}\check{\boldsymbol{\beta}}_t^{\mathsf{na}} = \underbrace{\mathbb{E}\check{\boldsymbol{\beta}}_{\mathsf{T}_1}^{\mathsf{na}}}_{=\bar{\lambda}\mathsf{T}_1} + \sum_{\tau=\mathsf{T}_1+1}^{t} \underbrace{\mathbb{E}\log\frac{\ell(\boldsymbol{x}_\tau|1)}{\ell(\boldsymbol{x}_\tau|2)}}_{=-\bar{\lambda}\text{ since }\vartheta^o=2} = \bar{\lambda}(2\mathsf{T}_1 - t), \quad t > \mathsf{T}_1. \tag{10.7}$$

We see from (10.7) that the earlier operation regime (i.e., for $t \le \mathsf{T}_1$) results in an initial bias term $\bar{\lambda}\mathsf{T}_1$ of positive sign. On the other hand, since the true hypothesis is now $\vartheta^o = 2$, we would like to observe a negative value for $\mathbb{E}\check{\boldsymbol{\beta}}_t^{\mathsf{na}}$. Accordingly, the adaptation time can be roughly identified by considering the time necessary to overcome the initial bias toward hypothesis 1 once the true hypothesis switches from 1 to 2 at instant T_1. In terms of our mean-value analysis, this is the time necessary for the expected log belief ratio $\mathbb{E}\check{\boldsymbol{\beta}}_t^{\mathsf{na}}$ to become nonpositive. In view of (10.7), this change happens at instant $t_0 = 2\mathsf{T}_1$. The adaptation time is computed as the difference between t_0 and T_1. Therefore, the adaptation time for the traditional, sequential update strategy (2.21) is on the order of

$$\mathsf{T}_{\mathsf{na}} = \mathsf{T}_1. \tag{10.8}$$

In other words, the time necessary to recover from an earlier wrong opinion is proportional to the stationarity interval during which that opinion was actually true! This behavior, illustrated in Figure 10.1, is clearly not admissible for an adaptive algorithm.

10.1.2 Adaptive Strategy

Let us switch to the adaptive strategy. Developing the recursion in (10.4) until time T_1 we get

$$\check{\boldsymbol{\beta}}_{\mathsf{T}_1}^{\mathsf{ad}} = \sum_{\tau=1}^{\mathsf{T}_1} (1-\delta)^{\mathsf{T}_1-\tau} \log \frac{\ell(\boldsymbol{x}_\tau|1)}{\ell(\boldsymbol{x}_\tau|2)}, \tag{10.9}$$

yielding

$$\mathbb{E}\check{\boldsymbol{\beta}}_{\mathsf{T}_1}^{\mathsf{ad}} = \bar{\lambda} \sum_{\tau=1}^{\mathsf{T}_1} (1-\delta)^{\tau-1} = \frac{\bar{\lambda}}{\delta}\left(1-(1-\delta)^{\mathsf{T}_1}\right) \approx \frac{\bar{\lambda}}{\delta}, \tag{10.10}$$

where the approximation assumes a sufficiently large T_1.

Likewise, developing the recursion in (10.4) from the time instant T_1 (i.e., from the initial state $\check{\boldsymbol{\beta}}_{\mathsf{T}_1}^{\mathsf{ad}}$) until an instant $t > \mathsf{T}_1$, where the true hypothesis becomes $\vartheta^o = 2$, we obtain

$$\check{\boldsymbol{\beta}}_t^{\mathsf{ad}} = (1-\delta)^{t-\mathsf{T}_1}\check{\boldsymbol{\beta}}_{\mathsf{T}_1}^{\mathsf{ad}} + \sum_{\tau=\mathsf{T}_1+1}^{t} (1-\delta)^{t-\tau} \log \frac{\ell(\boldsymbol{x}_\tau|1)}{\ell(\boldsymbol{x}_\tau|2)}, \quad t > \mathsf{T}_1, \tag{10.11}$$

yielding

$$
\begin{aligned}
\mathbb{E}\check{\boldsymbol{\beta}}_t^{\mathsf{ad}} &= (1-\delta)^{t-\mathsf{T}_1}\mathbb{E}\check{\boldsymbol{\beta}}_{\mathsf{T}_1}^{\mathsf{ad}} + \sum_{\tau=\mathsf{T}_1+1}^{t} (1-\delta)^{t-\tau}\,\mathbb{E}\log\frac{\ell(\boldsymbol{x}_\tau|1)}{\ell(\boldsymbol{x}_\tau|2)} \\
&\overset{\text{(a)}}{=} (1-\delta)^{t-\mathsf{T}_1}\frac{\bar{\lambda}}{\delta}\left(1-(1-\delta)^{\mathsf{T}_1}\right) - \bar{\lambda}\sum_{\tau=1}^{t-\mathsf{T}_1}(1-\delta)^{\tau-1} \\
&= (1-\delta)^{t-\mathsf{T}_1}\frac{\bar{\lambda}}{\delta}\left(1-(1-\delta)^{\mathsf{T}_1}\right) - \frac{\bar{\lambda}}{\delta}\left(1-(1-\delta)^{t-\mathsf{T}_1}\right) \\
&= \frac{\bar{\lambda}}{\delta}\left(2(1-\delta)^{t-\mathsf{T}_1} - 1 - (1-\delta)^t\right) \\
&\approx \frac{\bar{\lambda}}{\delta}\left(2(1-\delta)^{t-\mathsf{T}_1} - 1\right), \qquad t > \mathsf{T}_1,
\end{aligned} \tag{10.12}
$$

where in step (a) we apply (10.10) (actually, the final equality, not the approximation) to evaluate $\mathbb{E}\check{\boldsymbol{\beta}}_{\mathsf{T}_1}^{\mathsf{ad}}$ and we use the fact that $\mathbb{E}\log\frac{\ell(\boldsymbol{x}_\tau|1)}{\ell(\boldsymbol{x}_\tau|2)} = -\bar{\lambda}$ for $\tau > \mathsf{T}_1$. The last approximation holds for sufficiently large t. Equating (10.12) to 0 we obtain

$$
t_0 = \frac{\log 2}{\log(1-\delta)^{-1}} + \mathsf{T}_1 \approx \frac{\log 2}{\delta} + \mathsf{T}_1, \tag{10.13}
$$

where we used the fact that $1/\log(1-\delta)^{-1} \approx 1/\delta$ for small δ. Evaluating the adaptation time as $\mathsf{T}_{\mathsf{ad}} = t_0 - \mathsf{T}_1$, from (10.13) we get

$$
\mathsf{T}_{\mathsf{ad}} \approx \frac{\log 2}{\delta}. \tag{10.14}
$$

10.1.3 Comparison

A visual comparison between the nonadaptive and adaptive strategies is shown in Figure 10.1, where the expected log belief ratios $\mathbb{E}\check{\boldsymbol{\beta}}_t^{\mathsf{na}}$ and $\mathbb{E}\check{\boldsymbol{\beta}}_t^{\mathsf{ad}}$ are depicted as functions of t. Comparing (10.14) against (10.8), we see that for the nonadaptive strategy the adaptation time diverges as the duration T_1 of the stationarity interval increases, whereas for the adaptive strategy it is *independent* of T_1, and is controlled by the parameter δ, scaling roughly as $1/\delta$. One explanation for this difference is that the expected log belief ratio of the adaptive strategy given by (10.10) converges as $\mathsf{T}_1 \to \infty$, to the stable value $\bar{\lambda}/\delta$ that depends on δ. In contrast, for the nonadaptive strategy the expected log belief ratio in (10.5), increases linearly with T_1. This implies that, after a relatively long stationarity

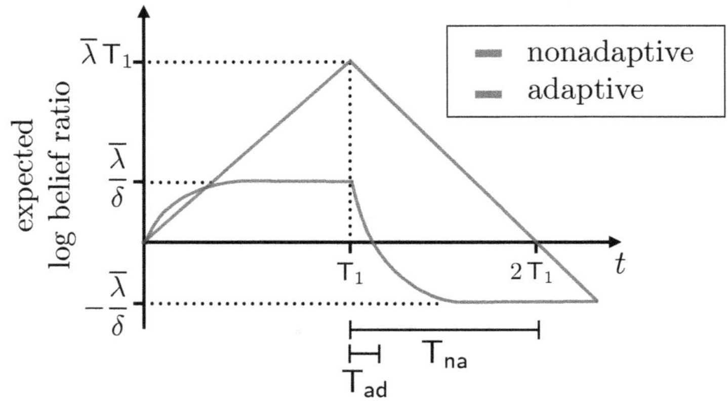

Figure 10.1: Single-agent learning under nonstationary conditions. Time evolution of the expected log belief ratios $\mathbb{E}\check{\boldsymbol{\beta}}_t^{\mathsf{na}}$ (traditional nonadaptive strategy (10.3), in blue) and $\mathbb{E}\check{\boldsymbol{\beta}}_t^{\mathsf{ad}}$ (adaptive strategy (10.4), in red).

interval, the nonadaptive strategy accumulates an initial bias that is more difficult to overcome.

In a nutshell, while the reaction capacity of traditional social learning given by (10.3) is not controlled by design and is severely affected by the duration of previous stationarity intervals, in the adaptive social learning update (10.4) the adaptation time is not affected by previous stationarity intervals, and the effective memory is controlled through the adaptation parameter δ. This adaptation ability comes at the expense of learning accuracy. In fact, as we have already established in the previous chapter, the steady-state error probability does not converge to 0 as time elapses, but converges to some stable value. However, this value vanishes exponentially fast as a function of $1/\delta$, highlighting the fundamental trade-off of adaptive social learning: The smaller the adaptation parameter δ is, the smaller the error probability will be (i.e., better learning accuracy) and the larger the adaptation time will be (i.e., slower adaptation).

10.2 Quantitative Transient Analysis

In this section we provide a more rigorous analysis to support the qualitative arguments of Section 10.1. We assume that the ASL strategy (8.13) has been in operation for a certain time t_0. All the knowledge accumulated by the agents until this time is summarized in the belief vectors $\{\mu_{k,t_0}\}_{k=1}^{K}$. We remark that the evolution of the statistical models from $t = 0$ to $t = t_0$ is left completely arbitrary, that is, the system could have experienced several drifts in the statistical conditions and/or other system parameters, e.g., the graph combination weights. From the viewpoint of the ASL algorithm, all these effects are summarized in the belief vectors $\{\mu_{k,t_0}\}_{k=1}^K$ that act as initial state at time t_0. In order to perform the transient analysis, we assume that from $t_0 + 1$ onward, some models $\{f_k(x)\}_{k=1}^K$ steadily govern the data of the different agents, with some target hypothesis equal to $\vartheta^\star$ — see Definition 8.1. We will establish how much time is necessary to get sufficiently close to the steady-state learning performance starting from the initial realization $\{\mu_{k,t_0}\}_{k=1}^K$. As done before, to simplify the notation we set $t_0 = 0$ and the initial state becomes $\{\mu_{k,0}\}_{k=1}^K$.

In the theory of adaptation and learning, the focus is typically on estimation of a continuous parameter, and the transient analysis is performed by characterizing the time evolution of suitable moments of an error variable, e.g., the second-order moment of a vector quantifying the difference between the estimated and true parameters. In this setting, the transient analysis ascertains how long it takes for the error to attain some small value [151, 154]. In comparison, in the social learning setting the adaptation time will be related to the time evolution of the instantaneous error probability introduced in (9.20), and specifically to the time necessary for this probability to approach the *steady-state* error probability.

The time evolution of the instantaneous error probability will be characterized in terms of the upper bound provided in Theorem 10.1. As we will see from the proof of the theorem, this bound relies on the logarithmic moment generating function of the log belief ratios. Compare this approach with the one adopted for the estimation of continuous parameters. In the latter case we examine the time evolution of *moments*, while, in adaptive social learning, it will be important to characterize the time evolution of *logarithmic moment generating functions*. This fact admits the following interesting interpretation: Since the logarithmic moment generating function of a random variable incorporates dependence on *all moments* of the variable, the transient analysis of adaptive social learning relies on all moments, while in problems addressing the estimation of continuous parameters we need only individual moments.

Theorem 10.1 (Bounds on the instantaneous error probability). Let Assumptions 5.1, 5.2, and 6.1 be satisfied. Let C and r be the constants defined in (4.25), which are determined by the combination matrix. Assume that, for all $\theta \neq \vartheta^\star$, $\lambda_{\mathsf{inf}}(\theta) \triangleq \inf\left(\mathsf{supp}_{\lambda_{\mathsf{net}}(\theta)}\right) < 0$, where $\mathsf{supp}_{\lambda_{\mathsf{net}}(\theta)}$ denotes the support of the distribution of the network average of log likelihood ratios $\boldsymbol{\lambda}_{\mathsf{net},t}(\theta)$ (see (6.7) and Definition E.1), and let $s_\theta^\star$ be the quantity introduced in Corollary 9.4. Define the scaled log belief ratios, for $k = 1, 2, \ldots, K$ and $t = 0, 1, \ldots$,

$$\boldsymbol{b}_{k,t}(\theta) \triangleq \delta \times \boldsymbol{\beta}_{k,t}(\theta), \tag{10.15}$$

and the following network average, corresponding to the (deterministic) initial values $b_{k,0}(\theta)$ weighted by the Perron vector entries $\{v_k\}$:

$$b_{\mathsf{net},0}(\theta) \triangleq \sum_{k=1}^{K} v_k b_{k,0}(\theta), \tag{10.16}$$

Let, for all $\theta \neq \vartheta^\star$,

$$\mathsf{K}_1(\theta) \triangleq |s_\theta^\star| \left[\bar{\lambda}_{\mathsf{net}}(\theta) - b_{\mathsf{net},0}(\theta)\right], \tag{10.17}$$

$$\mathsf{K}_2(\theta) \triangleq C\,|s_\theta^\star| \sum_{k=1}^{K} |b_{k,0}(\theta)|, \tag{10.18}$$

where $\bar{\lambda}_{\mathsf{net}}(\theta)$ is defined by (6.10). Then, for each agent k, the instantaneous error probability $p_{k,t}$ defined by (9.20) is upper bounded as

$$p_{k,t} \leq \sum_{\theta \neq \vartheta^\star} \exp\left\{\frac{1}{\delta}\Big[-\Phi(\theta) + \mathsf{K}_1(\theta)(1-\delta)^t + \mathsf{K}_2(\theta)(1-\delta)^t r^t + O(\delta)\Big]\right\}, \tag{10.19}$$

where $\Phi(\theta)$ is defined by (9.81).

Proof. Recalling the representation in (9.5), we can write

$$\boldsymbol{\beta}_{k,t}(\theta) = (1-\delta)^t \sum_{j=1}^{K} [A^t]_{jk} \beta_{j,0}(\theta) + \widehat{\boldsymbol{\beta}}_{k,t}(\theta). \tag{10.20}$$

Let us introduce the scaled quantity

$$\widehat{\boldsymbol{b}}_{k,t}(\theta) \triangleq \delta \times \widehat{\boldsymbol{\beta}}_{k,t}(\theta) = \delta \sum_{j=1}^{K} \sum_{\tau=1}^{t} (1-\delta)^{\tau-1} [A^\tau]_{jk}\, \boldsymbol{\lambda}_{j,t-\tau+1}(\theta), \tag{10.21}$$

where the equality follows from (9.6). Since $\boldsymbol{b}_{k,t}(\theta) = \delta \times \boldsymbol{\beta}_{k,t}(\theta)$ by definition, from

(10.20) and (10.21) we get

$$\begin{aligned}
\boldsymbol{b}_{k,t}(\theta) &= \widehat{\boldsymbol{b}}_{k,t}(\theta) + (1-\delta)^t \sum_{j=1}^{K} [A^t]_{jk} \boldsymbol{b}_{j,0}(\theta) \\
&= \widehat{\boldsymbol{b}}_{k,t}(\theta) + (1-\delta)^t \sum_{j=1}^{K} v_j \boldsymbol{b}_{j,0}(\theta) + (1-\delta)^t \sum_{j=1}^{K} \Big([A^t]_{jk} - v_j \Big) \boldsymbol{b}_{j,0}(\theta) \\
&\geq \widehat{\boldsymbol{b}}_{k,t}(\theta) + (1-\delta)^t \sum_{j=1}^{K} v_j \boldsymbol{b}_{j,0}(\theta) - C(1-\delta)^t r^t \sum_{j=1}^{K} |\boldsymbol{b}_{j,0}(\theta)| \\
&= \widehat{\boldsymbol{b}}_{k,t}(\theta) + (1-\delta)^t \boldsymbol{b}_{\mathsf{net},0}(\theta) - \frac{\mathsf{K}_2(\theta)}{|s_\theta^\star|}(1-\delta)^t r^t, \qquad (10.22)
\end{aligned}$$

where the inequality follows from (4.25), and in the last equality we used (10.16) and (10.18). In view of (10.22) we can write

$$\begin{aligned}
\mathbb{P}[\boldsymbol{b}_{k,t}(\theta) \leq 0] &\leq \mathbb{P}\left[\widehat{\boldsymbol{b}}_{k,t}(\theta) \leq -(1-\delta)^t \boldsymbol{b}_{\mathsf{net},0}(\theta) + \frac{\mathsf{K}_2(\theta)}{|s_\theta^\star|}(1-\delta)^t r^t\right] \\
&\stackrel{(a)}{=} \mathbb{P}\left[\frac{s_\theta^\star}{\delta}\widehat{\boldsymbol{b}}_{k,t}(\theta) \geq \frac{|s_\theta^\star|}{\delta}(1-\delta)^t \boldsymbol{b}_{\mathsf{net},0}(\theta) - \frac{\mathsf{K}_2(\theta)}{\delta}(1-\delta)^t r^t\right] \\
&\stackrel{(b)}{\leq} \frac{\mathbb{E}\exp\left\{\frac{s_\theta^\star}{\delta}\widehat{\boldsymbol{b}}_{k,t}(\theta)\right\}}{\exp\left\{\frac{|s_\theta^\star|}{\delta}(1-\delta)^t \boldsymbol{b}_{\mathsf{net},0}(\theta) - \frac{\mathsf{K}_2(\theta)}{\delta}(1-\delta)^t r^t\right\}} \\
&\stackrel{(c)}{=} \exp\left\{\frac{1}{\delta}\left[\delta\widehat{\Lambda}_{k,t}\left(\frac{s_\theta^\star}{\delta};\theta\right) - (1-\delta)^t |s_\theta^\star| \boldsymbol{b}_{\mathsf{net},0}(\theta) + \mathsf{K}_2(\theta)(1-\delta)^t r^t\right]\right\},
\end{aligned} \qquad (10.23)$$

where (a) follows by multiplying by $s_\theta^\star/\delta$ both sides of the inequality within the probability brackets and taking into account the fact that $s_\theta^\star < 0$ (see Corollary 9.4); (b) follows by applying Chernoff's bound (Theorem C.3); and in (c) we introduced the LMGF of $\widehat{\boldsymbol{b}}_{k,t}(\theta)$, defined as

$$\widehat{\Lambda}_{k,t}(s;\theta) \triangleq \log \mathbb{E}\exp\left\{s\,\widehat{\boldsymbol{b}}_{k,t}(\theta)\right\}. \qquad (10.24)$$

We now want to obtain a convenient expression for the LMGF $\widehat{\Lambda}_{k,t}(s;\theta)$ in (10.24). To this end, we will appeal to some results from Appendix F, which are more easily illustrated by introducing the following ad-hoc notation. Let us set, for $j = 1, 2, \ldots, K$

and $\tau \in \mathbb{N}$,

$$\boldsymbol{y}_{j,\tau} = \boldsymbol{\lambda}_{j,\tau}(\theta), \qquad \boldsymbol{y}_\tau = [\boldsymbol{y}_{1,\tau}, \boldsymbol{y}_{2,\tau}, \dots, \boldsymbol{y}_{K,\tau}], \tag{10.25}$$

$$\alpha_{j,\tau} = [A^\tau]_{jk}, \qquad \alpha_\tau = [\alpha_{1,\tau}, \alpha_{2,\tau}, \dots, \alpha_{K,\tau}], \qquad \alpha = v, \tag{10.26}$$

$$\boldsymbol{z}_t(\delta) = \delta \sum_{\tau=1}^{t} (1-\delta)^{\tau-1} \alpha_\tau^\top \boldsymbol{y}_\tau, \qquad \boldsymbol{z}(\delta) = \delta \sum_{\tau=1}^{\infty} (1-\delta)^{\tau-1} \alpha_\tau^\top \boldsymbol{y}_\tau, \tag{10.27}$$

$$\boldsymbol{y}_{\mathsf{ave},\tau} = v^\top \boldsymbol{y}_\tau = \boldsymbol{\lambda}_{\mathsf{net},\tau}(\theta), \qquad \Lambda_{\mathsf{ave}}(s) = \log \mathbb{E} \exp\left\{ s\, \boldsymbol{y}_{\mathsf{ave},\tau} \right\}, \tag{10.28}$$

$$\Lambda_y(u) = \log \mathbb{E} \exp\left\{ u^\top \boldsymbol{y}_t \right\}, \quad u \in \mathbb{R}^K, \tag{10.29}$$

$$\Lambda_{z_t}(s) = \log \mathbb{E} \exp\left\{ s\, \boldsymbol{z}_t(\delta) \right\}, \qquad \Lambda_\delta(s) = \log \mathbb{E} \exp\left\{ s\, \boldsymbol{z}(\delta) \right\}. \tag{10.30}$$

Now, observe that from (F.107) we have the representation

$$\begin{aligned} \Lambda_{z_t}(s) &= \sum_{\tau=1}^{t} \Lambda_{\mathsf{ave}}\Big(s\,\delta(1-\delta)^{\tau-1} \Big) \\ &+ \sum_{\tau=1}^{t} \Big[\Lambda_y\Big(s\,\delta(1-\delta)^{\tau-1} \alpha_\tau \Big) - \Lambda_y\Big(s\,\delta(1-\delta)^{\tau-1} \alpha \Big) \Big], \end{aligned} \tag{10.31}$$

which, in view of (F.119) and (F.120), implies that

$$\delta \Lambda_{z_t}(s/\delta) = \delta \sum_{\tau=1}^{t} \Lambda_{\mathsf{ave}}\Big(s(1-\delta)^{\tau-1} \Big) + O(\delta). \tag{10.32}$$

On the other hand, in view of Eqs. (F.97), (F.100), and (F.123), the summation on the RHS can be written as

$$\int_{s(1-\delta)^t}^{s} \frac{\Lambda_{\mathsf{ave}}(\varsigma)}{\varsigma} d\varsigma + O(\delta), \tag{10.33}$$

which combined with (10.32) yields

$$\delta \Lambda_{z_t}(s/\delta) = \int_{s(1-\delta)^t}^{s} \frac{\Lambda_{\mathsf{ave}}(\varsigma)}{\varsigma} d\varsigma + O(\delta). \tag{10.34}$$

Exploiting (10.21), (10.24), and the chain of definitions (10.25)–(10.30), we arrive at the identity

$$\Lambda_{z_t}(s) = \widehat{\Lambda}_{k,t}(s;\theta). \tag{10.35}$$

Likewise, using (10.28) and recalling from Table 6.1 that the LMGF of the average variable $\boldsymbol{\lambda}_{\mathsf{net},t}$ is denoted by $\Lambda_{\mathsf{net}}(s;\theta)$, we have

$$\Lambda_{\mathsf{ave}}(s) = \Lambda_{\mathsf{net}}(s;\theta). \tag{10.36}$$

Substituting (10.35) and (10.36) into (10.34), we obtain

$$\begin{aligned} \delta \widehat{\Lambda}_{k,t}(s/\delta;\theta) &= \int_{s(1-\delta)^t}^{s} \frac{\Lambda_{\mathsf{net}}(\varsigma)}{\varsigma} d\varsigma + O(\delta) \\ &= \int_0^{s} \frac{\Lambda_{\mathsf{net}}(\varsigma;\theta)}{\varsigma} d\varsigma - \int_0^{s(1-\delta)^t} \frac{\Lambda_{\mathsf{net}}(\varsigma;\theta)}{\varsigma} d\varsigma + O(\delta) \\ &= \phi(s;\theta) - \int_0^{s(1-\delta)^t} \frac{\Lambda_{\mathsf{net}}(\varsigma;\theta)}{\varsigma} d\varsigma + O(\delta), \end{aligned} \tag{10.37}$$

where the last equality follows from the definition of $\phi(s;\theta)$ in (9.79). Applying (10.37) with the choice $s = s_\theta^\star$, we get

$$\begin{aligned}
\delta\widehat{\Lambda}_{k,t}\left(s_\theta^\star/\delta;\theta\right) &= \phi(s_\theta^\star;\theta) - \int_0^{s_\theta^\star(1-\delta)^t} \frac{\Lambda_{\mathsf{net}}(\varsigma;\theta)}{\varsigma}d\varsigma + O(\delta)\\
&\overset{(a)}{=} -\Phi(\theta) - \int_0^{s_\theta^\star(1-\delta)^t} \frac{\Lambda_{\mathsf{net}}(\varsigma;\theta)}{\varsigma}d\varsigma + O(\delta)\\
&\overset{(b)}{=} -\Phi(\theta) + \int_{-|s_\theta^\star|(1-\delta)^t}^{0} \frac{\Lambda_{\mathsf{net}}(\varsigma;\theta)}{\varsigma}d\varsigma + O(\delta)\\
&\overset{(c)}{\leq} -\Phi(\theta) + (1-\delta)^t\,|s_\theta^\star|\,\bar{\lambda}_{\mathsf{net}}(\theta) + O(\delta),
\end{aligned} \tag{10.38}$$

where (a) follows by using the definition of $\Phi(\theta)$ from (9.104); (b) holds because $s_\theta^\star$ is negative; and (c) follows by observing that, in view of (A.3a) and the strict convexity of the LMGF Λ_{net}, for $\varsigma < 0$ we have

$$\frac{\Lambda_{\mathsf{net}}(\varsigma;\theta)}{\varsigma} < \Lambda'_{\mathsf{net}}(0;\theta) = \bar{\lambda}_{\mathsf{net}}(\theta). \tag{10.39}$$

Using (10.38) in (10.23) along with the definition of $\mathsf{K}_1(\theta)$ from (10.17), we get the upper bound in (10.19).

■

Theorem 10.1 reveals the main behavior of the *instantaneous* error probability. Examining the exponent of the upper bound in (10.19) we see, up to higher-order corrections embodied in the term $O(\delta)$, the emergence of three terms: the *steady-state* error exponent $\Phi(\theta)$ already identified in Theorem 9.4, and two other terms that characterize the *transient* behavior. The first transient term decays as $(1-\delta)^t$, and is thus influenced solely by the adaptation parameter δ. The second transient term decays as $(1-\delta)^t r^t$, which means it decays faster and is influenced also by the network through the combination-matrix parameter r. According to (4.24), this parameter is related to the second largest-magnitude eigenvalue of A, and is therefore related to the mixing properties of A (i.e., the convergence rate of $[A^t]_{jk}$ to the Perron vector entry v_j).

It is important to make a remark in relation to the terms $b_{\mathsf{net},0}$ and $b_{k,0}$ appearing in (10.17) and (10.18), respectively. In view of (10.15), we have

$$b_{k,0} = \delta \times \beta_{k,0} \tag{10.40}$$

and, from (10.16), also $b_{\mathsf{net},0}$ implicitly contains the multiplying factor δ. Accordingly, instead of including $b_{\mathsf{net},0}$ and $b_{k,0}$ in (10.17) and (10.18), it appears that we could have incorporated them into the $O(\delta)$ correction in (10.19). We now explain why this is not the best choice and why it is more useful to leave explicit the dependence on these two terms. Recall that the time instant $t = 0$ in our analysis represents an arbitrary time instant after which a stationary period begins. For example, $t = 0$ can correspond to the end of a previous stationary period (*learning cycle*) where the agents learned a certain model that has then changed at the beginning ($t = 1$) of the subsequent learning cycle. In other words, the "initial" state $b_{k,0}$ can correspond to the *steady state* of the previous learning cycle. In this case, $b_{k,0}$ would contain an implicit dependence on δ, since it would be the steady-state output of the ASL algorithm. More specifically, from Theorem 9.2 we know that in steady state this scaled log belief ratio approximates $\bar{\lambda}_{\mathsf{net}}$ for small δ. Therefore, when $b_{k,0}$ is interpreted as the steady-state vector of a previous learning cycle, from (9.28) we can write

$$b_{k,0} \approx \bar{\lambda}_{\mathsf{net}}^{\mathsf{prev}}, \qquad b_{\mathsf{net},0} \approx \bar{\lambda}_{\mathsf{net}}^{\mathsf{prev}}, \tag{10.41}$$

where we denote by $\bar{\lambda}_{\mathsf{net}}^{\mathsf{prev}}$ the limiting value characterizing the *previous* learning cycle. Note that (10.41) is *not* an $O(\delta)$ correction. For this reason, it is more appropriate to leave explicit the dependence on $b_{\mathsf{net},0}$ and $b_{k,0}$ and not to incorporate these terms into the $O(\delta)$ correction in (10.19).

10.3 Adaptation Time

In summary, Theorem 10.1 provides the upper bound in (10.19) on the instantaneous error probability. As $t \to \infty$, this bound converges to

$$\sum_{\theta \neq \vartheta^\star} \exp\left\{-\frac{1}{\delta}\Big[\Phi(\theta) + O(\delta)\Big]\right\} = \exp\left\{-\frac{1}{\delta}\Big[\Phi + O(\delta)\Big]\right\}, \tag{10.42}$$

where $\Phi = \min_{\theta \neq \vartheta^\star} \Phi(\theta)$ is the error exponent in (9.82).[2] In the theory of adaptation and learning [151], adaptation times are usually defined in

[2]The equality in (10.42) can be obtained as follows. Since there exists at least one value $\theta \neq \vartheta^\star$ such that $\Phi = \Phi(\theta)$, we have

$$\sum_{\theta \neq \vartheta^\star} \exp\left\{-\frac{1}{\delta}\Big[\Phi(\theta) + O(\delta)\Big]\right\} \geq \exp\left\{-\frac{1}{\delta}\Big[\Phi + O(\delta)\Big]\right\}. \tag{10.43}$$

On the other hand, we can write

$$\begin{aligned}
&\sum_{\theta \neq \vartheta^\star} \exp\left\{-\frac{1}{\delta}\Big[\Phi(\theta) + O(\delta)\Big]\right\} \leq (H-1)\exp\left\{-\frac{1}{\delta}\Big[\Phi + O(\delta)\Big]\right\} \\
&= \exp\left\{-\frac{1}{\delta}\Big[\Phi + O(\delta) + \delta\log(H-1)\Big]\right\} = \exp\left\{-\frac{1}{\delta}\Big[\Phi + O(\delta)\Big]\right\}.
\end{aligned} \tag{10.44}$$

terms of the number of iterations necessary to get "sufficiently" close to some limiting (i.e., steady-state) value. In our setting, we will apply this concept to the available upper bounds on the error probability, with the RHS of (10.42) being our limiting value. Specifically, we say that $\mathsf{T}_{\mathsf{ASL}}$ is a valid *adaptation time* when, for all $t > \mathsf{T}_{\mathsf{ASL}}$,

$$p_{k,t} \leq \exp\left\{-\frac{1}{\delta}\Big[(1-\varepsilon)\Phi + O(\delta)\Big]\right\}. \tag{10.45}$$

In other words, we require that, after $\mathsf{T}_{\mathsf{ASL}}$, the instantaneous error probability $p_{k,t}$ is upper bounded by a quantity that matches the exponent Φ on the RHS of (10.42), but for some small ε. This is made precise in the following corollary, where we determine expressions for the adaptation time by distinguishing the cases of "favorable" and "unfavorable" initial states.

The favorable scenario is identified by the condition $b_{\mathsf{net},0}(\theta) \geq \bar{\lambda}_{\mathsf{net}}(\theta)$ for all $\theta \neq \vartheta^\star$, which means that the initial states are larger than the limiting values $\bar{\lambda}_{\mathsf{net}}(\theta)$ to which the scaled log belief ratio converges in view of Theorem 9.2. We see from (10.17) that when $b_{\mathsf{net},0}(\theta) \geq \bar{\lambda}_{\mathsf{net}}(\theta)$, the terms $\mathsf{K}_1(\theta)$ are all nonpositive. Examining (10.19), we conclude that these terms contribute to reducing the value of the upper bound in (10.19) (or are irrelevant if they are equal to 0). For this reason, we say that when $b_{\mathsf{net},0}(\theta) \geq \bar{\lambda}_{\mathsf{net}}(\theta)$ for all $\theta \neq \vartheta^\star$ we are in a favorable scenario. The situation is reversed when $b_{\mathsf{net},0}(\theta) < \bar{\lambda}_{\mathsf{net}}(\theta)$ for at least one $\theta \neq \vartheta^\star$, since in this case at least one of the terms $\mathsf{K}_1(\theta)$ is positive, thus contributing to increase the value of the upper bound in (10.19).

Corollary 10.1 (Adaptation time). Under the same assumptions used in Theorem 10.1, let

$$\mathsf{K}_1 \triangleq \max_{\theta \neq \vartheta^\star} \mathsf{K}_1(\theta), \qquad \mathsf{K}_2 \triangleq \max_{\theta \neq \vartheta^\star} \mathsf{K}_2(\theta), \tag{10.46}$$

and let the *adaptation time* $\mathsf{T}_{\mathsf{ASL}}$ be a time instant such that, for all $t > \mathsf{T}_{\mathsf{ASL}}$,

$$p_{k,t} \leq e^{-\frac{1}{\delta}[(1-\varepsilon)\Phi + O(\delta)]} \tag{10.47}$$

for some small $\varepsilon > 0$. Then, we have the following two scenarios:

Favorable case (all initial states are good). If $b_{\mathsf{net},0}(\theta) \geq \bar{\lambda}_{\mathsf{net}}(\theta)$ for all $\theta \neq \vartheta^\star$, then for $\varepsilon < \mathsf{K}_2/\Phi$,

$$\mathsf{T}_{\mathsf{ASL}} = \frac{1}{\log r^{-1}} \log \frac{\mathsf{K}_2}{\varepsilon\,\Phi}. \tag{10.48}$$

Unfavorable case (at least one initial state is bad). If $b_{\mathsf{net},0}(\theta) < \bar{\lambda}_{\mathsf{net}}(\theta)$ for at least one $\theta \neq \vartheta^\star$, then for $\varepsilon < \mathsf{K}_1/\Phi$,

$$\mathsf{T}_{\mathsf{ASL}} = \frac{1}{\log(1-\delta)^{-1}} \log \frac{\mathsf{K}_1}{\varepsilon\,\Phi}. \tag{10.49}$$

Proof. We determine the adaptation time as the critical instant after which we stay close to the exponent Φ, in the precise sense specified by (10.47). For ease of reference, it is useful to report here (10.50), namely,

$$p_{k,t} \le \sum_{\theta\neq\vartheta^\star} \exp\left\{\frac{1}{\delta}\Big[-\Phi(\theta) + \mathsf{K}_1(\theta)(1-\delta)^t + \mathsf{K}_2(\theta)(1-\delta)^t r^t + O(\delta)\Big]\right\}. \tag{10.50}$$

Since $\Phi(\theta) \ge \Phi$ (see (9.82)), $\mathsf{K}_1(\theta) \le \mathsf{K}_1$ and $\mathsf{K}_2(\theta) \le \mathsf{K}_2$ (see (10.46)), and since $0 < \delta < 1$, from (10.50) we can write

$$\begin{aligned} p_{k,t} &\le \sum_{\theta\neq\vartheta^\star} \exp\left\{\frac{1}{\delta}\Big[-\Phi + \mathsf{K}_1(1-\delta)^t + \mathsf{K}_2 r^t + O(\delta)\Big]\right\} \\ &= (H-1)\exp\left\{\frac{1}{\delta}\Big[-\Phi + \mathsf{K}_1(1-\delta)^t + \mathsf{K}_2 r^t + O(\delta)\Big]\right\}. \end{aligned} \tag{10.51}$$

The constant factor $H-1$ can be incorporated into the $O(\delta)$ correction, yielding

$$p_{k,t} \le \exp\left\{\frac{1}{\delta}\Big[-\Phi + \mathsf{K}_1(1-\delta)^t + \mathsf{K}_2\, r^t + O(\delta)\Big]\right\}. \tag{10.52}$$

We now use (10.52) to evaluate the adaptation time in the favorable and unfavorable cases.

Let us consider first the favorable case, where $b_{\mathsf{net}}(\theta) \ge \bar{\lambda}_{\mathsf{net}}(\theta)$ for all $\theta \neq \vartheta^\star$, implying, in view of (10.17) and (10.46), that $\mathsf{K}_1 \le 0$, such that from (10.52) we have

$$p_{k,t} \le \exp\left\{\frac{1}{\delta}\Big[-\Phi + \mathsf{K}_2 r^t + O(\delta)\Big]\right\}. \tag{10.53}$$

On the other hand, with the choice of $\mathsf{T}_{\mathsf{ASL}}$ in (10.48) we have

$$t > \mathsf{T}_{\mathsf{ASL}} \iff t > \frac{1}{\log r^{-1}} \log\frac{\mathsf{K}_2}{\varepsilon\,\Phi} \iff \mathsf{K}_2\, r^t < \varepsilon\,\Phi, \tag{10.54}$$

which, when used in (10.53), yields (10.47). Thus, we have proved the claim for the case where $b_{\mathsf{net}}(\theta) \ge \bar{\lambda}_{\mathsf{net}}(\theta)$ for all $\theta \neq \vartheta^\star$.

We examine next the unfavorable case where $b_{\mathsf{net}}(\theta) < \bar{\lambda}_{\mathsf{net}}(\theta)$ for at least one value $\theta \neq \vartheta^\star$. Observe that in this case we have $\mathsf{K}_1 = \max_{\theta\neq\vartheta^\star} \mathsf{K}_1(\theta) > 0$. If we set the adaptation time $\mathsf{T}_{\mathsf{ASL}}$ according to the law in (10.49), we have

$$t > \mathsf{T}_{\mathsf{ASL}} \iff t > \frac{1}{\log(1-\delta)^{-1}} \log\frac{\mathsf{K}_1}{\varepsilon\,\Phi} \iff \mathsf{K}_1(1-\delta)^t < \varepsilon\Phi, \tag{10.55}$$

which, when used in (10.52), yields

$$p_{k,t} \le \exp\left\{\frac{1}{\delta}\Big[-(1-\varepsilon)\Phi + \mathsf{K}_2\, r^t + O(\delta)\Big]\right\}. \tag{10.56}$$

Moreover, from the known bound $\log x \le x - 1$, holding for all $x > 0$, we have the inequality

$$\log \frac{1}{1-\delta} \le \frac{1}{1-\delta} - 1 = \frac{\delta}{1-\delta}, \tag{10.57}$$

implying

$$\frac{1}{\log(1-\delta)^{-1}} \ge \frac{1-\delta}{\delta} = \frac{1}{\delta} - 1. \tag{10.58}$$

In the range $t > \mathsf{T}_{\mathsf{ASL}}$, from (10.55) and (10.58) we obtain (recall that $\varepsilon < \mathsf{K}_1/\Phi$)

$$t > \left(\frac{1}{\delta} - 1\right) \log \frac{\mathsf{K}_1}{\varepsilon\,\Phi}, \tag{10.59}$$

which, since $0 < r < 1$, also implies

$$r^t \le r^{\left(\frac{1}{\delta}-1\right) \log \frac{\mathsf{K}_1}{\varepsilon\,\Phi}}, \tag{10.60}$$

which implies that the quantity $\mathsf{K}_2\, r^t$ appearing in (10.56) can be incorporated into the term $O(\delta)$, yielding (10.47), and the proof is complete.

■

We are now ready to examine the main parameters and phenomena affecting the adaptation time $\mathsf{T}_{\mathsf{ASL}}$.

Memory. The memory from the past evolution of the algorithm is summarized in the starting belief vectors $\{\mu_{k,0}\}_{k=1}^K$, which determine the initial values $\{b_{k,0}\}_{k=1}^K$ and $b_{\mathsf{net},0}$.

As we observed before stating the corollary, when $b_{\mathsf{net},0}(\theta) \ge \bar{\lambda}_{\mathsf{net}}(\theta)$ we have $\mathsf{K}_1(\theta) \le 0$, and the transient term $\mathsf{K}_1(\theta)(1-\delta)^t$ reduces the value of the error probability (bound) or is irrelevant. Therefore, when $b_{\mathsf{net},0}(\theta) \ge \bar{\lambda}_{\mathsf{net}}(\theta)$ for all $\theta \neq \vartheta^\star$, we see from (10.19) that the dominant transient term is the one scaling as $(1-\delta)^t r^t$; the corresponding adaptation time in (10.48) is essentially determined by the mixing parameter r, i.e., by how fast the powers of the combination matrix converge to $v\,\mathbb{1}^\mathsf{T}$ — see Corollary 4.1. Under this regime, the adaptation time *does not depend critically on the adaptation parameter* δ. Moreover, the adaptation time in (10.48) increases for larger initial values $|b_{k,0}(\theta)|$ — see (10.18).

In comparison, when $b_{\mathsf{net},0}(\theta) < \bar{\lambda}_{\mathsf{net}}(\theta)$ for at least one $\theta \neq \vartheta^\star$, the dominant transient term is the one scaling as $(1-\delta)^t$; the corresponding adaptation time in (10.49) *scales with the adaptation parameter as* $1/\log(1-\delta)^{-1}$. For small δ, this scaling law can be approximated by $1/\delta$, which means that the adaptation time grows as the inverse of the adaptation parameter δ when $\delta \to 0$.

Within the unfavorable scenario, one particularly interesting case is when $b_{\mathsf{net},0}(\theta)$ is negative. This happens, for example, when the initial state comes from a previous learning cycle where the agents have converged to some hypothesis that has then changed at the beginning of the subsequent learning cycle. To see why in this situation we have $b_{\mathsf{net},0}(\theta) < 0$, let us examine a single learning cycle. Recall that we use the convention that the data of the learning cycle under examination are collected from time instant $t = 1$, and that the initial belief at time instant $t = 0$ collects all the knowledge stored until the beginning of the learning cycle, that is, the knowledge accumulated from previous learning cycles. Assume for instance that in the previous learning cycle the algorithm had been converging to some hypothesis θ, which has then switched to $\vartheta^\star$ at the beginning $(t = 1)$ of the successive learning cycle. This means that the log belief ratio between θ and $\vartheta^\star$ was positive at the end of the previous learning cycle. Actually, since in the new learning cycle we compute log belief ratios in the reverse direction, i.e., between $\vartheta^\star$ and θ, we have $b_{\mathsf{net},0}(\theta) < 0$. Therefore, the smaller $b_{\mathsf{net},0}(\theta)$ is, the worse the starting condition will be. We expect that a worse starting condition has a negative impact on the adaptation time. This is confirmed by (10.17), because smaller values of $b_{\mathsf{net},0}(\theta) < 0$ imply larger values of $\mathsf{K}_1(\theta)$, which in turn correspond to increasing the adaptation time in (10.49).

Finally, observe from (10.48) and (10.49) that the dependence of the adaptation time on K_1 and K_2 is logarithmic. Since we see from (10.17) and (10.18) that $\mathsf{K}_1(\theta)$ and $\mathsf{K}_2(\theta)$ embody the initial states, we conclude that the past algorithm evolution does not have a critical impact on the adaptation time.

Parameter $s_\theta^\star$. The parameter $s_\theta^\star$ influences the adaptation time through the constants $\mathsf{K}_1(\theta)$ and $\mathsf{K}_2(\theta)$ — see (10.17) and (10.18). Some insight into the role of $s_\theta^\star$ can be gained by focusing on the following setting. Consider the objective evidence model in Section 5.3 and assume statistical independence across the agents. Under this setting, the following inequalities were proved in [25]:

$$\frac{1}{v_{\mathsf{max}}} \leq |s_\theta^\star| \leq \frac{1}{v_{\mathsf{min}}}, \tag{10.61}$$

where v_{max} and v_{min} denote the maximum and minimum entries of the Perron vector of A. On the face of it, these bounds suggest a dependence of the adaptation times in (10.48) and (10.49) on the network parameters through the Perron vector.

However, we should recall that the network error exponent Φ, defined in (9.82) and appearing in (10.48) and (10.49), depends on the network as well. For example, when all likelihoods are equal across the agents and the combination matrix is doubly stochastic (yielding a uniform Perron vector), we showed in (9.130) that the network error exponent Φ is K times the exponent Φ_{ind} corresponding to an individual agent (i.e., to a standalone implementation of the ASL algorithm), namely,

$$\Phi = K\Phi_{\mathsf{ind}}. \tag{10.62}$$

Moreover, with a uniform Perron vector, Eq. (10.61) implies

$$s_\theta^\star = -K. \tag{10.63}$$

Using (10.62) and (10.63) in (10.48) or (10.49), we find that the network size appearing in the parameter $s_\theta^\star = -K$ is compensated for and canceled out by the network size embodied in the network exponent $\Phi = K\Phi_{\mathsf{ind}}$. Accordingly, we expect the network parameters to have a reduced impact on the adaptation time when the initial conditions are unfavorable — see (10.49). In comparison, we see from (10.48) that the adaptation time under favorable initial conditions would depend on the parameter r that is related to the second largest-magnitude eigenvalue of A, and, hence, embodies a dependence on the network features. However, note that the dependence of the adaptation time in (10.48) on r is logarithmic.

KL divergences and error exponents. As explained before, in the unfavorable scenario the dominant constant is $\mathsf{K}_1(\theta)$. We see from (10.17) that this constant depends on the quantity $\bar{\lambda}_{\mathsf{net}}(\theta)$ defined by (6.10), which in turn depends on the KL divergences relevant to the considered classification problem. To gain some insight into this dependence, we consider the following simplified setting. First, we focus on the objective evidence model in Section 5.3, where $\vartheta^\star = \vartheta^o$ for some true hypothesis ϑ^o, implying, in view of (6.10), that

$$\bar{\lambda}_{\mathsf{net}}(\theta) = D_{\mathsf{net}}(\theta) - \underbrace{D_{\mathsf{net}}(\vartheta^o)}_{=0} = D_{\mathsf{net}}(\theta). \tag{10.64}$$

Second, we neglect the initial state $b_{\mathsf{net},0}$, so that Eq. (10.49) becomes

$$\mathsf{T}_{\mathsf{ASL}} = \frac{1}{\log(1-\delta)^{-1}} \log \frac{\max_{\theta \neq \vartheta^\star} \left\{|s_\theta^\star| D_{\mathsf{net}}(\theta)\right\}}{\varepsilon\,\Phi}. \tag{10.65}$$

Under the objective evidence model, $D_{\mathsf{net}}(\theta)$ is a network average of the KL divergences between the true likelihoods $\ell_k(x|\vartheta^o)$ and the likelihoods $\ell_k(x|\theta)$ corresponding to a wrong hypothesis $\theta \neq \vartheta^o$. As a result, larger values of $D_{\mathsf{net}}(\theta)$ imply that the true hypothesis is more easily distinguishable, that is, the decision problem is easier. Now, expression (10.65) may suggest an *increase* of the adaptation time for larger $D_{\mathsf{net}}(\theta)$. This would imply that easier decision problems require more time to decide reliably, which is counterintuitive.

To see that this reasoning is incomplete, observe that the error exponent Φ is also related to the difficulty of the decision problem; it measures how fast the steady-state error probability converges to 0 as $\delta \to 0$. Therefore, when the decision problem becomes easier, the error probability is smaller, and Φ should also increase, resulting in a *reduction* of the adaptation time in (10.65). In summary, since the KL divergences and the error exponents have opposite effects on the adaptation time in (10.65), it is difficult to anticipate their combined impact. This impact can be quantified by evaluating the pertinent parameters for each particular learning problem. In any case, we remark that the effect of the KL divergences and error exponents is mitigated by the presence of the logarithm in (10.65).

Parameter ε**.** The adaptation time was defined as the time instant after which the error probability decays with an error exponent necessary to achieve The smaller ε is, the closer the error exponent to the steady-state exponent Φ will be. Remarkably, the impact of this parameter on the adaptation time is not critical, since we see from (10.48) and (10.49) that $\mathsf{T}_{\mathsf{ASL}}$ depends logarithmically on ε, namely, growing as $\log(1/\varepsilon)$.

Adaptation parameter δ**.** Observe that the adaptation time in (10.48) does not depend on δ. This means that if we reduce the adaptation parameter, we can achieve a higher learning accuracy (i.e., a smaller error probability), without incurring an expense in terms of learning time. This is due to the fact that the algorithm starts from a favorable initial condition where it is already inclined toward the target hypothesis.

It is clear that this is not the correct setting to evaluate a "genuine" adaptation time, i.e., the time needed by the algorithm to arrive at a correct determination that was not true at the beginning of the learning cycle. Let us focus instead on the case where the initial conditions are unfavorable, where the adaptation time scales, for small δ, as $1/\log(1-\delta)^{-1} \approx 1/\delta$ —

see (10.49). Comparing this result with (10.14), we see that this behavior matches well the qualitative analysis of Section 10.1.

The last expression in Corollary 10.1 shows that the adaptation time is reduced by increasing δ. However, this is not always desirable, since increasing δ also reduces the accuracy in the decision-making process (recall from Theorem 9.4 that we must instead reduce δ to obtain a small error probability). These contrasting effects represent two sides of the same coin; they show the trade-off between learning and adaptation that exists in the ASL strategy. This trade-off can be better summarized by combining Theorem 9.4 and Corollary 10.1 to conclude that the error probability decays exponentially with the adaptation time, roughly scaling as

$$\text{error prob.} \sim \exp\left\{-\frac{\Phi}{\log\Big(\mathsf{K}_1 \times (\varepsilon\,\Phi)^{-1}\Big)}\,\mathsf{T}_{\mathsf{ASL}}\right\} \quad \text{for all agents.} \tag{10.66}$$

Stability over successive learning cycles. The characterization of the transient phase provided by Theorem 10.1 and the related corollary are valid under an arbitrary choice of the initial state $b_{k,0}$. However, as we observed above, if we start from an unfavorable state, then the adaptation time is affected adversely. This gives rise to a fundamental issue that we now illustrate in detail. Assume that the time axis is divided into successive intervals (learning cycles) wherein the system evolves under stationary conditions. At the beginning of each learning cycle the statistical distributions and/or the likelihoods and/or other system parameters change, and a new cycle starts where the system evolves in a stationary manner, albeit under different conditions. Sufficient time to learn is given within each individual learning cycle, so that the system reaches the steady state for that learning cycle. Then, as explained before, the belief accumulated at the end of a learning cycle will become the initial belief for the subsequent learning cycle, which can be very different from the target belief for this new cycle. Thus, it makes sense to ask how "wrong" can the initial beliefs be at the beginning of a learning cycle. In particular, do errors accumulate over time as the algorithm progresses, impairing the learning process? These questions can be answered by combining the steady-state and transient analyses.

To see how, consider the beginning of a learning cycle. As observed in the last paragraph of Section 10.2, at the end of the previous learning cycle, the agents' belief vectors have converged to some vector $\bar{\lambda}_{\mathsf{net}}^{\mathsf{prev}}$ — see

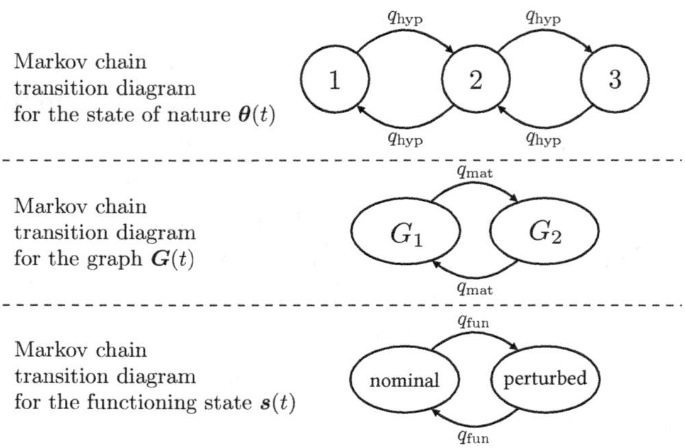

Figure 10.2: Illustration of the Markov chains corresponding to the sources of nonstationarity in Example 10.1. (*Top*) Transition diagram for the underlying state of nature $\boldsymbol{\theta}(t)$. (*Center*) Transition diagram for the graph $\boldsymbol{G}(t)$. (*Bottom*) Transition diagram for the functioning state $\boldsymbol{s}(t)$.

(10.41). Notably, this vector does not depend on the adaptation parameter δ and, in particular, it does not diverge as δ becomes small. That is, the initial conditions at the beginning of each learning cycle are not critically affected by the choice of δ. These arguments apply provided that sufficient time is given for learning within each cycle, namely, if the parameter δ guarantees a sufficient adaptation time that is smaller than the duration of the individual learning cycle. These aspects will be more quantitatively illustrated in the forthcoming example.

Example 10.1 (Evolution over successive learning cycles). In this example we focus on a specific nonstationary setting to illustrate the role of adaptation. We divide the time axis into successive *random* intervals (learning cycles) wherein the system conditions remain stationary. We examine an environment where there are three different sources of nonstationarity, which will be modeled as (mutually independent) homogeneous Markov chains, as specified below.

i) For $t \in \mathbb{N}$, let $\boldsymbol{\theta}(t)$ be a state of nature at time t, which is allowed to change over time. We assume $\boldsymbol{\theta}(t)$ follows a Markov chain with possible states in $\Theta = \{1, 2, 3\}$ and with transition probability q_{hyp} between any two different states, as represented by the finite-state diagram in the top panel of Figure 10.2 (where only transition probabilities are displayed, with the complementary probabilities of remaining in a state being omitted).

ii) The graph connectivity can drift over time. We assume that the agents can be connected according to two different undirected graphs G_1 and G_2, shown in Figure 10.3. All agents are assumed to have a self-loop, not shown in the figure.

Even if both graphs are strong, the former graph has high connectivity, while the latter has low connectivity. The combination matrix in both cases is designed using the Metropolis rule reported in Table 4.1. For the graph G_1, the resulting combination matrix has second largest-magnitude eigenvalue equal to 0.212, whereas for the graph G_2, the combination matrix has second largest-magnitude eigenvalue equal to 0.717. The graph in force at time t is denoted by $\boldsymbol{G}(t)$, and follows a Markov chain with transition probability q_{mat} — see the finite-state diagram shown in the center panel of Figure 10.2. Note that this type of drift is contemplated by our analysis. In fact, the characterization of the transient evolution for the error probability in Theorem 10.1 is very general. It summarizes all previous behavior until time instant $t = 0$ in the initial (scaled) log belief ratios $\{b_{k,0}\}$. Whatever the system parameters before that instant are (e.g., different combination matrices, different true distributions), the algorithm evolution for $t > 0$ will depend only on the initial beliefs (i.e., at $t = 0$) and on the system parameters ruling the *current* learning cycle (i.e., in force for $t > 0$).

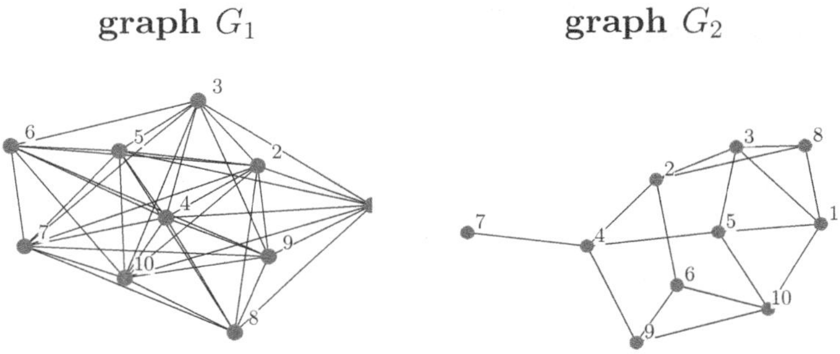

Figure 10.3: Network topologies used in Example 10.1. The graphs are undirected and all agents are assumed to have a self-loop (not shown in the figure).

iii) The system can be in one of two possible functioning states, namely, nominal (N) and perturbed (P). The functioning state at time t is denoted by $\boldsymbol{s}(t)$. Under state "$\boldsymbol{s}(t)$ = nominal" the data are generated according to the true likelihood corresponding to hypothesis $\theta(t)$, namely, we are under the objective evidence model of Section 5.3 (given the true hypothesis, the observations are generated as statistically independent across the agents). Specifically, the nominal likelihood models are chosen from the following family of Laplace distributions:

$$g_n(x) = \frac{1}{2} e^{-|x-n|}, \qquad n = 1, 2, 3, \tag{10.67}$$

in such a way that $\ell_k(x|\theta) = g_\theta(x)$, for $k = 1, 2, \dots, K$ and $\theta = 1, 2, 3$. Since the nominal functioning state corresponds to the objective evidence model, the target hypothesis $\vartheta^\star$ that minimizes the network average of KL divergences $D_{\mathsf{net}}(\theta)$ is equal to the underlying state of nature $\theta(t)$. Under state "$\boldsymbol{s}(t)$ = perturbed" we adopt the following construction. First, we generate samples from the nominal data model, and then contaminate them with iid zero-mean Gaussian noise having variance equal to 100. For the choice of the system parameters used in this example, the target hypothesis $\vartheta^\star$ that minimizes the network average of KL divergences

$D_{\text{net}}(\theta)$ (now computed under the modified data distributions corresponding to the perturbed state) is still equal to the true underlying state of nature $\theta(t)$. Transitions between the two functioning states occur according to a Markov chain ruled by a transition probability q_{fun} — see the finite-state diagram shown in the bottom panel of Figure 10.2.

Duration of a learning cycle. A learning cycle is identified by a time interval where all the conditions remain stationary. Let us evaluate the average duration of a learning cycle. Technically, in terms of the above Markov chain formulation, we need to identify the average time spent in each *joint* state $\{\boldsymbol{\theta}(t), \boldsymbol{G}(t), \boldsymbol{s}(t)\}$. In order to be conservative, we focus on the worst case, i.e., on the shortest average duration, which is obtained when the system is in the most unstable state (i.e., the state where transitions are more frequent). Examining Figure 10.2, the most unstable state is obtained when the hypothesis in force is $\boldsymbol{\theta}(t) = 2$ (since from such intermediate state we can move leftward or rightward, while from the other states it cannot), whereas for the combination matrix and the functioning state the particular choice is immaterial. Now, given that the overall system is in the joint state $\{\boldsymbol{\theta}(t) = 2, \boldsymbol{G}(t) = G_1, \boldsymbol{s}(t) = \text{nominal}\}$, the probability p_{min} that the system does *not* change state for a single step is equal to

$$p_{\text{min}} = (1 - 2\, q_{\text{hyp}})(1 - q_{\text{mat}})(1 - q_{\text{fun}}). \tag{10.68}$$

Likewise, the probability that the system remains in the considered state for exactly $t - 1$ steps (which means that the learning cycle has duration t) is equal to

$$p_{\text{min}}^{t-1}(1 - p_{\text{min}}), \qquad t \in \mathbb{N}, \tag{10.69}$$

which corresponds to the pmf of a geometric random variable. The expected value of a random variable following the distribution in (10.69) can be computed as

$$(1 - p_{\text{min}}) \sum_{t=1}^{\infty} p_{\text{min}}^{t-1}\, t = (1 - p_{\text{min}}) \underbrace{\frac{d}{dp_{\text{min}}} \underbrace{\left(\sum_{t=0}^{\infty} p_{\text{min}}^{t} \right)}_{\frac{1}{1-p_{\text{min}}}}}_{\frac{1}{(1-p_{\text{min}})^2}} = \frac{1}{1 - p_{\text{min}}}, \tag{10.70}$$

which means that the average duration for the worst-case (i.e., shortest) learning cycle is equal to

$$\mathsf{T}_{\text{LC}} = \frac{1}{1 - p_{\text{min}}}. \tag{10.71}$$

In order to model a nonstationary environment where the system parameters remain stable during the learning cycles, we take inspiration from the Gilbert-Elliott model, which is typically employed to describe random bursts of errors over communication channels [67, 82]. According to the Gilbert-Elliott model, the transition probabilities between states of the chain are kept small so as to ensure that the chain remains in the same state for several contiguous time samples.

Adaptation time. We consider the perspective of a network designer who wants to select the adaptation parameter δ. To make this choice in an informed manner, it is necessary to make an estimate of the adaptation time corresponding to a given δ. We assume that the network designer has no knowledge about the true underlying distributions when they differ from the nominal likelihoods, i.e., when the system is in the (unpredictable) perturbed functioning state. Accordingly, the following calculations are performed by using the nominal likelihoods.

Let us first focus on a particular true hypothesis $\vartheta^\star$. Then, we will repeat the computation for all possible choices of $\vartheta^\star$ and retain the highest, i.e., worst-case adaptation time. We assume that in a given learning cycle the system has evolved from a previous learning cycle where the agents converged to a hypothesis different from that in force during the current learning cycle. Under this setting, as was explained before, we are in the unfavorable case of Corollary 10.1, and thus we need to call upon (10.49) to evaluate the adaptation time. To this end, we can first evaluate numerically the exponent Φ as shown in Example 9.6. The evaluation of the constant K_1 in (10.46) requires a separate explanation.

We start by computing the constant $\mathsf{K}_1(\theta)$ in (10.17) for each $\theta \neq \vartheta^\star$. Observe that we need to evaluate the initial states $b_{k,0}(\theta)$ to obtain the network average $b_{\mathsf{net},0}$. Recalling that the initial states at $t = 0$ correspond to the final states at the end of the previous learning cycle, their values will obviously depend on the particular previous evolution. To compute $\mathsf{K}_1(\theta)$, we make a conservative choice and consider the worst-case initial state. The specific calculations are as follows.

We denote by $\vartheta^\diamond \neq \vartheta^\star$ the true hypothesis in force during the previous learning cycle, and by μ_k^{prev} the *steady-state* belief vector at the end of the previous learning cycle. Assuming that this learning cycle had a sufficiently long duration, so that the beliefs reached the steady state, for each agent k we can write

$$\delta \log \frac{\mu_k^{\mathsf{prev}}(\vartheta^\diamond)}{\mu_k^{\mathsf{prev}}(\theta)} \approx \frac{1}{K}\sum_{j=1}^{K} D(\ell_{j,\vartheta^\diamond}||\ell_{j,\theta}), \tag{10.72}$$

For $\theta = \vartheta^\diamond$, Eq. (10.72) is actually a trivial equality, whereas for $\theta \neq \vartheta^\diamond$ the approximation follows from Theorem 9.2, once we make explicit the expressions for the scaled steady-state log belief ratio $b_k(\theta)$ and the expected network average of log likelihood ratios $\bar{\lambda}_{\mathsf{net}}(\theta)$ appearing in (9.28), computed under the hypothesis pertaining to the previous learning cycle, and with uniform Perron vector entries $v_j = 1/K$ (since the combination matrix is doubly stochastic).

On the other hand, for all $\theta \neq \vartheta^\star$ we can write

$$b_{k,0}(\theta) = \delta \log \frac{\mu_{k,0}(\vartheta^\star)}{\mu_{k,0}(\theta)} = \delta \log \frac{\mu_{k,0}(\vartheta^\diamond)}{\mu_{k,0}(\theta)} - \delta \log \frac{\mu_{k,0}(\vartheta^\diamond)}{\mu_{k,0}(\vartheta^\star)}. \tag{10.73}$$

Since the initial belief vector $\mu_{k,0}$ coincides with the final value of the belief vector at the end of the previous learning cycle, from (10.72) and (10.73) we obtain, for all $\theta \neq \vartheta^\star$,

$$b_{k,0}(\theta) \approx \frac{1}{K}\sum_{j=1}^{K}\Big(D(\ell_{j,\vartheta^\diamond}||\ell_{j,\theta}) - D(\ell_{j,\vartheta^\diamond}||\ell_{j,\vartheta^\star})\Big). \tag{10.74}$$

Accordingly, since all agents converge to the same limit point, we can write

$$b_{\mathsf{net},0}(\theta) \approx b_{k,0}(\theta) \approx \frac{1}{K}\sum_{j=1}^{K}\Big(D(\ell_{j,\vartheta^\diamond}||\ell_{j,\theta}) - D(\ell_{j,\vartheta^\diamond}||\ell_{j,\vartheta^\star})\Big) \tag{10.75}$$

and introduce the minimum value

$$b^{\mathsf{min}}_{\mathsf{net},0}(\theta) = \min_{\vartheta^\diamond \neq \vartheta^\star} \frac{1}{K}\sum_{j=1}^{K}\Big(D(\ell_{j,\vartheta^\diamond}||\ell_{j,\theta}) - D(\ell_{j,\vartheta^\diamond}||\ell_{j,\vartheta^\star})\Big). \tag{10.76}$$

Note that this minimum is negative since the function to be minimized is negative for $\vartheta^\diamond = \theta$. The value $b^{\mathsf{min}}_{\mathsf{net},0}(\theta)$ can be inserted into (10.17) to compute the following upper bound:

$$\mathsf{K}_1(\theta) = |s^\star_\theta|\left[\bar{\lambda}_{\mathsf{net}}(\theta) - b_{\mathsf{net},0}(\theta)\right] \lessapprox |s^\star_\theta|\left[\bar{\lambda}_{\mathsf{net}}(\theta) - b^{\mathsf{min}}_{\mathsf{net},0}(\theta)\right]. \tag{10.77}$$

In view of (10.77), to compute an approximate upper bound on the maximum constant K_1 appearing in (10.46), we can maximize the RHS with respect to $\theta \neq \vartheta^\star$. The result would depend on $\vartheta^\star$ since all our calculations have been performed for a given $\vartheta^\star$. Therefore, to obtain a worst-case bound, we consider the constant $\mathsf{K}^{\mathsf{up}}_1$ that is obtained by further maximizing over all hypotheses $\vartheta^\star$ and use this upper bound to obtain the highest adaptation time

$$\mathsf{T}_{\mathsf{ASL}} \approx \frac{1}{\log(1-\delta)^{-1}}\log\frac{\mathsf{K}^{\mathsf{up}}_1}{\varepsilon\,\Phi}. \tag{10.78}$$

Inserting into this relation the numerical values corresponding to our simulation setup, the time necessary for the error exponent to reach half the value (i.e., we use $\varepsilon = 0.5$) of Φ is equal to

$$\mathsf{T}_{\mathsf{ASL}} \approx \frac{3.1916}{\log(1-\delta)^{-1}} \approx \frac{3.1916}{\delta}, \tag{10.79}$$

where in the last step we use the approximation $1/\log(1-\delta)^{-1} \approx 1/\delta$, holding for small δ. We now examine two scenarios that differ in the duration of the learning cycle.

"Short" learning cycles. First, we consider the following setting:

$$q_{\mathsf{hyp}} = 5 \times 10^{-3}, \quad q_{\mathsf{mat}} = 10^{-3}, \quad q_{\mathsf{fun}} = 10^{-3}, \tag{10.80}$$

yielding, in view of (10.68),

$$p_{\mathsf{min}} = 0.9880. \tag{10.81}$$

Using (10.71), the average duration of a learning cycle is approximated by

$$\mathsf{T}_{\mathsf{LC}} \approx 83 \text{ iterations}. \tag{10.82}$$

If we equate this value for T_{LC} to the adaptation time in (10.79), we get $\delta \approx 0.038$. To guarantee proper learning, we need an adaptation time sufficiently smaller than the average duration of a learning cycle. In the experiments shown in Figure 10.4 we made the choice

$$\delta = 0.1, \tag{10.83}$$

which, when substituted into (10.79), corresponds to the adaptation time

$$\mathsf{T}_{\mathsf{ASL}} \approx 32 \text{ iterations}. \tag{10.84}$$

This value is approximately one third of the average worst-case learning cycle in (10.82).

Figure 10.4 shows the simulation results pertaining to the considered setup. The first (top) row shows the transitions for the three sources of nonstationarity illustrated in Figure 10.2, namely, true state of nature, functioning state, and network graph. In the second row we display the time evolution of the beliefs of agent 1 obtained by running the ASL strategy, whereas the third row shows the error probability achieved by this

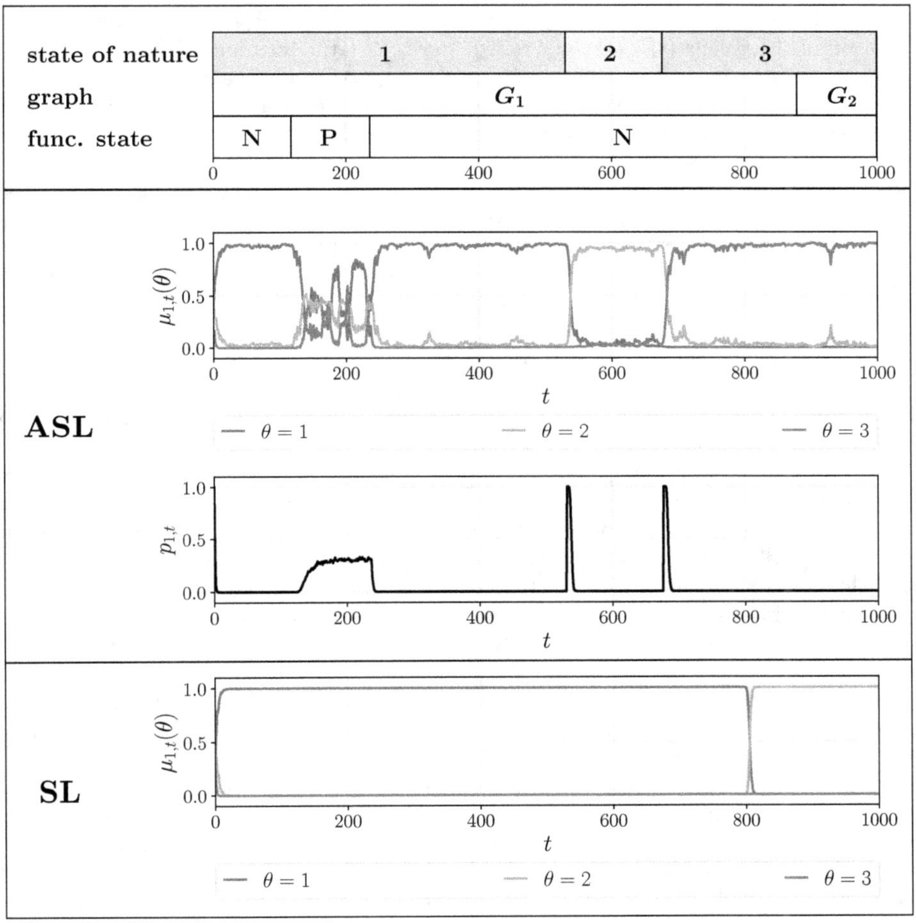

Figure 10.4: Evolution over successive learning cycles (Example 10.1), with adaptation parameter $\delta = 0.1$ and average cycle duration $\mathsf{T}_{\mathsf{LC}} \approx 83$. (*First (top) row*) Observed transitions for the three sources of nonstationarity illustrated in Figure 10.2, namely, true state of nature, functioning state, and network graph. (*Second row*) Time evolution of the belief of agent 1 for the *adaptive* social learning (ASL) strategy from listing (8.13). (*Third row*) Time evolution of the error probability of agent 1 for the adaptive social learning strategy. (*Fourth row*) Time evolution of the belief of agent 1 for the *nonadaptive* social learning (SL) strategy from listing (3.16).

agent, estimated empirically from 1000 Monte Carlo runs. For comparison purposes, in the fourth row we report the time evolution of the beliefs of agent 1 for the nonadaptive social learning strategy from listing (3.16), which is labeled as SL.

First, we observe that, except for the learning cycle corresponding to the perturbed functioning state, the ASL strategy exhibits good performance after a relatively short transient at the beginning of each cycle. The learning ability is revealed by the time evolution of the beliefs (second row), which shows how the maximum belief corresponds to the true hypothesis, after relatively short adaptation intervals necessary to react in the face of nonstationarities. More quantitatively, the learning ability is reflected by the time evolution of the error probabilities (third row), where we see some peaks (error probability close to 1) that clearly correspond to the changes, and that have a short duration dictated by the adaptation times. In sharp contrast, the traditional social learning strategy loses its learning ability after the first learning cycle.

Zooming in on Figure 10.4, we see that nonstationarities in the hypotheses induce a perceivable change in the learning performance, whereas nonstationarities in the network graph or in the functioning state deserve a separate analysis.

For what concerns the graph, we see that the learning ability is preserved in the face of changes, i.e., the system does *not* undergo an interval of poor performance. This behavior makes sense, since from the theoretical analysis we know that the ASL strategy must learn consistently provided that the graph is primitive; this is the case for both graphs G_1 and G_2 considered in our example.

Regarding the functioning state, we see that when "$\boldsymbol{s}(t)$ = perturbed" the system undergoes an interval of worse performance (error probability $\approx 1/3$), which sounds reasonable since the observations are very noisy and thus provide unreliable information. Remarkably, the adaptation capacity of the ASL strategy allows the agents to recover from this failure state in the successive learning cycles.

In summary, we have seen that the log belief ratios at the beginning of each learning cycle are stable, since they arise as steady-state limiting values at the end of the previous learning cycle. In other words, the log belief ratios do not diverge as time elapses. Contrast this behavior with what happens for traditional social learning, where, at the end of a learning cycle, the log belief ratio tends to diverge with the length of the learning cycle. As a result, the initial log belief ratio of the subsequent learning cycle becomes very distant from the log belief ratio that should correspond to the new model, and the system becomes unable to track variations over successive learning cycles. In comparison, with adaptive social learning, the number of variations of the underlying statistical conditions occurring during the entire algorithm evolution does not impair successful learning by the ASL strategy. What really matters is that the duration of the learning cycle is sufficiently large to allow a sufficiently small value of δ to enable accurate learning.

"Long" learning cycles. In Figure 10.5 we consider the more favorable situation where the average duration of the learning cycle is increased by one order of magnitude, using the following transition probabilities for the pertinent Markov chains:

$$q_{\mathsf{hyp}} = 5 \times 10^{-4}, \quad q_{\mathsf{mat}} = 10^{-4}, \quad q_{\mathsf{fun}} = 10^{-4}. \tag{10.85}$$

Accordingly, we expect that the adaptation properties of the system will be preserved if we reduce the adaptation parameter by one order of magnitude, yielding

$$\delta = 0.01. \tag{10.86}$$

Comparing Figure 10.5 against Figure 10.4, we see that the general behavior is perfectly confirmed, and two notable effects emerge. First, the adaptation properties are preserved,

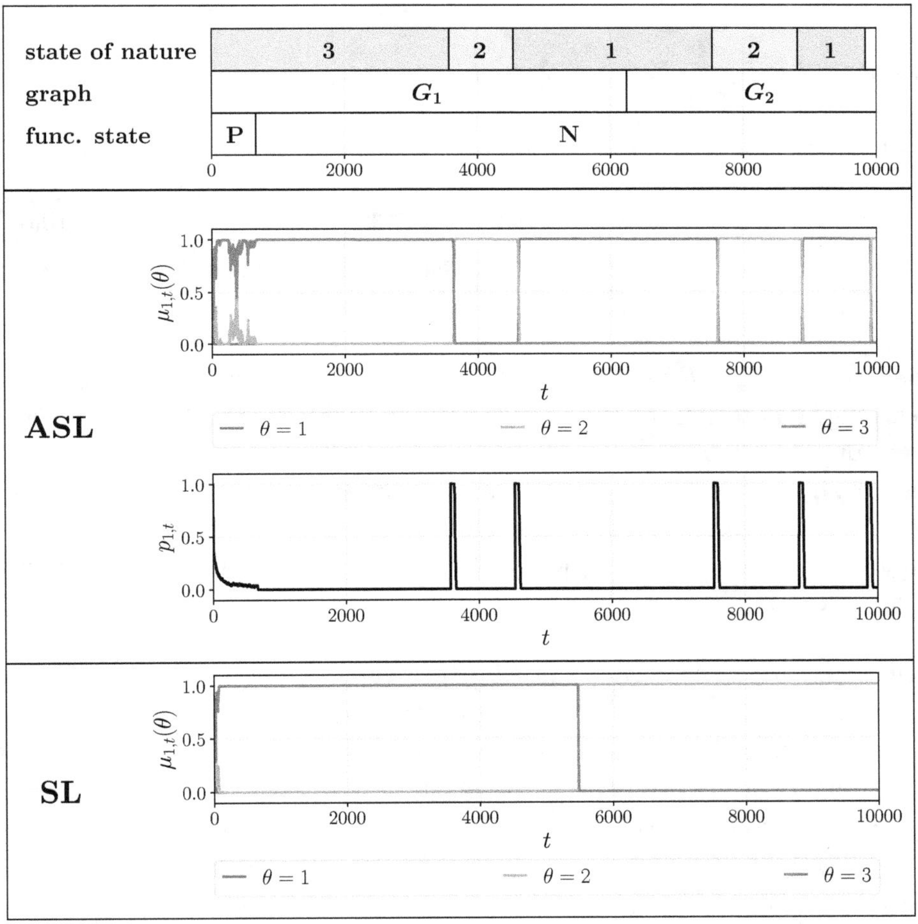

Figure 10.5: Evolution over successive learning cycles (Example 10.1), with adaptation parameter $\delta = 0.01$ and average cycle duration $\mathsf{T}_{\mathsf{LC}} \approx 833$. (*First (top) row*) Observed transitions for the three sources of nonstationarity illustrated in Figure 10.2, namely, true state of nature, functioning state, and network graph. (*Second row*) Time evolution of the belief of agent 1 for the *adaptive* social learning (ASL) strategy from listing (8.13). (*Third row*) Time evolution of the error probability of agent 1 for the adaptive social learning strategy. (*Fourth row*) Time evolution of the belief of agent 1 for the *traditional* social learning (SL) strategy from listing (3.16).

i.e., the system is able to adapt to the changes sufficiently fast to guarantee a stable evolution over successive learning cycles. Second, the fluctuations around the limiting steady-state are reduced with respect to Figure 10.4, yielding a smaller error probability. This confirms the theoretical analysis carried out in the previous sections, since we are now using a smaller adaptation parameter $\delta = 0.01$.

10.4 Summary: Learning and Adaptation under ASL

Equation (10.66) reveals the universal scaling law for adaptive social learning:

$$\text{error probability} \sim e^{-\text{adaptation time}} \tag{10.87}$$

We will now comment on this fundamental result in relation to different aspects.

Learning and adaptation trade-off. Equation (10.87) summarizes the learning/adaptation trade-off, since it implies that a better learning quality, i.e., lower error probability, requires a reduced adaptation capacity, i.e., larger adaptation times. The trade-off between learning and adaptation arises (albeit with different scaling laws, as discussed in the next paragraph) in other research domains, such as adaptive filtering [154] or inference over networks [155].

Adaptive social learning vs. adaptive distributed estimation. In the distributed estimation or regression context the goal is to learn the value of a continuous parameter. For this type of inference problem, adaptive implementations based on distributed stochastic gradient approximations have been shown to provide a mean-square estimation error that scales proportionally to the inverse of the adaptation time [151, 152]. In the social learning context, we observe that the error probabilities decay exponentially with the adaptation time. These scaling laws represent the *universal* scaling laws governing errors of adaptive social learning and adaptive distributed estimation.

Scaling laws for inference problems. We have observed that the scaling laws governing adaptive social learning and adaptive distributed estimation are rather different. The significance of this result emerges more fully through an analogy with other traditional inference problems. As a first example, consider a *classic* (i.e., centralized, nonadaptive) inference setting with N iid data samples. If these samples are used to solve a

Table 10.1: Fundamental scaling laws of the learning performance with respect to the cost of information for different types of inference problems. The symbol $\sim$ means "scales as".

Scheme	**Cost**	**Error probability**	**Estimation error (MSE)**
Centralized	No. of samples N	$\sim e^{-N}$	$\sim 1/N$
Fusion center	Bit-rate R	$\sim e^{-R}$	$\sim 1/R$
Adaptive	Adapt. time T	$\sim e^{-T}$	$\sim 1/T$

classification problem (e.g., a binary detection problem), it is known that the error probability of the best classifier decays exponentially with N [59], whereas if we have to solve an estimation problem, the optimal mean-square estimation error decays as $1/N$ [159]. As a second example, consider a distributed (nonadaptive) inference problem with a fusion center. The fundamental limits for such problem have been examined in the context of rate-constrained multiterminal inference, and, more specifically, with reference to the so-called CEO problem [18, 169]. In this setting, given a bit-rate R, the error probability decays exponentially with R [18], whereas the mean-square estimation error vanishes as $1/R$ [169]. Comparing the scaling laws characterizing these two problems with the laws for adaptive social learning and adaptive distributed estimation, we see that increasing the adaptation time corresponds to increasing the number of independent samples in the first inference problem, or increasing the bit-rate in the second problem. This makes perfect sense, since the adaptation time represents the *cost of information* used by the network for inference purposes, much as the number of samples N or the bit-rate R in the considered examples. A summary of the aforementioned comparisons is provided in Table 10.1.

Chapter 11

Partial Information Sharing

As explained in the previous chapters, the fundamental learning mechanism of social learning is activated by the exchange of beliefs between neighboring agents. In this chapter we examine the case where the agents are constrained to share only *partial* beliefs. This limitation arises in practice for different reasons.

For example, consider the following social dynamics. Some agents within a group collect reviews and share opinions about a certain commercial product. Assume that this product is released by brands $1, 2$, or 3. During their interactions, the agents focus on a specific brand of interest, say brand 1. They share positive or negative impressions only about 1, without sharing information regarding the other two brands. Despite this limited exchange of information, the agents inherently update their opinions also about the other brands. One fundamental question arising under *partial information sharing* is the following: Would the agents be able to establish whether brand 1 is the best among the three brands by exchanging opinions concerning only brand 1?

Another motivation for partial information sharing relates to communication constraints. In fact, the growing interest in distributed learning architectures has motivated the search for communication-efficient distributed algorithms for optimization and learning [6, 40, 41, 102, 127, 134]. This issue has been recently addressed also in the context of social learning [89, 128, 129, 149, 164]. Two main approaches have been considered to guarantee communication efficiency in social learning: belief quantization, where the belief vectors are represented with a prescribed number of bits to cope with the communication constraints; and belief sparsification, where the agents transmit a subset of the belief-vector entries or communicate only when there is sufficient innovation in the beliefs.

Motivated by these considerations, we examine in this chapter the effect of partial information sharing in social learning, and the impact it has on the learning ability of the agents. In particular, we will constrain the agents to share their belief about a single *hypothesis of interest.*

11.1 Partial Information Framework

Social learning under partial information sharing was briefly introduced in Example 3.3, as a special case of the unifying framework for non-Bayesian social learning shown in (3.77a)–(3.77d).

Figure 11.1 illustrates a block diagram of social learning under partial information sharing, in terms of the four steps described by (3.77a)–(3.77d). The core parts where partial information sharing acts are the encoding step (3.77b) and the decoding step (3.77c). Let us examine in greater detail these steps.

In the partial information framework, we assume that each agent k shares its opinion regarding a single hypothesis of interest $\vartheta^\bullet \in \Theta$, which means that only the entry $\psi_{k,t}(\vartheta^\bullet)$ of the intermediate belief vector $\psi_{k,t}$ is shared. In terms of step (3.77b), this corresponds to saying that each agent k *encodes* its intermediate belief into a single scalar value $\psi_{k,t}(\vartheta^\bullet)$, namely,

$$\psi_{k,t} \overset{\text{encode}}{\longrightarrow} \psi_{k,t}(\vartheta^\bullet). \tag{11.1}$$

Then, the agents must perform a *decoding* operation to turn the available information into *full* belief vectors. To this end, each agent k can use its own belief vector $\psi_{k,t}$ and the intermediate beliefs $\psi_{j,t}(\vartheta^\bullet)$ (i.e., the beliefs about the hypothesis of interest) received from the neighbors $j \in \mathcal{N}_k \backslash \{k\}$ to build an estimate $\widehat{\psi}_{j,t}^{(k)}$ of the full belief vectors for all agents $j \in \mathcal{N}_k$ (including $j = k$ if $a_{kk} > 0$). When $a_{kk} > 0$ and $\widehat{\psi}_{k,t}^{(k)} = \psi_{k,t}$, we say that the strategy is *self-aware.*

According to the aforementioned description, the decoding step (3.77c) can be specialized to

$$\Big(\psi_{k,t}, \{\psi_{j,t}(\vartheta^\bullet)\}_{j \in \mathcal{N}_k \backslash \{k\}}\Big) \overset{\text{decode}}{\longrightarrow} \Big\{\widehat{\psi}_{j,t}^{(k)}\Big\}_{j \in \mathcal{N}_k}. \tag{11.2}$$

Finally, in the combination step (3.77d) the reconstructed beliefs can be combined using one of the pooling rules (e.g., geometric or arithmetic averaging) introduced in Section 3.3.

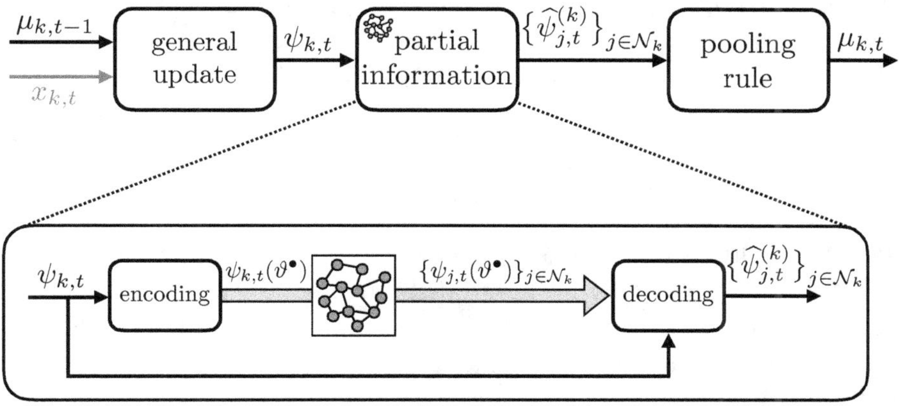

Figure 11.1: Diagram of social learning under partial information sharing. In comparison with Figure 3.3, the encoding/decoding operations are specialized as follows. For each agent k: *i)* the encoding step outputs only $\psi_{k,t}(\vartheta^\bullet)$, i.e., the belief about the hypothesis of interest; and *ii)* the decoding step is applied to the information available to agent k, i.e., to its own (entire) belief vector $\psi_{k,t}$ and the beliefs $\psi_{j,t}(\vartheta^\bullet)$ received from its neighbors $j \in \mathcal{N}_k\backslash\{k\}$.

It is clear how the agent must perform the encoding step in (11.1): It should extract the entry corresponding to hypothesis $\vartheta^\bullet$ from its intermediate belief vector. However, the decoding step in (11.2) is a design choice and can therefore be tailored to different applications. The reasoning behind this step is that, upon receiving the belief $\psi_{j,t}(\vartheta^\bullet)$, agent k will seek to fill in the missing entries using some decoding strategy, thereby reconstructing a complete belief vector $\widehat{\psi}_{j,t}^{(k)}$ to approximate the unknown intermediate belief vector of its neighbor j.

The described partial information framework is valid under arbitrary choices for the first and last blocks in the top panel of Figure 11.1. That is, we are free to choose a general update and pooling rules. For the analysis in this chapter, we will choose in particular a Bayesian update rule and a geometric-average pooling rule.

11.2 Decoding Strategies

In this section we show how the decoding strategy can be derived from a Bayesian approach. To avoid confusion, we remark that in the following development, when we refer to a neighboring agent $j \in \mathcal{N}_k$, the case $j = k$ is included whenever $a_{kk} > 0$.

For each agent $j \in \mathcal{N}_k$, agent k possesses the intermediate belief $\psi_{j,t}(\vartheta^\bullet)$.

While reconstructing the full belief vector $\widehat{\psi}_{j,t}^{(k)}$, agent k trusts agent j and, hence, it sets

$$\widehat{\psi}_{j,t}^{(k)}(\vartheta^\bullet) = \psi_{j,t}(\vartheta^\bullet) \quad \forall j \in \mathcal{N}_k. \tag{11.3}$$

Consider now the set of *unshared hypotheses*

$$\mathcal{U} \triangleq \Theta \backslash \{\vartheta^\bullet\}. \tag{11.4}$$

Once we assume the equality in (11.3), the remaining mass assigned to the set $\mathcal{U}$ must necessarily be $1 - \psi_{j,t}(\vartheta^\bullet)$ in order to ensure that $\widehat{\psi}_{j,t}^{(k)}$ is a valid belief vector, i.e., that its entries add up to 1. From Bayes' rule, this implies that $\widehat{\psi}_{j,t}^{(k)}$ must satisfy, for all $\theta \neq \vartheta^\bullet$, the equation[1]

$$\widehat{\psi}_{j,t}^{(k)}(\theta) = \mathsf{p}_k(\theta|\mathcal{U})\Big(1 - \psi_{j,t}(\vartheta^\bullet)\Big), \tag{11.6}$$

where $\mathsf{p}_k(\theta|\mathcal{U})$ is the belief about θ *conditioned* on the set $\mathcal{U}$, computed by agent k. To complete the decoding strategy, it is necessary to choose the form of $\mathsf{p}_k(\theta|\mathcal{U})$.

An agnostic, maximum-entropy choice for $\mathsf{p}_k(\theta|\mathcal{U})$ is given by

$$\mathsf{p}_k(\theta|\mathcal{U}) = \frac{1}{H-1}, \tag{11.7}$$

where agent k assumes no knowledge available to determine $\mathsf{p}_k(\theta|\mathcal{U})$ and thus splits the remaining belief mass $1 - \psi_{k,i}(\vartheta^\bullet)$ uniformly across the $H-1$ hypotheses belonging to $\mathcal{U}$.

An alternative approach consists of leveraging the most up-to-date knowledge that agent k has accumulated up to time t. As a matter of fact, the most up-to-date belief vector available to agent k at time t is $\psi_{k,t}$, which leads to the *conditional* belief given $\mathcal{U}$:

$$\mathsf{p}_k(\theta|\mathcal{U}) = \frac{\psi_{k,t}(\theta)}{1 - \psi_{k,t}(\vartheta^\bullet)}. \tag{11.8}$$

We see that (11.8) diversifies the allocation of the conditional-belief mass across the unshared hypotheses, based on the available knowledge stored in the intermediate belief vector $\psi_{k,t}$. In contrast, strategy (11.7) opts for a

[1]To interpret (11.6), consider a random variable $\boldsymbol{\theta} \in \Theta$. For all $\theta \in \mathcal{U}$,

$$\mathbb{P}[\boldsymbol{\theta} = \theta] = \mathbb{P}[\boldsymbol{\theta} = \theta, \boldsymbol{\theta} \in \mathcal{U}] = \mathbb{P}[\boldsymbol{\theta} = \theta | \boldsymbol{\theta} \in \mathcal{U}]\, \mathbb{P}[\boldsymbol{\theta} \in \mathcal{U}], \tag{11.5}$$

where the first equality holds since $\theta \in \mathcal{U}$, while the second equality is Bayes' rule. We see from (11.5) that the probability of a particular value θ can be expressed as the product of a conditional probability (the term $\mathsf{p}_k(\theta|\mathcal{U})$ in (11.6)) and the total probability assigned to the set $\mathcal{U}$ (the term $1 - \psi_{j,t}(\vartheta^\bullet)$ in (11.6)).

uniform allocation, thus forgetting any evidence that agent k accumulated in the past. We refer to (11.7) as the *memoryless* strategy, and to (11.8) as the *memory-aware* strategy.

Note that with strategy (11.8), when $a_{kk} > 0$ agent k is automatically *self-aware*, in the sense that $\widehat{\psi}_{k,t}^{(k)} = \psi_{k,t}$. Self-awareness is a compelling property, which arises naturally from our Bayesian interpretation of the decoding strategy once we allow it to incorporate the information contained in $\psi_{k,t}$. In comparison, note that in strategy (11.7) agent k is *not* self-aware.

The two decoding strategies proposed in this chapter are summarized as follows.

Memoryless decoding strategy.

$$\widehat{\psi}_{j,t}^{(k)}(\theta) = \begin{cases} \psi_{j,t}(\vartheta^\bullet) & \text{if } \theta = \vartheta^\bullet, \\ \dfrac{1}{H-1}\Big(1 - \psi_{j,t}(\vartheta^\bullet)\Big) & \text{if } \theta \neq \vartheta^\bullet. \end{cases} \tag{11.9}$$

Memory-aware decoding strategy.

$$\widehat{\psi}_{j,t}^{(k)}(\theta) = \begin{cases} \psi_{j,t}(\vartheta^\bullet) & \text{if } \theta = \vartheta^\bullet, \\ \dfrac{\psi_{k,t}(\theta)}{1 - \psi_{k,t}(\vartheta^\bullet)}\Big(1 - \psi_{j,t}(\vartheta^\bullet)\Big) & \text{if } \theta \neq \vartheta^\bullet. \end{cases} \tag{11.10}$$

Observe that both decoding strategies act as *filling strategies*, where the unshared entries of $\psi_{j,t}$ are filled in according to different approaches. Another property of the two strategies is that the resulting belief vector of agent j estimated by agent k depends only on the partial information $\psi_{j,t}(\vartheta^\bullet)$ and (for the memory-aware approach) on the full belief vector $\psi_{k,t}$. More general decoding strategies can be considered, e.g., taking into account the information received from *all* neighbors and not only from agent j. Note also that, in the binary case, for both filling strategies we have $\widehat{\psi}_{j,t}^{(k)} = \psi_{j,t}$ for $k = 1, 2, \ldots, K$ and $j \in \mathcal{N}_k$, which means that the agents recover the exact intermediate beliefs from their neighbors. Therefore, in the partial information setting the binary case corresponds to a trivial case that boils down to traditional social learning under full information sharing.

The social learning strategy with partial information is summarized in listing (11.11). The decoding step is either (11.9) or (11.10), whereas for the combination step we focus on the geometric-averaging rule.

Social learning with partial information

set variable memory=0 or memory=1
start from the prior belief vectors $\mu_{k,0}$ for $k = 1, 2, \ldots, K$
choose the hypothesis of interest $\vartheta^\bullet \in \Theta$
for $t = 1, 2, \ldots$
  for $k = 1, 2, \ldots, K$
    agent k observes $x_{k,t}$
    for $\theta = 1, 2, \ldots, H$

$$\psi_{k,t}(\theta) = \frac{\mu_{k,t-1}(\theta)\ell_k(x_{k,t}|\theta)}{\sum\limits_{\theta' \in \Theta} \mu_{k,t-1}(\theta')\ell_k(x_{k,t}|\theta')} \qquad \text{(self-learning)}$$

    end
  end

  for $k = 1, 2, \ldots, K$
    for each $j \in \mathcal{N}_k$ (decoding) (11.11)

$$\widehat{\psi}^{(k)}_{j,t}(\vartheta^\bullet) = \psi_{j,t}(\vartheta^\bullet)$$

      for each $\theta \neq \vartheta^\bullet$
        if memory=0

$$\widehat{\psi}^{(k)}_{j,t}(\theta) = \frac{1}{H-1}\Big(1 - \psi_{j,t}(\vartheta^\bullet)\Big)$$

        elseif memory=1

$$\widehat{\psi}^{(k)}_{j,t}(\theta) = \frac{\psi_{k,t}(\theta)}{1 - \psi_{k,t}(\vartheta^\bullet)}\Big(1 - \psi_{j,t}(\vartheta^\bullet)\Big)$$

        end
      end
    end

    for $\theta = 1, 2, \ldots, H$

$$\mu_{k,t}(\theta) = \frac{\prod_{j \in \mathcal{N}_k}\left[\widehat{\psi}^{(k)}_{j,t}(\theta)\right]^{a_{jk}}}{\sum\limits_{\theta' \in \Theta}\prod_{j \in \mathcal{N}_k}\left[\widehat{\psi}^{(k)}_{j,t}(\theta')\right]^{a_{jk}}} \qquad \text{(cooperation)}$$

    end
  end
end

Before going further, it is important to remark that, under Assumptions 5.1 and 5.3, the beliefs $\boldsymbol{\psi}_{k,t}(\theta)$ and $\boldsymbol{\mu}_{k,t}(\theta)$ resulting from (11.11) are almost-surely positive for all k, t, and θ. This property has already been established in Chapters 5 and 7 for traditional social learning. To show that the same property holds for strategy (11.11), one can proceed as follows: *i)* repeat the same arguments used in these chapters to establish that, starting from a belief $\boldsymbol{\mu}_{j,t-1}(\theta)$ that is nonzero at any θ, the update rule yields intermediate beliefs that satisfy in particular the inequalities $\psi_{j,t}(\vartheta^\bullet) < 1$, $\psi_{k,t}(\vartheta^\bullet) < 1$, and $\psi_{k,t}(\theta) > 0$; *ii)* observe that, once fed with these intermediate beliefs, the reconstructed beliefs $\widehat{\psi}_{j,t}^{(k)}(\theta)$ preserve the positivity property; and *iii)* finally note that, as was explained in in Chapter 5, the geometric-average pooling rule also preserves positivity.

11.3 Asymptotic Learning Objectives

In this chapter we focus on the objective evidence model described in Section 5.3, where, as $t \to \infty$, the learning system observes an infinite amount of data supporting the true hypothesis ϑ^o. This increasing knowledge should hopefully correspond to an increasing confidence gained about the true hypothesis. Since the confidence about the veracity of a hypothesis is quantified by the belief about that hypothesis, we expect that the learning system ultimately places full mass on the true hypothesis. We will refer to this type of truth learning as being *traditional.*

Definition 11.1 (Traditional truth learning). We say that traditional truth learning is achieved when

$$\mu_{k,t}(\vartheta^o) \xrightarrow[t\to\infty]{\text{a.s.}} 1 \quad \forall k = 1, 2, \ldots, K. \tag{11.12}$$

Consider for example the sequential Bayesian update (2.21), where the belief is constructed as a posterior probability given the knowledge originating from a data stream. Under correct likelihood models, we know this is the optimal construction. Lemma 2.2 showed that the resulting belief is asymptotically concentrated on the true hypothesis. We conclude that the optimal Bayesian strategy achieves traditional truth learning. Even in the social learning setting, the agents observe increasing evidence over time in the form of streaming data, and therefore it would be desirable to reach the same kind of asymptotic certainty as in the optimal Bayesian strategy. Notably, we showed before in Chapters 5 and 7 that, over connected graphs, traditional truth learning is also achieved in non-Bayesian social learning. In fact, it is guaranteed by Corollary 5.1 under geometric averaging and by Theorem 7.1 under arithmetic averaging.

Within the partial information setting, one relevant objective is to establish the validity of the hypothesis of interest $\vartheta^\bullet$. Referring back to the example mentioned at the beginning of the chapter, we note that the agents there exchange opinions about a product by brand 1. That is, the hypothesis of interest is $\vartheta^{\bullet} = 1$. The best product in the market is represented by hypothesis ϑ^{o}, which could be a different brand such as $\vartheta^{o} = 2$. By collecting multiple data over time (e.g., reviews about the product of interest) and repeated exchanges of opinions, the agents are interested in deciding whether or not brand 1 manufactures the best product, without exchanging any opinions regarding brands 2 or 3. In other words, upon exchanging information regarding $\vartheta^{\bullet}$, the agents are interested in deciding whether or not this hypothesis corresponds to the truth ϑ^{o}. If an agent is successful in doing that, we say that this agent achieves *partial truth learning*.

Definition 11.2 (Partial truth learning). We say that partial truth learning is achieved when the nature of the hypothesis of interest $\vartheta^{\bullet}$ is correctly identified. This entails two different definitions depending on whether or not the hypothesis of interest is equal to the true hypothesis.

i) If $\vartheta^{\bullet} = \vartheta^{o}$, partial truth learning is achieved when

$$\boldsymbol{\mu}_{k,t}(\vartheta^{\bullet}) \xrightarrow[t\to\infty]{\text{a.s.}} 1 \quad \forall k = 1, 2, \ldots, K. \tag{11.13}$$

ii) If $\vartheta^{\bullet} \neq \vartheta^{o}$, partial truth learning is achieved when

$$\boldsymbol{\mu}_{k,t}(\vartheta^{\bullet}) \xrightarrow[t\to\infty]{\text{a.s.}} 0 \quad \forall k = 1, 2, \ldots, K. \tag{11.14}$$

Note that traditional truth learning implies partial truth learning for any $\vartheta^{\bullet}$. However, the converse statement depends on $\vartheta^{\bullet}$. If $\vartheta^{\bullet} = \vartheta^{o}$, traditional truth learning is implied by partial truth learning. However, if $\vartheta^{\bullet} \neq \vartheta^{o}$, Eq. (11.14) only reveals that the hypothesis of interest will be discarded, and traditional truth learning is not guaranteed. In summary,

$$\begin{cases} \text{traditional truth learning} \quad \Longrightarrow \quad \text{partial truth learning} \\ \text{traditional truth learning} \quad \Longleftarrow \quad \begin{array}{c}\text{partial truth learning} \\ \text{with } \vartheta^{\bullet} = \vartheta^{o}\end{array} \end{cases} \tag{11.15}$$

11.4 Memoryless Strategy

In this section we investigate the convergence behavior of the partial information algorithm in listing (11.11) when the decoding step is specialized to the memoryless filling strategy in (11.9). In this case, at each instant t, each agent k performs the following three steps for each $\theta \in \Theta$:

$$\boldsymbol{\psi}_{k,t}(\theta) = \frac{\boldsymbol{\mu}_{k,t-1}(\theta)\ell_k(\boldsymbol{x}_{k,t}|\theta)}{\sum_{\theta' \in \Theta} \boldsymbol{\mu}_{k,t-1}(\theta')\ell_k(\boldsymbol{x}_{k,t}|\theta')}, \tag{11.16a}$$

$$\widehat{\boldsymbol{\psi}}_{j,t}^{(k)}(\theta) = \begin{cases} \boldsymbol{\psi}_{j,t}(\vartheta^\bullet) & \text{if } \theta = \vartheta^\bullet, \\ \dfrac{1}{H-1}\Big(1 - \boldsymbol{\psi}_{j,t}(\vartheta^\bullet)\Big) & \text{if } \theta \neq \vartheta^\bullet, \end{cases} \qquad j \in \mathcal{N}_k, \tag{11.16b}$$

$$\boldsymbol{\mu}_{k,t}(\theta) = \frac{\prod_{j \in \mathcal{N}_k} \left[\widehat{\boldsymbol{\psi}}_{j,t}^{(k)}(\theta)\right]^{a_{jk}}}{\sum_{\theta' \in \Theta} \prod_{j \in \mathcal{N}_k} \left[\widehat{\boldsymbol{\psi}}_{j,t}^{(k)}(\theta')\right]^{a_{jk}}}. \tag{11.16c}$$

It is useful to observe that, in the memoryless approach, the entries of the belief vector corresponding to the unshared hypotheses *evolve equally* over time. To see this, we substitute the decoding step (11.16b) into the cooperation step (11.16c) and write the log belief ratios for any two hypotheses $\theta, \theta' \in \mathcal{U}$ as follows:

$$\log \frac{\boldsymbol{\mu}_{k,t}(\theta)}{\boldsymbol{\mu}_{k,t}(\theta')} = \sum_{j \in \mathcal{N}_k} a_{jk} \log \frac{\widehat{\boldsymbol{\psi}}_{j,t}(\theta)}{\widehat{\boldsymbol{\psi}}_{j,t}(\theta')} = \sum_{j \in \mathcal{N}_k} a_{jk} \log \frac{1 - \boldsymbol{\psi}_{j,t}(\vartheta^\bullet)}{1 - \boldsymbol{\psi}_{j,t}(\vartheta^\bullet)} = 0, \tag{11.17}$$

which implies that

$$\boldsymbol{\mu}_{k,t}(\theta) = \boldsymbol{\mu}_{k,t}(\theta'). \tag{11.18}$$

Since the entries of the vector $\boldsymbol{\mu}_{k,t}$ add up to 1, we have

$$\sum_{\theta \in \mathcal{U}} \boldsymbol{\mu}_{k,t}(\theta) = 1 - \boldsymbol{\mu}_{k,t}(\vartheta^\bullet), \tag{11.19}$$

and we can use (11.18) to conclude that

$$\boldsymbol{\mu}_{k,t}(\theta) = \frac{1 - \boldsymbol{\mu}_{k,t}(\vartheta^\bullet)}{H-1} \tag{11.20}$$

for any unshared hypothesis $\theta \in \mathcal{U}$.

Before delving into the analysis of the learning performance under partial information sharing, we consider the following assumptions. First, we assume that the observations are generated under objective evidence, i.e., under Assumption 5.3. Second, we assume that the network graph is primitive according to Definition 4.5.

Next, we introduce some definitions that will be useful in the forthcoming analysis. We start with the *aggregate likelihood*

$$\ell_k(x|\mathcal{U}) \triangleq \frac{1}{H-1} \sum_{\theta \in \mathcal{U}} \ell_k(x|\theta), \tag{11.21}$$

which averages uniformly the likelihoods corresponding to the hypotheses different from $\vartheta^\bullet$. Then we consider the KL divergence between the true likelihood and the aggregate likelihood, namely,

$$D(\ell_{k,\vartheta^o} || \ell_{k,\mathcal{U}}) \triangleq \mathbb{E} \log \frac{\ell_k(\boldsymbol{x}_{k,t}|\vartheta^o)}{\ell_k(\boldsymbol{x}_{k,t}|\mathcal{U})}, \tag{11.22}$$

where, according to our usual notation, the symbol $\ell_{k,\mathcal{U}}$ is used to refer to the entire pdf or pmf in (11.21), not to a particular value x. We now show that this divergence is finite. We have that

$$\begin{aligned}
\log \frac{\ell_k(x|\vartheta^o)}{\ell_k(x|\mathcal{U})} &= \log \frac{\ell_k(x|\vartheta^o)}{\frac{1}{H-1} \sum_{\theta \in \mathcal{U}} \ell_k(x|\theta)} \\
&= \log \ell_k(x|\vartheta^o) - \log \left(\frac{1}{H-1} \sum_{\theta \in \mathcal{U}} \ell_k(x|\theta) \right) \\
&\overset{(a)}{\leq} \frac{1}{H-1} \sum_{\theta \in \mathcal{U}} \log \frac{\ell_k(x|\vartheta^o)}{\ell_k(x|\theta)},
\end{aligned} \tag{11.23}$$

where (a) follows from Jensen's inequality (see Theorem C.5 and in particular (C.10) with uniform weights $1/(H-1)$) applied to the convex function $-\log$. Then, taking expectations in (11.23) with respect to the true likelihood model $\ell_k(x|\vartheta^o)$ allows us to bound the KL divergence $D(\ell_{k,\vartheta^o} || \ell_{k,\mathcal{U}})$ in terms of the KL divergences relative to the unshared hypotheses as follows:

$$D(\ell_{k,\vartheta^o} || \ell_{k,\mathcal{U}}) \leq \frac{1}{H-1} \sum_{\theta \in \mathcal{U}} D(\ell_{j,\vartheta^o} || \ell_{j,\theta}) < \infty, \tag{11.24}$$

where the last inequality follows from (5.37). We have in fact proved that $D(\ell_{k,\vartheta^o} || \ell_{k,\mathcal{U}})$ is finite.

Finally, we introduce the following weighted quantity:

$$D_{\mathsf{net}}(\mathcal{U}) \triangleq \sum_{k=1}^{K} v_k D(\ell_{k,\vartheta^o} || \ell_{k,\mathcal{U}}), \tag{11.25}$$

which extends the definition in (5.24) to the aggregate likelihood in (11.21).

11.4.1 Convergence Results

The results illustrated in this section were originally presented in [24, 26]. The next theorem reveals that different types of convergence behavior arise, depending on two quantities: the network average of KL divergences relative to $\vartheta^\bullet$, which, from (5.24), is defined by

$$D_{\mathsf{net}}(\vartheta^\bullet) = \sum_{k=1}^{K} v_k D(\ell_{k,\vartheta^\circ} || \ell_{k,\vartheta^\bullet}), \tag{11.26}$$

and the network average of KL divergences relative to the set of unshared hypotheses, which is the quantity $D_{\mathsf{net}}(\mathcal{U})$ defined by (11.25). When $D_{\mathsf{net}}(\vartheta^\bullet)$ is larger than $D_{\mathsf{net}}(\mathcal{U})$, then hypothesis $\vartheta^\bullet$ will be discarded. If it is smaller, the belief will be concentrated on $\vartheta^\bullet$.

Theorem 11.1 (Memoryless strategy: Belief convergence). Let Assumptions 5.1 and 5.3 be satisfied. If the network graph is primitive, then for $k = 1, 2, \dots, K$,

i) If $D_{\mathsf{net}}(\vartheta^\bullet) > D_{\mathsf{net}}(\mathcal{U})$,

$$\boldsymbol{\mu}_{k,t}(\vartheta^\bullet) \xrightarrow[t\to\infty]{\text{a.s.}} 0 \quad \text{and} \quad \boldsymbol{\mu}_{k,t}(\theta) \xrightarrow[t\to\infty]{\text{a.s.}} \frac{1}{H-1} \quad \forall \theta \neq \vartheta^\bullet. \tag{11.27}$$

ii) If $D_{\mathsf{net}}(\vartheta^\bullet) < D_{\mathsf{net}}(\mathcal{U})$,

$$\boldsymbol{\mu}_{k,t}(\vartheta^\bullet) \xrightarrow[t\to\infty]{\text{a.s.}} 1. \tag{11.28}$$

Proof. From (11.16b) it follows that, for all $\theta \in \mathcal{U}$,

$$\begin{aligned}
\log \frac{\widehat{\boldsymbol{\psi}}^{(k)}_{j,t}(\theta)}{\widehat{\boldsymbol{\psi}}^{(k)}_{j,t}(\vartheta^\bullet)} &= \log \frac{\frac{1}{H-1}\Big(1-\boldsymbol{\psi}_{j,t}(\vartheta^\bullet)\Big)}{\boldsymbol{\psi}_{j,t}(\vartheta^\bullet)} \\
&= \log \frac{\frac{1}{H-1}\sum_{\theta'\in\mathcal{U}} \boldsymbol{\psi}_{j,t}(\theta')}{\boldsymbol{\psi}_{j,t}(\vartheta^\bullet)}.
\end{aligned} \tag{11.29}$$

Using (11.16a) in (11.29) we obtain

$$\log \frac{\widehat{\boldsymbol{\psi}}^{(k)}_{j,t}(\theta)}{\widehat{\boldsymbol{\psi}}^{(k)}_{j,t}(\vartheta^\bullet)} = \log \frac{\frac{1}{H-1}\sum_{\theta'\in\mathcal{U}} \boldsymbol{\mu}_{j,t-1}(\theta')\ell_j(\boldsymbol{x}_{j,t}|\theta')}{\boldsymbol{\mu}_{j,t-1}(\vartheta^\bullet)\ell_j(\boldsymbol{x}_{j,t}|\vartheta^\bullet)}, \tag{11.30}$$

and exploiting (11.18) we can write

$$\begin{aligned}
\log \frac{\widehat{\psi}_{j,t}^{(k)}(\theta)}{\widehat{\psi}_{j,t}^{(k)}(\vartheta^\bullet)} &= \log \frac{\boldsymbol{\mu}_{j,t-1}(\theta)\frac{1}{H-1}\sum_{\theta'\in\mathcal{U}} \ell_j(\boldsymbol{x}_{j,t}|\theta')}{\boldsymbol{\mu}_{j,t-1}(\vartheta^\bullet)\ell_j(\boldsymbol{x}_{j,t}|\vartheta^\bullet)} \\
&= \log \frac{\boldsymbol{\mu}_{j,t-1}(\theta)}{\boldsymbol{\mu}_{j,t-1}(\vartheta^\bullet)} + \log \frac{\frac{1}{H-1}\sum_{\theta'\in\mathcal{U}} \ell_j(\boldsymbol{x}_{j,t}|\theta')}{\ell_j(\boldsymbol{x}_{j,t}|\vartheta^\bullet)} \\
&= \log \frac{\boldsymbol{\mu}_{j,t-1}(\theta)}{\boldsymbol{\mu}_{j,t-1}(\vartheta^\bullet)} + \log \frac{\ell_j(\boldsymbol{x}_{j,t}|\mathcal{U})}{\ell_j(\boldsymbol{x}_{j,t}|\vartheta^\bullet)},
\end{aligned} \tag{11.31}$$

where in the last equality we used the aggregate likelihood defined in (11.21). Substituting (11.31) into (11.16c), we obtain the following recursion for the log belief ratios:

$$\begin{aligned}
\log \frac{\boldsymbol{\mu}_{k,t}(\theta)}{\boldsymbol{\mu}_{k,t}(\vartheta^\bullet)} &= \sum_{j\in\mathcal{N}_k} a_{jk} \left[\log \frac{\boldsymbol{\mu}_{j,t-1}(\theta)}{\boldsymbol{\mu}_{j,t-1}(\vartheta^\bullet)} + \log \frac{\ell_j(\boldsymbol{x}_{j,t}|\mathcal{U})}{\ell_j(\boldsymbol{x}_{j,t}|\vartheta^\bullet)}\right] \\
&= \sum_{j=1}^{K} a_{jk} \left[\log \frac{\boldsymbol{\mu}_{j,t-1}(\theta)}{\boldsymbol{\mu}_{j,t-1}(\vartheta^\bullet)} + \log \frac{\ell_j(\boldsymbol{x}_{j,t}|\mathcal{U})}{\ell_j(\boldsymbol{x}_{j,t}|\vartheta^\bullet)}\right],
\end{aligned} \tag{11.32}$$

where the last equality follows from the definition of neighborhood in (4.1). Next, to establish the claim of the theorem, we follow similar steps to those used in the proof of Theorem 5.1, by exploiting Lemma D.3.

We start by noting that (11.32) can be cast in the vector form (D.57), namely,

$$\boldsymbol{z}_t = A^\mathsf{T}(\boldsymbol{z}_{t-1} + \boldsymbol{y}_t), \tag{11.33}$$

by setting

$$\boldsymbol{y}_t = \left[\log \frac{\ell_1(\boldsymbol{x}_{1,t}|\mathcal{U})}{\ell_1(\boldsymbol{x}_{1,t}|\vartheta^\bullet)}, \log \frac{\ell_2(\boldsymbol{x}_{2,t}|\mathcal{U})}{\ell_2(\boldsymbol{x}_{2,t}|\vartheta^\bullet)}, \ldots, \log \frac{\ell_K(\boldsymbol{x}_{K,t}|\mathcal{U})}{\ell_K(\boldsymbol{x}_{K,t}|\vartheta^\bullet)}\right], \tag{11.34}$$

$$\boldsymbol{z}_t = \left[\log \frac{\boldsymbol{\mu}_{1,t}(\theta)}{\boldsymbol{\mu}_{1,t}(\vartheta^\bullet)}, \log \frac{\boldsymbol{\mu}_{2,t}(\theta)}{\boldsymbol{\mu}_{2,t}(\vartheta^\bullet)}, \ldots, \log \frac{\boldsymbol{\mu}_{K,t}(\theta)}{\boldsymbol{\mu}_{K,t}(\vartheta^\bullet)}\right], \tag{11.35}$$

where we recall that in our notation all vectors are column vectors. Next, we note that the network graph is assumed to be primitive, implying, in view of Corollary 4.1, that (D.58) holds with $A^\bullet = v\,\mathbb{1}^\mathsf{T}$.

Now, to use the results from Lemma D.3, we must verify that the sequence $\{\boldsymbol{y}_t\}$ is formed by iid vectors whose entries have finite mean. The first condition is satisfied under Assumption 5.3. Regarding the second condition, consider the jth entry of $\boldsymbol{y}_t$ expressed as follows:

$$\log \frac{\ell_j(\boldsymbol{x}_{j,t}|\mathcal{U})}{\ell_j(\boldsymbol{x}_{j,t}|\vartheta^\bullet)} = \log \frac{\ell_j(\boldsymbol{x}_{j,t}|\vartheta^o)}{\ell_j(\boldsymbol{x}_{j,t}|\vartheta^\bullet)} - \log \frac{\ell_j(\boldsymbol{x}_{j,t}|\vartheta^o)}{\ell_j(\boldsymbol{x}_{j,t}|\mathcal{U})}, \tag{11.36}$$

where we recall that ϑ^o denotes the true hypothesis. Under Assumption 5.3, the first term on the RHS of (11.36) has finite mean. The same property holds for the second term in view of (11.23). Thus, the jth entry of the vector $\boldsymbol{y}_t$ has finite mean. We conclude that the sequence $\{\boldsymbol{y}_t\}$ satisfies the conditions required by Lemma D.3, where the vector $\bar{y}$

used in Lemma D.3 corresponds to $\mathbb{E}\boldsymbol{y}_t$. We can therefore apply the result of Lemma D.3 with $A^{\bullet} = v\mathbb{1}^{\mathsf{T}}$ to conclude that

$$\frac{1}{t}\,\boldsymbol{z}_t \xrightarrow[t\to\infty]{\text{a.s.}} \mathbb{1}\,v^{\mathsf{T}}\,\mathbb{E}\boldsymbol{y}_t. \tag{11.37}$$

In view of (11.34) and (11.35), we can rewrite (11.37) in terms of the kth entry as follows:

$$\begin{aligned}\frac{1}{t}\log\frac{\boldsymbol{\mu}_{k,t}(\theta)}{\boldsymbol{\mu}_{k,t}(\vartheta^{\bullet})} &\xrightarrow[t\to\infty]{\text{a.s.}} \sum_{j=1}^{K} v_j \mathbb{E}\log\frac{\ell_j(\boldsymbol{x}|\mathcal{U})}{\ell_j(\boldsymbol{x}|\vartheta^{\bullet})}\\ &= \sum_{j=1}^{K} v_j\Big[D(\ell_{j,\vartheta^{\circ}}||\ell_{j,\vartheta^{\bullet}}) - D(\ell_{j,\vartheta^{\circ}}||\ell_{j,\mathcal{U}})\Big]\\ &= D_{\mathsf{net}}(\vartheta^{\bullet}) - D_{\mathsf{net}}(\mathcal{U}),\end{aligned} \tag{11.38}$$

where $D_{\mathsf{net}}(\mathcal{U})$ and $D_{\mathsf{net}}(\vartheta^{\bullet})$ are defined by (11.25) and (11.26), respectively. The above result holds for all $\theta \in \mathcal{U}$. The sign of the quantity on the RHS of (11.38) will dictate different convergence behaviors.

Consider first case i), namely,

$$D_{\mathsf{net}}(\vartheta^{\bullet}) > D_{\mathsf{net}}(\mathcal{U}), \tag{11.39}$$

under which the RHS of (11.38) becomes positive. This implies that

$$\log\frac{\boldsymbol{\mu}_{k,t}(\theta)}{\boldsymbol{\mu}_{k,t}(\vartheta^{\bullet})} \xrightarrow[t\to\infty]{\text{a.s.}} \infty \quad \forall\theta\in\mathcal{U}. \tag{11.40}$$

Since $\boldsymbol{\mu}_{k,t}(\theta)$ is bounded by 1 for any $\theta \in \Theta$, it follows from (11.40) that

$$\boldsymbol{\mu}_{k,t}(\vartheta^{\bullet}) \xrightarrow[t\to\infty]{\text{a.s.}} 0, \tag{11.41}$$

which, in view of (11.20), implies that

$$\boldsymbol{\mu}_{k,t}(\theta) \xrightarrow[t\to\infty]{\text{a.s.}} \frac{1}{H-1} \quad \forall\theta\in\mathcal{U}, \tag{11.42}$$

thus concluding the proof for case i).

Second, consider case ii), namely,

$$D_{\mathsf{net}}(\vartheta^{\bullet}) < D_{\mathsf{net}}(\mathcal{U}), \tag{11.43}$$

under which the RHS of (11.38) becomes negative. Using similar arguments as before, we deduce that

$$\log\frac{\boldsymbol{\mu}_{k,t}(\theta)}{\boldsymbol{\mu}_{k,t}(\vartheta^{\bullet})} \xrightarrow[t\to\infty]{\text{a.s.}} -\infty \quad \forall\theta\in\mathcal{U}, \tag{11.44}$$

implying that

$$\boldsymbol{\mu}_{k,t}(\theta) \xrightarrow[t\to\infty]{\text{a.s.}} 0 \quad \forall\theta\in\mathcal{U}. \tag{11.45}$$

This, in turn, implies that

$$\boldsymbol{\mu}_{k,t}(\vartheta^{\bullet}) \xrightarrow[t\to\infty]{\text{a.s.}} 1, \tag{11.46}$$

which concludes the proof for case ii).

■

Theorem 11.1 shows two types of convergence behavior for the beliefs across the network. The critical condition necessary to establish which behavior is activated involves the comparison between two network KL divergences, specifically, $D_{\sf net}(\vartheta^\bullet)$ (which quantifies the difference between the true likelihood and the likelihood relative to the hypothesis of interest $\vartheta^\bullet$) and $D_{\sf net}(\mathcal{U})$ (which quantifies the difference between the true likelihood and the fictitious likelihood from (11.21) corresponding to the ensemble of unshared hypotheses).

Condition i) in Theorem 11.1, $D_{\sf net}(\vartheta^\bullet) > D_{\sf net}(\mathcal{U})$, means that the likelihood relative to $\vartheta^\bullet$ is sufficiently distinct from the true one in comparison with the aggregate likelihood. This relatively high difference from ϑ^o drives the agents to believe that $\vartheta^\bullet$ is not the true hypothesis — see the first convergence result in (11.27).

Conversely, condition ii) in Theorem 11.1, $D_{\sf net}(\vartheta^\bullet) < D_{\sf net}(\mathcal{U})$, means that the likelihood relative to $\vartheta^\bullet$ is closer to the true one than the aggregate likelihood. As a result, in this case the agents tend to accept $\vartheta^\bullet$ as the true hypothesis — see (11.28).

To gain further insight, it is useful to examine how the convergence behavior changes under the possible choices of $\vartheta^\bullet$. In particular, in the next two sections we will consider the two possible scenarios: *truth sharing*, i.e., $\vartheta^\bullet = \vartheta^o$, and *false-hypothesis sharing*, i.e., $\vartheta^\bullet \neq \vartheta^o$.

11.4.2 Truth Sharing

When $\vartheta^\bullet = \vartheta^o$, from (11.26) we have

$$D_{\sf net}(\vartheta^\bullet) = D_{\sf net}(\vartheta^o) = 0, \tag{11.47}$$

which implies that condition i) in Theorem 11.1 never holds, due to the nonnegativity of the KL divergence $D_{\sf net}(\mathcal{U})$. It is instead possible that condition ii) is verified. This happens when $D_{\sf net}(\mathcal{U}) > 0$. If this is the case, Eq. (11.28) holds, which means that traditional truth learning is achieved.

Therefore, we are interested in establishing when $D_{\sf net}(\mathcal{U}) > 0$. To this end, we introduce the next assumption. Preliminarily, it is useful to recall from (7.2) the definition of an *indistinguishable set*:

$$\mathcal{I}_k \triangleq \Big\{\theta \in \Theta\backslash\{\vartheta^o\} \text{ such that } D(\ell_{k,\vartheta^o}||\ell_{k,\theta}) = 0\Big\}, \tag{11.48}$$

and from (7.3) the definition of a *distinguishable set*:

$$\mathcal{D}_k \triangleq \Theta\backslash\Big(\mathcal{I}_k \cup \{\vartheta^o\}\Big). \tag{11.49}$$

Assumption 11.1 (Average likelihood of distinguishable hypotheses). There exists an agent k whose distinguishable set $\mathcal{D}_k$ is nonempty, and whose true likelihood ℓ_{k,ϑ^o} is not a *uniform* combination of the likelihoods $\{\ell_{k,\theta}\}_{\theta\in\mathcal{D}_k}$ of the distinguishable hypotheses. That is, we have

$$\ell_{k,\vartheta^o} \neq \frac{1}{|\mathcal{D}_k|}\sum_{\theta\in\mathcal{D}_k}\ell_{k,\theta}. \tag{11.50}$$

Note that Assumption 11.1 is a relaxed version of Assumption 7.1. While Assumption 7.1 requires that the true likelihood cannot take the form of *any* convex combination of likelihoods $\{\ell_{k,\theta}\}_{\theta\in\mathcal{D}_k}$, in Assumption 11.1 the combination to be avoided is the one with *uniform* weights. This is not a strong assumption, since the case where the true likelihood matches *exactly* a mixture of the likelihoods relative to the distinguishable hypotheses with uniform weights is deemed to be a rare coincidence. Moreover, for Assumption 11.1 to hold, the existence of a single agent k satisfying (11.50) is sufficient, while in Assumption 7.1 the required condition of convex independence must be satisfied by all agents with nonempty distinguishable sets.

The next corollary shows that Assumption 11.1 is equivalent to the condition $D_{\mathsf{net}}(\mathcal{U}) > 0$, and, hence, that when $\vartheta^\bullet = \vartheta^o$ the memoryless filling strategy achieves traditional truth learning.

Corollary 11.1 (Memoryless strategy: Truth sharing implies traditional truth learning). Under the same assumptions used in Theorem 11.1 and under Assumption 11.1, if $\vartheta^\bullet = \vartheta^o$, then for $k = 1, 2, \ldots, K$,

$$\boldsymbol{\mu}_{k,t}(\vartheta^o) \xrightarrow[t\to\infty]{\text{a.s.}} 1. \tag{11.51}$$

Proof. It suffices to show that condition ii) of Theorem 11.1 holds, namely, that

$$D_{\mathsf{net}}(\vartheta^\bullet) < D_{\mathsf{net}}(\mathcal{U}) \tag{11.52}$$

when $\vartheta^\bullet = \vartheta^o$. First, under $\vartheta^\bullet = \vartheta^o$, it follows from (11.26) that

$$D_{\mathsf{net}}(\vartheta^\bullet) = \sum_{k=1}^{K} v_k D(\ell_{k,\vartheta^o}||\ell_{k,\vartheta^o}) = 0. \tag{11.53}$$

Therefore, condition (11.52) will be verified if we establish that

$$D_{\mathsf{net}}(\mathcal{U}) > 0. \tag{11.54}$$

In view of the positivity of the Perron vector entries and the nonnegativity of the KL divergence, we see from (11.25) that condition (11.54) is violated if, and only if,

$$D(\ell_{k,\vartheta^o}||\ell_{k,\mathcal{U}}) = 0 \quad \forall k = 1, 2, \ldots, K. \tag{11.55}$$

We now proceed to show that, under Assumption 11.1, Eq. (11.55) is not verified for at least one agent k. Specifically, it is not verified for the agent k mentioned in Assumption 11.1, which has as nonempty distinguishable set $\mathcal{D}_k$ and satisfies (11.50). To this end, observe that by using the definition of $\ell_{k,\mathcal{U}}$ from (11.21), Eq. (11.55) is equivalent to

$$\ell_k(x|\vartheta^o) = \frac{1}{H-1}\sum_{\theta\in\mathcal{U}} \ell_k(x|\theta). \tag{11.56}$$

We recall that when $\boldsymbol{x}_{k,t}$ happens to be a continuous random vector, then the equality between pdfs in (11.56) is intended to hold for all $x \in \mathcal{X}_k$, except possibly for sets with zero Lebesgue measure.

Now note that, when $\vartheta^\bullet = \vartheta^o$, we have

$$\mathcal{U} = \Theta\backslash\{\vartheta^o\} = \mathcal{I}_k \cup \mathcal{D}_k, \tag{11.57}$$

namely, $\mathcal{U}$ is equivalent to the union of the disjoint sets of indistinguishable and distinguishable hypotheses defined by (11.48) and (11.49), respectively. Thus, we can rewrite (11.56) as

$$(H-1)\,\ell_k(x|\vartheta^o) = \sum_{\theta\in\mathcal{D}_k} \ell_k(x|\theta) + \sum_{\theta\in\mathcal{I}_k} \ell_k(x|\theta), \tag{11.58}$$

which in turn is equivalent to

$$(H-1)\,\ell_k(x|\vartheta^o) = \sum_{\theta\in\mathcal{D}_k} \ell_k(x|\theta) + |\mathcal{I}_k|\,\ell_k(x|\vartheta^o), \tag{11.59}$$

or

$$\ell_k(x|\vartheta^o) = \frac{1}{|\mathcal{D}_k|}\sum_{\theta\in\mathcal{D}_k} \ell_k(x|\theta), \tag{11.60}$$

where in the last step we used the fact that $|\mathcal{D}_k| = H - 1 - |\mathcal{I}_k|$. Since (11.60) violates Assumption 11.1, we conclude that (11.55) cannot hold, which in turn implies (11.52), and the proof is complete.

■

It is interesting to draw a parallel between memoryless partial information sharing and traditional social learning. In view of (11.50), Assumption 11.1 has the interpretation that at least one agent k has the capability to discount $\mathcal{D}_k$ as a whole, i.e., to discount all $\theta \in \mathcal{D}_k$. On the other hand, in (11.20) we showed that, under the memoryless approach, the beliefs about the unshared hypotheses *evolve equally*, i.e., $\boldsymbol{\mu}_{k,t}(\theta)$ takes on the same value for all $\theta \neq \vartheta^\bullet$ during the algorithm evolution. Accordingly, once agent k is able to discount all $\theta \in \mathcal{D}_k$, it is also able to discount all $\theta \neq \vartheta^\bullet$.

Finally, this possibility is extended to all the other agents by propagation of information across the primitive graph.

We can now compare the described learning mechanism with the one occurring in traditional social learning. There we required global identifiability, which implies that, for each $\theta \neq \vartheta^o$ there must be an agent that distinguish θ from ϑ^o. This condition is stronger than Assumption 11.1. In other words, Assumption 11.1 can be satisfied even if global identifiability is violated. For example, consider $\Theta = \{1, 2, 3\}$ with $\vartheta^o = 1$. Assume that $\theta = 2$ is indistinguishable from $\vartheta^o = 1$ for all agents. Then it follows that global identifiability cannot hold. Consider furthermore that no agent can distinguish hypothesis $\theta = 3$ from $\vartheta^o = 1$, *except* for agent k, for which $\ell_{k,3} \neq \ell_{k,1}$. Therefore, the distinguishable set of agent k is $\mathcal{D}_k = \{3\}$. Then, Assumption 11.1 requires $\ell_{k,3} \neq \ell_{k,1}$, which is true.[2] Then, due to this single informative agent k, Assumption 11.1 holds, even though global identifiability does not.

This phenomenon might appear puzzling, since it seems to imply that learning under partial information sharing is easier! However, we must keep in mind that we are considering only the truth sharing scenario. In practice, the agents of course cannot decide to share the true hypothesis. Therefore, the learning performance must always be examine by taking also into account what happens under false-hypothesis sharing. In the next section, we will in fact argue that the *advantage* (of memoryless partial information sharing over traditional social learning) observed when $\vartheta^\bullet = \vartheta^o$ occurs *at the expense* of a *disadvantage* in the case $\vartheta^\bullet \neq \vartheta^o$.

11.4.3 False-Hypothesis Sharing

When $\theta \neq \vartheta^\bullet$, both conditions i) and ii) in Theorem 11.1 *can* occur, as we illustrate later in Example 11.1, depending on the choice of $\vartheta^\bullet$ among the wrong hypotheses. If condition i) is satisfied, then partial truth learning is achieved, since the first convergence result in (11.27) reveals that the hypothesis of interest is correctly discarded. Moreover, from the second convergence result in (11.27) we see that $\boldsymbol{\mu}_{k,t}(\vartheta^o) \xrightarrow[t\to\infty]{\text{a.s.}} 1/(H-1)$, which implies that traditional truth learning can only take place in the binary case (which, as already discussed, is a trivial case within the partial information setting).

Alternatively, if condition ii) is satisfied, we see from (11.28) that the full

[2]This example also shows that Assumption 11.1 is automatically satisfied whenever $\mathcal{D}_k$ contains a single element.

belief mass is ultimately concentrated on the hypothesis of interest, which however is wrong and, hence, partial truth learning cannot be achieved.

To sum up, under false-hypothesis sharing partial truth learning is achieved when condition i) holds. One *necessary* condition for it to hold is that

$$D_{\mathsf{net}}(\vartheta^\bullet) > 0, \tag{11.61}$$

which implies that some agent must be able to distinguish $\vartheta^\bullet$ from ϑ^o. Since the true hypothesis can be any hypothesis, condition (11.61) should be required for any hypothesis $\vartheta^\bullet \neq \vartheta^o$, which is equivalent global identifiability — see Assumption 5.4. We conclude that, under false-hypothesis sharing, global identifiability is necessary but not sufficient and it can only guarantee partial truth learning. Therefore, as anticipated in the previous section, under false-hypothesis sharing we end up with a disadvantage with respect to traditional social learning.

Example 11.1 (Memoryless strategy: False-hypothesis sharing). Consider $K = 10$ agents interested in solving a social learning problem. The agents operate under partial information sharing, and adopt a memoryless filling strategy. They are connected according to the strong graph shown in the left panel of Figure 11.2 (the graph is undirected and all agents have a self-loop, not shown in the figure). On top of this graph, a combination matrix is designed using the uniform-averaging rule — see Table 4.1. The set of hypotheses is $\Theta = \{1, 2, 3\}$ and the true hypothesis is $\vartheta^o = 1$. All agents possess the same family of likelihood models, denoted by $\ell(x|\theta)$, and illustrated in the right panel of Figure 11.2. These are Gaussian models with unit variance and means $1, 2$, and 5. Moreover, in the simulations the observations are drawn as statistically independent across the agents.

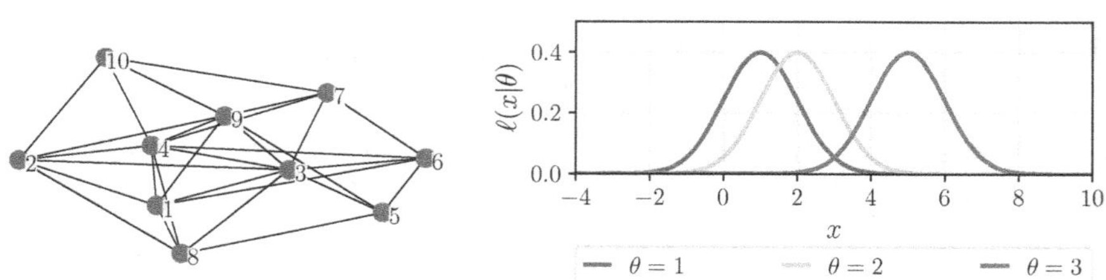

Figure 11.2: (*Left*) Network topology used in Example 11.1. The graph is undirected and all agents are assumed to have a self-loop, not shown in the figure. (*Right*) Family of Gaussian likelihood models used in the example.

Evaluating (5.24) with $\ell_{k,\theta} = \ell_\theta$ for all $\theta \in \Theta$ (because the likelihood models are the same for all agents) and with $f_k = \ell_{\vartheta^o}$ (because in this chapter we are focusing on the

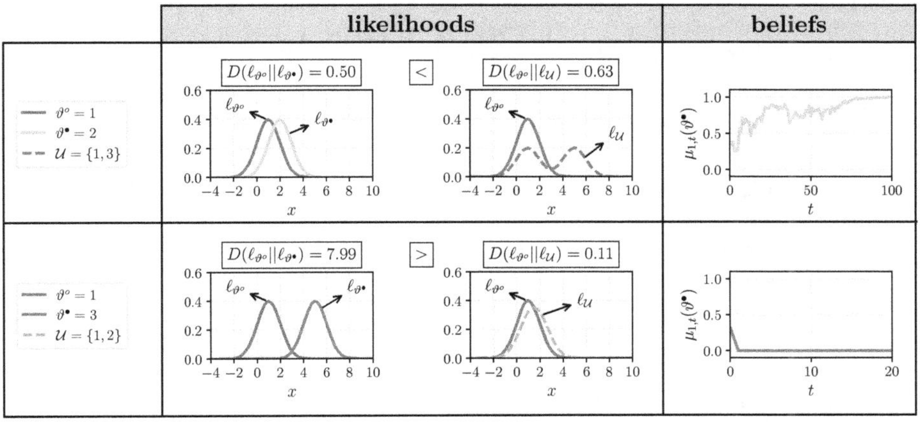

Figure 11.3: Likelihood models and belief evolution for agent 1 in Example 11.1. (*Left and center*) Actual likelihood models (solid line) and aggregate likelihood models defined in (11.21) (dashed line). (*Right*) Time evolution of the belief about the hypothesis of interest for agent 1.

objective evidence model) it follows that

$$D_{\text{net}}(\vartheta^\bullet) = \sum_{k=1}^{K} v_k D(\ell_{\vartheta^o}||\ell_{\vartheta^\bullet}) = D(\ell_{\vartheta^o}||\ell_{\vartheta^\bullet}). \tag{11.62}$$

Likewise, evaluating (11.25) we have

$$D_{\text{net}}(\mathcal{U}) = \sum_{k=1}^{K} v_k D(\ell_{\vartheta^o}||\ell_{\mathcal{U}}) = D(\ell_{\vartheta^o}||\ell_{\mathcal{U}}), \tag{11.63}$$

where $\ell_{\mathcal{U}}$ is the aggregate likelihood defined in (11.21), with subscript k omitted since the likelihoods are equal across the agents. Using (11.62) and (11.63), we see that in the example under consideration the Perron vector does not play a role in the convergence behavior in Theorem 11.1, and only the following two quantities will determine the behavior of all agents:

$$D(\ell_{\vartheta^o}||\ell_{\vartheta^\bullet}) \quad \text{and} \quad D(\ell_{\vartheta^o}||\ell_{\mathcal{U}}). \tag{11.64}$$

We now examine how the learning behavior changes depending on the particular choice of the hypothesis of interest from among the wrong hypotheses. Consider first the case $\vartheta^\bullet = 2$. This case is examined in the top panels of Figure 11.3. We see that $D(\ell_{\vartheta^o}||\ell_{\vartheta^\bullet}) < D(\ell_{\vartheta^o}||\ell_{\mathcal{U}})$, which means that the likelihood relative to the hypothesis of interest is closer to the true likelihood in comparison with the aggregate likelihood. Accordingly, condition ii) in Theorem 11.1 is satisfied, which implies that all agents are fooled into believing that $\vartheta^\bullet$ is the true state. This behavior is confirmed by the experiment (obtained by running the algorithm in (11.11) with the memoryless filling strategy in (11.9)) shown in the top right panel of Figure 11.3, where we display in particular the beliefs of agent 1.

We switch to the case $\vartheta^\bullet = 3$, examined in the bottom panels of Figure 11.3. Now we have $D(\ell_{\vartheta^o}||\ell_{\vartheta^\bullet}) > D(\ell_{\vartheta^o}||\ell_{\mathcal{U}})$, i.e., the likelihood relative to the hypothesis of interest is farther from the true likelihood in comparison with the aggregate likelihood. Accordingly,

Table 11.1: Identifiability setup for the learning problem in Example 11.2. We highlight in lilac the distributions corresponding to the distinguishable hypotheses, and in pink the distributions corresponding to the indistinguishable hypotheses.

Likelihood model: $\ell_k(x\|\theta)$										
k \\ θ	$\vartheta^o = 1$	2	3	4	5	6	7	8	9	10
1	g_1	g_2	g_3	g_4	g_5	g_6	g_7	g_8	g_9	g_{10}
2	g_1	g_1	g_3	g_4	g_5	g_6	g_7	g_8	g_9	g_{10}
3	g_1	g_1	g_1	g_4	g_5	g_6	g_7	g_8	g_9	g_{10}
4	g_1	g_1	g_1	g_1	g_5	g_6	g_7	g_8	g_9	g_{10}
5	g_1	g_1	g_1	g_1	g_1	g_6	g_7	g_8	g_9	g_{10}
6	g_1	g_1	g_1	g_1	g_1	g_1	g_7	g_8	g_9	g_{10}
7	g_1	g_1	g_1	g_1	g_1	g_1	g_1	g_8	g_9	g_{10}
8	g_1	g_1	g_1	g_1	g_1	g_1	g_1	g_1	g_9	g_{10}
9	g_1	g_1	g_1	g_1	g_1	g_1	g_1	g_1	g_1	g_{10}
10	g_1	g_1	g_1	g_1	g_1	g_1	g_1	g_1	g_1	g_1
11	g_1	g_2	g_1	g_1	g_1	g_1	g_1	g_1	g_1	g_1
12	g_1	g_2	g_3	g_1	g_1	g_1	g_1	g_1	g_1	g_1

condition i) in Theorem 11.1 is satisfied, which implies that the agents correctly classify the hypothesis of interest as being false. This behavior is confirmed by the experiment shown in the bottom right panel of Figure 11.3, where we focus again on the beliefs of agent 1.

Example 11.2 (Memoryless strategy: Convergence behavior). Consider a network of $K = 12$ agents connected according to the strong graph displayed in the top left panel of Figure 11.4. The graph is undirected and all agents are assumed to have a self-loop (not shown in the figure). The combination matrix A is designed according to the Metropolis rule (see Table 4.1), resulting in a doubly stochastic matrix, and, hence, the entries of the Perron vector are $v_k = (1/12)$ — see (4.18). The set of hypotheses is $\Theta = \{1, 2, \ldots, 10\}$, and the true hypothesis is $\vartheta^o = 1$. The observations are statistically independent over time and across the agents.

Before describing the likelihoods of each agent, let us consider the following family of unit-variance Gaussian pdfs with different means:

$$g_n(x) = \frac{1}{\sqrt{2\pi}} \exp\left\{-\frac{1}{2}\Big(x - 0.5(n-1)\Big)^2\right\}, \qquad n = 1, 2, \ldots, 10. \tag{11.65}$$

The distributions are depicted in the top right panel of Figure 11.4. We assume that the likelihoods are taken from this family of distributions as detailed in Table 11.1.

Note from Table 11.1 that

$$\begin{cases} \mathcal{I}_k = \emptyset & \text{if } k = 1, \\ \mathcal{I}_k \neq \emptyset & \text{if } k = 2, 3, \ldots, 12, \end{cases} \tag{11.66}$$

which means that only agent 1 is able to solve the inference problem alone, while for the other agents there always exist some hypotheses indistinguishable from ϑ^o. For example, for agent 4 we have

$$\mathcal{I}_4 = \{2, 3, 4\} \text{ and } \mathcal{D}_4 = \{5, 6, 7, 8, 9, 10\}, \tag{11.67}$$

whereas for agent 10 we have

$$\mathcal{I}_{10} = \Theta \backslash \{\vartheta^o\} \text{ and } \mathcal{D}_{10} = \emptyset. \tag{11.68}$$

To determine the belief convergence for different hypotheses, we can use Theorem 11.1 and Corollary 11.1. To illustrate how these results can be used, we consider three different cases.

Case $\vartheta^\bullet = \vartheta^o = 1$. In the truth sharing case, we resort to Corollary 11.1 to claim that all agents learn the truth in the traditional sense. To be able to do so, we must show that Assumption 11.1 holds, i.e., for each agent k such that $\mathcal{D}_k$ is nonempty, it should hold that

$$\ell_{k,1} \neq \frac{1}{|\mathcal{D}_k|} \sum_{\theta \in \mathcal{D}_k} \ell_{k,\theta}. \tag{11.69}$$

Let us focus on one agent, for example agent 11. We see from Table 11.1 that $\mathcal{D}_{11} = \{2\}$, so that condition (11.69) becomes

$$\ell_{11,1} \neq \ell_{11,2}. \tag{11.70}$$

Observing from the table that $\ell_{11,1} = g_1$ and $\ell_{11,2} = g_2$, we conclude that agent 11 satisfies (11.69). If we consider now agent 12, we see that $\mathcal{D}_{12} = \{2, 3\}$ and condition (11.69) becomes

$$\ell_{12,1} \neq \frac{1}{2}\big(\ell_{12,2} + \ell_{12,3}\big). \tag{11.71}$$

Observing from the table that $\ell_{12,2} = g_2$ and $\ell_{12,3} = g_3$, we see that Eq. (11.71) is equivalent to

$$g_1 \neq \frac{1}{2}\big(g_2 + g_3\big), \tag{11.72}$$

which holds since the Gaussian mixture on the RHS cannot be equal to the Gaussian pdf on the LHS. The above reasoning can be extended to all other agents for which $\mathcal{D}_k$ is nonempty. This implies that Assumption 11.1 holds. Therefore, in view of Corollary 11.1, all agents must place their full belief mass on the true hypothesis. This behavior is confirmed by the experiment shown in the bottom left panel of Figure 11.4, where we focus on the beliefs of agent 1. This experiment, as well as the other experiments shown in the bottom panels, are obtained by running the algorithm in (11.11) with the memoryless filling strategy in (11.9).

Case $\vartheta^\bullet = 4$. To determine the asymptotic behavior of the memoryless strategy when $\vartheta^\bullet \neq 1$, we resort to Theorem 11.1 and compute the quantities $D_{\mathsf{net}}(\vartheta^\bullet)$ and $D_{\mathsf{net}}(\mathcal{U})$. The first quantity can be evaluated as follows:

$$\begin{aligned} D_{\mathsf{net}}(4) &= \sum_{k=1}^{12} v_k D(\ell_{k,1} || \ell_{k,4}) \\ &\overset{\text{(a)}}{=} \sum_{k=1}^{3} v_k D(g_1 || g_4) + \sum_{k=4}^{12} v_k D(g_1 || g_1) \\ &= \sum_{k=1}^{3} v_k D(g_1 || g_4) \\ &\overset{\text{(b)}}{=} \frac{3}{12} \frac{(0 - 0.5 \times 3)^2}{2} = 0.2812, \end{aligned} \tag{11.73}$$

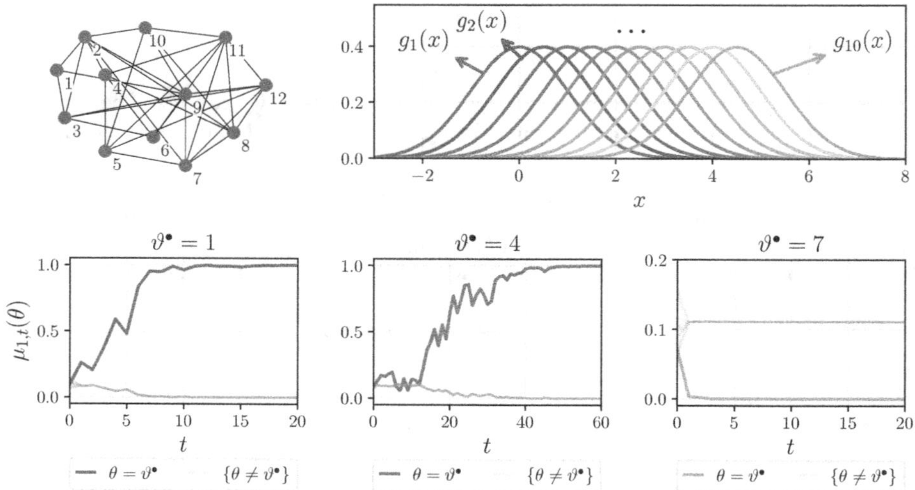

Figure 11.4: (*Top left*) Network topology used in Example 11.2. The graph is undirected and all agents are assumed to have a self-loop, not shown in the figure. (*Top right*) Family of Gaussian distributions used in the example. (*Bottom*) Belief evolution over time for agent 1. The bottom panels correspond to different hypotheses of interest $\vartheta^\bullet$.

where in (a) we used the information from Table 11.1, and in (b) we used the formula for the KL divergence between Gaussian distributions with different means reported in (2.45).

The second quantity, namely $D_{\mathsf{net}}(\mathcal{U})$, is given by

$$D_{\mathsf{net}}(\mathcal{U}) = \frac{1}{12}\sum_{k=1}^{12} D\left(\ell_{k,1}\|\ell_{k,\mathcal{U}}\right), \tag{11.74}$$

where $\mathcal{U} = \Theta\backslash\{4\}$ and

$$\ell_{k,\mathcal{U}} = \frac{1}{9}\sum_{\theta\in\Theta\backslash\{4\}} \ell_{k,\theta}. \tag{11.75}$$

Using Table 11.1, we find numerically that

$$D_{\mathsf{net}}(\mathcal{U}) = 0.4521, \tag{11.76}$$

which is smaller than the value of $D_{\mathsf{net}}(4)$ obtained in (11.73). It follows that condition ii) in Theorem 11.1 holds. Therefore, all agents must mistakenly place their full belief mass on the hypothesis of interest $\vartheta^\bullet = 4$. This behavior is in fact observed in the bottom center panel of Figure 11.4, with reference to agent 1.

Case $\vartheta^\bullet = 7$. Similarly to the previous case, we will compute the quantities $D_{\mathsf{net}}(\vartheta^\bullet)$

and $D_{\text{net}}(\mathcal{U})$, obtaining

$$\begin{aligned} D_{\text{net}}(7) &= \sum_{k=1}^{12} v_k D(\ell_{k,1}||\ell_{k,7}) \\ &\stackrel{(a)}{=} \sum_{k=1}^{6} v_k D(g_1||g_7) + \sum_{k=7}^{12} v_k D(g_1||g_1) \\ &= \sum_{k=1}^{6} v_k D(g_1||g_7) \\ &\stackrel{(b)}{=} \frac{6}{12}\frac{(0-0.5\times 6)^2}{2} = 2.25, \end{aligned} \tag{11.77}$$

and (now $\mathcal{U} = \Theta\backslash\{7\}$)

$$D_{\text{net}}(\mathcal{U}) = 0.3817. \tag{11.78}$$

We see that in this case condition i) in Theorem 11.1 holds and therefore all agents must correctly discard the hypothesis of interest $\vartheta^\bullet = 7$. This behavior is confirmed by the experiment shown in the bottom right panel of Figure 11.4, where we focus again on the beliefs of agent 1.

11.5 Memory in Partial Information

We next study the asymptotic behavior of algorithm (11.11) under the memory-aware filling strategy in (11.10). Preliminarily, we introduce the following notation:

$$\boldsymbol{\mu}_{k,t}(\mathcal{S}) \triangleq \sum_{\theta\in\mathcal{S}} \boldsymbol{\mu}_{k,t}(\theta), \qquad \boldsymbol{\psi}_{k,t}(\mathcal{S}) \triangleq \sum_{\theta\in\mathcal{S}} \boldsymbol{\psi}_{k,t}(\theta), \tag{11.79}$$

for any subset of hypotheses $\mathcal{S} \subseteq \Theta$, with the convention

$$\boldsymbol{\mu}_{k,t}(\emptyset) = \boldsymbol{\psi}_{k,t}(\emptyset) = 0. \tag{11.80}$$

Furthermore, we define for each agent k the *prior confusion ratio*

$$\Gamma_k \triangleq \frac{\mu_{k,0}(\mathcal{I}_k)}{\mu_{k,0}(\vartheta^o)}, \tag{11.81}$$

namely, the ratio between the mass assigned by agent k to the indistinguishable set and the mass assigned to the true hypothesis. Observe that Γ_k increases as agent k assigns more mass to the indistinguishable set and/or less mass to the true hypothesis, i.e., when it is more confused at the beginning of the learning process.

Exploiting (11.79), we see that the uniform prior assignment, $\mu_{k,0}(\theta) = 1/H$ for all $\theta \in \Theta$, leads to

$$\Gamma_k = \frac{\mu_{k,0}(\mathcal{I}_k)}{\mu_{k,0}(\vartheta^o)} = \frac{\sum_{\theta\in\mathcal{I}_k}(1/H)}{(1/H)} = |\mathcal{I}_k| \triangleq J_k. \tag{11.82}$$

In other words, with a flat prior, the ratio Γ_k coincides with the cardinality of the indistinguishable set $|\mathcal{I}_k|$, which we denote by J_k. In this case, higher values of J_k correspond to agents whose local inference abilities are worse, i.e., whose indistinguishable set is larger.

In the *social* learning framework, it is also useful to consider a measure of confusion at the *network* level. We accordingly introduce the network average of prior confusion ratios,

$$\Gamma \triangleq \prod_{k=1}^{K} \Gamma_k^{v_k}, \tag{11.83}$$

which is a weighted geometric average of the individual confusion ratios $\{\Gamma_k\}$, with weights given by the entries of the Perron vector. Likewise, we introduce the network average of the cardinalities of the indistinguishable sets,

$$J \triangleq \prod_{k=1}^{K} J_k^{v_k}. \tag{11.84}$$

11.5.1 Convergence Results

The results illustrated in this section originally appeared in [46]. Our first theorem examines the learning behavior of the memory-aware strategy when $\vartheta^\bullet \neq \vartheta^o$. In this case, we note that for each agent k the hypothesis of interest $\vartheta^\bullet$ can belong to either the subset $\mathcal{I}_k$ or the subset $\mathcal{D}_k$. In view of this fact and before introducing the forthcoming result, we define the sets of indistinguishable and distinguishable hypotheses *excluding* the hypothesis of interest, namely,

$$\mathcal{I}_k^\bullet \triangleq \mathcal{I}_k \backslash \{\vartheta^\bullet\}, \quad \mathcal{D}_k^\bullet \triangleq \mathcal{D}_k \backslash \{\vartheta^\bullet\}. \tag{11.85}$$

Clearly, if $\vartheta^\bullet$ belongs to $\mathcal{I}_k$, then $\mathcal{D}_k^\bullet = \mathcal{D}_k$, whereas if $\vartheta^\bullet$ belongs to $\mathcal{D}_k$, then $\mathcal{I}_k^\bullet = \mathcal{I}_k$. It is also convenient to introduce the set

$$\mathcal{I}_k^o \triangleq \mathcal{I}_k^\bullet \cup \{\vartheta^o\}. \tag{11.86}$$

Theorem 11.2 (Memory-aware strategy: Convergence when $\vartheta^\bullet \neq \vartheta^o$). Let Assumptions 5.1, 5.3, 5.4, and 7.1 be satisfied. Let $\vartheta^\bullet \neq \vartheta^o$ and assume that the network graph is connected. Then, for $k = 1, 2, \ldots, K$, we have the following three behaviors depending on the particular hypothesis:

i) **Hypothesis of interest $\vartheta^\bullet$:**

$$\boldsymbol{\mu}_{k,t}(\vartheta^\bullet) \xrightarrow[t\to\infty]{\text{a.s.}} 0. \tag{11.87}$$

ii) **Hypotheses $\theta \in \mathcal{D}_k^\bullet$:**

$$\boldsymbol{\mu}_{k,t}(\theta) \xrightarrow[t\to\infty]{\text{a.s.}} 0. \tag{11.88}$$

iii) **Hypotheses $\theta \in \mathcal{I}_k^o$:**
For all $t \in \mathbb{N}$, the conditional belief given that $\theta \in \mathcal{I}_k^o$ remains equal to the same conditional belief at $t = 0$, namely,

$$\frac{\boldsymbol{\mu}_{k,t}(\theta)}{\boldsymbol{\mu}_{k,t}(\mathcal{I}_k^o)} = \frac{\mu_{k,0}(\theta)}{\mu_{k,0}(\mathcal{I}_k^o)}. \tag{11.89}$$

Since $\boldsymbol{\mu}_{k,t}(\mathcal{I}_k^o) \xrightarrow[t\to\infty]{\text{a.s.}} 1$ in view of (11.87) and (11.88), we also have

$$\boldsymbol{\mu}_{k,t}(\theta) \xrightarrow[t\to\infty]{\text{a.s.}} \frac{\mu_{k,0}(\theta)}{\mu_{k,0}(\mathcal{I}_k^o)}. \tag{11.90}$$

Proof. See Appendix 11.B. ■

The fundamental message from Theorem 11.2 is that all agents are able to learn well when $\vartheta^\bullet \neq \vartheta^o$, since they ultimately place zero mass on the (false) hypothesis $\vartheta^\bullet$. That is, the algorithm achieves partial truth learning for $\vartheta^\bullet \neq \vartheta^o$ — see Definition 11.2.

In addition, the theorem shows that all distinguishable hypotheses are discarded. Indeed, when $\vartheta^\bullet \in \mathcal{D}_k$, then Eqs. (11.87) and (11.88) imply that both $\vartheta^\bullet$ and the set $\mathcal{D}_k^\bullet$ are discarded, which corresponds to rejecting all the distinguishable hypotheses. If otherwise $\vartheta^\bullet \in \mathcal{I}_k$, then not only $\mathcal{D}_k$ is rejected, but also $\vartheta^\bullet$.

The remaining belief mass is distributed over the set $\mathcal{I}_k^o$ (i.e., over the true hypothesis and the indistinguishable hypotheses different from $\vartheta^\bullet$) according to the ratio $\mu_{k,0}(\theta)/\mu_{k,0}(\mathcal{I}_k^o)$ in (11.90). This ratio represents a *conditional* belief given that θ belongs to $\mathcal{I}_k^o$. Specifically, it is the *prior* conditional belief, i.e., corresponding to $t = 0$. The fact that for $\theta \in \mathcal{I}_k^o$ the asymptotic beliefs depend exclusively on this prior conditional belief makes perfect sense since: *i)* the observations cannot help agent k distinguish between the true and the indistinguishable hypotheses; and *ii)* no information about the unshared hypotheses is diffused across the network. Therefore, the information available to agent k about the true and the indistinguishable *unshared* hypotheses does not increase over time, i.e., it is the same information available *at the beginning of the learning process.*

Equation (11.90) also reveals that, for $\vartheta^\bullet \neq \vartheta^o$, traditional truth learning is possible if, and only if, $\mathcal{I}_k^o = \{\vartheta^o\}$ for all k. In view of (11.86), this condition is equivalent to the condition $\mathcal{I}_k^\bullet = \emptyset$, which, in view of (11.85), is satisfied when

$$\mathcal{I}_k = \{\vartheta^\bullet\} \text{ or } \mathcal{I}_k = \emptyset \quad \forall k = 1, 2, \ldots, K. \tag{11.91}$$

In other words, for $\vartheta^\bullet \neq \vartheta^o$, traditional truth learning is achieved if, and only if, for each agent k either the hypothesis of interest is the only indistinguishable hypothesis (i.e., $\mathcal{I}_k = \{\vartheta^\bullet\}$) or the problem is locally identifiable (i.e., $\mathcal{I}_k = \emptyset$).

Let us switch to the case $\vartheta^\bullet = \vartheta^o$, which is covered by the next theorem.

Theorem 11.3 (Memory-aware strategy: Convergence when $\vartheta^\bullet = \vartheta^o$). Let Assumptions 5.1, 5.3, 5.4, and 7.1 be satisfied. Let $\vartheta^\bullet = \vartheta^o$ and assume that the network graph is primitive. Then, for $k = 1, 2, \ldots, K$, we have the following three behaviors depending on the particular hypothesis:

i) **Hypothesis of interest $\vartheta^\bullet$:**

$$\boldsymbol{\mu}_{k,t}(\vartheta^\bullet) \xrightarrow[t\to\infty]{\text{a.s.}} \frac{1}{1+\Gamma}, \tag{11.92}$$

where Γ is the network confusion ratio defined in (11.83).

ii) **Hypotheses $\theta \in \mathcal{D}_k$:**

$$\boldsymbol{\mu}_{k,t}(\theta) \xrightarrow[t\to\infty]{\text{a.s.}} 0. \tag{11.93}$$

iii) **Hypotheses $\theta \in \mathcal{I}_k$:**
For all $t \in \mathbb{N}$, the conditional belief given that $\theta \in \mathcal{I}_k^o$ remains equal to the same conditional belief at $t = 0$, namely,

$$\frac{\boldsymbol{\mu}_{k,t}(\theta)}{\boldsymbol{\mu}_{k,t}(\mathcal{I}_k)} = \frac{\mu_{k,0}(\theta)}{\mu_{k,0}(\mathcal{I}_k)}. \tag{11.94}$$

Since $\boldsymbol{\mu}_{k,t}(\mathcal{I}_k) \xrightarrow[t\to\infty]{\text{a.s.}} \Gamma/(1+\Gamma)$ in view of (11.92) and (11.93), we also have

$$\boldsymbol{\mu}_{k,t}(\theta) \xrightarrow[t\to\infty]{\text{a.s.}} \frac{\Gamma}{1+\Gamma} \frac{\mu_{k,0}(\theta)}{\mu_{k,0}(\mathcal{I}_k)}. \tag{11.95}$$

Proof. See Appendix 11.C. ∎

The results in Theorem 11.3 can be summarized as follows. From (11.93) we see that all the distinguishable hypotheses are correctly discarded. Moreover, from (11.92) we see that the belief about the true hypothesis converges to 1 if, and only if, the network confusion ratio Γ is zero. In other words, for $\vartheta^{\bullet} = \vartheta^{o}$, the memory-aware strategy achieves traditional truth learning (see Definition 11.1) if, and only if, $\Gamma = 0$.

Since Γ is a weighted geometric average of the single-agent confusion ratios $\{\Gamma_k\}$ defined in (11.81), then $\Gamma = 0$ provided that at least one agent k has $\Gamma_k = 0$. Moreover, since the initial beliefs are assumed to be positive (see Assumptions 5.1), Γ_k can only admit value 0 when the indistinguishable set $\mathcal{I}_k$ is empty. Therefore, we conclude that traditional truth learning is achieved if, and only if,

$$\exists k \in \{1, 2, \ldots, K\} \text{ such that } \mathcal{I}_k = \emptyset. \tag{11.96}$$

It is important to note that (11.96) does not mean that the problem must be locally identifiable for *all* agents; this must be the case for *at least one* powerful agent k. In other words, when such an agent exists, we can also have a problem that is locally unidentifiable for the other $K - 1$ agents, which would be therefore unable to discriminate ϑ^{o} from some of the other hypotheses if they worked in isolation. However, by exploiting cooperation across the network, they can profit from the powerful agent and overcome their individual limitations.

When the problem is locally unidentifiable for all agents, we have instead $\Gamma > 0$. In this case, while (11.93) reveals that zero mass is still assigned to the distinguishable hypotheses, the residual mass is now split between ϑ^{o} *and* the indistinguishable hypotheses, since the belief about the true hypothesis converges to a value strictly less than 1. This splitting is ruled by (11.95), implying that the behavior of the memory-aware strategy *depends on the initial beliefs* and in particular that if two agents start to learn with different initial beliefs, they can also learn differently in the long run. This conclusion is in contrast with traditional social learning, whose asymptotic behavior is instead independent of the initial beliefs.

Example 11.3 (Memory-aware strategy: Convergence behavior). In this example we consider the same setting used in Example 11.2.

In Table 11.2 we report the cardinality of the indistinguishable sets for each agent according to the likelihoods in Table 11.1. From Table 11.2 we can compute the network average of cardinalities according to (11.84), resulting in $J = 0$.

Table 11.2: Cardinality of the indistinguishable set $\mathcal{I}_k$ for each $k = 1, 2, \ldots, 12$ according to the likelihoods in Table 11.1.

Agent k	1	2	3	4	5	6	7	8	9	10	11	12
J_k	0	1	2	3	4	5	6	7	8	9	8	7

Let us examine first the truth sharing scenario ($\vartheta^\bullet = \vartheta^o = 1$). Observe from the table that the indistinguishable set of agent 1 is empty, i.e., the problem is locally identifiable for agent 1. In this case (see the discussion that led to (11.96)), it follows from Theorem 11.3 that all agents achieve traditional truth learning. This is confirmed by the experiments (obtained by running the algorithm in (11.11) with the memory-aware approach in (11.10)) shown in Figure 11.5, where we focus on the beliefs of agents 1, 4, and 10.

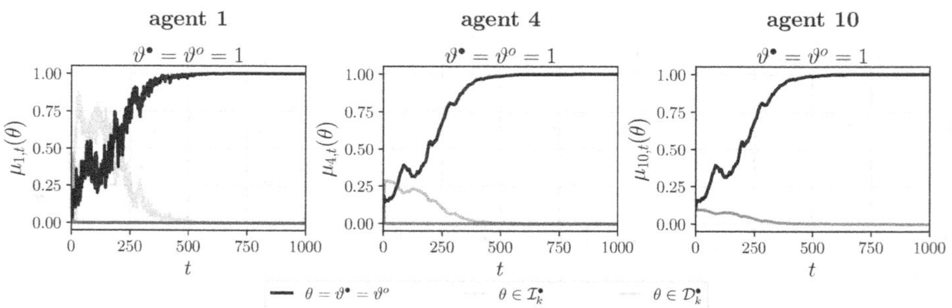

Figure 11.5: Truth sharing. Belief evolution over time for agents 1, 4, and 10 in Example 11.3.

Note that agents 1, 4, and 10 have different characteristics: For agent 1 the decision problem is locally identifiable; agent 4 is able to distinguish some hypotheses from ϑ^o, but not all of them, so that the problem is locally unidentifiable for it; for agent 10 all hypotheses $\theta \neq \vartheta^o$ are indistinguishable from ϑ^o, i.e., agent 10 has totally uninformative data. Despite these differences, under truth sharing they all achieve traditional truth learning.

We will now see that the situation changes under false-hypothesis sharing ($\vartheta^\bullet \neq \vartheta^o$). More specifically, Eqs. (11.87) and (11.93) reveal that the hypothesis of interest and the distinguishable hypotheses are all discarded. The asymptotic behavior of the beliefs about the remaining hypotheses (i.e., about $\theta \in \mathcal{I}_k^o$) is instead governed by (11.90). Since this behavior depends in general on the particular agent k, it is convenient to examine agents 1, 4, and 10 separately. Moreover, since the behavior is also dependent on the particular value chosen for $\vartheta^\bullet$, we will examine two cases, namely, $\vartheta^\bullet = 4$ and $\vartheta^\bullet = 7$.

Agent 1. As already observed, the indistinguishable set of agent 1 is empty. Then, from (11.86) we have $\mathcal{I}_1^o = \{1\}$, and from (11.90) we conclude that, whatever the choice of $\vartheta^\bullet \neq \vartheta^o$,

$$\mu_{1,t}(\vartheta^o) \xrightarrow[t\to\infty]{\text{a.s.}} 1. \tag{11.97}$$

This is confirmed by the evolution of the beliefs of agent 1, for $\vartheta^\bullet = 4$ and $\vartheta^\bullet = 7$, shown in Figure 11.5. We conclude that, thanks to local identifiability, agent 1 is able to learn

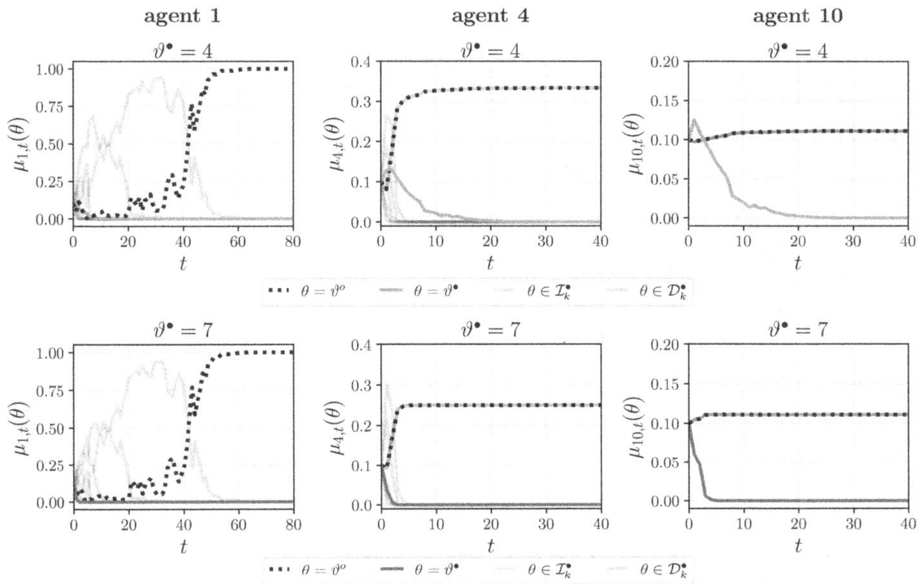

Figure 11.6: False-hypothesis sharing. Belief evolution over time for agents 1, 4, and 10 in Example 11.3. We consider two cases for the hypothesis of interest $\vartheta^\bullet$: The top panels refer to $\vartheta^\bullet = 4$, whereas the bottom panels refer to $\vartheta^\bullet = 7$.

properly even under false-hypothesis sharing.

Agent 4. Let us consider the case $\vartheta^\bullet = 4$. Recalling that $\vartheta^o = 1$, from Table 11.1 we see that $\mathcal{I}_4 = \{2, 3, 4\}$, and thus we have

$$\mathcal{I}_4^\bullet = \mathcal{I}_4 \backslash \{4\} = \{2, 3\}, \quad \mathcal{I}_4^o = \mathcal{I}_4^\bullet \cup \{1\} = \{1, 2, 3\}. \tag{11.98}$$

Using (11.90), and since in the experiments we considered uniform initial beliefs for all agents, we can write

$$\mu_{4,t}(\theta) \xrightarrow[t\to\infty]{\text{a.s.}} \frac{\mu_{4,0}(\theta)}{\mu_{4,0}(\mathcal{I}_4^o)} = \frac{1}{|\mathcal{I}_4^o|} = \frac{1}{3} \quad \forall \theta \in \mathcal{I}_4^o = \{1, 2, 3\}. \tag{11.99}$$

Examining Figure 11.6 (top center), we see that the black dashed curve, which corresponds to hypothesis 1, converges to 1/3. Moreover, there are also some pink curves that converge to 1/3. We have verified that these curves correspond to hypotheses 2 and 3. Thus, the numerical experiments confirm the behavior predicted by (11.99). We see that, differently from agent 1, agent 4 is not able to place full belief mass on the true hypothesis. Due to the lack of local identifiability, the belief mass is instead uniformly distributed over the true hypothesis and the indistinguishable hypotheses different from $\vartheta^\bullet$.

Consider next the choice $\vartheta^\bullet = 7$. Since this hypothesis is distinguishable for agent 4, we have

$$\mathcal{I}_4^\bullet = \mathcal{I}_4 \backslash \{7\} = \mathcal{I}_4 = \{2, 3, 4\}, \quad \mathcal{I}_4^o = \mathcal{I}_4^\bullet \cup \{1\} = \{1, 2, 3, 4\}, \tag{11.100}$$

yielding

$$\mu_{4,t}(\theta) \xrightarrow[t\to\infty]{\text{a.s.}} \frac{\mu_{4,0}(\theta)}{\mu_{4,0}(\mathcal{I}_4^o)} = \frac{1}{|\mathcal{I}_4^o|} = \frac{1}{4} \quad \forall\theta \in \mathcal{I}_4^o = \{1,2,3,4\}. \tag{11.101}$$

This convergence result matches the experiments in Figure 11.6 (bottom center), where the black dashed curve (hypothesis 1) and some other pink curves (hypotheses 2, 3, and 4) all converge to 1/4.

Agent 10. Consider again the case $\vartheta^\bullet = 4$. From Table 11.1 we see that $\mathcal{I}_{10} = \{2,3,\dots,10\}$, and thus we have

$$\mathcal{I}_{10}^\bullet = \mathcal{I}_{10}\backslash\{4\} = \{2,3,5,6,\dots,10\}, \quad \mathcal{I}_{10}^o = \mathcal{I}_{10}^\bullet \cup \{1\} = \{1,2,3,5,6,\dots,10\}, \tag{11.102}$$

and using (11.90) we obtain

$$\mu_{10,t}(\theta) \xrightarrow[t\to\infty]{\text{a.s.}} \frac{\mu_{10,0}(\theta)}{\mu_{10,0}(\mathcal{I}_{10}^o)} = \frac{1}{|\mathcal{I}_{10}^o|} = \frac{1}{9} \quad \forall\theta \in \mathcal{I}_{10}^o = \{1,2,3,5,6,\dots,10\}. \tag{11.103}$$

This convergence result is confirmed by the pertinent experiment in Figure 11.6 (top right), where we see that the beliefs of agent 10 pertaining to all hypotheses different from $\vartheta^\bullet = 4$ converge to 1/9. In comparison with agent 4, we see that agent 10 has a higher degree of uncertainty. In fact, the belief mass is uniformly distributed on more hypotheses. This happens because the indistinguishable set of agent 10 is larger than the indistinguishable set of agent 4. However, note that agent 10 is initially in a state of complete ignorance, since all hypotheses are indistinguishable from the true hypothesis. Thanks to social learning, agent 10 is able to understand that $\vartheta^\bullet = 4$ is false, and then it distributes the residual mass evenly across the remaining hypotheses. Similar consideration apply to the experiment referring to $\vartheta^\bullet = 7$ in Figure 11.6 (bottom right).

We will discuss next the case where the agents initialize their beliefs in a uniform manner, from which we can derive interesting conclusions regarding the learning mechanism of the memory-aware approach.

11.5.2 Unbiased Initialization

An interesting scenario that captures the learning mechanism of the memory-aware strategy is the *unbiased* case, i.e., when the initial beliefs are all uniform. It is useful to summarize the results for this case in the following corollary.

Corollary 11.2 (Memory-aware strategy: Uniform initial beliefs). Let the same assumptions used in Theorem 11.2 be satisfied and consider, for each agent $k = 1, 2, \dots, K$, the uniform prior assignment

$$\mu_{k,0}(\theta) = \frac{1}{H} \quad \forall\theta \in \Theta. \tag{11.104}$$

When $\vartheta^\bullet \neq \vartheta^o$, Eqs. (11.87) and (11.88) hold as they are, whereas in Eqs. (11.89) and (11.90) we must set, for $k = 1, 2, \ldots, K$,

$$\frac{\mu_{k,0}(\theta)}{\mu_{k,0}(\mathcal{I}_k^o)} = \frac{1}{|\mathcal{I}_k^o|}. \tag{11.105}$$

In particular, Eq. (11.90) becomes

$$\boldsymbol{\mu}_{k,t}(\theta) \xrightarrow[t\to\infty]{\text{a.s.}} \frac{1}{|\mathcal{I}_k^o|}, \tag{11.106}$$

which means that the mass is asymptotically equipartitioned over the set comprising ϑ^o and the indistinguishable hypotheses different from $\vartheta^\bullet$.
When $\vartheta^\bullet = \vartheta^o$, we have the following three behaviors depending on the particular hypothesis:

i) **Hypothesis of interest $\vartheta^\bullet$:**

$$\boldsymbol{\mu}_{k,t}(\vartheta^o) \xrightarrow[t\to\infty]{\text{a.s.}} \frac{1}{1+J}. \tag{11.107}$$

ii) **Hypotheses $\theta \in \mathcal{D}_k$:**

$$\boldsymbol{\mu}_{k,t}(\theta) \xrightarrow[t\to\infty]{\text{a.s.}} 0. \tag{11.108}$$

iii) **Hypotheses $\theta \in \mathcal{I}_k$:**
For all $t \in \mathbb{N}$, the conditional belief given that $\theta \in \mathcal{I}_k$ remains equal to the same conditional belief at $t = 0$, namely,

$$\frac{\boldsymbol{\mu}_{k,t}(\theta)}{\boldsymbol{\mu}_{k,t}(\mathcal{I}_k)} = \frac{1}{J_k}. \tag{11.109}$$

Since $\boldsymbol{\mu}_{k,t}(\mathcal{I}_k) \xrightarrow[t\to\infty]{\text{a.s.}} J/(1+J)$ in view of (11.107) and (11.108), we also have

$$\boldsymbol{\mu}_{k,t}(\theta) \xrightarrow[t\to\infty]{\text{a.s.}} \left(\frac{J}{J_k}\right) \frac{1}{1+J}. \tag{11.110}$$

Proof. The claims follow from Theorems 11.2 and 11.3 by setting $\mu_{k,0}(\theta) = 1/H$ for $k = 1, 2, \ldots, K$ and for all $\theta \in \Theta$. ∎

Under false-hypothesis sharing, Eq. (11.106) shows that the belief mass is asymptotically equipartitioned over the set $\mathcal{I}_k^o$, namely, over the true hypothesis and the indistinguishable hypotheses different from $\vartheta^\bullet$. Such asymptotic equipartition is perfectly coherent with the uniform prior assignment and the fact that no information about these hypotheses propagates across the network.

Consider now the truth sharing scenario. The particular case $J = 0$, which, in view of (11.82), (11.83), and (11.84), corresponds to $\Gamma = 0$, has been discussed in the comments on Theorem 11.2. We have seen that in this case the memory-aware strategy achieves traditional truth learning. Let us now focus on the case $J > 0$, where, as we are going to show, Corollary 11.2 allows us to investigate more closely the role of cooperation in the memory-aware strategy.

Observe that the geometric average of a set of numbers is bounded by the minimum and maximum values in the set. Since, in view of (11.84), J is a geometric average of the individual cardinalities J_k, for each agent k we have either $J_k > J$ or $J_k \leq J$. Consider first the agents with a number of indistinguishable hypotheses J_k larger than the network average J, i.e., with $(J/J_k) < 1$. In view of (11.108) and (11.110), it follows that, with probability 1,

$$J_k > J \implies \lim_{t\to\infty} \boldsymbol{\mu}_{k,t}(\theta) < \frac{1}{1+J} \quad \forall \theta \in \Theta \backslash \{\vartheta^o\}. \tag{11.111}$$

Combining this result with (11.107), we conclude that for a sufficiently large t, with probability 1,

$$J_k > J \implies \boldsymbol{\mu}_{k,t}(\vartheta^o) > \boldsymbol{\mu}_{k,t}(\theta) \quad \forall \theta \in \Theta \backslash \{\vartheta^o\}. \tag{11.112}$$

Conversely, for agents with $(J/J_k) > 1$, it follows that, with probability 1,

$$J_k < J \implies \lim_{t\to\infty} \boldsymbol{\mu}_{k,t}(\theta) > \frac{1}{1+J} \quad \forall \theta \in \mathcal{I}_k, \tag{11.113}$$

which, along with (11.107) and (11.108), implies that for a sufficiently large t, with probability 1,

$$J_k < J \implies \begin{cases} \boldsymbol{\mu}_{k,t}(\vartheta^o) < \boldsymbol{\mu}_{k,t}(\theta) & \forall \theta \in \mathcal{I}_k, \\ \boldsymbol{\mu}_{k,t}(\vartheta^o) > \boldsymbol{\mu}_{k,t}(\theta) & \forall \theta \in \mathcal{D}_k. \end{cases} \tag{11.114}$$

This behavior has an interesting implication for the role of cooperation. From (11.112) we see that, after cooperation, the agents that were *individually more confused* at time $t = 0$ (i.e., the agents featuring $J_k > J$) truly benefit from cooperation and end up with a belief that is maximized at the true hypothesis. However, Eq. (11.114) reveals that the situation is reversed for the agents that were *individually less confused* (i.e., with $J_k < J$); they end up with a belief that is no longer maximized at the true hypothesis since it is smaller than the belief about any indistinguishable hypothesis.

In summary, in the memory-aware strategy *altruism is not rewarding.* In other words, cooperation is not beneficial for all agents, since agents do not necessarily increase their confidence about the true hypothesis. However, the aforementioned discussion is based on the implicit assumption that maximization of the beliefs is what one should aim for in order to discover the true state of nature. Is maximization of the belief necessary to classify correctly $\vartheta^\bullet$? The next section provides an unexpected answer to this question.

11.5.3 Correct Decision under the Memory-Aware Approach

From Theorems 11.2 and 11.3 we obtain the following corollary, which reveals a fundamental dichotomy arising between the cases $\vartheta^\bullet \neq \vartheta^o$ and $\vartheta^\bullet = \vartheta^o$.

Corollary 11.3 (Memory-aware strategy: Asymptotic classification of $\vartheta^\bullet$). Under the same assumptions used in Theorem 11.3, for $k = 1, 2, \dots, K$,

$$\begin{cases} \boldsymbol{\mu}_{k,t}(\vartheta^\bullet) \xrightarrow[t\to\infty]{\text{a.s.}} 0 & \text{if } \vartheta^\bullet \neq \vartheta^o, \\ \boldsymbol{\mu}_{k,t}(\vartheta^\bullet) \xrightarrow[t\to\infty]{\text{a.s.}} \dfrac{1}{1+\Gamma} & \text{if } \vartheta^\bullet = \vartheta^o. \end{cases} \tag{11.115}$$

Proof. The claim follows from (11.87) and (11.92). ■

We see from (11.115) that the belief about $\vartheta^\bullet$ converges to 0 when $\vartheta^\bullet \neq \vartheta^o$ and to a positive number when $\vartheta^\bullet = \vartheta^o$. The gap between these limiting values suggests that it is possible to devise a decision rule that makes each agent k capable of classifying $\vartheta^\bullet$ correctly (with probability 1 as $t \to \infty$). More precisely, we need to define a decision rule for each time t, and examine the online behavior of the resulting decisions as $t \to \infty$. Note that, when the belief about $\vartheta^\bullet$ converges to 1 if $\vartheta^\bullet = \vartheta^o$ and to 0 otherwise, correct classification of the hypothesis of interest is obviously achieved by the standard rule that selects the hypothesis maximizing the belief.

However, we have observed that if $\boldsymbol{\mu}_{k,t}(\vartheta^\bullet)$ does not converge to 1 when $\vartheta^\bullet = \vartheta^o$, the rule maximizing the belief can fail since the maximum belief may correspond to one of the indistinguishable hypotheses. To overcome this issue, one can employ the following threshold test:

$$\begin{cases} \mu_{k,t}(\vartheta^\bullet) \le \eta \implies \text{reject } \vartheta^\bullet, \\ \mu_{k,t}(\vartheta^\bullet) > \eta \implies \text{accept } \vartheta^\bullet, \end{cases} \qquad 0 < \eta < \frac{1}{1+\Gamma}, \tag{11.116}$$

which, in view of (11.115), guarantees that the probability of classifying correctly $\vartheta^\bullet$ converges to 1 as $t \to \infty$.

Note that the decision rule (11.116) requires $\eta < 1/(1+\Gamma)$. From (11.83) we see that, to compute Γ, each agent needs to know the initial belief assignments of all agents, as well as the Perron vector. When this knowledge is available, the threshold can be surely set. However, there are several situations where this knowledge is not available. We now show that it is possible to set a threshold $\eta < 1/(1+\Gamma)$ with a much coarser prior information. To this end, we start by observing from (11.81) that we can write

$$\Gamma_k = \frac{\mu_{k,0}(\mathcal{I}_k)}{\mu_{k,0}(\vartheta^o)} \le \frac{1-\mu_{k,0}(\vartheta^o)}{\mu_{k,0}(\vartheta^o)} = \frac{1}{\mu_{k,0}(\vartheta^o)} - 1 \le \frac{1}{\mu_{\mathsf{min},0}} - 1, \tag{11.117}$$

where

$$\mu_{\mathsf{min},0} = \min_{\substack{k\in\{1,2,\dots,K\} \\ \theta\in\Theta}} \mu_{k,0}(\theta). \tag{11.118}$$

The network average of prior confusion ratios Γ is upper bounded by the maximum prior confusion ratio across the agents, which in view of (11.117) yields

$$\Gamma \le \frac{1}{\mu_{\mathsf{min},0}} - 1, \tag{11.119}$$

which is equivalent to

$$\mu_{\mathsf{min},0} \le \frac{1}{1+\Gamma}, \tag{11.120}$$

further implying that the choice

$$\eta = \mu_{\mathsf{min},0} - \varepsilon, \qquad \text{with } 0 < \varepsilon < \mu_{\mathsf{min},0}, \tag{11.121}$$

guarantees that $\eta < 1/(1+\Gamma)$. Accordingly, with (11.121) the hypothesis of interest is accepted provided that the observed belief exceeds (but for a small ε) the smallest initial belief across all agents and all hypotheses. In particular, in the case of unbiased initialization, Eq. (11.121) becomes

$$\eta = \frac{1}{H} - \varepsilon, \qquad \text{with } 0 < \varepsilon < \frac{1}{H}, \tag{11.122}$$

which essentially means that a belief larger than the uniform belief is sufficient to accept the hypothesis of interest.

Note that to implement (11.121), the agents must know $\mu_{\text{min},0}$. This requires, for example, that the agents share their initial beliefs in a preliminary phase of the algorithm, or that the initial beliefs are assigned with a protocol known to all agents beforehand. Remarkably, from (11.122) we see that, with an unbiased initialization, the only quantity necessary to set the threshold is the number of hypotheses, which is obviously known to all agents.

Before concluding this section, it is useful to contrast the truth-learning concept employed in traditional social learning with the decision rule (11.116). In both cases, each agent is able to make the right choice with probability 1 as $t \to \infty$. However, there is a difference that can emerge depending on the particular application context. Following traditional social learning, we might require that the belief $\boldsymbol{\mu}_{k,t}(\vartheta^\bullet)$ converges to 1 or 0 if the hypothesis of interest is true or false, respectively. This viewpoint is important, e.g., in applications where the agents are humans, since it reflects the natural behavior by which individuals express the strength of their opinions, and this strength is expected to increase as more evidence is collected. In particular, the choice of accepting the hypothesis of interest can be naturally formulated in terms of selecting the maximum belief.

On the other hand, when allowing $\Gamma > 0$ in Theorem 11.3, the situation changes, since the limiting belief about the hypothesis of interest is allowed to be even smaller than the belief about some indistinguishable hypothesis. This notwithstanding, we showed that the decision rule (11.116) allows to achieve correct decisions. This is because to accept $\vartheta^\bullet$ this rule neither requires $\boldsymbol{\mu}_{k,t}(\vartheta^\bullet)$ to converge to 1, nor that it is the maximum belief! We showed that it is sufficient to fulfill the milder requirement that $\boldsymbol{\mu}_{k,t}(\vartheta^\bullet)$ exceeds the minimum initial belief. One explanation for this behavior is as follows. When $\vartheta^\bullet = \vartheta^o$, the hypothesis of interest is by definition not statistically different from the *indistinguishable* hypotheses. However, what makes the hypothesis of interest different from the indistinguishable hypotheses is the way it is treated by the social learning algorithm, since it is the only hypothesis the agents exchange information about. This induces the agents to treat $\vartheta^\bullet$ in a "privileged" way. In other words, by using the decision rule (11.116) in place of the maximum-belief rule, the agents introduce a bias in favor of $\vartheta^\bullet$, which is used to overcome the limitations of partial information sharing. This requires that the agents are aware of how the underlying algorithm works, since to learn correctly they must combine this additional knowledge with their beliefs. From a practical viewpoint, this is definitely possible when the agents are programmable devices.

11.6 Comparing Strategies

In the previous sections we have characterized the learning behavior of social learning under partial information sharing, for both the memoryless and the memory-aware filling strategies. In this section we want to exploit the obtained results to examine two aspects. First, we discuss the role of the social exchange of information as opposed to a standalone implementation. Second, we consider the advantages of leveraging the agents' memory.

Standalone vs. social algorithms. One fundamental aspect is to establish how social collaboration influences the agents' beliefs. To this end, we study next the standalone algorithm, i.e., the sequential Bayesian scheme presented in Chapter 2, where there is no information exchange and the agents iteratively update their beliefs as

$$\boldsymbol{\mu}_{k,t}(\theta) \propto \boldsymbol{\mu}_{k,t-1}(\theta)\ell_k(\boldsymbol{x}_{k,t}|\theta). \tag{11.123}$$

The following theorem characterizing the standalone scheme is proved by using similar analytical tools to those used in Lemma 2.2. However, the results in Lemma 2.2 are obtained by assuming that the decision problem is identifiable, which in the multi-agent setting would correspond to assuming local identifiability for any agent. In contrast, the following theorem assumes an arbitrary identifiability setup for each agent k.

Theorem 11.4 (Standalone learning algorithm). Let Assumption 5.3 be satisfied, and assume initial belief vectors with positive entries for all agents. Then, with the standalone learning algorithm (11.123), all agents asymptotically discard the distinguishable hypothesis. Regarding the other hypotheses, for all $t \in \mathbb{N}$, the conditional belief given that $\theta \notin \mathcal{D}_k$ remains equal to the same conditional belief at $t = 0$, namely, for $k = 1, 2, \ldots, K$ and for all $\theta \in \mathcal{I}_k \cup \{\vartheta^o\}$,

$$\frac{\boldsymbol{\mu}_{k,t}(\theta)}{\boldsymbol{\mu}_{k,t}(\vartheta^o) + \boldsymbol{\mu}_{k,t}(\mathcal{I}_k)} = \frac{\mu_{k,0}(\theta)}{\mu_{k,0}(\vartheta^o) + \mu_{k,0}(\mathcal{I}_k)}. \tag{11.124}$$

Since $\boldsymbol{\mu}_{k,t}(\vartheta^o) + \boldsymbol{\mu}_{k,t}(\mathcal{I}_k) \xrightarrow[t\to\infty]{\text{a.s.}} 1$ because the distinguishable hypotheses are asymptotically discarded, we also have

$$\boldsymbol{\mu}_{k,t}(\theta) \xrightarrow[t\to\infty]{\text{a.s.}} \frac{\mu_{k,0}(\theta)}{\mu_{k,0}(\vartheta^o) + \mu_{k,0}(\mathcal{I}_k)} \quad \forall \theta \in \mathcal{I}_k \cup \{\vartheta^o\}. \tag{11.125}$$

By using the definition of Γ_k in (11.81), the above results can be schematically summarized as follows.

i) **True hypothesis ϑ^o:**

$$\boldsymbol{\mu}_{k,t}(\vartheta^o) \xrightarrow[t\to\infty]{\text{a.s.}} \frac{1}{1+\Gamma_k}. \tag{11.126}$$

ii) **Hypotheses $\theta \in \mathcal{D}_k$:**

$$\boldsymbol{\mu}_{k,t}(\theta) \xrightarrow[t\to\infty]{\text{a.s.}} 0. \tag{11.127}$$

iii) **Hypotheses $\theta \in \mathcal{I}_k$:**

$$\boldsymbol{\mu}_{k,t}(\theta) \xrightarrow[t\to\infty]{\text{a.s.}} \frac{\Gamma_k}{1+\Gamma_k} \frac{\mu_{k,0}(\theta)}{\mu_{k,0}(\mathcal{I}_k)}. \tag{11.128}$$

Proof. Under Assumption 5.3, we can follow the same steps used in the proof of Lemma 2.2 up to (2.39), obtaining, for all $\theta \in \Theta$,

$$\frac{1}{t} \log \frac{\boldsymbol{\mu}_{k,t}(\vartheta^o)}{\boldsymbol{\mu}_{k,t}(\theta)} \xrightarrow[t\to\infty]{\text{a.s.}} D(\ell_{k,\vartheta^o} || \ell_{k,\theta}). \tag{11.129}$$

For $\theta \in \mathcal{D}_k$ the RHS of (11.129) is positive. This implies that

$$\log \frac{\boldsymbol{\mu}_{k,t}(\vartheta^o)}{\boldsymbol{\mu}_{k,t}(\theta)} \xrightarrow[t\to\infty]{\text{a.s.}} \infty \quad \forall \theta \in \mathcal{D}_k, \tag{11.130}$$

further yielding

$$\boldsymbol{\mu}_{k,t}(\theta) \xrightarrow[t\to\infty]{\text{a.s.}} 0 \quad \forall \theta \in \mathcal{D}_k, \tag{11.131}$$

and condition (11.127) is proved. Now, Eq. (11.127) implies that

$$\boldsymbol{\mu}_{k,t}(\mathcal{I}_k \cup \{\vartheta^o\}) \xrightarrow[t\to\infty]{\text{a.s.}} 1. \tag{11.132}$$

Taking, for any pair of hypotheses $\theta, \theta' \in \mathcal{I}_k \cup \{\vartheta^o\}$, the ratio between $\boldsymbol{\mu}_{k,t}(\theta')$ and $\boldsymbol{\mu}_{k,t}(\theta)$, and using (11.123), we obtain

$$\frac{\boldsymbol{\mu}_{k,t}(\theta')}{\boldsymbol{\mu}_{k,t}(\theta)} = \frac{\boldsymbol{\mu}_{k,t-1}(\theta')}{\boldsymbol{\mu}_{k,t-1}(\theta)} = \ldots = \frac{\mu_{k,0}(\theta')}{\mu_{k,0}(\theta)}. \tag{11.133}$$

Summing over $\theta' \in \mathcal{I}_k \cup \{\vartheta^o\}$, from the first definition in (11.79) we have

$$\frac{\boldsymbol{\mu}_{k,t}(\mathcal{I}_k \cup \{\vartheta^o\})}{\boldsymbol{\mu}_{k,t}(\theta)} = \frac{\mu_{k,0}(\mathcal{I}_k \cup \{\vartheta^o\})}{\mu_{k,0}(\theta)}, \tag{11.134}$$

which, in view of (11.132), implies

$$\boldsymbol{\mu}_{k,t}(\theta) \xrightarrow[t\to\infty]{\text{a.s.}} \frac{\mu_{k,0}(\theta)}{\mu_{k,0}(\mathcal{I}_k \cup \{\vartheta^o\})} = \frac{\mu_{k,0}(\theta)}{\mu_{k,0}(\mathcal{I}_k) + \mu_{k,0}(\vartheta^o)} \tag{11.135}$$

for all $\theta \in \mathcal{I}_k \cup \{\vartheta^o\}$. From the definition of Γ_k in (11.81), we can rewrite (11.135) as

$$\boldsymbol{\mu}_{k,t}(\theta) \xrightarrow[t\to\infty]{\text{a.s.}} \frac{\Gamma_k}{1+\Gamma_k} \frac{\mu_{k,0}(\theta)}{\mu_{k,0}(\mathcal{I}_k)} \tag{11.136}$$

for all $\theta \in \mathcal{I}_k \cup \{\vartheta^o\}$, thus corresponding to (11.128). When $\theta = \vartheta^o$, the result specializes to (11.126).

■

We see a nice symmetry between Theorems 11.3 and 11.4, with the *network* confusion ratio Γ (in memory-aware social learning) being replaced by the *individual* confusion ratio Γ_k (in standalone learning). Despite this symmetry, there exist substantial differences between the standalone algorithm and the social learning algorithms with either the memoryless or the memory-aware filling strategy, as we now explain.

Theorem 11.4 reveals that, as $t \to \infty$, agent k in isolation can place full mass on ϑ^o only if $\Gamma_k = 0$, i.e., if the problem is *locally* identifiable for agent k. When there are instead indistinguishable hypotheses, the observations collected by a standalone agent do not convey useful information to discriminate between ϑ^o and the indistinguishable hypotheses. The only ability given to a standalone agent is to discard the distinguishable hypotheses. Other than that, the agent can redistribute the belief mass over the true and the indistinguishable hypotheses according to the only information it has to discriminate among them, that is, the initial belief vector. This behavior is summarized by (11.124). In particular, if the initial beliefs are uniform, from (11.124) we conclude that the belief about ϑ^o and the belief about the indistinguishable hypotheses converge to the same value $1/(1 + J_k)$, i.e., indistinguishability persists in the long term. No matter which decision rule is used, the standalone algorithm will be unable to discern in this case.

The situation is different for the memoryless or memory-aware social learning algorithms, where, thanks to cooperation, proper learning can be attained in different circumstances examined in detail in the previous sections.

Memoryless vs. memory-aware. For the memoryless strategy (11.9), it was shown in Theorem 11.1 that the belief about the hypothesis of interest converges either to 0 or 1. The conditions under which any of the two asymptotic behaviors take place depend on the comparison of the network KL divergences $D_{\text{net}}(\vartheta^\bullet)$ and $D_{\text{net}}(\mathcal{U})$, whose role was discussed in Section 11.4.1. The following behavior was established: Under truth sharing, traditional truth learning is always achieved; under false-hypothesis sharing, partial truth learning is achieved when $D_{\text{net}}(\vartheta^\bullet) > D_{\text{net}}(\mathcal{U})$, but an undesirable behavior emerges when $D_{\text{net}}(\vartheta^\bullet) < D_{\text{net}}(\mathcal{U})$, since the belief about $\vartheta^\bullet$ converges to 1, i.e., the agents are completely fooled and end up placing full mass on the wrong hypothesis.

For the memory-aware strategy (11.10), we ascertained that: Under false-hypothesis sharing, partial truth learning is always guaranteed; under truth sharing, traditional truth learning is achieved when there exists at least one powerful agent in the network that can solve the problem on its own, i.e., an agent endowed with local identifiability. Moreover, in Section 11.5.3 we showed that, even if no agent in the network satisfies local identifiability, there exists a decision rule, namely (11.116), under which agents are always able to classify correctly the hypothesis of interest.

11.A Appendix: Preliminary Results

The appendices at the end of this chapter are devoted to the proof of Theorems 11.2 and 11.3. We start in this section by presenting a series of auxiliary results.

For ease of reference, we report here the algorithm from listing (11.11), specialized to the memory-aware approach. In this case, at each instant t, each agent k performs the following three steps for each $\theta \in \Theta$:

$$\boldsymbol{\psi}_{k,t}(\theta) = \frac{\boldsymbol{\mu}_{k,t-1}(\theta)\ell_k(\boldsymbol{x}_{k,t}|\theta)}{\sum\limits_{\theta' \in \Theta} \boldsymbol{\mu}_{k,t-1}(\theta')\ell_k(\boldsymbol{x}_{k,t}|\theta')}, \tag{11.137a}$$

$$\widehat{\boldsymbol{\psi}}_{j,t}^{(k)}(\theta) = \begin{cases} \boldsymbol{\psi}_{j,t}(\vartheta^\bullet) & \text{if } \theta = \vartheta^\bullet, \\ \dfrac{\boldsymbol{\psi}_{k,t}(\theta)}{1 - \boldsymbol{\psi}_{k,t}(\vartheta^\bullet)}\Big(1 - \boldsymbol{\psi}_{j,t}(\vartheta^\bullet)\Big) & \text{if } \theta \neq \vartheta^\bullet, \end{cases} \qquad j \in \mathcal{N}_k, \tag{11.137b}$$

$$\boldsymbol{\mu}_{k,t}(\theta) = \frac{\prod\limits_{j \in \mathcal{N}_k} \left[\widehat{\boldsymbol{\psi}}_{j,t}^{(k)}(\theta)\right]^{a_{jk}}}{\sum\limits_{\theta' \in \Theta} \prod\limits_{j \in \mathcal{N}_k} \left[\widehat{\boldsymbol{\psi}}_{j,t}^{(k)}(\theta')\right]^{a_{jk}}}. \tag{11.137c}$$

The first auxiliary result introduces two submartingales related to the belief vectors.

Lemma 11.1 (Useful submartingales). Let Assumptions 5.1 and 5.3 be satisfied, and assume that the network graph is connected, with a combination matrix A having Perron vector v. Let $\mathcal{S}_k$ be any nonempty agent-dependent set of hypotheses satisfying

$$\mathcal{S}_k \subseteq \big(\mathcal{I}_k \cup \{\vartheta^o\}\big) \backslash \{\vartheta^\bullet\}, \tag{11.138}$$

and let $\mathcal{S} \triangleq \{\mathcal{S}_1, \mathcal{S}_2, \ldots, \mathcal{S}_K\}$. Define the random variables, for $t = 0, 1, \ldots,$

$$\boldsymbol{m}_t \triangleq \sum_{k=1}^{K} v_k \log \boldsymbol{\mu}_{k,t}(\vartheta^o), \quad \boldsymbol{n}_t(\mathcal{S}) \triangleq \sum_{k=1}^{K} v_k \log \boldsymbol{\mu}_{k,t}(\mathcal{S}_k), \tag{11.139}$$

where the notation for beliefs computed over subsets, like $\boldsymbol{\mu}_{k,t}(\mathcal{S}_k)$, was introduced in (11.79). Moreover, recall the definition of $d_k(q)$ from (7.7),

$$d_k(q) \triangleq \mathbb{E} \log \frac{\ell_k(\boldsymbol{x}_{k,t}|\vartheta^o)}{\sum\limits_{\theta \in \Theta} q(\theta)\ell_k(\boldsymbol{x}_{k,t}|\theta)}, \tag{11.140}$$

where q is a convex combination vector (i.e., it has nonnegative entries that add up to 1) of dimension H. The following properties hold for any choice of $\vartheta^\bullet$:

i) For $t = 1, 2, \ldots$,

$$\mathbb{E}[\boldsymbol{m}_t|\mathcal{F}_{t-1}] \geq \boldsymbol{m}_{t-1} + \sum_{k=1}^{K} v_k d_k(\boldsymbol{\mu}_{k,t-1}), \tag{11.141}$$

$$\mathbb{E}[\boldsymbol{n}_t(\mathcal{S})|\mathcal{F}_{t-1}] \geq \boldsymbol{n}_{t-1}(\mathcal{S}) + \sum_{k=1}^{K} v_k d_k(\boldsymbol{\mu}_{k,t-1}), \tag{11.142}$$

where, given the underlying probability space $(\Omega, \mathscr{F}, \mathbb{P})$, we introduce the filtration (see Definition D.5) generated by the belief vectors of all agents, namely, the sequence of sub-σ-fields

$$\mathcal{F}_t \triangleq \sigma\Big(\{\boldsymbol{\mu}_{k,0}\}_{k=1}^K, \{\boldsymbol{\mu}_{k,1}\}_{k=1}^K, \ldots, \{\boldsymbol{\mu}_{k,t}\}_{k=1}^K\Big), \quad t = 0, 1, \ldots \tag{11.143}$$

Note that $\mathcal{F}_0 = \sigma\left(\{\boldsymbol{\mu}_{k,0}\}_{k=1}^K\right) = \{\emptyset, \Omega\}$ is the trivial σ-field, since we are modeling the initial beliefs as deterministic.

ii) Both sequences $\{\boldsymbol{m}_t\}_{t=0}^{\infty}$ and $\{\boldsymbol{n}_t(\mathcal{S})\}_{t=0}^{\infty}$ are negative submartingales with respect to $\{\mathcal{F}_t\}_{t=0}^{\infty}$, and there exist random variables $\boldsymbol{m}_\infty$ and $\boldsymbol{n}_\infty(\mathcal{S})$ such that

$$\boldsymbol{m}_t \xrightarrow[t\to\infty]{\text{a.s.}} \boldsymbol{m}_\infty, \qquad \boldsymbol{n}_t(\mathcal{S}) \xrightarrow[t\to\infty]{\text{a.s.}} \boldsymbol{n}_\infty(\mathcal{S}). \tag{11.144}$$

iii) The sequences of expected values $\mathbb{E}\boldsymbol{m}_t$ and $\mathbb{E}\boldsymbol{n}_t(\mathcal{S})$ have finite limits.

Proof. We first prove the claims for $\boldsymbol{m}_t$. Applying the arithmetic/geometric mean inequality we can write [37]

$$\sum_{\theta\in\Theta}\prod_{j\in\mathcal{N}_k}\left[\widehat{\boldsymbol{\psi}}_{j,t}^{(k)}(\theta)\right]^{a_{jk}} \leq \sum_{\theta\in\Theta}\sum_{j\in\mathcal{N}_k} a_{jk}\widehat{\boldsymbol{\psi}}_{j,t}^{(k)}(\theta) = 1. \tag{11.145}$$

Grouping (11.137c) and (11.145) we get

$$\boldsymbol{\mu}_{k,t}(\vartheta^o) = \frac{\prod_{j\in\mathcal{N}_k}\left[\widehat{\boldsymbol{\psi}}_{j,t}^{(k)}(\vartheta^o)\right]^{a_{jk}}}{\sum_{\theta\in\Theta}\prod_{j\in\mathcal{N}_k}\left[\widehat{\boldsymbol{\psi}}_{j,t}^{(k)}(\theta)\right]^{a_{jk}}} \geq \prod_{j\in\mathcal{N}_k}\left[\widehat{\boldsymbol{\psi}}_{j,t}^{(k)}(\vartheta^o)\right]^{a_{jk}}, \tag{11.146}$$

which, using the definition of $\widehat{\boldsymbol{\psi}}_{j,t}^{(k)}$ from (11.137b), yields

$$\boldsymbol{\mu}_{k,t}(\vartheta^o) \geq \begin{cases} \prod_{j\in\mathcal{N}_k}\left[\boldsymbol{\psi}_{j,t}(\vartheta^o)\right]^{a_{jk}} & \text{if } \vartheta^\bullet = \vartheta^o, \\ \boldsymbol{\psi}_{k,t}(\vartheta^o)\,\dfrac{\prod_{j\in\mathcal{N}_k}\left[1-\boldsymbol{\psi}_{j,t}(\vartheta^\bullet)\right]^{a_{jk}}}{1-\boldsymbol{\psi}_{k,t}(\vartheta^\bullet)} & \text{if } \vartheta^\bullet \neq \vartheta^o. \end{cases} \tag{11.147}$$

Taking the logarithm we get

$$\log \boldsymbol{\mu}_{k,t}(\vartheta^o) \geq \begin{cases} \sum\limits_{j\in\mathcal{N}_k} a_{jk} \log \boldsymbol{\psi}_{j,t}(\vartheta^o) & \text{if } \vartheta^\bullet = \vartheta^o, \\ \log \boldsymbol{\psi}_{k,t}(\vartheta^o) + \sum\limits_{j\in\mathcal{N}_k} a_{jk} \log \dfrac{1-\boldsymbol{\psi}_{j,t}(\vartheta^\bullet)}{1-\boldsymbol{\psi}_{k,t}(\vartheta^\bullet)} & \text{if } \vartheta^\bullet \neq \vartheta^o, \end{cases} \tag{11.148}$$

which, from the definition of neighborhood in (4.1), is equivalently written as

$$\log \boldsymbol{\mu}_{k,t}(\vartheta^o) \geq \begin{cases} \sum\limits_{j=1}^{K} a_{jk} \log \boldsymbol{\psi}_{j,t}(\vartheta^o) & \text{if } \vartheta^\bullet = \vartheta^o, \\ \log \boldsymbol{\psi}_{k,t}(\vartheta^o) + \sum\limits_{j=1}^{K} a_{jk} \log \dfrac{1-\boldsymbol{\psi}_{j,t}(\vartheta^\bullet)}{1-\boldsymbol{\psi}_{k,t}(\vartheta^\bullet)} & \text{if } \vartheta^\bullet \neq \vartheta^o. \end{cases} \tag{11.149}$$

Since the network graph is assumed to be connected, with a left stochastic combination matrix A, it follows from Definition 4.6 and Lemma 4.3 that A is an irreducible matrix with spectral radius $\rho(A) = 1$. From the Perron-Frobenius theorem (Theorem 4.1) it follows that we can define the Perron vector v, which has positive entries adding up to 1, and satisfies

$$Av = v. \tag{11.150}$$

Equation (11.150) can be expanded in terms of the individual entries of Av and v, yielding the following identities:

$$\sum_{k=1}^{K} a_{jk} v_k = v_j, \qquad j = 1, 2, \ldots, K. \tag{11.151}$$

Let us now focus on the RHS of (11.149). Consider first the term corresponding to $\vartheta^\bullet = \vartheta^o$, namely, $\sum_{j=1}^{K} a_{jk} \log \boldsymbol{\psi}_{j,t}(\vartheta^o)$. Multiplying this quantity by v_k and summing over k, we obtain the following expression:

$$\begin{aligned} &\sum_{k=1}^{K} v_k \sum_{j=1}^{K} a_{jk} \log \boldsymbol{\psi}_{j,t}(\vartheta^o) \\ &= \sum_{j=1}^{K} \underbrace{\sum_{k=1}^{K} a_{jk} v_k}_{=v_j \text{ from (11.151)}} \log \boldsymbol{\psi}_{j,t}(\vartheta^o) = \sum_{k=1}^{K} v_k \log \boldsymbol{\psi}_{k,t}(\vartheta^o). \end{aligned} \tag{11.152}$$

Consider next the term corresponding to $\vartheta^\bullet \neq \vartheta^o$. Multiplying this term by v_k and summing over k, we get

$$\begin{aligned}
&\sum_{k=1}^{K} v_k \log \boldsymbol{\psi}_{k,t}(\vartheta^o) + \sum_{k=1}^{K} v_k \sum_{j=1}^{K} a_{jk} \log \frac{1-\boldsymbol{\psi}_{j,t}(\vartheta^\bullet)}{1-\boldsymbol{\psi}_{k,t}(\vartheta^\bullet)} \\
&= \sum_{k=1}^{K} v_k \log \boldsymbol{\psi}_{k,t}(\vartheta^o) \\
&+ \sum_{j=1}^{K} \underbrace{\sum_{k=1}^{K} v_k a_{jk}}_{= v_j \text{ from (11.151)}} \log\left(1-\boldsymbol{\psi}_{j,t}(\vartheta^\bullet)\right) - \underbrace{\sum_{j=1}^{K} a_{jk}}_{=1} \sum_{k=1}^{K} v_k \log\left(1-\boldsymbol{\psi}_{k,t}(\vartheta^\bullet)\right) \\
&= \sum_{k=1}^{K} v_k \log \boldsymbol{\psi}_{k,t}(\vartheta^o) + \sum_{j=1}^{K} v_j \log\left(1-\boldsymbol{\psi}_{j,t}(\vartheta^\bullet)\right) - \sum_{k=1}^{K} v_k \log\left(1-\boldsymbol{\psi}_{k,t}(\vartheta^\bullet)\right) \\
&= \sum_{k=1}^{K} v_k \log \boldsymbol{\psi}_{k,t}(\vartheta^o).
\end{aligned} \tag{11.153}$$

Multiplying both sides of (11.149) by v_k and summing over k, from (11.152) and (11.153) we obtain the following inequality:

$$\boldsymbol{m}_t = \sum_{k=1}^{K} v_k \log \boldsymbol{\mu}_{k,t}(\vartheta^o) \geq \sum_{k=1}^{K} v_k \log \boldsymbol{\psi}_{k,t}(\vartheta^o), \tag{11.154}$$

where we used the definition of $\boldsymbol{m}_t$ from (11.139). Furthermore, substituting (11.137a) into (11.154) yields

$$\boldsymbol{m}_t \geq \boldsymbol{m}_{t-1} + \sum_{k=1}^{K} v_k \log \frac{\ell_k(\boldsymbol{x}_{k,t}|\vartheta^o)}{\sum\limits_{\theta\in\Theta} \boldsymbol{\mu}_{k,t-1}(\theta)\ell_k(\boldsymbol{x}_{k,t}|\theta)}. \tag{11.155}$$

Now, taking the conditional expectation $\mathbb{E}\left[\cdot|\mathcal{F}_{t-1}\right]$ of both sides of (11.155), we obtain, for $t = 1, 2, \dots,$

$$\mathbb{E}\left[\boldsymbol{m}_t|\mathcal{F}_{t-1}\right] \geq \boldsymbol{m}_{t-1} + \sum_{k=1}^{K} v_k d_k(\boldsymbol{\mu}_{k,t-1}), \tag{11.156}$$

where $d_k(q)$, for a convex combination vector q of dimension H, is defined in (11.140). This proves part i) for $\boldsymbol{m}_t$. Observe from (11.140) that $d_k(q)$ corresponds to the KL divergence between the true model ℓ_{k,ϑ^o} and the mixture model $\sum_{\theta\in\Theta} q(\theta)\ell_{k,\theta}$. From the nonnegativity of the KL divergence it follows that $d_k(\boldsymbol{\mu}_{k,t-1}) \geq 0$, which, in view of (11.156), implies

$$\mathbb{E}\left[\boldsymbol{m}_t|\mathcal{F}_{t-1}\right] \geq \boldsymbol{m}_{t-1}. \tag{11.157}$$

Observe that $\boldsymbol{m}_t$ is a negative random variable since the entries of the Perron vector are positive and all the beliefs are almost surely strictly less than 1 — see the discussion at the end of Section 11.2. Taking the expectation of both sides of (11.157), we can further write

$$0 > \mathbb{E}\boldsymbol{m}_t \geq \mathbb{E}\boldsymbol{m}_{t-1} \geq \cdots \geq m_0, \tag{11.158}$$

which shows that $\boldsymbol{m}_t$ has finite mean for $t = 0, 1, \ldots$ (note that m_0 is finite since the initial beliefs are nonzero in view of point ii) in Assumption 5.1). In view of (11.157), we see that the sequence $\{\boldsymbol{m}_t\}_{t=0}^{\infty}$ is a negative submartingale. Thus, we can call upon the martingale convergence theorem (in particular, Corollary D.1) to conclude that $\boldsymbol{m}_t$ converges almost surely. This proves part ii) for $\boldsymbol{m}_t$. Finally, part iii) for $\boldsymbol{m}_t$ follows from (11.158), which implies that the sequence of expectations converges (since it is nondecreasing and bounded from above).

Now we focus on $\boldsymbol{n}_t(\mathcal{S})$. In view of (11.137c), (11.79), and (11.145),

$$\boldsymbol{\mu}_{k,t}(\mathcal{S}_k) = \sum_{\theta\in\mathcal{S}_k} \frac{\prod\limits_{j\in\mathcal{N}_k}\left[\widehat{\boldsymbol{\psi}}_{j,t}^{(k)}(\theta)\right]^{a_{jk}}}{\sum\limits_{\theta'\in\Theta}\prod\limits_{j\in\mathcal{N}_k}\left[\widehat{\boldsymbol{\psi}}_{j,t}^{(k)}(\theta')\right]^{a_{jk}}} \geq \sum_{\theta\in\mathcal{S}_k}\prod_{j\in\mathcal{N}_k}\left[\widehat{\boldsymbol{\psi}}_{j,t}^{(k)}(\theta)\right]^{a_{jk}}. \tag{11.159}$$

From the definition of $\mathcal{S}_k$ in (11.138) we see that $\vartheta^{\bullet} \notin \mathcal{S}_k$ for any agent k. Therefore, the term $\widehat{\boldsymbol{\psi}}_{j,t}^{(k)}(\theta)$ must be evaluated by applying the expression in (11.137b) that is valid for the case $\theta \neq \vartheta^{\bullet}$. Substituting this expression into the RHS of (11.159) yields

$$\begin{aligned}\boldsymbol{\mu}_{k,t}(\mathcal{S}_k) &\geq \sum_{\theta\in\mathcal{S}_k}\frac{\boldsymbol{\psi}_{k,t}(\theta)}{1-\boldsymbol{\psi}_{k,t}(\vartheta^{\bullet})}\prod_{j\in\mathcal{N}_k}\left[1-\boldsymbol{\psi}_{j,t}(\vartheta^{\bullet})\right]^{a_{jk}}\\ &= \frac{\boldsymbol{\psi}_{k,t}(\mathcal{S}_k)}{1-\boldsymbol{\psi}_{k,t}(\vartheta^{\bullet})}\prod_{j=1}^{K}\left[1-\boldsymbol{\psi}_{j,t}(\vartheta^{\bullet})\right]^{a_{jk}},\end{aligned} \tag{11.160}$$

where in the equality we apply the definition of $\boldsymbol{\psi}_{k,t}(\mathcal{S}_k)$ from (11.79) and extend the product to all j by exploiting the definition of neighborhood from (4.1). Taking the logarithm in (11.160), multiplying by v_k, and summing over k gives

$$\begin{aligned}\sum_{k=1}^{K} v_k \log \boldsymbol{\mu}_{k,t}(\mathcal{S}_k) &\geq \sum_{k=1}^{K} v_k \log \boldsymbol{\psi}_{k,t}(\mathcal{S}_k) + \sum_{k=1}^{K} v_k \sum_{j=1}^{K} a_{jk}\log\frac{1-\boldsymbol{\psi}_{j,t}(\vartheta^{\bullet})}{1-\boldsymbol{\psi}_{k,t}(\vartheta^{\bullet})}\\ &= \sum_{k=1}^{K} v_k \log \boldsymbol{\psi}_{k,t}(\mathcal{S}_k),\end{aligned} \tag{11.161}$$

where the equality follows by applying the same steps used to obtain (11.153). Substituting the definition of $\boldsymbol{n}_t(\mathcal{S})$ from (11.139) into (11.161), we obtain

$$\begin{aligned}\boldsymbol{n}_t(\mathcal{S}) &\geq \sum_{k=1}^{K} v_k \log \boldsymbol{\psi}_{k,t}(\mathcal{S}_k)\\ &= \boldsymbol{n}_{t-1}(\mathcal{S}) + \sum_{k=1}^{K} v_k \log \frac{\ell_k(\boldsymbol{x}_{k,t}|\vartheta^{o})}{\sum\limits_{\theta\in\Theta}\boldsymbol{\mu}_{k,t-1}(\theta)\ell_k(\boldsymbol{x}_{k,t}|\theta)},\end{aligned} \tag{11.162}$$

where the equality follows from (11.137a) and the fact that for $\theta \in \mathcal{S}_k$ we have $\ell_{k,\theta} = \ell_{k,\vartheta^{o}}$. The proof can be completed by repeating the same steps used to prove (11.156)–(11.158) starting from (11.155), replacing $\boldsymbol{m}_t$ with the submartingale $\boldsymbol{n}_t(\mathcal{S})$ defined in (11.139). ■

From Lemma 11.1 we can derive the following corollary, which provides bounds on the expectation of the logarithm of the intermediate beliefs.

Corollary 11.4 (Expectation of log beliefs). Consider the same assumptions used in Lemma 11.1. Then, for $k = 1, 2, \dots, K$ and for all $t \in \mathbb{N}$,

$$\mathbb{E} \log \frac{1}{\psi_{k,t}(\vartheta^o)} < \frac{|m_0|}{v_k}, \tag{11.163}$$

$$\mathbb{E} \log \frac{1}{\psi_{k,t}(\mathcal{S}_k)} < \frac{|n_0(\mathcal{S})|}{v_k}. \tag{11.164}$$

Proof. Let us first prove (11.163). Using (11.137a) and (11.140), we have that

$$\begin{aligned}\mathbb{E} \log \frac{1}{\psi_{k,t}(\vartheta^o)} &= \mathbb{E} \log \frac{1}{\boldsymbol{\mu}_{k,t-1}(\vartheta^o)} - \mathbb{E} d_k(\boldsymbol{\mu}_{k,t-1}) \\ &\leq \mathbb{E} \log \frac{1}{\boldsymbol{\mu}_{k,t-1}(\vartheta^o)},\end{aligned} \tag{11.165}$$

where the inequality follows from the fact that $d_k(\mu_{k,t-1}) \geq 0$ in view of the nonnegativity of the KL divergence. Moreover, from the explanation at the end of Section 11.2, we know that $\log \boldsymbol{\mu}_{k,t-1}(\vartheta^o)$ is a negative random variable. On the other hand, since $v_k > 0$, it follows that

$$v_k \log \boldsymbol{\mu}_{k,t-1}(\vartheta^o) > \sum_{k=1}^{K} v_k \log \boldsymbol{\mu}_{k,t-1}(\vartheta^o), \tag{11.166}$$

which, from definition (11.139), is equivalent to

$$\log \boldsymbol{\mu}_{k,t-1}(\vartheta^o) > \frac{\boldsymbol{m}_{t-1}}{v_k}. \tag{11.167}$$

Then, taking expectations and using (11.158), we obtain

$$\mathbb{E} \log \boldsymbol{\mu}_{k,t-1}(\vartheta^o) > \frac{\mathbb{E} \boldsymbol{m}_{t-1}}{v_k} \geq \frac{m_0}{v_k}, \tag{11.168}$$

or

$$\mathbb{E} \log \frac{1}{\boldsymbol{\mu}_{k,t-1}(\vartheta^o)} < -\frac{m_0}{v_k} = \frac{|m_0|}{v_k}, \tag{11.169}$$

where in the last step we use the fact that $m_0 < 0$. Using (11.169) in (11.165) we get (11.163). It remains to prove (11.164).

To this end, observe that $\ell_{k,\theta} = \ell_{k,\vartheta^o}$ for $\theta \in \mathcal{S}_k$. As a result, from (11.137a) and (11.79) we have

$$\mathbb{E} \log \frac{1}{\psi_{k,t}(\mathcal{S}_k)} = \mathbb{E} \log \frac{1}{\boldsymbol{\mu}_{k,t-1}(\mathcal{S}_k)} - \mathbb{E} d_k(\boldsymbol{\mu}_{k,t-1}). \tag{11.170}$$

Then, Eq. (11.164) is obtained by repeating the same steps used to get (11.169) from (11.165), replacing $\boldsymbol{m}_t$ with the submartingale $\boldsymbol{n}_t(\mathcal{S})$ defined in (11.139). ∎

Lemma 11.1 is also useful to prove the following intermediate result, where we show that all agents in the network discard the distinguishable hypotheses in probability.

Lemma 11.2 (All agents discard the distinguishable hypotheses). Let Assumptions 5.1, 5.3, and 7.1 be satisfied, and assume that the network graph is connected, with a combination matrix A having Perron vector v. Then, for $k = 1, 2, \dots, K$ and for all $\theta \in \mathcal{D}_k$,

$$\boldsymbol{\mu}_{k,t}(\theta) \xrightarrow[t\to\infty]{\text{p}} 0. \tag{11.171}$$

Proof. Taking expectations in (11.141), we can write

$$0 \le \sum_{k=1}^{K} v_k \mathbb{E} d_k(\boldsymbol{\mu}_{k,t-1}) \le \mathbb{E}\boldsymbol{m}_t - \mathbb{E}\boldsymbol{m}_{t-1}. \tag{11.172}$$

Then, in view of part iii) of Lemma 11.1, and using the squeeze theorem [144, Thm. 3.19], we have

$$\lim_{t\to\infty} \sum_{k=1}^{K} v_k \mathbb{E} d_k(\boldsymbol{\mu}_{k,t-1}) = 0. \tag{11.173}$$

From Theorem 4.1, which can be invoked since the network graph is connected (hence, the associated combination matrix A is irreducible) it follows that the entries of the Perron vector are positive, i.e., $v_k > 0$ for $k = 1, 2, \dots K$. Since $d_k(\boldsymbol{\mu}_{k,t-1})$ is nonnegative, the positivity of v_k implies that each individual summand in (11.173) must converge to 0. This means that, for each k, $d_k(\boldsymbol{\mu}_{k,t-1})$ converges to 0 in the 1st mean — see Definition D.3. In view of (D.17), convergence in the 1st mean implies convergence in probability. Therefore, for all agents we have

$$d_k(\boldsymbol{\mu}_{k,t-1}) \xrightarrow[t\to\infty]{\text{p}} 0. \tag{11.174}$$

We will use the above result to conclude that (11.171) holds. Applying Pinsker's inequality (Theorem C.7), we can lower bound the KL divergence $d_k(\boldsymbol{\mu}_{k,t-1})$ as follows:

$$d_k(\boldsymbol{\mu}_{k,t-1}) \ge \frac{1}{2} D_{\mathsf{TV}}^2 \left(\ell_{k,\vartheta^o}\,,\, \sum_{\theta\in\Theta} \boldsymbol{\mu}_{k,t-1}(\theta) \ell_{k,\theta} \right), \tag{11.175}$$

where the symbol D_{TV} denotes the total variation distance, whose expression is provided in Definition C.1.

Consider now an agent k for which $|\mathcal{D}_k| > 0$. We can write

$$\begin{aligned}
&\ell_k(x|\vartheta^o) - \sum_{\theta\in\Theta} \boldsymbol{\mu}_{k,t-1}(\theta)\ell_k(x|\theta)\\
&= \ell_k(x|\vartheta^o) - \sum_{\theta\in\mathcal{I}_k\cup\{\vartheta^o\}} \boldsymbol{\mu}_{k,t-1}(\theta)\ell_k(x|\theta) - \sum_{\theta\in\mathcal{D}_k} \boldsymbol{\mu}_{k,t-1}(\theta)\ell_k(x|\theta)\\
&= \underbrace{\left(1 - \sum_{\theta\in\mathcal{I}_k\cup\{\vartheta^o\}} \boldsymbol{\mu}_{k,t-1}(\theta)\right)}_{=\sum_{\theta\in\mathcal{D}_k}\boldsymbol{\mu}_{k,t-1}(\theta)}\ell_k(x|\vartheta^o) - \sum_{\theta\in\mathcal{D}_k} \boldsymbol{\mu}_{k,t-1}(\theta)\ell_k(x|\theta)\\
&= \left(\ell_k(x|\vartheta^o) - \sum_{\theta\in\mathcal{D}_k} \boldsymbol{q}(\theta)\ell_k(x|\theta)\right)\sum_{\theta'\in\mathcal{D}_k}\boldsymbol{\mu}_{k,t-1}(\theta'), \qquad (11.176)
\end{aligned}$$

where in the second equality we used the fact that within the indistinguishable set we have $\ell_k(x|\theta) = \ell_k(x|\vartheta^o)$, whereas in the last equality we introduced the notation

$$\boldsymbol{q}(\theta) = \frac{\boldsymbol{\mu}_{k,t-1}(\theta)}{\sum_{\theta'\in\mathcal{D}_k}\boldsymbol{\mu}_{k,t-1}(\theta')}. \qquad (11.177)$$

Then, in view of the formulas for the total variation distance in Definition C.1, Eq. (11.176) implies that

$$\begin{aligned}
&D_{\mathsf{TV}}\left(\ell_{k,\vartheta^o}, \sum_{\theta\in\Theta}\boldsymbol{\mu}_{k,t-1}(\theta)\ell_{k,\theta}\right)\\
&= \left|\sum_{\theta\in\mathcal{D}_k}\boldsymbol{\mu}_{k,t-1}(\theta)\right| \times D_{\mathsf{TV}}\left(\ell_{k,\vartheta^o}, \sum_{\theta\in\mathcal{D}_k}\boldsymbol{q}(\theta)\ell_{k,\theta}\right). \qquad (11.178)
\end{aligned}$$

By repeating the same arguments used in the proof of Lemma 7.2 (see the discussion following (7.30)), we arrive at the inequality

$$D_{\mathsf{TV}}\left(\ell_{k,\vartheta^o}, \sum_{\theta\in\Theta}\boldsymbol{\mu}_{k,t-1}(\theta)\ell_{k,\theta}\right) \ge d_{\mathsf{min}}\left|\sum_{\theta\in\mathcal{D}_k}\boldsymbol{\mu}_{k,t-1}(\theta)\right|, \qquad (11.179)$$

for a certain $d_{\mathsf{min}} > 0$. Using this result in (11.175) yields

$$d_k(\boldsymbol{\mu}_{k,t-1}) \ge \frac{d^2_{\mathsf{min}}}{2}\left(\sum_{\theta\in\mathcal{D}_k}\boldsymbol{\mu}_{k,t-1}(\theta)\right)^2. \qquad (11.180)$$

In view of (11.174), the above result implies (11.171).

■

Using Eqs. (11.137a)–(11.137c), we can establish the following additional result, which provides a condition under which an agent is able to discard the distinguishable unshared hypotheses *almost surely*.

Lemma 11.3 (Belief convergence for the distinguishable unshared hypotheses). Let Assumptions 5.1 and 5.3 be satisfied. If $\vartheta^\bullet \neq \vartheta^o$ or $\mathcal{I}_k^\bullet \neq \emptyset$, then

$$\boldsymbol{\mu}_{k,t}(\theta) \xrightarrow[t\to\infty]{\text{a.s.}} 0 \quad \forall \theta \in \mathcal{D}_k^\bullet. \tag{11.181}$$

Proof. By assumption, ϑ^o is unshared or there exists at least one indistinguishable unshared hypothesis. Let θ' be equal to ϑ^o if ϑ^o is unshared, or to an indistinguishable unshared hypothesis. Let θ be distinguishable and unshared. Using (11.137a)–(11.137c), we can write

$$\begin{aligned}
\frac{\boldsymbol{\mu}_{k,t}(\theta')}{\boldsymbol{\mu}_{k,t}(\theta)} &\overset{(11.137c)}{=} \frac{\prod_{j\in\mathcal{N}_k}\left[\widehat{\boldsymbol{\psi}}_{j,t}^{(k)}(\theta')\right]^{a_{jk}}}{\prod_{j\in\mathcal{N}_k}\left[\widehat{\boldsymbol{\psi}}_{j,t}^{(k)}(\theta)\right]^{a_{jk}}} \\
&\overset{(11.137b)}{=} \frac{\prod_{j\in\mathcal{N}_k}\left[\dfrac{\boldsymbol{\psi}_{k,t}(\theta')}{1-\boldsymbol{\psi}_{k,t}(\vartheta^\bullet)}\Big(1-\boldsymbol{\psi}_{j,t}(\vartheta^\bullet)\Big)\right]^{a_{jk}}}{\prod_{j\in\mathcal{N}_k}\left[\dfrac{\boldsymbol{\psi}_{k,t}(\theta)}{1-\boldsymbol{\psi}_{k,t}(\vartheta^\bullet)}\Big(1-\boldsymbol{\psi}_{j,t}(\vartheta^\bullet)\Big)\right]^{a_{jk}}} = \frac{\boldsymbol{\psi}_{k,t}(\theta')}{\boldsymbol{\psi}_{k,t}(\theta)} \\
&\overset{(11.137a)}{=} \frac{\boldsymbol{\mu}_{k,t-1}(\theta')}{\boldsymbol{\mu}_{k,t-1}(\theta)}\,\frac{\ell_k(\boldsymbol{x}_{k,t}|\theta')}{\ell_k(\boldsymbol{x}_{k,t}|\theta)} \\
&= \frac{\boldsymbol{\mu}_{k,t-1}(\theta')}{\boldsymbol{\mu}_{k,t-1}(\theta)}\,\frac{\ell_k(\boldsymbol{x}_{k,t}|\vartheta^o)}{\ell_k(\boldsymbol{x}_{k,t}|\theta)},
\end{aligned} \tag{11.182}$$

where in the step that applies (11.137b) we use the fact that both θ' and θ are unshared hypotheses, while in the last equality we replace $\ell_k(\boldsymbol{x}_{k,t}|\theta')$ with $\ell_k(\boldsymbol{x}_{k,t}|\vartheta^o)$ since θ' is equal to ϑ^o or is indistinguishable from ϑ^o. Taking the logarithm of the LHS and the RHS of (11.182), we obtain the recursion

$$\log\frac{\boldsymbol{\mu}_{k,t}(\theta')}{\boldsymbol{\mu}_{k,t}(\theta)} = \log\frac{\boldsymbol{\mu}_{k,t-1}(\theta')}{\boldsymbol{\mu}_{k,t-1}(\theta)} + \log\frac{\ell_k(\boldsymbol{x}_{k,t}|\vartheta^o)}{\ell_k(\boldsymbol{x}_{k,t}|\theta)}, \tag{11.183}$$

and unfolding it we get

$$\log\frac{\boldsymbol{\mu}_{k,t}(\theta')}{\boldsymbol{\mu}_{k,t}(\theta)} = \log\frac{\mu_{k,0}(\theta')}{\mu_{k,0}(\theta)} + \sum_{\tau=1}^{t}\log\frac{\ell_k(\boldsymbol{x}_{k,\tau}|\vartheta^o)}{\ell_k(\boldsymbol{x}_{k,\tau}|\theta)}. \tag{11.184}$$

Under Assumption 5.3, the random variables $\ell_k(\boldsymbol{x}_{k,\tau}|\vartheta^o)/\ell_k(\boldsymbol{x}_{k,\tau}|\theta)$ are iid, and their expectation is given by the KL divergence $D(\ell_{k,\vartheta^o}||\ell_{k,\theta}) < \infty$. Therefore, after dividing (11.184) by t we can call upon the strong law of large numbers (Theorem D.7) to conclude that

$$\frac{1}{t}\log\frac{\boldsymbol{\mu}_{k,t}(\theta')}{\boldsymbol{\mu}_{k,t}(\theta)} \xrightarrow[t\to\infty]{\text{a.s.}} \mathbb{E}\log\frac{\ell_k(\boldsymbol{x}_{k,\tau}|\vartheta^o)}{\ell_k(\boldsymbol{x}_{k,\tau}|\theta)} = D(\ell_{k,\vartheta^o}||\ell_{k,\theta}) \quad \forall\theta\in\mathcal{D}_k^\bullet. \tag{11.185}$$

Note that $D(\ell_{k,\vartheta^o}||\ell_{k,\theta})$ is positive because θ is a distinguishable hypothesis. Therefore, Eq. (11.185) implies

$$\log\frac{\boldsymbol{\mu}_{k,t}(\theta')}{\boldsymbol{\mu}_{k,t}(\theta)} \xrightarrow[t\to\infty]{\text{a.s.}} \infty \quad \forall\theta\in\mathcal{D}_k^\bullet, \tag{11.186}$$

from which we conclude that

$$\mu_{k,t}(\theta) \xrightarrow[t\to\infty]{\text{a.s.}} 0 \quad \forall \theta \in \mathcal{D}_k^\bullet, \tag{11.187}$$

since the entries of the belief vector are bounded. Thus, we have established the claim of the lemma.

■

It is worth commenting on the differences between Lemmas 11.2 and 11.3. Lemma 11.2 allows us to conclude that all agents discard the distinguishable hypotheses in probability, irrespective of the choice of $\vartheta^\bullet$. Meanwhile, Lemma 11.3 ensures that an agent k can discard the distinguishable *unshared* hypotheses *almost-surely*, under the additional assumption that $\vartheta^\bullet \neq \vartheta^o$ or $\mathcal{I}_k^\bullet \neq \emptyset$. In order to prove almost-sure convergence of the beliefs according to Theorems 11.2 and 11.3, we will resort to a combination of both results.

Before we present the proofs of Theorems 11.2 and 11.3, we characterize in the following lemma the ratios between intermediate beliefs corresponding to distinguishable and indistinguishable hypotheses.

Lemma 11.4 (Intermediate belief ratios between distinguishable and indistinguishable hypotheses). Let Assumptions 5.1, 5.3, and 7.1 be satisfied. Let $\vartheta^\bullet = \vartheta^o$, $\mathcal{D}_k \neq \emptyset$, and $\mathcal{I}_k \neq \emptyset$, and assume that the network graph is connected, with a combination matrix A having Perron vector v. Then, for $k = 1, 2, \ldots, K$:

i) The sequence defined, for $t \in \mathbb{N}$, by the intermediate belief ratios

$$\frac{\psi_{k,t}(\mathcal{D}_k)}{\psi_{k,t}(\mathcal{I}_k)} \tag{11.188}$$

is a positive martingale with respect to the filtration (see Definition D.5) formed by the sub-σ-fields

$$\mathcal{F}_t \triangleq \sigma\left(\{\mu_{k,0}\}_{k=1}^K, \{\mu_{k,1}\}_{k=1}^K, \ldots, \{\mu_{k,t}\}_{k=1}^K\right), \quad t \in \mathbb{N}. \tag{11.189}$$

ii) This martingale vanishes almost surely.

Proof. To begin with, observe that $\psi_{k,t}(\mathcal{D}_k)$ and $\psi_{k,t}(\mathcal{I}_k)$ are positive random variables, since the sets $\mathcal{D}_k$ and $\mathcal{I}_k$ are nonempty and the intermediate beliefs are almost surely positive — see the discussion at the end of Section 11.2.

We first establish property i). The following chain of identities holds:

$$\frac{\boldsymbol{\psi}_{k,t}(\mathcal{D}_k)}{\boldsymbol{\psi}_{k,t}(\mathcal{I}_k)} \overset{(a)}{=} \frac{\sum_{\theta\in\mathcal{D}_k} \boldsymbol{\psi}_{k,t}(\theta)}{\sum_{\theta\in\mathcal{I}_k} \boldsymbol{\psi}_{k,t}(\theta)} \overset{(b)}{=} \frac{\sum_{\theta\in\mathcal{D}_k} \boldsymbol{\mu}_{k,t-1}(\theta)\ell_k(\boldsymbol{x}_{k,t}|\theta)}{\sum_{\theta\in\mathcal{I}_k} \boldsymbol{\mu}_{k,t-1}(\theta)\ell_k(\boldsymbol{x}_{k,t}|\theta)} \overset{(c)}{=} \frac{\sum_{\theta\in\mathcal{D}_k} \boldsymbol{\mu}_{k,t-1}(\theta)\ell_k(\boldsymbol{x}_{k,t}|\theta)}{\ell_k(\boldsymbol{x}_{k,t}|\vartheta^o)\sum_{\theta\in\mathcal{I}_k} \boldsymbol{\mu}_{k,t-1}(\theta)} \overset{(d)}{=} \frac{1}{\boldsymbol{\mu}_{k,t-1}(\mathcal{I}_k)}\sum_{\theta\in\mathcal{D}_k} \boldsymbol{\mu}_{k,t-1}(\theta)\frac{\ell_k(\boldsymbol{x}_{k,t}|\theta)}{\ell_k(\boldsymbol{x}_{k,t}|\vartheta^o)}, \tag{11.190}$$

where in (a) and (d) we apply definitions (11.79) for beliefs computed over subsets; (b) follows from (11.137a); and (c) holds since, within the indistinguishable set $\mathcal{I}_k$, all likelihood models are equal to ℓ_{k,ϑ^o}. Since $\mathbb{E}[\ell_k(\boldsymbol{x}_{k,t}|\theta)/\ell_k(\boldsymbol{x}_{k,t}|\vartheta^o)] = 1$ for each $\theta \in \Theta$, from (11.190) we can write, for $t = 2, 3, \dots,$

$$\begin{aligned}
&\mathbb{E}\left[\frac{\boldsymbol{\psi}_{k,t}(\mathcal{D}_k)}{\boldsymbol{\psi}_{k,t}(\mathcal{I}_k)}\middle|\mathcal{F}_{t-1}\right] \\
&= \frac{1}{\boldsymbol{\mu}_{k,t-1}(\mathcal{I}_k)}\sum_{\theta\in\mathcal{D}_k} \boldsymbol{\mu}_{k,t-1}(\theta)\,\mathbb{E}\left[\frac{\ell_k(\boldsymbol{x}_{k,t}|\theta)}{\ell_k(\boldsymbol{x}_{k,t}|\vartheta^o)}\right] \overset{(11.79)}{=} \frac{\boldsymbol{\mu}_{k,t-1}(\mathcal{D}_k)}{\boldsymbol{\mu}_{k,t-1}(\mathcal{I}_k)} \\
&\overset{(11.137c)}{=} \frac{\sum_{\theta\in\mathcal{D}_k}\prod_{j\in\mathcal{N}_k}\left[\widehat{\boldsymbol{\psi}}^{(k)}_{j,t-1}(\theta)\right]^{a_{jk}}}{\sum_{\theta\in\mathcal{I}_k}\prod_{j\in\mathcal{N}_k}\left[\widehat{\boldsymbol{\psi}}^{(k)}_{j,t-1}(\theta)\right]^{a_{jk}}} \\
&\overset{(11.137b)}{=} \frac{\sum_{\theta\in\mathcal{D}_k}\prod_{j\in\mathcal{N}_k}\left[\frac{\boldsymbol{\psi}_{k,t-1}(\theta)}{1-\boldsymbol{\psi}_{k,t-1}(\vartheta^\bullet)}\left(1-\boldsymbol{\psi}_{j,t-1}(\vartheta^\bullet)\right)\right]^{a_{jk}}}{\sum_{\theta\in\mathcal{I}_k}\prod_{j\in\mathcal{N}_k}\left[\frac{\boldsymbol{\psi}_{k,t-1}(\theta)}{1-\boldsymbol{\psi}_{k,t-1}(\vartheta^\bullet)}\left(1-\boldsymbol{\psi}_{j,t-1}(\vartheta^\bullet)\right)\right]^{a_{jk}}} \\
&= \frac{\sum_{\theta\in\mathcal{D}_k}\prod_{j\in\mathcal{N}_k}\left[\boldsymbol{\psi}_{k,t-1}(\theta)\right]^{a_{jk}}}{\sum_{\theta\in\mathcal{I}_k}\prod_{j\in\mathcal{N}_k}\left[\boldsymbol{\psi}_{k,t-1}(\theta)\right]^{a_{jk}}} = \frac{\sum_{\theta\in\mathcal{D}_k}\left[\boldsymbol{\psi}_{k,t-1}(\theta)\right]^{\sum_{j\in\mathcal{N}_k}a_{jk}}}{\sum_{\theta\in\mathcal{I}_k}\left[\boldsymbol{\psi}_{k,t-1}(\theta)\right]^{\sum_{j\in\mathcal{N}_k}a_{jk}}} = \frac{\sum_{\theta\in\mathcal{D}_k}\boldsymbol{\psi}_{k,t-1}(\theta)}{\sum_{\theta\in\mathcal{I}_k}\boldsymbol{\psi}_{k,t-1}(\theta)} \\
&\overset{(11.79)}{=} \frac{\boldsymbol{\psi}_{k,t-1}(\mathcal{D}_k)}{\boldsymbol{\psi}_{k,t-1}(\mathcal{I}_k)}.
\end{aligned} \tag{11.191}$$

Note that since $\mathcal{D}_k$ and $\mathcal{I}_k$ do not contain ϑ^o, and since $\vartheta^\bullet = \vartheta^o$, the hypotheses belonging to $\mathcal{D}_k$ or $\mathcal{I}_k$ are all unshared hypotheses. Accordingly, in (11.191), the reconstructed beliefs $\widehat{\psi}_{j,t}^{(k)}(\theta)$ have been computed from (11.137b) using the formula corresponding to $\theta \neq \vartheta^\bullet$. In summary, we have shown that

$$\mathbb{E}\left[\frac{\psi_{k,t}(\mathcal{D}_k)}{\psi_{k,t}(\mathcal{I}_k)}\middle|\mathcal{F}_{t-1}\right] = \frac{\psi_{k,t-1}(\mathcal{D}_k)}{\psi_{k,t-1}(\mathcal{I}_k)}. \tag{11.192}$$

Moreover, specializing (11.190) to the case $t = 1$ we get

$$\begin{aligned}\mathbb{E}\left[\frac{\psi_{k,1}(\mathcal{D}_k)}{\psi_{k,1}(\mathcal{I}_k)}\right] &= \frac{1}{\mu_{k,0}(\mathcal{I}_k)}\sum_{\theta\in\mathcal{D}_k}\mu_{k,0}(\theta)\underbrace{\mathbb{E}\left[\frac{\ell_k(\boldsymbol{x}_{k,t}|\theta)}{\ell_k(\boldsymbol{x}_{k,t}|\vartheta^o)}\right]}_{=1}\\ &= \frac{\mu_{k,0}(\mathcal{D}_k)}{\mu_{k,0}(\mathcal{I}_k)} < \infty, \end{aligned}\tag{11.193}$$

where the last inequality holds because $\mu_{k,0}(\mathcal{I}_k) > 0$ in view of point ii) in Assumption 5.1. Taking the expectation of both sides of (11.192), we obtain

$$\mathbb{E}\left[\frac{\psi_{k,t}(\mathcal{D}_k)}{\psi_{k,t}(\mathcal{I}_k)}\right] = \mathbb{E}\left[\frac{\psi_{k,t-1}(\mathcal{D}_k)}{\psi_{k,t-1}(\mathcal{I}_k)}\right] = \cdots = \mathbb{E}\left[\frac{\psi_{k,1}(\mathcal{D}_k)}{\psi_{k,1}(\mathcal{I}_k)}\right] = \frac{\mu_{k,0}(\mathcal{D}_k)}{\mu_{k,0}(\mathcal{I}_k)}, \tag{11.194}$$

which shows that $\psi_{k,t}(\mathcal{D}_k)/\psi_{k,t}(\mathcal{I}_k)$ has finite mean for all $t \in \mathbb{N}$. We conclude from (11.192) that the sequence $\{\psi_{k,t}(\mathcal{D}_k)/\psi_{k,t}(\mathcal{I}_k)\}_{t\in\mathbb{N}}$ is a nonnegative martingale. We can therefore call upon the martingale convergence theorem (in particular, Corollary D.1), to establish that $\psi_{k,t}(\mathcal{D}_k)/\psi_{k,t}(\mathcal{I}_k)$ converges almost surely. This means that, to prove property ii) in the lemma, it suffices to show the following convergence in probability:

$$\frac{\psi_{k,t}(\mathcal{D}_k)}{\psi_{k,t}(\mathcal{I}_k)} \xrightarrow[t\to\infty]{\mathrm{p}} 0. \tag{11.195}$$

The convergence in (11.195) results from Lemma D.2, applied with the choices

$$\boldsymbol{w}_t = \psi_{k,t}(\mathcal{D}_k), \qquad \boldsymbol{y}_t = \frac{1}{\psi_{k,t}(\mathcal{I}_k)}. \tag{11.196}$$

We now verify that with these choices the conditions of Lemma D.2 are satisfied.

Specifically, condition (D.27) holds because $\boldsymbol{w}_t = \psi_{k,t}(\mathcal{D}_k)$ converges to 0 in probability in view of Lemma 11.2 (actually, the lemma refers to beliefs $\boldsymbol{\mu}_{k,t}(\theta)$, but the claim of the lemma can be readily extended to the intermediate beliefs $\psi_{k,t}(\theta)$ by using (11.137a)).

Conditions (D.28) and (D.29) are satisfied since, by using Markov's inequality (Theorem C.1), we have the following upper bounds holding for any $y > 1$:

$$\begin{aligned}\mathbb{P}\left[\frac{1}{\psi_{k,t}(\mathcal{I}_k)} > y\right] &= \mathbb{P}\left[\log\frac{1}{\psi_{k,t}(\mathcal{I}_k)} > \log y\right]\\ &\le \frac{1}{\log y}\,\mathbb{E}\log\frac{1}{\psi_{k,t}(\mathcal{I}_k)}\\ &< \frac{1}{\log y}\,\frac{|n_0(\mathcal{I})|}{v_k} \overset{y\to\infty}{\longrightarrow} 0, \end{aligned}\tag{11.197}$$

where $\mathcal{I} = \{\mathcal{I}_1, \mathcal{I}_2, \ldots, \mathcal{I}_K\}$, and the last inequality follows from Corollary 11.4 applied with the choice $\mathcal{S}_k = \mathcal{I}_k$ for all k (note that this choice satisfies condition (11.138) since, for $\vartheta^\bullet = \vartheta^o$, $(\mathcal{I}_k \cup \{\vartheta^o\})\backslash\vartheta^\bullet = \mathcal{I}_k$). Thus, Lemma D.2 allows us to claim (11.195), which in turn implies that the martingale $\{\boldsymbol{\psi}_{k,t}(\mathcal{D}_k)/\boldsymbol{\psi}_{k,t}(\mathcal{I}_k)\}_{t\in\mathbb{N}}$ vanishes almost surely. ∎

11.B Appendix: Proof of Theorem 11.2

Proof. We need to establish (11.87), (11.88), and (11.89).

Equation (11.88) follows directly from Lemma 11.3, once we observe that $\vartheta^\bullet \neq \vartheta^o$.

Let us focus on (11.89). Observe that, in the considered case $\vartheta^\bullet \neq \vartheta^o$, the set $\mathcal{I}_k^o$ defined by (11.86) contains only unshared hypotheses. If $\mathcal{I}_k = \emptyset$, then $\mathcal{I}_k^o$ contains only ϑ^o, and Eq. (11.89) trivially yields the identity $1 = 1$. Consider then the case $\mathcal{I}_k \neq \emptyset$, and let $\theta, \theta' \in \mathcal{I}_k^o$. Note that these hypotheses are the true hypothesis or indistinguishable hypotheses. We have the following chain of identities:

$$\begin{aligned}
\frac{\boldsymbol{\mu}_{k,t}(\theta')}{\boldsymbol{\mu}_{k,t}(\theta)} &\overset{\text{(a)}}{=} \prod_{j\in\mathcal{N}_k}\left[\frac{\widehat{\boldsymbol{\psi}}_{j,t}^{(k)}(\theta')}{\widehat{\boldsymbol{\psi}}_{j,t}^{(k)}(\theta)}\right]^{a_{jk}} \overset{\text{(b)}}{=} \prod_{j\in\mathcal{N}_k}\left[\frac{\boldsymbol{\psi}_{k,t}(\theta')}{\boldsymbol{\psi}_{k,t}(\theta)}\right]^{a_{jk}} \overset{\text{(c)}}{=} \frac{\boldsymbol{\psi}_{k,t}(\theta')}{\boldsymbol{\psi}_{k,t}(\theta)} \\
&\overset{\text{(d)}}{=} \frac{\boldsymbol{\mu}_{k,t-1}(\theta')\ell_k(\boldsymbol{x}_{k,t}|\theta')}{\boldsymbol{\mu}_{k,t-1}(\theta)\ell_k(\boldsymbol{x}_{k,t}|\theta)} \overset{\text{(e)}}{=} \frac{\boldsymbol{\mu}_{k,t-1}(\theta')\ell_k(\boldsymbol{x}_{k,t}|\vartheta^o)}{\boldsymbol{\mu}_{k,t-1}(\theta)\ell_k(\boldsymbol{x}_{k,t}|\vartheta^o)} \\
&= \frac{\boldsymbol{\mu}_{k,t-1}(\theta')}{\boldsymbol{\mu}_{k,t-1}(\theta)},
\end{aligned} \tag{11.198}$$

where (a) follows from (11.137c); in (b) we use the expression for the reconstructed beliefs from (11.137b) holding for unshared hypotheses; (c) follows from the fact that A is left stochastic; (d) from (11.137a); and (e) holds because θ and θ' are indistinguishable. From (11.198) we see that

$$\frac{\boldsymbol{\mu}_{k,t}(\theta')}{\boldsymbol{\mu}_{k,t}(\theta)} = \frac{\boldsymbol{\mu}_{k,t-1}(\theta')}{\boldsymbol{\mu}_{k,t-1}(\theta)} = \cdots = \frac{\mu_{k,0}(\theta')}{\mu_{k,0}(\theta)}. \tag{11.199}$$

Summing over $\theta' \in \mathcal{I}_k^o$, from the first definition in (11.79) we obtain, for any $\theta \in \mathcal{I}_k^o$,

$$\frac{\boldsymbol{\mu}_{k,t}(\mathcal{I}_k^o)}{\boldsymbol{\mu}_{k,t}(\theta)} = \frac{\mu_{k,0}(\mathcal{I}_k^o)}{\mu_{k,0}(\theta)}, \tag{11.200}$$

which is equivalent to (11.89).

It remains to prove (11.87). To this end, we first establish that $\boldsymbol{\mu}_{k,t}(\vartheta^\bullet)$ vanishes *in probability*, and then use this result to show that it vanishes almost surely. Under global identifiability (Assumption 5.4) we have $\vartheta^\bullet \in \mathcal{D}_h$ for some agent h, and since the network is assumed to be connected, we can use Lemma 11.2 to conclude that

$$\boldsymbol{\mu}_{h,t}(\vartheta^\bullet) \xrightarrow[t\to\infty]{\text{p}} 0. \tag{11.201}$$

We want to show that the same result holds for all agents in the network. To this end, we start by showing that Eq. (11.201) implies that the intermediate belief $\boldsymbol{\psi}_{h,t}(\vartheta^\bullet)$ converges to 0 in probability.

From (11.137a) we can write (note that, as usual, the likelihood ratios are well defined with probability 1 due to Assumption 5.3)

$$\begin{aligned}
\boldsymbol{\psi}_{h,t}(\vartheta^\bullet) &= \frac{\boldsymbol{\mu}_{h,t-1}(\vartheta^\bullet)}{\displaystyle\sum_{\theta\in\Theta}\boldsymbol{\mu}_{h,t-1}(\theta)\frac{\ell_h(\boldsymbol{x}_{h,t}|\theta)}{\ell_h(\boldsymbol{x}_{h,t}|\vartheta^o)}}\,\frac{\ell_h(\boldsymbol{x}_{h,t}|\vartheta^\bullet)}{\ell_h(\boldsymbol{x}_{h,t}|\vartheta^o)} \\
&= \frac{\boldsymbol{\mu}_{h,t-1}(\vartheta^\bullet)}{\displaystyle\sum_{\theta\in\mathcal{D}_h}\boldsymbol{\mu}_{h,t-1}(\theta)\frac{\ell_h(\boldsymbol{x}_{h,t}|\theta)}{\ell_h(\boldsymbol{x}_{h,t}|\vartheta^o)} + \underbrace{\sum_{\theta\in\mathcal{I}_h\cup\{\vartheta^o\}}\boldsymbol{\mu}_{h,t-1}(\theta)}_{=1-\boldsymbol{\mu}_{h,t-1}(\mathcal{D}_h)}}\,\frac{\ell_h(\boldsymbol{x}_{h,t}|\vartheta^\bullet)}{\ell_h(\boldsymbol{x}_{h,t}|\vartheta^o)} \\
&= \frac{\boldsymbol{\mu}_{h,t-1}(\vartheta^\bullet)}{1+\displaystyle\sum_{\theta\in\mathcal{D}_h}\boldsymbol{\mu}_{h,t-1}(\theta)\left(\frac{\ell_h(\boldsymbol{x}_{h,t}|\theta)}{\ell_h(\boldsymbol{x}_{h,t}|\vartheta^o)}-1\right)}\,\frac{\ell_h(\boldsymbol{x}_{h,t}|\vartheta^\bullet)}{\ell_h(\boldsymbol{x}_{h,t}|\vartheta^o)},
\end{aligned} \tag{11.202}$$

where in the second equality we used the fact that $\ell_{h,\theta} = \ell_{h,\vartheta^o}$ for all $\theta \in \mathcal{I}_h \cup \{\vartheta^o\}$. By introducing the definitions

$$\boldsymbol{s}_t' = \frac{\boldsymbol{\mu}_{h,t-1}(\vartheta^\bullet)}{1+\displaystyle\sum_{\theta\in\mathcal{D}_h}\boldsymbol{\mu}_{h,t-1}(\theta)\left(\frac{\ell_h(\boldsymbol{x}_{h,t}|\theta)}{\ell_h(\boldsymbol{x}_{h,t}|\vartheta^o)}-1\right)}, \tag{11.203}$$

$$\boldsymbol{s}_t'' = \frac{\ell_h(\boldsymbol{x}_{h,t}|\vartheta^\bullet)}{\ell_h(\boldsymbol{x}_{h,t}|\vartheta^o)}, \tag{11.204}$$

we see from (11.202) that we have the identity

$$\boldsymbol{\psi}_{h,t}(\vartheta^\bullet) = \boldsymbol{s}_t'\,\boldsymbol{s}_t''. \tag{11.205}$$

We want to show that $\boldsymbol{\psi}_{h,t}(\vartheta^\bullet)$ vanishes in probability and, for this purpose, we start by examining the first term in the denominator of $\boldsymbol{s}_t'$, namely,

$$\sum_{\theta\in\mathcal{D}_h}\boldsymbol{\mu}_{h,t-1}(\theta)\left(\frac{\ell_h(\boldsymbol{x}_{h,t}|\theta)}{\ell_h(\boldsymbol{x}_{h,t}|\vartheta^o)}-1\right). \tag{11.206}$$

Observe that (11.88) and (11.201) imply that $\boldsymbol{\mu}_{h,t-1}(\theta)$ vanishes almost surely (hence, in probability) for all $\theta \in \mathcal{D}_h$. Moreover, the term $\ell_h(\boldsymbol{x}_{h,t}|\theta)/\ell_h(\boldsymbol{x}_{h,t}|\vartheta^o)$ has constant distribution over time. Calling upon Slutsky's theorem (in particular, Eq. (D.38) in Theorem D.4), we obtain the following convergence holding for all summands in (11.206):

$$\boldsymbol{\mu}_{h,t-1}(\theta)\left(\frac{\ell_h(\boldsymbol{x}_{h,t}|\theta)}{\ell_h(\boldsymbol{x}_{h,t}|\vartheta^o)}-1\right) \xrightarrow[t\to\infty]{\mathrm{P}} 0, \tag{11.207}$$

which implies that the denominator of $\boldsymbol{s}_t'$ converges to 1 in probability. Since the numerator of $\boldsymbol{s}_t'$ vanishes in probability in view of (11.201), $\boldsymbol{s}_t'$ vanishes in probability.[3]

[3]When we have two random sequences converging in probability, Theorem D.3 implies that their ratio converges in probability to the ratio between the limiting variables, provided that the limiting variable in the denominator is zero with zero probability.

From this convergence result and from the fact that the term s_t'' in (11.204) has constant distribution over time, by invoking again Slutsky's theorem we obtain

$$\psi_{h,t}(\vartheta^\bullet) = s_t' \, s_t'' \xrightarrow[t\to\infty]{\text{p}} 0. \tag{11.208}$$

We now proceed to show that, if agent h is a neighbor of agent k, then the condition $\psi_{h,t}(\vartheta^\bullet) \xrightarrow[t\to\infty]{\text{p}} 0$ implies that $\mu_{k,t}(\vartheta^\bullet) \xrightarrow[t\to\infty]{\text{p}} 0$. Consider then an agent k for which $a_{hk} > 0$, that is, an agent k such that $h \in \mathcal{N}_k$. In view of (11.137c) we can write

$$\mu_{k,t}(\vartheta^\bullet) = \frac{\prod\limits_{j\in\mathcal{N}_k} \left[\psi_{j,t}(\vartheta^\bullet)\right]^{a_{jk}}}{\sum\limits_{\theta\in\Theta}\prod\limits_{j\in\mathcal{N}_k}\left[\widehat{\psi}_{j,t}^{(k)}(\theta)\right]^{a_{jk}}} \leq \frac{\prod\limits_{j\in\mathcal{N}_k} \left[\psi_{j,t}(\vartheta^\bullet)\right]^{a_{jk}}}{\sum\limits_{\theta\in\mathcal{U}}\prod\limits_{j\in\mathcal{N}_k}\left[\widehat{\psi}_{j,t}^{(k)}(\theta)\right]^{a_{jk}}}, \tag{11.209}$$

where we recall that $\mathcal{U}$ is the set collecting the unshared hypotheses. Let us examine the denominator on the RHS of (11.209). Exploiting (11.137b) we obtain the following equalities:

$$\begin{aligned}
\sum_{\theta\in\mathcal{U}}\prod_{j\in\mathcal{N}_k}\left[\widehat{\psi}_{j,t}^{(k)}(\theta)\right]^{a_{jk}} &= \sum_{\theta\in\mathcal{U}}\prod_{j\in\mathcal{N}_k}\left[\frac{\psi_{k,t}(\theta)}{1-\psi_{k,t}(\vartheta^\bullet)}\right]^{a_{jk}}\left[1-\psi_{j,t}(\vartheta^\bullet)\right]^{a_{jk}}\\
&= \sum_{\theta\in\mathcal{U}}\left[\frac{\psi_{k,t}(\theta)}{1-\psi_{k,t}(\vartheta^\bullet)}\right]^{\sum\limits_{j\in\mathcal{N}_k} a_{jk}}\prod_{j\in\mathcal{N}_k}\left[1-\psi_{j,t}(\vartheta^\bullet)\right]^{a_{jk}}\\
&= \frac{\sum\limits_{\theta\in\mathcal{U}}\psi_{k,t}(\theta)}{1-\psi_{k,t}(\vartheta^\bullet)}\prod_{j\in\mathcal{N}_k}\left[1-\psi_{j,t}(\vartheta^\bullet)\right]^{a_{jk}}\\
&= \prod_{j\in\mathcal{N}_k}\left[1-\psi_{j,t}(\vartheta^\bullet)\right]^{a_{jk}}.
\end{aligned} \tag{11.210}$$

Substituting (11.210) into (11.209), we obtain

$$\begin{aligned}
\mu_{k,t}(\vartheta^\bullet) &\leq \prod_{j\in\mathcal{N}_k}\left[\frac{\psi_{j,t}(\vartheta^\bullet)}{1-\psi_{j,t}(\vartheta^\bullet)}\right]^{a_{jk}}\\
&\overset{\text{(a)}}{\leq} \left(\prod_{j\in\mathcal{N}_k}\left[\psi_{j,t}(\vartheta^\bullet)\right]^{a_{jk}}\right)\times\left(\prod_{j\in\mathcal{N}_k}\left[\frac{1}{\psi_{j,t}(\vartheta^o)}\right]^{a_{jk}}\right)\\
&\overset{\text{(b)}}{\leq} \left[\psi_{h,t}(\vartheta^\bullet)\right]^{a_{hk}}\prod_{j\in\mathcal{N}_k}\left[\frac{1}{\psi_{j,t}(\vartheta^o)}\right]^{a_{jk}},
\end{aligned} \tag{11.211}$$

where (a) follows because (recall that $\vartheta^\bullet \neq \vartheta^o$, i.e., $\vartheta^o \in \mathcal{U}$)

$$1-\psi_{j,t}(\vartheta^\bullet) = \sum_{\theta\in\mathcal{U}}\psi_{j,t}(\theta) \geq \psi_{j,t}(\vartheta^o) \tag{11.212}$$

and from the fact that the beliefs and combination weights are nonnegative; whereas (b) holds because the beliefs are bounded by 1 and, hence,

$$\prod_{\substack{j\in\mathcal{N}_k\\ j\neq h}}\left[\psi_{j,t}(\vartheta^\bullet)\right]^{a_{jk}} \leq 1. \tag{11.213}$$

Now we apply Lemma D.2 to the RHS of (11.211), where: *i)* $[\boldsymbol{\psi}_{h,t}(\vartheta^\bullet)]^{a_{hk}}$ plays the role of $\boldsymbol{w}_t$ since it converges in probability to 0 in view of (11.208) (recall that $a_{hk} > 0$); and *ii)* $\prod_{j\in\mathcal{N}_k}[\boldsymbol{\psi}_{j,t}(\vartheta^o)]^{-a_{jk}}$ plays the role of $\boldsymbol{y}_t$ since, by exploiting Markov's inequality (Theorem C.1) and Corollary 11.4, for any $y > 1$ we have

$$\begin{aligned}\mathbb{P}\left[\prod_{j\in\mathcal{N}_k}\left[\frac{1}{\boldsymbol{\psi}_{j,t}(\vartheta^o)}\right]^{a_{jk}} > y\right] &= \mathbb{P}\left[\sum_{j\in\mathcal{N}_k} a_{jk}\log\frac{1}{\boldsymbol{\psi}_{j,t}(\vartheta^o)} > \log y\right]\\ &\le \frac{1}{\log y}\sum_{j\in\mathcal{N}_k} a_{jk}\mathbb{E}\log\frac{1}{\boldsymbol{\psi}_{j,t}(\vartheta^o)}\\ &< \frac{|m_0|}{\log y}\sum_{j\in\mathcal{N}_k}\frac{a_{jk}}{v_j}\xrightarrow{y\to\infty} 0,\end{aligned} \tag{11.214}$$

which means that conditions (D.28) and (D.29) are satisfied. Thus, from Lemma D.2 we conclude that the RHS of (11.211) vanishes in probability, implying that $\boldsymbol{\mu}_{k,t}(\vartheta^\bullet) \xrightarrow[t\to\infty]{\mathrm{p}} 0$ for any agent k such that $h \in \mathcal{N}_k$.

In summary, we have shown that $\boldsymbol{\mu}_{h,t} \xrightarrow[t\to\infty]{\mathrm{p}} 0$ implies $\boldsymbol{\mu}_{k,t} \xrightarrow[t\to\infty]{\mathrm{p}} 0$ when h is a neighbor of k. Since the network is assumed to be connected, we can iterate the reasoning so as to extend the result to all agents in the network. In other words, we have established that $\boldsymbol{\mu}_{k,t}(\vartheta^\bullet) \xrightarrow[t\to\infty]{\mathrm{p}} 0$ for all k. Note that this result does not correspond to (11.87) since we only established convergence in probability. We now show how to extend the result to almost-sure convergence.

To this end, observe that the belief vector is a probability vector, and, hence,

$$\boldsymbol{\mu}_{k,t}(\vartheta^\bullet) + \boldsymbol{\mu}_{k,t}(\mathcal{I}_k^o) + \boldsymbol{\mu}_{k,t}(\mathcal{D}_k^\bullet) = 1, \tag{11.215}$$

which, combined with (11.88), yields

$$\boldsymbol{\mu}_{k,t}(\mathcal{I}_k^o) \xrightarrow[t\to\infty]{\mathrm{p}} 1 \qquad \text{for } k = 1, 2, \ldots, K. \tag{11.216}$$

Consider now the submartingale $\boldsymbol{n}_t(\mathcal{S})$ defined in (11.139). By choosing the set $\mathcal{S}$ as $\mathcal{S} = \{\mathcal{I}_1^o, \mathcal{I}_2^o, \ldots, \mathcal{I}_K^o\}$, from (11.216) we get

$$\boldsymbol{n}_t(\mathcal{S}) = \sum_{k=1}^{K} v_k \log\boldsymbol{\mu}_{k,t}(\mathcal{I}_k^o) \xrightarrow[t\to\infty]{\mathrm{p}} 0. \tag{11.217}$$

From (11.217) and part ii) of Lemma 11.1, we conclude that the convergence of $\boldsymbol{n}_t(\mathcal{S})$ must take place almost surely, that is,

$$\sum_{k=1}^{K} v_k \log\boldsymbol{\mu}_{k,t}(\mathcal{I}_k^o) \xrightarrow[t\to\infty]{\mathrm{a.s.}} 0, \tag{11.218}$$

Since $v_k > 0$ and $\log\boldsymbol{\mu}_{k,t}(\mathcal{I}_k^o) < 0$ for all k, we conclude that $\boldsymbol{\mu}_{k,t}(\mathcal{I}_k^o) \xrightarrow[t\to\infty]{\mathrm{a.s.}} 1$, which, in view of (11.215), finally implies (11.87), and the proof is complete.

■

11.C Appendix: Proof of Theorem 11.3

Proof. We start by focusing on point iii). If $\mathcal{I}_k = \emptyset$, there are no indistinguishable hypotheses, and this point is not present. We consider then the case $\mathcal{I}_k \neq \emptyset$ and prove that (11.94) holds. If $\mathcal{I}_k$ contains only one hypothesis, Eq. (11.94) trivially yields the identity $1 = 1$. It remains to examine the case where $\mathcal{I}_k$ contains at least two hypotheses. Observe that all the hypotheses belonging to $\mathcal{I}_k$ are indistinguishable by definition, and are also unshared since they are distinct from $\vartheta^\bullet$ (recall that we are considering the case $\vartheta^\bullet = \vartheta^o$). Accordingly, Eq. (11.199) holds for any $\theta, \theta' \in \mathcal{I}_k$. Summing over $\theta' \in \mathcal{I}_k$ the LHS and the RHS of (11.199), and applying the first definition in (11.79), we conclude that

$$\frac{\boldsymbol{\mu}_{k,t}(\mathcal{I}_k)}{\boldsymbol{\mu}_{k,t}(\theta)} = \frac{\mu_{k,0}(\mathcal{I}_k)}{\mu_{k,0}(\theta)}, \tag{11.219}$$

which is equivalent to (11.94).

Next, we prove (11.93). Since we are considering the case $\vartheta^\bullet = \vartheta^o$, from (11.85) we have

$$\mathcal{I}_k^\bullet = \mathcal{I}_k \backslash \{\vartheta^o\} = \mathcal{I}_k, \qquad \mathcal{D}_k^\bullet = \mathcal{D}_k \backslash \{\vartheta^o\} = \mathcal{D}_k, \tag{11.220}$$

which hold because $\mathcal{I}_k$ and $\mathcal{D}_k$ do not contain ϑ^o.

Now, if $\mathcal{I}_k = \emptyset$, then $\Gamma = 0$ in view of (11.81) and (11.83). Under this condition, the limit in (11.92) would be equal to 1, which would in turn imply (11.93). Thus, if $\mathcal{I}_k = \emptyset$ it would suffice to prove (11.92).

On the other hand, if $\mathcal{I}_k \neq \emptyset$, we can invoke Lemma 11.3 (along with the fact that $\mathcal{I}_k^\bullet = \mathcal{I}_k$ and $\mathcal{D}_k^\bullet = \mathcal{D}_k$) to see that (11.93) holds. Also in this case, to complete the proof we need to establish (11.92). To this end, it is convenient to consider separately the cases $\Gamma = 0$ and $\Gamma > 0$.

Case $\Gamma = 0$. In view of (11.83), when $\Gamma = 0$ we must have an agent h with $\Gamma_h = 0$, a condition that, from (11.81) and the assumption that the initial beliefs are positive, is equivalent to $\mathcal{I}_h = \emptyset$. This means that all hypotheses $\theta \neq \vartheta^o$ are distinguishable, implying, in view of Lemma 11.2,

$$\boldsymbol{\mu}_{h,t}(\vartheta^o) \xrightarrow[t\to\infty]{\mathrm{p}} 1. \tag{11.221}$$

We now want to show that the same convergence result holds for all agents in the network. To this end, we will first establish that the *intermediate* belief $\boldsymbol{\psi}_{h,t}(\vartheta^o)$ also converges to 1 in probability. Observe that from (11.137a) we can write

$$\boldsymbol{\psi}_{h,t}(\vartheta^o) = \frac{\boldsymbol{\mu}_{h,t-1}(\vartheta^o)}{\boldsymbol{\mu}_{h,t-1}(\vartheta^o) + \sum_{\theta\neq\vartheta^o} \boldsymbol{\mu}_{h,t-1}(\theta) \dfrac{\ell_h(\boldsymbol{x}_{h,t}|\theta)}{\ell_h(\boldsymbol{x}_{h,t}|\vartheta^o)}}. \tag{11.222}$$

Since the ratios $\ell_h(\boldsymbol{x}_{h,t}|\theta)/\ell_h(\boldsymbol{x}_{h,t}|\vartheta^o)$ are identically distributed over time, and $\boldsymbol{\mu}_{h,t-1}(\theta)$ vanishes in probability for $\theta \neq \vartheta^o$ in view of (11.221), from Slutsky's theorem (see in particular (D.38) in Theorem D.4) we have

$$\sum_{\theta\in\mathcal{U}} \boldsymbol{\mu}_{h,t-1}(\theta) \frac{\ell_h(\boldsymbol{x}_{h,t}|\theta)}{\ell_h(\boldsymbol{x}_{h,t}|\vartheta^o)} \xrightarrow[t\to\infty]{\mathrm{p}} 0, \tag{11.223}$$

which, when used in (11.222) along with (11.221), yields

$$\boldsymbol{\psi}_{h,t}(\vartheta^o) \xrightarrow[t\to\infty]{\mathrm{p}} 1. \tag{11.224}$$

The next step is to show that (11.224) implies that $\boldsymbol{\mu}_{k,t}(\vartheta^o) \xrightarrow[t\to\infty]{\text{p}} 1$ if h is a neighbor of k, i.e., if $a_{hk} > 0$. Obviously, we already know that $\boldsymbol{\mu}_{k,t}(\vartheta^o) \xrightarrow[t\to\infty]{\text{p}} 1$ if $\Gamma_k = 0$. Therefore, it suffices to focus on an agent k with $\Gamma_k > 0$. The following chain of relations holds:

$$
\begin{aligned}
\boldsymbol{\mu}_{k,t}(\mathcal{I}_k) &\overset{\text{(a)}}{=} \frac{\sum_{\theta\in\mathcal{I}_k}\prod_{j\in\mathcal{N}_k}\left[\widehat{\boldsymbol{\psi}}^{(k)}_{j,t}(\theta)\right]^{a_{jk}}}{\sum_{\theta'\in\Theta}\prod_{j\in\mathcal{N}_k}\left[\widehat{\boldsymbol{\psi}}^{(k)}_{j,t}(\theta')\right]^{a_{jk}}}\\
&\overset{\text{(b)}}{=} \frac{\sum_{\theta\in\mathcal{I}_k}\dfrac{\boldsymbol{\psi}_{k,t}(\theta)}{1-\boldsymbol{\psi}_{k,t}(\vartheta^\bullet)}\prod_{j\in\mathcal{N}_k}\left[1-\boldsymbol{\psi}_{j,t}(\vartheta^\bullet)\right]^{a_{jk}}}{\sum_{\theta'\in\Theta}\prod_{j\in\mathcal{N}_k}\left[\widehat{\boldsymbol{\psi}}^{(k)}_{j,t}(\theta')\right]^{a_{jk}}}\\
&\le \frac{\sum_{\theta\in\mathcal{I}_k}\dfrac{\boldsymbol{\psi}_{k,t}(\theta)}{1-\boldsymbol{\psi}_{k,t}(\vartheta^\bullet)}\prod_{j\in\mathcal{N}_k}\left[1-\boldsymbol{\psi}_{j,t}(\vartheta^\bullet)\right]^{a_{jk}}}{\prod_{j\in\mathcal{N}_k}\left[\widehat{\boldsymbol{\psi}}^{(k)}_{j,t}(\vartheta^o)\right]^{a_{jk}}}\\
&\overset{\text{(c)}}{=} \frac{\sum_{\theta\in\mathcal{I}_k}\dfrac{\boldsymbol{\psi}_{k,t}(\theta)}{1-\boldsymbol{\psi}_{k,t}(\vartheta^o)}\prod_{j\in\mathcal{N}_k}\left[1-\boldsymbol{\psi}_{j,t}(\vartheta^o)\right]^{a_{jk}}}{\prod_{j\in\mathcal{N}_k}\left[\boldsymbol{\psi}^{(k)}_{j,t}(\vartheta^o)\right]^{a_{jk}}}\\
&\overset{\text{(d)}}{=} \frac{\boldsymbol{\psi}_{k,t}(\mathcal{I}_k)}{1-\boldsymbol{\psi}_{k,t}(\vartheta^o)}\prod_{j\in\mathcal{N}_k}\left(\frac{1-\boldsymbol{\psi}_{j,t}(\vartheta^o)}{\boldsymbol{\psi}_{j,t}(\vartheta^o)}\right)^{a_{jk}}\\
&\overset{\text{(e)}}{\le} \prod_{j\in\mathcal{N}_k}\left(\frac{1-\boldsymbol{\psi}_{j,t}(\vartheta^o)}{\boldsymbol{\psi}_{j,t}(\vartheta^o)}\right)^{a_{jk}}\\
&\overset{\text{(f)}}{\le} \left[1-\boldsymbol{\psi}_{h,t}(\vartheta^o)\right]^{a_{hk}}\prod_{j\in\mathcal{N}_k}\left[\frac{1}{\boldsymbol{\psi}_{j,t}(\vartheta^o)}\right]^{a_{jk}},
\end{aligned}
\tag{11.225}
$$

where (a) follows from (11.79) and (11.137c); in (b) we apply (11.137b); (c) holds because $\vartheta^\bullet = \vartheta^o$; (d) follows from (11.79); (e) follows from

$$
\frac{\boldsymbol{\psi}_{k,t}(\mathcal{I}_k)}{1-\boldsymbol{\psi}_{k,t}(\vartheta^o)} = \frac{\boldsymbol{\psi}_{k,t}(\mathcal{I}_k)}{\boldsymbol{\psi}_{k,t}(\mathcal{I}_k)+\boldsymbol{\psi}_{k,t}(\mathcal{D}_k)} \le 1, \tag{11.226}
$$

whereas (f) holds since the beliefs are bounded by 1 and, hence,

$$
\prod_{\substack{j\in\mathcal{N}_k\\ j\neq h}}\left[1-\boldsymbol{\psi}_{j,t}(\vartheta^o)\right]^{a_{jk}} \le 1. \tag{11.227}
$$

Next, we show that the RHS of (11.225) vanishes in probability. This conclusion follows from Lemma D.2, where the sequence $[1-\boldsymbol{\psi}_{h,t}(\vartheta^o)]^{a_{hk}}$ plays the role of $\boldsymbol{w}_t$, since it converges in probability to 0 according to (11.224); and where the sequence

$\prod_{j\in\mathcal{N}_k}[\boldsymbol{\psi}_{j,t}(\vartheta^o)]^{-a_{jk}}$ plays the role of $\boldsymbol{y}_t$, since this sequence satisfies (11.214) and, hence, satisfies conditions (D.28) and (D.29).

Since the RHS of (11.225) converges in probability to 0, we have that $\boldsymbol{\mu}_{k,t}(\mathcal{I}_k) \xrightarrow[t\to\infty]{\text{p}} 0$. Recalling that $\boldsymbol{\mu}_{k,t}(\vartheta^o) = 1 - \boldsymbol{\mu}_{k,t}(\mathcal{I}_k) - \boldsymbol{\mu}_{k,t}(\mathcal{D}_k)$, and that we have already shown that $\boldsymbol{\mu}_{k,t}(\mathcal{D}_k) \xrightarrow[t\to\infty]{\text{p}} 0$, we conclude that $\boldsymbol{\mu}_{k,t}(\vartheta^o) \xrightarrow[t\to\infty]{\text{p}} 1$. In summary, we have shown that $\boldsymbol{\mu}_{h,t}(\vartheta^o) \xrightarrow[t\to\infty]{\text{p}} 1$ implies $\boldsymbol{\mu}_{k,t}(\vartheta^o) \xrightarrow[t\to\infty]{\text{p}} 1$ when h is a neighbor of k. Since the network is connected, by iterating the above reasoning we conclude that $\boldsymbol{\mu}_{k,t}(\vartheta^o) \xrightarrow[t\to\infty]{\text{p}} 1$ for all k, which implies

$$\boldsymbol{m}_t = \sum_{k=1}^{K} v_k \log \boldsymbol{\mu}_{k,t}(\vartheta^o) \xrightarrow[t\to\infty]{\text{p}} 0. \tag{11.228}$$

From part ii) of Lemma 11.1, this convergence must take place almost surely, i.e.,

$$\sum_{k=1}^{K} v_k \log \boldsymbol{\mu}_{k,t}(\vartheta^o) \xrightarrow[t\to\infty]{\text{a.s.}} 0. \tag{11.229}$$

Since $v_k > 0$ and $\log \boldsymbol{\mu}_{k,t}(\vartheta^o) < 0$ for all k, we must have $\boldsymbol{\mu}_{k,t}(\vartheta^o) \xrightarrow[t\to\infty]{\text{a.s.}} 1$ for any agent k, which concludes the proof for the case $\Gamma = 0$.

Case $\Gamma > 0$. For $k = 1, 2, \dots, K$, we have the following chain of equalitites

$$\begin{aligned}
\frac{\boldsymbol{\mu}_{k,t}(\mathcal{I}_k)}{\boldsymbol{\mu}_{k,t}(\vartheta^o)} &\overset{(11.79)}{=} \frac{\sum_{\theta\in\mathcal{I}_k}\boldsymbol{\mu}_{k,t}(\theta)}{\boldsymbol{\mu}_{k,t}(\vartheta^o)} \overset{(11.137\text{c})}{=} \frac{\sum_{\theta\in\mathcal{I}_k}\prod_{j\in\mathcal{N}_k}\left[\widehat{\boldsymbol{\psi}}^{(k)}_{j,t}(\theta)\right]^{a_{jk}}}{\prod_{j\in\mathcal{N}_k}\left[\boldsymbol{\psi}^{(k)}_{j,t}(\vartheta^o)\right]^{a_{jk}}} \\
&\overset{(11.137\text{b})}{=} \frac{\sum_{\theta\in\mathcal{I}_k}\dfrac{\boldsymbol{\psi}_{k,t}(\theta)}{1-\boldsymbol{\psi}_{k,t}(\vartheta^o)}\prod_{j\in\mathcal{N}_k}\left[1-\boldsymbol{\psi}_{j,t}(\vartheta^o)\right]^{a_{jk}}}{\prod_{j\in\mathcal{N}_k}\left[\boldsymbol{\psi}^{(k)}_{j,t}(\vartheta^o)\right]^{a_{jk}}} \\
&= \frac{\boldsymbol{\psi}_{k,t}(\mathcal{I}_k)}{1-\boldsymbol{\psi}_{k,t}(\vartheta^o)}\prod_{j\in\mathcal{N}_k}\left[\frac{1-\boldsymbol{\psi}_{j,t}(\vartheta^o)}{\boldsymbol{\psi}_{j,t}(\vartheta^o)}\right]^{a_{jk}} \\
&= \frac{\boldsymbol{\psi}_{k,t}(\mathcal{I}_k)}{1-\boldsymbol{\psi}_{k,t}(\vartheta^o)}\left(\prod_{j=1}^{K}\left[\frac{\boldsymbol{\psi}_{j,t}(\mathcal{I}_j)}{\boldsymbol{\psi}_{j,t}(\vartheta^o)}\right]^{a_{jk}}\right)\times\left(\prod_{j=1}^{K}\left[\frac{1-\boldsymbol{\psi}_{j,t}(\vartheta^o)}{\boldsymbol{\psi}_{j,t}(\mathcal{I}_j)}\right]^{a_{jk}}\right),
\end{aligned} \tag{11.230}$$

where in the last step we multiplied and divided by $\prod_{j=1}^{K}[\boldsymbol{\psi}_{j,t}(\mathcal{I}_j)]^{a_{jk}}$, and applied the definition of neighborhood from (4.1). Using (11.137a) and (11.48) we get

$$\frac{\boldsymbol{\psi}_{j,t}(\mathcal{I}_j)}{\boldsymbol{\psi}_{j,t}(\vartheta^o)} = \frac{\boldsymbol{\mu}_{j,t-1}(\mathcal{I}_j)\ell_j(\boldsymbol{x}_{j,t}|\vartheta^o)}{\boldsymbol{\mu}_{j,t-1}(\vartheta^o)\ell_j(\boldsymbol{x}_{j,t}|\vartheta^o)} = \frac{\boldsymbol{\mu}_{j,t-1}(\mathcal{I}_j)}{\boldsymbol{\mu}_{j,t-1}(\vartheta^o)}. \tag{11.231}$$

Likewise, exploiting (11.48) and (11.49) we have

$$\frac{1-\boldsymbol{\psi}_{j,t}(\vartheta^o)}{\boldsymbol{\psi}_{j,t}(\mathcal{I}_j)} = \frac{\boldsymbol{\psi}_{j,t}(\mathcal{I}_j)+\boldsymbol{\psi}_{j,t}(\mathcal{D}_j)}{\boldsymbol{\psi}_{j,t}(\mathcal{I}_j)} = 1 + \frac{\boldsymbol{\psi}_{j,t}(\mathcal{D}_j)}{\boldsymbol{\psi}_{j,t}(\mathcal{I}_j)}. \tag{11.232}$$

Subtituting (11.231) and (11.232) into (11.230) and taking the logarithm we obtain

$$\log \frac{\mu_{k,t}(\mathcal{I}_k)}{\mu_{k,t}(\vartheta^o)} = \sum_{j=1}^{K} a_{jk} \log \frac{\mu_{j,t-1}(\mathcal{I}_j)}{\mu_{j,t-1}(\vartheta^o)} + \log\left(\left[1+\frac{\psi_{k,t}(\mathcal{D}_k)}{\psi_{k,t}(\mathcal{I}_k)}\right]^{-1} \prod_{j=1}^{K}\left[1+\frac{\psi_{j,t}(\mathcal{D}_j)}{\psi_{j,t}(\mathcal{I}_j)}\right]^{a_{jk}}\right). \tag{11.233}$$

It is now convenient to introduce the vectors

$$\boldsymbol{z}_t \triangleq \text{col}\left\{\log \frac{\mu_{k,t}(\mathcal{I}_k)}{\mu_{k,t}(\vartheta^o)}\right\}_{k=1}^{K}, \tag{11.234}$$

$$\boldsymbol{y}_t \triangleq \text{col}\left\{\log\left(\left[1+\frac{\psi_{k,t}(\mathcal{D}_k)}{\psi_{k,t}(\mathcal{I}_k)}\right]^{-1} \prod_{j=1}^{K}\left[1+\frac{\psi_{j,t}(\mathcal{D}_j)}{\psi_{j,t}(\mathcal{I}_j)}\right]^{a_{jk}}\right)\right\}_{k=1}^{K}, \tag{11.235}$$

where $\text{col}\{x_k\}_{k=1}^{K}$ denotes the $K \times 1$ vector obtained by stacking into a single column the entries $x_1, x_2, \ldots, x_K$. Using (11.234) and (11.235), we can recast (11.233) in the vector form

$$\boldsymbol{z}_t = A^{\mathsf{T}} \boldsymbol{z}_{t-1} + \boldsymbol{y}_t. \tag{11.236}$$

Unfolding the recursion we get

$$\boldsymbol{z}_t = (A^t)^{\mathsf{T}} z_0 + \sum_{\tau=0}^{t-1} (A^{\tau})^{\mathsf{T}} \boldsymbol{y}_{t-\tau}. \tag{11.237}$$

Let V be the $K \times K$ matrix whose columns are all equal to the Perron vector, i.e., $V = v\mathbb{1}^{\mathsf{T}}$. We now show that the vectors $\boldsymbol{y}_t$ in (11.235) are in the null space of V^{T}:

$$V^{\mathsf{T}} \boldsymbol{y}_t = 0. \tag{11.238}$$

Since V has equal columns, Eq. (11.238) is equivalent to the relation $\sum_{k=1}^{K} v_k \boldsymbol{y}_{k,t} = 0$, where $\boldsymbol{y}_{k,t}$ corresponds to the kth entry of the vector $\boldsymbol{y}_t$. Exploiting the definition of $\boldsymbol{y}_t$ from (11.235), we have

$$\sum_{k=1}^{K} v_k \boldsymbol{y}_{k,t} = -\sum_{k=1}^{K} v_k \log\left(1+\frac{\psi_{k,t}(\mathcal{D}_k)}{\psi_{k,t}(\mathcal{I}_k)}\right) + \sum_{k=1}^{K} v_k \sum_{j=1}^{K} a_{jk} \log\left(1+\frac{\psi_{j,t}(\mathcal{D}_j)}{\psi_{j,t}(\mathcal{I}_j)}\right) = 0, \tag{11.239}$$

where the final equality follows from the identity $\sum_{k=1}^{K} a_{jk} v_k = v_j$.

In view of (11.238) we can rewrite (11.237) as

$$\boldsymbol{z}_t = (A^t)^{\mathsf{T}} z_0 + \sum_{\tau=0}^{t-1} (A^{\tau} - V)^{\mathsf{T}} \boldsymbol{y}_{t-\tau} + \sum_{\tau=0}^{t-1} \underbrace{V^{\mathsf{T}} \boldsymbol{y}_{t-\tau}}_{=0} (A^t)^{\mathsf{T}} z_0 + \sum_{\tau=0}^{t-1} F_{\tau}^{\mathsf{T}} \boldsymbol{y}_{t-\tau}, \tag{11.240}$$

where $F_\tau \triangleq A^\tau - V$. The next step of the proof is to establish that the vector defined by the sum on the RHS of (11.240) vanishes almost surely as $t \to \infty$, namely, in terms of the kth entry of this vector, we want to show that

$$\sum_{\tau=0}^{t-1}\sum_{j=1}^{K}[F_\tau]_{jk}\,\boldsymbol{y}_{j,t-\tau} \xrightarrow[t\to\infty]{\text{a.s.}} 0. \tag{11.241}$$

To this end, observe that since the graph is primitive, in view of Corollary 4.1 there exist two constants $C > 0$ and $r \in (0,1)$ such that, for $t = 0, 1, \ldots$, the following condition is satisfied:[4]

$$\max_{j,k\in\{1,2,\ldots,K\}}\left|[A^t - V]_{jk}\right| \le Cr^t. \tag{11.243}$$

We can write

$$\left|\sum_{\tau=0}^{t-1}\sum_{j=1}^{K}[F_\tau]_{jk}\,\boldsymbol{y}_{j,t-\tau}\right| \le \sum_{\tau=0}^{t-1}\left|[F_\tau]_{jk}\right|\sum_{j=1}^{K}|\boldsymbol{y}_{j,t-\tau}| \le \sum_{\tau=0}^{t-1} r^\tau \boldsymbol{w}_{t-\tau}, \tag{11.244}$$

where in the last step we used the bound from (11.243) and introduced the definition

$$\boldsymbol{w}_t = C\sum_{j=1}^{K}|\boldsymbol{y}_{j,t}|. \tag{11.245}$$

When $\mathcal{D}_j = \emptyset$, we see from (11.235) that $\boldsymbol{y}_{j,t} = 0$. Consider then the agents j for which $\mathcal{D}_j \neq \emptyset$ (there must be at least one such an agent in view of global identifiability). For these agents, part ii) of Lemma 11.4 implies that $\boldsymbol{y}_{j,t} \xrightarrow[t\to\infty]{\text{a.s.}} 0$ for any j, and, hence, $\boldsymbol{w}_t \xrightarrow[t\to\infty]{\text{a.s.}} 0$. This implies that, almost surely, for any $\varepsilon > 0$ there exists a (random) value $\boldsymbol{t}_\varepsilon$ such that for all $t > \boldsymbol{t}_\varepsilon$ we have $\boldsymbol{w}_t < \varepsilon(1-r)$. Therefore, for $t > \boldsymbol{t}_\varepsilon$, the following relations hold (almost surely):

$$\begin{aligned}\sum_{\tau=0}^{t-1} r^\tau \boldsymbol{w}_{t-\tau} &= \sum_{\tau=0}^{t-\boldsymbol{t}_\varepsilon-1} r^\tau \boldsymbol{w}_{t-\tau} + \sum_{\tau=t-\boldsymbol{t}_\varepsilon}^{t-1} r^\tau \boldsymbol{w}_{t-\tau}\\ &< \varepsilon(1-r)\sum_{\tau=0}^{t-\boldsymbol{t}_\varepsilon-1} r^\tau + \sum_{\tau=t-\boldsymbol{t}_\varepsilon}^{t-1} r^\tau \boldsymbol{w}_{t-\tau}\\ &< \varepsilon(1-r)\sum_{\tau=0}^{\infty} r^\tau + \sum_{\tau=t-\boldsymbol{t}_\varepsilon}^{t-1} r^\tau \boldsymbol{w}_{t-\tau}\\ &= \varepsilon + \sum_{\tau=t-\boldsymbol{t}_\varepsilon}^{t-1} r^\tau \boldsymbol{w}_{t-\tau},\end{aligned} \tag{11.246}$$

[4]Actually, Corollary 4.1 does not consider the case $t = 0$. However, including the case $t = 0$ in (11.243) is possible by defining the constant C as the maximum between the constant valid for $t > 1$ and the value

$$\max_{j,k\in\{1,2,\ldots,K\}}\left|[I_K - V]_{jk}\right|, \tag{11.242}$$

which corresponds to $t = 0$.

where in the final equality we used the fact that $\sum_{\tau=0}^{\infty} r^{\tau} = 1/(1-r)$. Moreover, since r^{τ} is decreasing and $\boldsymbol{w}_{t-\tau}$ is nonnegative, we get

$$\sum_{\tau=t-t_{\varepsilon}}^{t-1} r^{\tau} \boldsymbol{w}_{t-\tau} \le r^{t-t_{\varepsilon}} \sum_{\tau=t-t_{\varepsilon}}^{t-1} \boldsymbol{w}_{t-\tau} = r^{t} \left(\frac{1}{r^{t_{\varepsilon}}} \sum_{\tau=1}^{t_{\varepsilon}} \boldsymbol{w}_{\tau} \right). \tag{11.247}$$

We argue now that the quantity within brackets is almost-surely finite. Recalling the discussion following (11.137a)–(11.137c), we know that, almost surely, $\boldsymbol{\psi}_{k,t}(\theta)$ is positive for all k and θ. From the definition of $\boldsymbol{y}_t$ in (11.235), it follows that all entries $\boldsymbol{y}_{k,t}$ are almost-surely finite. This implies that $\boldsymbol{w}_t$ defined in (11.245) is almost-surely finite, and so is the quantity within brackets appearing in (11.247).

Thus, from (11.246) we obtain

$$\limsup_{t\to\infty} \sum_{\tau=0}^{t-1} r^{\tau} \boldsymbol{w}_{t-\tau} \le \varepsilon \qquad \text{almost surely}, \tag{11.248}$$

which proves (11.241) in view of (11.244) and the arbitrariness of ε.

We have therefore shown that the second term on the RHS of (11.240) vanishes almost surely as $t \to \infty$. It remains to characterize the first term on the RHS of (11.240). To this end, observe that (11.243) implies

$$\lim_{t\to\infty} (A^{t})^{\mathsf{T}} z_0 = V^{\mathsf{T}} z_0. \tag{11.249}$$

Considering the explicit form of the entries of the vector z_0 available from (11.234), we can write

$$\left[V^{\mathsf{T}} z_0\right]_k = \sum_{j=1}^{K} v_j \log \frac{\mu_{j,0}(\mathcal{I}_j)}{\mu_{j,0}(\vartheta^o)} = \sum_{j=1}^{K} v_j \log \Gamma_j = \log \Gamma, \tag{11.250}$$

where, in the last equality, we further apply the definition of Γ_j and Γ from (11.81) and (11.83), respectively. Using (11.241), (11.249), and (11.250) in (11.240) and exploiting the definition of $\boldsymbol{z}_t$ from (11.234), we get

$$\frac{\boldsymbol{\mu}_{k,t}(\mathcal{I}_k)}{\boldsymbol{\mu}_{k,t}(\vartheta^o)} \xrightarrow[t\to\infty]{\text{a.s.}} \Gamma. \tag{11.251}$$

On the other hand, from (11.93) we have

$$\boldsymbol{\mu}_{k,t}(\vartheta^o) + \boldsymbol{\mu}_{k,t}(\mathcal{I}_k) \xrightarrow[t\to\infty]{\text{a.s.}} 1, \tag{11.252}$$

and since we can write

$$\boldsymbol{\mu}_{k,t}(\vartheta^o) + \boldsymbol{\mu}_{k,t}(\mathcal{I}_k) = \boldsymbol{\mu}_{k,t}(\vartheta^o) \left(1 + \frac{\boldsymbol{\mu}_{k,t}(\mathcal{I}_k)}{\boldsymbol{\mu}_{k,t}(\vartheta^o)}\right), \tag{11.253}$$

by using (11.251) and (11.252), we conclude that

$$\boldsymbol{\mu}_{k,t}(\vartheta^o) \xrightarrow[t\to\infty]{\text{a.s.}} \frac{1}{1+\Gamma}, \tag{11.254}$$

which corresponds to (11.92), and the proof is complete.

■

Chapter 12

Social Machine Learning

We have seen in the previous chapters that in social learning the agents employ some locally available statistical models represented by the likelihoods, which are meant to approximate the possible true models. However, in many cases the likelihoods are not known beforehand, and the analysis so far in the text has not addressed the problem of how the agents can select them. We explain in this chapter how the agents can address this important issue by learning from training data, and carry out a detailed performance analysis to show that, under reasonable conditions, correct decision-making continues to be attained. We start with a motivating example.

Assume a certain classification problem must be solved to distinguish between some hypotheses or classes. It is useful to describe first a single-agent setting. To solve the classification problem, the agent would resort to some standard machine learning strategy, for example, a logistic regression classifier or a neural network [155]. Under supervised learning, the operation of these structures would involve two distinct phases. One is the *training phase*, where the agent (i.e., the classifier) learns how to construct the decision statistics necessary to perform the classification task. The agent is trained over a dataset containing several examples, which provide *clues* on the statistical relation between the observation x and the corresponding hypothesis θ. Borrowing a standard terminology from machine learning, in the following we refer to the observation x as the *feature*[1] and to the hypothesis θ as the *label*. Each clue in the training data therefore consists of a feature x marked with a label θ that denotes the particular hypothesis that gave rise to x. At the end of the training phase, the classifier would

[1]Note that a feature x can be a vector collecting different attributes.

have learned some decision statistics that allow it to construct a decision rule, i.e., a mapping from the feature space to the label space. Thus, the classifier can now switch to the *prediction phase*, where it uses the learned decision rule to predict the label θ' of any new feature x', i.e., to classify new (unlabeled) observations.

In the social machine learning (SML) paradigm to be discussed in this chapter, we will be dealing with a group of agents, rather than with a single agent. We will have a *distributed* ensemble of spatially dispersed datasets. Each dataset will be used to train a local learning machine at the corresponding agent location. Once training is concluded, the individual learning machines switch into a social learning mode where they cooperate with each other over a graph topology. For example, we can have an ensemble of mobile phones distributed over a certain geographic area. These phones would have embedded into them some local routines to learn *individual* weather forecasting models from training data consisting, for instance, of air humidity, atmospheric pressure, or temperature data collected at their respective locations. These routines are fixed, and can be chosen from among classic machine learning procedures. Subsequently, the phones would be able to interact with each other by means of a certain app that allows them to use their individual knowledge to accomplish a *social* weather forecasting task with enhanced accuracy.

To avoid confusion, it is important to compare this new setting with the settings described in the previous chapters. There, the local statistical models $\ell_k(x|\theta)$ for each agent k were taken for granted and there was no need for a training phase. In other words, in the context of social learning, the term "learning" referred to prediction only, while in social *machine* learning we have *two* learning stages, which are conveniently represented by the two concentric circles in future Figure 12.1. The outer circle corresponds to the *memory* layer, where each agent builds its individual memory by storing the likelihood[2] models learned during training; and the inner circle corresponds to the *processing* layer, where the agents cooperatively solve the classification problem over the graph by using streaming observations during the prediction phase.

The material presented in this chapter is based on [28, 29]. The study of social learning under uncertain likelihood models was also addressed in [87, 88], albeit from a different perspective. The approach in the latter

[2]Technically, as we will see from Lemma 12.1 further ahead, to construct the beliefs it would be sufficient to learn likelihood *ratios*, rather than the likelihoods themselves.

references requires the selection of suitable families of likelihood models (e.g., Gaussian, multinomial, or Poisson) that must be amenable to analytical manipulations, such as the construction of conjugate priors. Moreover, the chosen family must also match the underlying physics of the observed phenomenon.

In contrast, in the SML approach we abandon the idea of relying on analytical models and rely instead on a *data-driven* approach by using some arbitrary machine learning architecture at each node. Machine learning architectures are particularly appropriate when the designer has limited knowledge about the statistical models that describe the data distributions and even when these models exhibit a high degree of complexity. For example, neural networks involving the concatenation of nonlinear functions with millions of parameters have been proved to learn efficiently from training data the "shape" of very sophisticated models. This property is important for several learning applications, e.g., in the distributed classification of images or videos where it is hard to encode the data distribution into some classic statistical distributions. We will present useful instances of this type of problems in Section 12.6.

12.1 Social Machine Learning Model

We now introduce the details of the SML model.

The prediction phase corresponds to the belief formation problem that the agents want to solve. In other words, it corresponds to the same type of problem addressed in earlier chapters in the text. We conveniently collect in the next assumption the details and conditions for the prediction phase that will be used in this chapter.

Assumption 12.1 (Prediction phase). The feature observed by agent k at instant t of the prediction phase is denoted by $\boldsymbol{x}_{k,t} \in \mathcal{X}_k$. We will work under the following conditions:

i) *During the prediction phase* we are under the objective evidence model considered in Section 5.3, i.e., there exists one true underlying hypothesis $\vartheta^o \in \Theta$ that gives rise to the collections of observations $\{\boldsymbol{x}_{k,t}\}_{k=1}^{K}$ across the K agents, which are iid over time.

ii) We consider a family of likelihood models $\ell_{k,\theta}$, for $k = 1, 2, \ldots, K$ and $\theta \in \Theta$. The agents do not know these models and they should learn them during a training phase, as described next. The feature observed by agent k at time t, $\boldsymbol{x}_{k,t}$, is (marginally) distributed according to the model ℓ_{k,ϑ^o} that corresponds to the true underlying hypothesis ϑ^o.

iii) The likelihoods are assumed to satisfy Assumption 5.4, so that the classification problem to be solved in the prediction phase is globally identifiable.

iv) The likelihoods are assumed to satisfy

$$D(\ell_{k,\theta}||\ell_{k,\theta'}) < \infty, \tag{12.1}$$

for $k = 1, 2, \ldots, K$ and for all θ and θ' belonging to Θ.

v) *Prediction is performed cooperatively by all agents*, and cooperation takes place over a *primitive* graph — see Definition 4.5.

Since the likelihoods are assumed to be unknown, the agents must be trained *before* the prediction phase takes place. The aim of the training phase is to learn suitable decision models based on a set of available examples. The next assumption introduces the notation and describes the conditions relative to the training phase.

Assumption 12.2 (Training phase). During the training phase, each agent k has access to E_k clues consisting of (feature, label) pairs and forming the *training set*

$$\mathcal{T}_k \triangleq \left\{\widehat{\boldsymbol{x}}_{k,n}, \widehat{\boldsymbol{\theta}}_{k,n}\right\}_{n=1}^{E_k}, \quad \text{with } \widehat{\boldsymbol{x}}_{k,n} \in \mathcal{X}_k \text{ and } \widehat{\boldsymbol{\theta}}_{k,n} \in \Theta. \tag{12.2}$$

The pairs from (12.2) are assumed to be iid over n and distributed as follows. Given a label $\widehat{\boldsymbol{\theta}}_{k,n} = \theta$, the feature $\widehat{\boldsymbol{x}}_{k,n}$ is generated according to some (unknown) model $\ell_k(x|\theta)$. We further assume that, *during training*, the labels $\widehat{\boldsymbol{\theta}}_{k,n}$ are uniformly distributed:

$$\mathbb{P}\left[\widehat{\boldsymbol{\theta}}_{k,n} = \theta\right] = \frac{1}{H} \quad \forall \theta \in \Theta. \tag{12.3}$$

Training is performed *individually* by each agent. Furthermore, the mechanisms governing the training and prediction phases are statistically independent.

Each agent k is trained to approximate the unknown local likelihood model $\ell_k(x|\theta)$. More precisely, as explained in the forthcoming sections, agent k will instead learn some decision statistics, denoted by $h_k(x;\theta)$, to approximate *log likelihood ratios*.

Note that the data samples in the training sets are topped with the symbol $\widehat{\ }$. This is done to avoid confusion between features used in the training phase and features used in the prediction phase. We also emphasize that subscripts n and t have a different meaning.

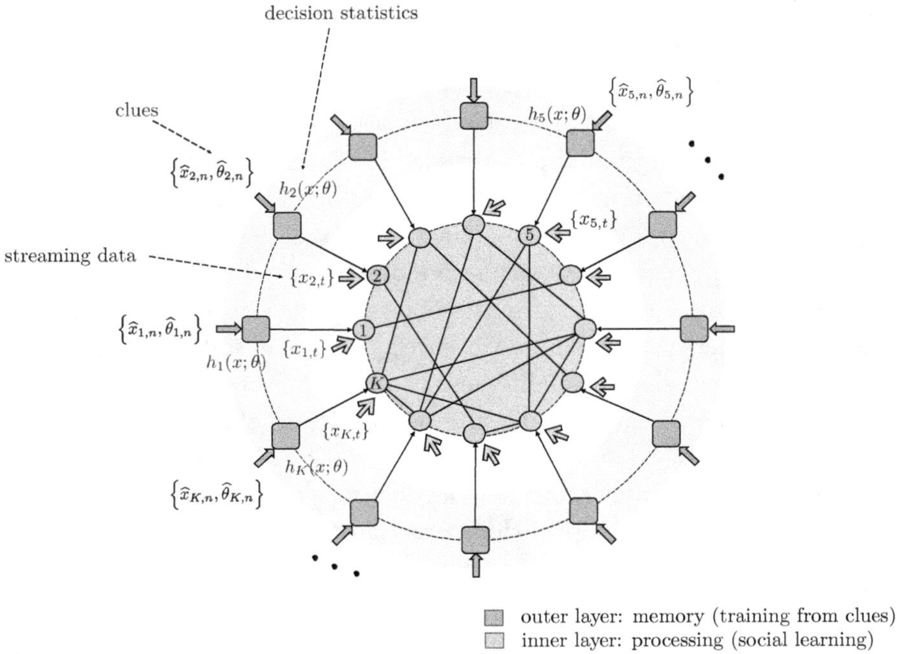

Figure 12.1: Schematic illustration of the social machine learning problem. The outer layer corresponds to the memory of the network, where each agent k uses the clues $\{\widehat{x}_{k,n}, \widehat{\theta}_{k,n}\}$ in its training set to learn some decision statistics $h_k(x;\theta)$, as described in Section 12.2. Once training is completed, the agents enter the inner layer, i.e., the processing stage where they perform cooperatively the social learning task by applying the learned decision statistics to the streaming observations $\{x_{k,t}\}$ during the prediction phase.

Subscript n refers to a *training* observation $\widehat{\boldsymbol{x}}_{k,n}$ that was generated under hypothesis $\widehat{\boldsymbol{\theta}}_{k,n}$. As a result, the observations aggregated over different values of n correspond to *different* hypotheses. In particular, condition (12.3) means that the hypotheses in the training set are drawn uniformly, i.e., the training set is balanced so that all classes are sufficiently explored.

In comparison, subscript t is a time index that refers to a *prediction* observation $\boldsymbol{x}_{k,t}$ arising from the *true* hypothesis ϑ^o. This means that the observations aggregated over time are generated under *one and the same* hypothesis.

Figure 12.1 summarizes the description of the SML paradigm in terms of the two "concentric" layers of *memory* and *processing*. During the training phase, each individual learning machine k uses the *clues*, i.e., (feature, label) pairs, available in its local training set to build its individual *memory*, where information about the learned decision statistics is stored. Once training is performed, the learning machines enter the *processing* layer, where they are fed by the streaming observations collected during the prediction phase, and apply the learned models to these observations in a *social* manner by cooperating over a network.

It is worth pointing out the distinguishing attributes of *social* machine learning, as opposed to more traditional machine learning problems. In a nutshell, the SML architecture is *dispersed in both space and time* and is capable of handling heterogeneous data more directly. Regarding dispersion in space, it results from the simultaneous presence of multiple remotely dispersed classifiers (i.e., agents). Moreover, the features at these agents can be of different type, size, or quality, and therefore heterogeneous. Regarding dispersion in time, it arises from the possibility in the prediction phase to base the classification decision on observations streaming over time, which enables increasingly more reliable decisions as time elapses (as would happen, for instance, in applications involving image or video sequences).

12.2 General Decision Statistics

In order to describe the social machine learning problem, it is necessary to introduce a framework and some notation to deal with classification under general decision statistics. For clarity of presentation, in this chapter the set of hypotheses will be represented as follows:

$$\Theta = \{\theta_1, \theta_2, \ldots, \theta_H\}. \tag{12.4}$$

We start with a useful observation summarized in the next lemma, where we show that, in general, social learning algorithms do not require knowledge of the individual likelihood models, but only of likelihood *ratios* relative to an arbitrary hypothesis.

Since in the following treatment we will deal with both the nonadaptive and adaptive social learning updates, it is convenient to treat them in a unified manner. To this end, recall that the adaptive update strategy in listing (8.13) is given by, for $0 < \delta < 1$,

$$\psi_{k,t}(\theta) = \frac{\mu_{k,t-1}^{1-\delta}(\theta)\ell_k(x_{k,t}|\theta)}{\sum\limits_{\theta' \in \Theta} \mu_{k,t-1}^{1-\delta}(\theta')\ell_k(x_{k,t}|\theta')}. \tag{12.5}$$

If we set $\delta = 0$ in (12.5), we recover the classic nonadaptive Bayesian update in listing (3.16). Accordingly, in the following we will use (12.5) with $0 < \delta < 1$ when we refer to the adaptive strategy, and with $\delta = 0$ for the nonadaptive strategy.

Lemma 12.1 (Sufficiency of log likelihood ratios). Assume that $\ell_k(x_{k,t}|\theta) > 0$ for all $\theta \in \Theta$ and consider the belief update (12.5) for $0 \le \delta < 1$. Then, this update requires only knowledge of the log likelihood ratios $\log \frac{\ell_k(x_{k,t}|\theta)}{\ell_k(x_{k,t}|\theta_H)}$. The choice of θ_H in the denominator is customary, and the same result continues to hold if θ_H is replaced by any other hypothesis belonging to Θ.

Proof. If we divide the numerator and denominator in (12.5) by the term $\ell_k(x_{k,t}|\theta_H)$, we obtain

$$\psi_{k,t}(\theta) = \frac{\mu_{k,t-1}^{1-\delta}(\theta)\dfrac{\ell_k(x_{k,t}|\theta)}{\ell_k(x_{k,t}|\theta_H)}}{\sum\limits_{\theta' \in \Theta} \mu_{k,t-1}^{1-\delta}(\theta')\dfrac{\ell_k(x_{k,t}|\theta')}{\ell_k(x_{k,t}|\theta_H)}}$$

$$= \frac{\mu_{k,t-1}^{1-\delta}(\theta)\exp\left\{\log \dfrac{\ell_k(x_{k,t}|\theta)}{\ell_k(x_{k,t}|\theta_H)}\right\}}{\sum\limits_{\theta' \in \Theta} \mu_{k,t-1}^{1-\delta}(\theta')\exp\left\{\log \dfrac{\ell_k(x_{k,t}|\theta')}{\ell_k(x_{k,t}|\theta_H)}\right\}}, \tag{12.6}$$

implying that the intermediate belief $\psi_{k,t}(\theta)$ can be computed only from knowledge of the (log) likelihood ratios appearing in the numerator and denominator. ■

To avoid confusion, observe that in previous chapters (e.g., in (6.3)) we used likelihood ratios taken with respect to the target hypothesis $\vartheta^\star$. This hypothesis is obviously unknown at the design stage, so that the social learning algorithms cannot depend on it in their computations; we only use it in our analytical developments to carry out performance analysis. In contrast, in Lemma 12.1 we take likelihood ratios with respect to a reference hypothesis that has no special meaning; it is set to θ_H for concreteness, but can be any arbitrary hypothesis. As a result, these ratios can be used by the agents during the implementation of the social learning algorithms. Note also that, under condition iv) from Assumption 12.1, the numerator or denominator of the likelihood ratios are nonzero with probability 1, and, hence, the log likelihood ratios are well defined.

The next step to specify the SML procedure is to generalize the social learning strategy by replacing the exact, unknown log likelihood ratios with general decision statistics learned during the training phase.

Accordingly, in place of (12.6), the SML algorithm will compute the intermediate belief $\psi_{k,t}(\theta)$ through the following update step:

$$\psi_{k,t}(\theta) = \frac{\mu_{k,t-1}^{1-\delta}(\theta) e^{h_k(x_{k,t};\theta)}}{\sum\limits_{\theta' \in \Theta} \mu_{k,t-1}^{1-\delta}(\theta') e^{h_k(x_{k,t};\theta')}}, \tag{12.7}$$

where the exact log likelihood ratio $\log \frac{\ell_k(x|\theta)}{\ell_k(x|\theta_H)}$ is replaced by a general decision statistic denoted by $h_k(x;\theta)$. As we explain later, the agents select some optimized decision statistic by learning from training data. For convenience, we set by definition

$$h_k(x;\theta_H) = 0 \quad \forall x \in \mathcal{X}_k. \tag{12.8}$$

Listing (12.11) describes the social learning algorithm that results from using the general decision statistics $h_k(x;\theta)$ in place of the true log likelihood ratios.

For later use, it is convenient to collect the functions $h_k(x;\theta)$ for all $\theta \neq \theta_H$ into a vector-valued function

$$h_k : \mathcal{X}_k \mapsto \mathbb{R}^{H-1}, \tag{12.9}$$

namely,

$$h_k(x) = [h_k(x;\theta_1), h_k(x;\theta_2), \ldots, h_k(x;\theta_{H-1})]. \tag{12.10}$$

For brevity, sometimes we will simply refer to $h_k(x)$ as decision statistic or function.

Social learning with general decision statistics $h_k(x;\theta)$

start from the prior belief vectors $\mu_{k,0}$ for $k = 1, 2, \ldots, K$
choose $\delta = 0$ for a nonadaptive update and $0 < \delta < 1$ otherwise
for $t = 1, 2, \ldots$
  for $k = 1, 2, \ldots, K$
    agent k observes $x_{k,t}$
    for $\theta = \theta_1, \theta_2, \ldots, \theta_H$

$$\psi_{k,t}(\theta) = \frac{\mu_{k,t-1}^{1-\delta}(\theta) e^{h_k(x_{k,t};\theta)}}{\sum_{\theta' \in \Theta} \mu_{k,t-1}^{1-\delta}(\theta') e^{h_k(x_{k,t};\theta')}} \quad \text{(self-learning)}$$

    end
  end

  for $k = 1, 2, \ldots, K$
    for $\theta = \theta_1, \theta_2, \ldots, \theta_H$

$$\mu_{k,t}(\theta) = \frac{\prod_{j \in \mathcal{N}_k} [\psi_{j,t}(\theta)]^{a_{jk}}}{\sum_{\theta' \in \Theta} \prod_{j \in \mathcal{N}_k} [\psi_{j,t}(\theta')]^{a_{jk}}} \quad \text{(cooperation)}$$

    end
  end
end

(12.11)

12.2.1 Conditions for Consistent Learning

Recall first that for the nonadaptive implementation ($\delta = 0$), consistency means that the belief of any agent places unit mass on the true hypothesis ϑ^o, almost surely as $t \to \infty$; for the adaptive implementation ($0 < \delta < 1$), consistency refers to the fact that, at any agent, the steady-state belief about the true hypothesis converges in probability to 1 as the adaptation parameter δ approaches zero.

In order to examine under which conditions the functions $h_k(x; \theta)$ achieve consistent learning, we start with the nonadaptive setting. For the case where the likelihood models are known, Corollary 5.1 establishes that the belief about the true hypothesis converges to 1. To establish this result we assumed finite KL divergences between the true model and the likelihood models (see Assumption 5.3) and positivity of the *network average of KL divergences* (see Assumption 5.4). The former condition can be written as

$$\mathbb{E}_{\ell_{k,\vartheta^o}} \log \frac{\ell_k(\boldsymbol{x}_{k,t}|\vartheta^o)}{\ell_k(\boldsymbol{x}_{k,t}|\theta)} < \infty, \tag{12.12}$$

which is verified in view of Assumption 12.1, point iv).

The latter condition can be written, for all $\theta \neq \vartheta^o$, as

$$\sum_{k=1}^{K} v_k \, \mathbb{E}_{\ell_{k,\vartheta^o}} \log \frac{\ell_k(\boldsymbol{x}_{k,t}|\vartheta^o)}{\ell_k(\boldsymbol{x}_{k,t}|\theta)} > 0, \tag{12.13}$$

where v_k is the kth entry of the Perron vector associated with the combination matrix A (the Perron vector exists because the combination matrix is irreducible in view of Assumption 12.1, point v). Condition (12.13) is verified since from Assumption 12.1, point iii), we have global identifiability.

We now observe that

$$\log \frac{\ell_k(\boldsymbol{x}_{k,t}|\vartheta^o)}{\ell_k(\boldsymbol{x}_{k,t}|\theta)} = \log \frac{\ell_k(\boldsymbol{x}_{k,t}|\vartheta^o)}{\ell_k(\boldsymbol{x}_{k,t}|\theta_H)} - \log \frac{\ell_k(\boldsymbol{x}_{k,t}|\theta)}{\ell_k(\boldsymbol{x}_{k,t}|\theta_H)} \tag{12.14}$$

and the proof of Theorem 5.1 remains unaltered if we replace the log likelihood ratios

$$\log \frac{\ell_k(\boldsymbol{x}_{k,t}|\theta)}{\ell_k(\boldsymbol{x}_{k,t}|\theta_H)} \tag{12.15}$$

with a general decision statistic

$$h_k(\boldsymbol{x}_{k,t}; \theta) \tag{12.16}$$

for all $\theta \in \Theta$. Along with this replacement, we also need to rephrase accordingly conditions (12.12) and (12.13). Specifically, it is sufficient to substitute condition (12.12) with the assumption that $h_k(\boldsymbol{x};\theta)$ has finite mean under ℓ_{k,ϑ^o}. Likewise, in view of (12.14), condition (12.13) becomes

$$\sum_{k=1}^{K} v_k \, \mathbb{E}_{\ell_{k,\vartheta^o}} \left[h_k(\boldsymbol{x}_{k,t};\vartheta^o) - h_k(\boldsymbol{x}_{k,t};\theta)\right] > 0 \quad \forall \theta \neq \vartheta^o. \tag{12.17}$$

Applying the same argument to Corollary 9.2 and considering the adaptive social learning strategy with general decision statistics $h_k(x;\theta)$, we conclude that under the same condition (12.17), the steady-state belief about the true hypothesis converges to 1 as the adaptation parameter vanishes. These results are summarized in the next lemma without proof.

Lemma 12.2 (Consistent learning under general decision statistics). Let Assumptions 5.1 and 12.1 be satisfied and let v be the Perron vector associated with the combination matrix A. Consider, for $k = 1, 2, \ldots, K$, the vector-valued decision statistic $h_k(x)$ defined by (12.10), along with condition (12.8). Assume that, for all $\theta \in \Theta$, the mean of $h_k(\boldsymbol{x}_{k,t};\theta)$ is finite. If the following condition is satisfied:

$$\sum_{k=1}^{K} v_k \, \mathbb{E}_{\ell_{k,\vartheta^o}} \left[h_k(\boldsymbol{x}_{k,t};\vartheta^o) - h_k(\boldsymbol{x}_{k,t};\theta)\right] > 0 \ \text{ for all pairs } (\vartheta^o,\theta), \ \ \theta \neq \vartheta^o, \tag{12.18}$$

then, whatever the true hypothesis ϑ^o is, consistent learning is achieved under both nonadaptive and adaptive social learning, in the following precise sense:

i) The nonadaptive social learning algorithm ($\delta = 0$) learns consistently in the sense that, for each agent, the belief about ϑ^o converges almost surely to 1 as $t \to \infty$.

ii) The adaptive social learning algorithm ($0 < \delta < 1$) learns consistently in the sense that, for each agent, the steady-state belief about ϑ^o converges in probability to 1 as $\delta \to 0$.

12.3 Training Phase

We now explain how the decision statistics $h_k(x;\theta)$ necessary to implement the social learning algorithm from listing (12.11) are selected by the agents. Since training is performed individually by each agent, we do not need to refer to a particular agent k in this section. Therefore, the subscript k will be omitted for now.

In classification problems, there are two main training paradigms. In the *generative* paradigm, the agent first learns a generative model, i.e., an approximation for the likelihood $\ell(x|\theta)$, and then constructs a posterior distribution for θ given x from this learned model. In comparison, in the *discriminative* paradigm (which we consider in our treatment), the agent learns *directly posterior probabilities.* That is, the agent constructs some posterior

$$q(\theta|x), \qquad \theta \in \Theta = \{\theta_1, \theta_2, \ldots, \theta_H\}, \tag{12.19}$$

to approximate the true (unknown) posterior $p(\theta|x)$.

From (12.3) we know that the labels in the training set are uniformly distributed, which in view of Bayes' rule implies that the *true* posterior satisfies

$$p(\theta|x) \propto \ell(x|\theta) \tag{12.20}$$

and, hence,

$$\log \frac{p(\theta|x)}{p(\theta_H|x)} = \log \frac{\ell(x|\theta)}{\ell(x|\theta_H)}. \tag{12.21}$$

Recall now that the social learning algorithm from listing (12.11) was constructed by replacing the log likelihood ratios with general decision statistics. Accordingly, if we replace the true (unknown) posterior $p(\theta|x)$ in (12.21) with its approximation $q(\theta|x)$, we obtain the following decision statistic:

$$h(x;\theta) \triangleq \log \frac{q(\theta|x)}{q(\theta_H|x)}. \tag{12.22}$$

Using (12.22), we can map $q(\theta|x)$ and $h(x;\theta)$ into each other by using the softmax expression

$$q(\theta|x) = \frac{e^{h(x;\theta)}}{\sum\limits_{\theta' \in \Theta} e^{h(x;\theta')}}, \qquad \theta \in \Theta. \tag{12.23}$$

Note that, while $q(\theta|x)$ is constrained to the interval $[0,1]$, the function $h(x;\theta)$ plays the role of a decision statistic whose domain can be the entire real axis.

In practice, the manner in which the decision statistics are learned is as follows. First, the designer chooses some admissible family of functions for $h(x;\theta)$. Then, an optimal function $\widehat{h}(x;\theta)$ is selected from this family in accordance with suitable criteria that incorporate information contained in the training data. For example, as we will see later, one could select the decision statistic that maximizes the similarity between the candidate posterior $q(\theta|x)$ and the true posterior $p(\theta|x)$. Two popular families of decision statistics are illustrated in the next examples.

Example 12.1 (Logistic multiclass regression). In logistic regression with multiple classes, we have a feature $x \in \mathbb{R}^d$, and the family of admissible functions $h(x;\theta)$, with $\theta \neq \theta_H$, is chosen to consist of linear regression functions parameterized by some weight vectors $w_\theta \in \mathbb{R}^d$:

$$h(x;\theta) = w_\theta^\mathsf{T} x, \qquad \theta \neq \theta_H. \tag{12.24}$$

An intercept or bias can be added in the regression model by extending the feature vector x to incorporate an additional unit entry. Using (12.24) in (12.23) allows us to parameterize the posterior probabilities in the following manner:

$$q(\theta|x) = \begin{cases} \dfrac{e^{w_\theta^\mathsf{T} x}}{1 + \sum\limits_{\theta' \neq \theta_H} e^{w_{\theta'}^\mathsf{T} x}} & \text{if } \theta \neq \theta_H, \\[2ex] \dfrac{1}{1 + \sum\limits_{\theta' \neq \theta_H} e^{w_{\theta'}^\mathsf{T} x}} & \text{if } \theta = \theta_H. \end{cases} \tag{12.25}$$

Example 12.2 (Multilayer perceptron). Consider a basic neural network architecture, that is, a multilayer perceptron (MLP) deployed to solve an H-ary classification problem. This architecture is illustrated in Figure 12.2. The input feature $x \in \mathbb{R}^d$ feeds the cascade of L layers, followed by the last layer that applies the softmax function (12.23) to compute the posterior probabilities from the decision statistics. Each layer l consists of n_l nodes. At each node $m = 1, 2, \ldots, n_l$ of layer $l = 2, 3, \ldots, L$, the following function $g_m^{(l)}(x)$ is implemented:

$$g_m^{(l)}(x) = \sum_{i=1}^{n_{l-1}} w_{im}^{(l)} \, \sigma_a\left(g_i^{(l-1)}(x)\right), \tag{12.26}$$

where σ_a is an *activation function.* The parameters $w_{im}^{(l)}$ correspond to the elements of a weight matrix W_l of dimension $n_{l-1} \times n_l$. For layer $l = 1$, the function implemented at node m is of the form

$$g_m^{(1)}(x) = \sum_{i=1}^{d} w_{im}^{(1)} \, x(i), \tag{12.27}$$

where $x(i)$ denotes the ith entry of x. Bias variables can be incorporated at one or more layers by adding one node at the pertinent layer and placing a dummy feature equal to 1 on this node.

The final layer, i.e., layer L, is deployed to produce the decision statistic $h(x)$. Accordingly, it has $n_L = H - 1$ nodes, with

$$g_m^{(L)}(x) = h(x;\theta_m), \qquad m = 1, 2, \ldots, H-1. \tag{12.28}$$

The final output of the classifier must be a posterior distribution $q(\theta|x)$ over the H classes, and is accordingly obtained from (12.28) by applying a softmax function, namely, by applying (12.23), with the usual convention $h(x;\theta_H) = 0$. This convention motivates the addition of the dummy input 0 in Figure 12.2.

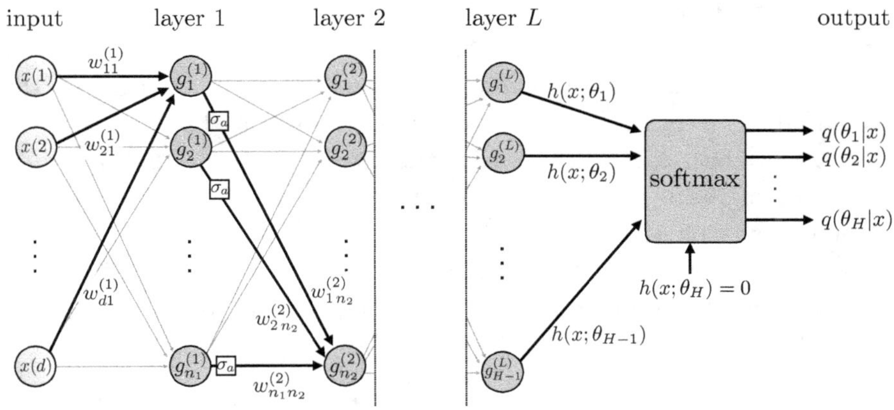

Figure 12.2: Illustration of the neural network architecture from Example 12.2.

We see that the MLP architecture is determined by the number of layers L, the number of nodes n_l for each layer $l = 1, 2, \ldots, L$, the activation function σ_a, and the number of hypotheses H. Once an architecture is chosen, the space of possible outputs of the MLP classifier is spanned by varying the matrices $\{W_l\}$ within some admissible family. Therefore, learning the final posterior or, equivalently, the final decision statistic $h(x)$, amounts to learning the matrices $\{W_l\}$ that minimize a suitable risk function, as we will see in the forthcoming sections.

Note that, compared with Example 12.1, here the family of functions allows to explore a variety of models significantly more general than a linear regression model.

The design of the classifier structure (i.e., the choice of its parameters $\{w_\theta\}$ in the logistic regression case or $\{W_l\}$ in the neural network case) is guided by the desire to minimize the "distance" between the true posterior $p(\theta|x)$ and the approximate posterior $q(\theta|x)$ (which depends on the decision statistic $h(x;\theta)$ via (12.23)). This discrepancy is usually measured by the KL divergence

$$\sum_{\theta\in\Theta} p(\theta|x) \log \frac{p(\theta|x)}{q(\theta|x)}. \tag{12.29}$$

Averaging this sum over the distribution of the feature data $\boldsymbol{x}$, we get the *conditional* KL divergence (see Definition B.6)

$$\begin{aligned} D_{\theta|x}(p||q) &= \mathbb{E} \log \frac{p(\boldsymbol{\theta}|\boldsymbol{x})}{q(\boldsymbol{\theta}|\boldsymbol{x})} \\ &= \mathbb{E} \log \frac{1}{q(\boldsymbol{\theta}|\boldsymbol{x})} - \mathbb{E} \log \frac{1}{p(\boldsymbol{\theta}|\boldsymbol{x})} \\ &= H_{\theta|x}(p, q) - H_{\theta|x}(p), \end{aligned} \tag{12.30}$$

where we introduced the conditional cross-entropy $H_{\theta|x}(p,q)$ and the conditional entropy $H_{\theta|x}(p)$ — see Definitions B.5 and B.2, respectively. We remark that the expectations are relative to all bold quantities, i.e., they are computed under the *true joint* distribution of $\boldsymbol{x}$ and $\boldsymbol{\theta}$.

A critical observation here is that the second term in (12.30), namely, the conditional entropy $H_{\theta|x}(p)$, does not depend on the classifier structure; it depends only on the true distribution, which cannot be controlled by the designer. This implies that minimizing the conditional KL divergence over q amounts to minimizing the conditional cross-entropy

$$H_{\theta|x}(p,q) = \mathbb{E}\log\frac{1}{q(\boldsymbol{\theta}|\boldsymbol{x})}. \tag{12.31}$$

Substituting (12.23) into (12.31), the conditional cross-entropy can be expressed as a function of the decision statistic $h(x,\theta)$ in the following form:

$$H_{\theta|x}(p,q) = \mathbb{E}\log\frac{\sum_{\theta'\in\Theta}\exp\left\{h(\boldsymbol{x};\theta')\right\}}{\exp\left\{h(\boldsymbol{x};\boldsymbol{\theta})\right\}}. \tag{12.32}$$

Since from now on we return to examining the training of the individual agents, we restore the subscript k. We recall that the decision statistic of agent k is denoted by $h_k(x,\theta)$. Choosing as risk function the conditional cross-entropy and exploiting (12.32), the risk function of agent k is given by

$$R_k(h_k) \triangleq \mathbb{E}\log\frac{\sum_{\theta\in\Theta}\exp\left\{h_k(\widehat{\boldsymbol{x}}_{k,n};\theta)\right\}}{\exp\left\{h_k(\widehat{\boldsymbol{x}}_{k,n};\widehat{\boldsymbol{\theta}}_{k,n})\right\}}, \tag{12.33}$$

where the expectation in (12.33) is computed over the distribution of the $(\widehat{\boldsymbol{x}}_{k,n},\widehat{\boldsymbol{\theta}}_{k,n})$ pairs belonging to the training set of agent k.[3] Since this distribution is unknown, the exact risk value is in practice replaced by the *empirical* risk function

$$\widehat{\boldsymbol{R}}_k(h_k) = \frac{1}{E_k}\sum_{n=1}^{E_k}\log\frac{\sum_{\theta\in\Theta}\exp\left\{h_k(\widehat{\boldsymbol{x}}_{k,n};\theta)\right\}}{\exp\left\{h_k(\widehat{\boldsymbol{x}}_{k,n};\widehat{\boldsymbol{\theta}}_{k,n})\right\}}. \tag{12.34}$$

That is, the expectation in (12.33) is replaced by an empirical average computed over the E_k samples available in the training set of agent k.

[3]Due to the identical distribution across the clues (i.e., across n), the risk does not depend on n, but only on the agent index k.

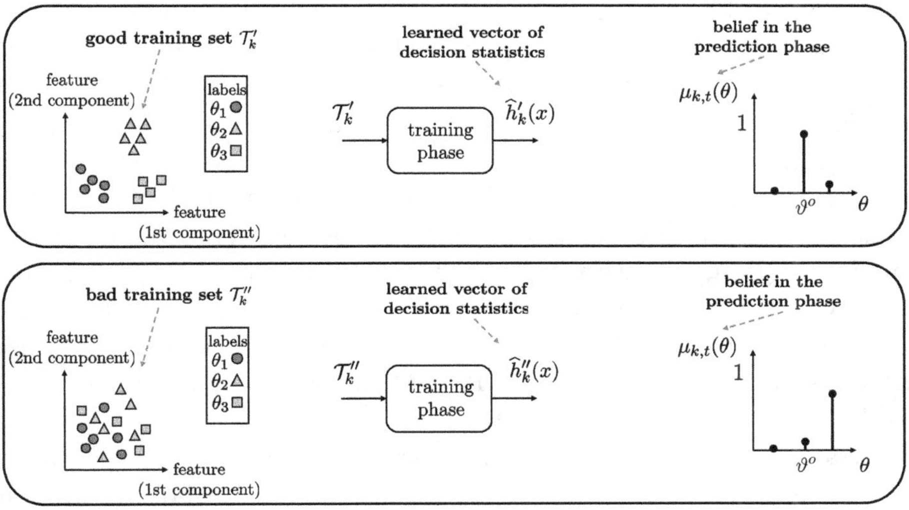

Figure 12.3: The optimized decision statistics learned during the training phase depend on the particular realization of the training set.

Note that the empirical risk function depends on the training set $\boldsymbol{\mathcal{T}}_k = \{\widehat{\boldsymbol{x}}_{k,n}, \widehat{\boldsymbol{\theta}}_{k,n}\}_{n=1}^{E_k}$, which explains the bold notation for $\widehat{\boldsymbol{R}}_k(h_k)$.

Now, during training, each agent k collects a training set $\boldsymbol{\mathcal{T}}_k$ and chooses an admissible family for the vector-valued decision statistic h_k in (12.10). After training, an optimal decision statistic $\widehat{\boldsymbol{h}}_k$ is selected from this family:

$$\boldsymbol{\mathcal{T}}_k \overset{\text{training}}{\longrightarrow} \widehat{\boldsymbol{h}}_k. \tag{12.35}$$

Note that the learned function $\widehat{\boldsymbol{h}}_k$ is written in bold since it embodies the randomness of the training set $\boldsymbol{\mathcal{T}}_k$.

One way to learn a decision statistic is by minimizing the empirical risk from (12.34):

$$\boldsymbol{h}_k^o = \arg\min_{h_k \in \mathcal{H}_k} \widehat{\boldsymbol{R}}_k(h_k), \tag{12.36}$$

where $\mathcal{H}_k$ denotes the function family where the search is performed. We can use the minimizer $\boldsymbol{h}_k^o$ as the learned decision statistic, i.e., $\widehat{\boldsymbol{h}}_k = \boldsymbol{h}_k^o$. However, this is not the only way in which a decision statistic can be learned. In some cases it is convenient to replace the risk function $\widehat{\boldsymbol{R}}_k(h_k)$ with a regularized version thereof, by adding a suitable regularization term [155]. Another possibility useful for binary classification problems, which is suggested in [29] and exploited in the next section, is to perform a de-biasing operation after minimizing the risk.

The dependence of the learned decision statistics on the training set has important implications, as illustrated in Figure 12.3. This is because, depending on the particular realization of the training sets, the learned functions $\widehat{\boldsymbol{h}}_k$ may or may not satisfy the conditions for consistent learning from Lemma 12.2. For instance, the agents may have access to some "good" realizations $\mathcal{T}_k'$ of the training sets (see the top panel of Figure 12.3), for which the learned functions $\widehat{h}_k'$ satisfy condition (12.18). Under this condition, all agents learn well and place their full belief mass on the true hypothesis ϑ^o. However, the agents may also observe some "bad" realizations $\mathcal{T}_k''$ (see the bottom panel of Figure 12.3), for which the learned functions $\widehat{h}_k''$ would not satisfy (12.18) and the prediction performance will not be satisfactory.

As a result, in the social machine learning framework, the occurrence of consistent learning during the prediction phase depends on the randomness of the agents' training sets. Therefore, a proper way to assess the learning guarantees of the system is to evaluate the *probability* of consistent learning [62, 155, 167], that is, the probability that the decision statistics produced at the end of the training phase allow all agents to classify consistently the underlying hypothesis through the algorithm in listing (12.11). According to Lemma 12.2, this probability of consistent learning can be formulated as

$$P_c \triangleq \mathbb{P}\left[\text{the functions } \left\{\widehat{\boldsymbol{h}}_k\right\}_{k=1}^{K} \text{ satisfy (12.18)}\right], \tag{12.37}$$

where the probability is computed with respect to the randomness in the training sets.

12.4 Performance Guarantees

In this section we characterize the performance of the social machine learning strategy for *binary* classification problems, for which $H = 2$. Preliminarily, we introduce a convenient representation for the risk functions in the binary case, and four useful quantities: the target risks, the complexity of the decision statistics, a descriptor quantifying the role of the training set sizes, and the de-biased decision statistics.

Risk representation in the binary case. When $H = 2$, the vector-valued decision statistic $h_k(x)$ becomes a scalar, namely, we have

$$h_k(x) = h_k(x;\theta_1) \qquad \text{[binary case, } H = 2]. \tag{12.38}$$

Using (12.22), the relation between the function $h_k(x)$ and the posterior $q_k(\theta|x)$ of agent k is

$$h_k(x) = \log \frac{q_k(\theta_1|x)}{1 - q_k(\theta_1|x)}. \tag{12.39}$$

The function $h_k(x)$ written in the form (12.39) is often referred to as the *logit statistic.* For convenience, we adopt the convention

$$\theta_1 = +1, \qquad \theta_2 = -1, \tag{12.40}$$

and from (12.39) we conclude that

$$q_k(+1|x) = \frac{e^{h_k(x)}}{e^{h_k(x)} + 1} = \frac{1}{1 + e^{-h_k(x)}}, \tag{12.41}$$

$$q_k(-1|x) = 1 - q_k(+1|x) = \frac{1}{1 + e^{h_k(x)}}. \tag{12.42}$$

These two relations can be combined into the following single equation, for $\theta \in \{+1, -1\}$:

$$q_k(\theta|x) = \frac{1}{1 + e^{-\theta\, h_k(x)}}. \tag{12.43}$$

Using (12.43) in (12.33), we find that the cross-entropy risk of agent k reduces to

$$R_k(h_k) = \mathbb{E} \log\left(1 + \exp\left\{-\widehat{\boldsymbol{\theta}}_{k,n}\, h_k(\widehat{\boldsymbol{x}}_{k,n})\right\}\right), \tag{12.44}$$

where the expectation is computed over the distribution characterizing the iid (feature, label) pairs $(\widehat{\boldsymbol{x}}_{k,n}, \widehat{\boldsymbol{\theta}}_{k,n})$ in the training set. The corresponding *empirical* risk (12.34) becomes

$$\widehat{\boldsymbol{R}}_k(h_k) = \frac{1}{E_k} \sum_{n=1}^{E_k} \log\left(1 + \exp\left\{-\widehat{\boldsymbol{\theta}}_{k,n}\, h_k(\widehat{\boldsymbol{x}}_{k,n})\right\}\right). \tag{12.45}$$

Target risks. As happens in classic statistical learning frameworks (e.g., in the Vapnik-Chervonenkis theory), the interplay between empirical and exact risks is critical to ascertain the learning and prediction ability of the classifiers [62, 155, 167]. In particular, one summary descriptor is the *target* risk, which is defined as the infimum of the exact risk over all possible decision statistics. However, differently from what is obtained in classic statistical learning theory, our results will depend on the *graph* properties. In particular, a major role will be played by *weighted combinations of the* *individual risks.* The combination weights turn out to be the entries of the Perron vector associated with the combination matrix that governs the social learning interactions between the agents. This property leads to phenomena that are not observed in traditional machine learning. For example, consistent classification can be achieved even if some of the agents learn bad models, but the plurality of the agents is able to reach a satisfying *aggregate* risk value. The next definition introduces the target risks of every agent k and the aggregate target risk of the entire network.

Definition 12.1 (Target risks). Given a family $\mathcal{H}_k$ from which agent k can pick its decision statistic h_k, we introduce the individual *target* risk

$$\mathsf{R}_k^o \triangleq \inf_{h_k \in \mathcal{H}_k} R_k(h_k) \tag{12.46}$$

and the *network* target risk

$$\mathsf{R}_{\mathsf{net}}^o \triangleq \sum_{k=1}^{K} v_k \mathsf{R}_k^o, \tag{12.47}$$

where v is the Perron vector associated with the combination matrix A.

We will assume that the network target risk $\mathsf{R}_{\mathsf{net}}^o$ is strictly smaller than $\log 2$. This condition is in a sense the counterpart of global identifiability in terms of risk functions. To understand why, consider the following *uninformative* posterior at agent k:

$$q_k(\theta|x) = \frac{1}{2} \quad \forall x \in \mathcal{X}_k, \quad \forall \theta \in \Theta. \tag{12.48}$$

In this case we have $R_k(h_k) = \log 2$, since the cross-entropy between *any* pmf and a binary uniform pmf is equal to $\log 2$, as can be immediately verified from Definition B.3. A posterior in the form (12.48) would not be useful for classification, since it corresponds to randomly assigning labels $+1$ and -1 with equal probability. Situations of this type occur in practice when the features do not carry information about the labels, or the classifier structure is not complex enough to address the classification task at hand. Requiring $\mathsf{R}_k^o < \log 2$ rules out the possibility that the risk is minimized by uninformative decision statistics of this type. Requiring the *network* target risk to satisfy $\mathsf{R}_{\mathsf{net}}^o < \log 2$ is a weaker assumption, since it imposes this bound on the risk values averaged over the graph. For example, the global condition $\mathsf{R}_{\mathsf{net}}^o < \log 2$ can be achieved even if $K-1$

uninformed agents have target risks equal to $\log 2$, while one informed agent k fulfills the inequality $\mathsf{R}_k^o < \log 2$.

Complexity of the decision statistics. The complexity of the decision structure, namely, of the family $\mathcal{H}_k$ of decision statistics h_k, will be seen to play an important role in the performance of the SML strategy. This complexity will be quantified through a statistical descriptor called *Rademacher complexity*, introduced in Definition G.1. Specifically, we will denote by ρ_k the Rademacher complexity associated with the kth agent, and by

$$\rho_{\mathsf{net}} \triangleq \sum_{k=1}^{K} v_k \rho_k \tag{12.49}$$

the *network* Rademacher complexity obtained as an average of the individual Rademacher complexities, weighted by the Perron vector entries.

Training set sizes. Assume that all agents have at least one clue in their training set, i.e., $E_k > 0$ for all k, and define the ratios

$$e_k \triangleq \frac{E_{\mathsf{max}}}{E_k}, \tag{12.50}$$

with

$$E_{\mathsf{max}} \triangleq \max_{k\in\{1,2,\ldots,K\}} E_k. \tag{12.51}$$

The *individual imbalance penalty* e_k quantifies the dissimilarity between the number of training samples of agent k and the maximum number of training samples.

De-biased decision statistics. To prove our consistency result, we will construct the decision statistics with a two-step procedure. First, we will minimize the empirical risk $\widehat{\boldsymbol{R}}_k(h_k)$ to obtain an intermediate function $\boldsymbol{h}_k^o$. Then we will obtain a decision statistic $\widetilde{\boldsymbol{h}}_k$ through a de-biasing operation that subtracts from $\boldsymbol{h}_k^o$ its empirical average computed over the training set. This operation is useful to favor consistency in the binary case, as we explain in Appendix 12.A.

Definition 12.2 (De-biased decision statistics). The learned decision statistic is computed as follows. First, an intermediate decision statistic $\boldsymbol{h}_k^o(x)$ is obtained by minimizing the empirical risk:

$$\boldsymbol{h}_k^o = \arg\min_{h_k \in \mathcal{H}_k} \widehat{\boldsymbol{R}}_k(h_k). \tag{12.52}$$

Then, a de-biased decision statistic $\widetilde{\boldsymbol{h}}_k(x)$ is computed by subtracting the empirical average:

$$\widetilde{\boldsymbol{h}}_k(x) = \boldsymbol{h}_k^o(x) - \frac{1}{E_k}\sum_{n=1}^{E_k} \boldsymbol{h}_k^o(\boldsymbol{x}_{k,n}). \tag{12.53}$$

The learned decision statistic is then chosen as $\widehat{\boldsymbol{h}}_k(x) = \widetilde{\boldsymbol{h}}_k(x)$.

12.4.1 Consistency with High Probability

The next theorem characterizes the consistency of the SML strategy in terms of a lower bound on the probability of consistent learning (12.37).

Theorem 12.1 (SML consistency). Let Assumptions 5.1, 12.1, and 12.2 be satisfied. For $k = 1, 2, \ldots, K$, consider a family $\mathcal{H}_k$ of bounded functions $h_k : \mathcal{X}_k \mapsto \mathbb{R}$:

$$|h_k(x)| \le h_{k,\mathsf{max}} \quad \forall x \in \mathcal{X}_k, \quad \text{with } 0 < h_{k,\mathsf{max}} < \infty, \tag{12.54}$$

and assume that each agent k employs as learned decision statistic $\widehat{\boldsymbol{h}}_k(x)$ the de-biased[4] decision statistic $\widetilde{\boldsymbol{h}}_k(x)$ introduced in Definition 12.2. Assume that the network target risk from (12.47) fulfills the inequality $\mathsf{R}_{\mathsf{net}}^o < \log 2$ and that the network Rademacher complexity from (12.49) is bounded as

$$\rho_{\mathsf{net}} < \mathscr{E}(\mathsf{R}_{\mathsf{net}}^o), \tag{12.55}$$

where the function $\mathscr{E}(\mathsf{R}_{\mathsf{net}}^o)$ is computed exactly in Appendix 12.C (see (12.146)) and can be approximated as (see Figure 12.10)

$$\mathscr{E}(\mathsf{R}_{\mathsf{net}}^o) \approx 0.1406\left(1 - \frac{\mathsf{R}_{\mathsf{net}}^o}{\log 2}\right). \tag{12.56}$$

Then, we have the following lower bound for the probability of consistent learning defined in (12.37):

$$P_c \ge 1 - 2\exp\left\{-2\,E_{\mathsf{max}}\left(\frac{\mathscr{E}(\mathsf{R}_{\mathsf{net}}^o) - \rho_{\mathsf{net}}}{h_{\mathsf{net}}}\right)^2\right\}, \tag{12.57}$$

where the parameter

$$h_{\mathsf{net}} \triangleq \sum_{k=1}^{K} v_k\, e_k\, h_{k,\mathsf{max}} \tag{12.58}$$

[4]Note that the de-biased function in (12.53) satisfies the looser constraint $\left|\widetilde{\boldsymbol{h}}_k(x)\right| \le 2\,h_{k,\mathsf{max}}$ and, hence, the final family to which the learned functions $\widetilde{\boldsymbol{h}}_k(x)$ belong is different from the original family $\mathcal{H}_k$ over which the risk was minimized.

globally accounts for the graph structure (through the Perron vector entries $\{v_k\}$), the imbalance penalties $\{e_k\}$ from (12.50), and the function families $\mathcal{H}_k$ of the individual agents (through the bounding constants $\{h_{k,\mathsf{max}}\}$).

Proof. See Appendix 12.C. ∎

Theorem 12.1 reveals that, if the network Rademacher complexity ρ_{net} is smaller than $\mathscr{E}(\mathsf{R}^o_{\mathsf{net}})$, then the probability of consistent learning converges to 1 exponentially with the number of training samples, i.e., as $E_{\mathsf{max}} \to \infty$ (with the proportion between E_{max} and E_k kept fixed, i.e., e_k kept constant). The exponent ruling this convergence (actually, the convergence of the bound in (12.57)) is

$$2\left(\frac{\mathscr{E}(\mathsf{R}^o_{\mathsf{net}}) - \rho_{\mathsf{net}}}{h_{\mathsf{net}}}\right)^2. \tag{12.59}$$

Larger values of this exponent are preferable, since they imply that the probability of consistent learning converges faster.

We see from (12.59) that three main factors determine how fast the probability of consistent learning approaches 1, namely, $\mathscr{E}(\mathsf{R}^o_{\mathsf{net}})$, ρ_{net}, and h_{net}. Let us examine their meaning separately.

Term $\mathscr{E}(\mathsf{R}^o_{\mathsf{net}})$ is a function of the network target risk $\mathsf{R}^o_{\mathsf{net}}$ — see the definition in (12.146). A good approximation for this function is given by (12.56), which reveals that $\mathscr{E}(\mathsf{R}^o_{\mathsf{net}})$ quantifies the difference between $\mathsf{R}^o_{\mathsf{net}}$ and the value $\log 2$. As already discussed, the value $\log 2$ corresponds to a "blind" decision system that classifies the observed features by randomly assigning labels $+1$ and -1 with equal probability. The closer the network target risk is to $\log 2$, the smaller the value of $\mathscr{E}(\mathsf{R}^o_{\mathsf{net}})$ will be. In other words, smaller values of $\mathscr{E}(\mathsf{R}^o_{\mathsf{net}})$ are symptomatic of more difficult classification problems. In fact, the error exponent in (12.59) decreases when $\mathscr{E}(\mathsf{R}^o_{\mathsf{net}})$ decreases. More precisely, what matters is *the difference* between $\mathscr{E}(\mathsf{R}^o_{\mathsf{net}})$ and the network Rademacher complexity ρ_{net}. Therefore, Eq. (12.59) reveals a remarkable interplay between the inherent difficulty of the classification problem, quantified inversely by $\mathscr{E}(\mathsf{R}^o_{\mathsf{net}})$, and the complexity of the decision statistics, quantified by ρ_{net}. Ideally, we would like to have simple classification problems (i.e., high values of $\mathscr{E}(\mathsf{R}^o_{\mathsf{net}})$) and low classifier complexity (i.e., low values of ρ_{net}). Notably, both indices are *network* indices, that is, they embody the graph structure.

The third parameter appearing in (12.59) is the constant h_{net} defined by (12.58), which consists of a weighted average (with weights given by the Perron vector entries v_k) of the product $h_{k,\text{max}}\, e_k$. Constant $h_{k,\text{max}}$ is a bound on the admissible values for the decision statistic of agent k.[5] Thus, this constant reflects the "breadth" of the decision-statistic family employed by agent k. Constant e_k quantifies the relative level of "ignorance" of agent k, in the sense that agents with a small number of training examples E_k with respect to the maximum number E_{max} exhibit large values of e_k. Accordingly, the product $h_{k,\text{max}}\, e_k$ is another measure of the complexity of the classification problem, in terms of the family of decision statistics and the training set sizes. The role of v_k, as usual, is to obtain a network index where the contribution of the individual agents is weighted according to their centrality in the network. As a result, the global constant h_{net} is an average measure of complexity across the network. Accordingly, we see from (12.59) that large values of h_{net} reduce the exponent, i.e., they slow down the convergence of the probability of consistent learning to 1.

12.5 Sample Complexity

It is useful to evaluate the sample complexity of the SML strategy, namely, how many training examples are sufficient to achieve a desired value for the probability of consistent learning. To this end, we can exploit (12.57). However, it is necessary to account for the fact that the quantity ρ_{net} itself depends on the number of training examples.

For typical families of decision functions, the Rademacher complexity ρ_k is upper bounded by $C_k/\sqrt{E_k}$ for some positive constant C_k [137]. One popular structure satisfying this property is a norm-constrained MLP, as we show in Lemma G.2.

Now, assuming that $\rho_k \le C_k/\sqrt{E_k}$, the network Rademacher complexity from (12.49) will be bounded as

$$\rho_{\text{net}} \le \sum_{k=1}^{K} v_k \frac{C_k}{\sqrt{E_k}} = \frac{1}{\sqrt{E_{\text{max}}}} \underbrace{\sum_{k=1}^{K} v_k C_k \sqrt{e_k}}_{\triangleq C_{\text{net}}}, \tag{12.60}$$

[5] The assumption of bounded decision statistics is met in several relevant cases. For instance, consider the logistic regression formulation from Example 12.1. In many practical applications, the values that the feature x can take are bounded. In this case, as seen from (12.24), the decision statistic is bounded if the weight vectors w_θ are bounded. Similarly, with bounded features, the multilayer perceptron from Example 12.2 meets the condition $|h_k(x)| \le h_{k,\text{max}}$ for norm-constrained neural networks [137], where the weight matrices W_l are bounded.

where in the last step we used (12.50). The global constant C_{net} mixes the individual complexity constants C_k, the Perron vector entries v_k, and the imbalance penalties e_k. Assuming that $(C_{\mathsf{net}}/\sqrt{E_{\mathsf{max}}}) < \mathscr{E}(\mathsf{R}^o_{\mathsf{net}})$ to guarantee condition (12.55), and substituting (12.60) into (12.57), we obtain the bound

$$P_c \geq 1 - 2\exp\left\{-\frac{2E_{\mathsf{max}}}{h^2_{\mathsf{net}}}\left(\mathscr{E}(\mathsf{R}^o_{\mathsf{net}}) - \frac{C_{\mathsf{net}}}{\sqrt{E_{\mathsf{max}}}}\right)^2\right\}. \tag{12.61}$$

which can be used to carry out a sample-complexity analysis of the SML strategy, as stated in the forthcoming theorem.

Theorem 12.2 (SML sample complexity). Let the same assumptions used in Theorem 12.1 be satisfied. Assume, for $k = 1, 2, \ldots, K$, that $\rho_k \leq C_k/\sqrt{E_k}$ for some constants $C_k > 0$ and let

$$C_{\mathsf{net}} \triangleq \sum_{k=1}^{K} v_k C_k \sqrt{e_k}. \tag{12.62}$$

If the maximum number of training samples across the agents satisfies the condition

$$E_{\mathsf{max}} \geq \left(\frac{C_{\mathsf{net}}}{\mathscr{E}(\mathsf{R}^o_{\mathsf{net}})}\right)^2 \left(1 + \frac{h_{\mathsf{net}}}{C_{\mathsf{net}}}\sqrt{\frac{1}{2}\log\left(\frac{2}{\varepsilon}\right)}\right)^2, \tag{12.63}$$

then consistent learning takes place with probability at least $1 - \varepsilon$.

Proof. See Appendix 12.D. ∎

We now examine how the relevant system parameters appearing in (12.63) influence the sample complexity.

Target performance. The desired probability of consistent learning, $1-\varepsilon$, influences the bound in (12.63) through the logarithmic term $\log(2/\varepsilon)$ and, hence, has a mild effect on the number of training samples.

Imbalance penalties. The imbalance penalty e_k appearing in the global parameter h_{net} in (12.58) quantifies how far the individual agent k is from the maximum size E_{max}. Larger values for e_k imply that agent k has less training data, and thus require that E_{max} be increased to compensate for this deficiency.

Decision statistic bounds. The term $h_{k,\mathsf{max}}$ corresponds to the bound on the output of the decision statistic $h_k(x)$ and, hence, other conditions being equal, increasing $h_{k,\mathsf{max}}$ corresponds to increasing the possible functions to choose from. Accordingly, from (12.63) we see that the larger $h_{k,\mathsf{max}}$ is, the larger the number of training samples must be to guarantee a probability of consistent learning $P_c \geq 1 - \varepsilon$.

Term C_{net}. The constant C_{net} quantifies the average complexity of the decision statistics across the network. The number of training samples grows quadratically with C_{net}.

Term $\mathscr{E}(\mathsf{R}^o_{\mathsf{net}})$. As explained before, the term $\mathscr{E}(\mathsf{R}^o_{\mathsf{net}})$ quantifies (inversely) the difficulty of the classification problem. Smaller values of $\mathscr{E}(\mathsf{R}^o_{\mathsf{net}})$ are representative of more difficult classification problems, and accordingly necessitate the use of more training samples.

Role of the network. Given the networked nature of our inference problem, it is expected that the network structure plays a significant role in the results of Theorems 12.1 and 12.2. The network influence is captured through the terms $\mathsf{R}^o_{\mathsf{net}}$, ρ_{net}, and h_{net} appearing in (12.57). All these terms contain the Perron vector entries v_k.

The Perron vector entries reveal the influence of each agent. For example, we see from (12.47) that an agent k with higher weight v_k has more power to steer the value of the network target risk $\mathsf{R}^o_{\mathsf{net}}$ toward its own private target risk R^o_k. We recall that v_k is an index of the centrality of agent k — see the discussion following Theorem 4.4. In the previous chapters, we have already observed how the agent centrality plays a role in social learning. For example, in traditional social learning (Chapter 5) we encountered the network average of KL divergences $D_{\mathsf{net}}(\theta)$ defined in Table 6.1. Likewise, to characterize the performance of both traditional (Chapter 6) and adaptive (Chapter 9) strategies, we worked with statistical descriptors (e.g., the covariance matrix or the logarithmic moment generating function) of the network average of log likelihood ratios $\boldsymbol{\lambda}_{\mathsf{net},t}(\theta)$ — see again Table 6.1. Notably, the dependence on the graph structure is generally not found in the literature on statistical bounds for ensembles of classifiers [30, 50], while we see that in social machine learning the graph (in particular, the Perron vector) matters.

12.6 Illustrative Examples

In this section we illustrate the application of social machine learning to practical classification problems, and compare it against a traditional learning approach employed to aggregate multiple classifiers.

Example 12.3 (MNIST dataset). We consider the MNIST dataset [108], which contains several realizations of images representing digits $0, 1, \ldots, 9$. We focus on the first two digits, and build a binary classification problem aimed at distinguishing digits 0 and 1. In terms of our notation, we have a hypothesis $\theta \in \Theta = \{+1, -1\}$, where we map digit 1 into hypothesis $\theta = +1$ and digit 0 into $\theta = -1$. We employ a network of $K = 9$ spatially distributed agents, where each agent observes only a part of the image (see Figure 12.4). These agents wish to collaborate and discover which digit corresponds to the image they are collectively observing.

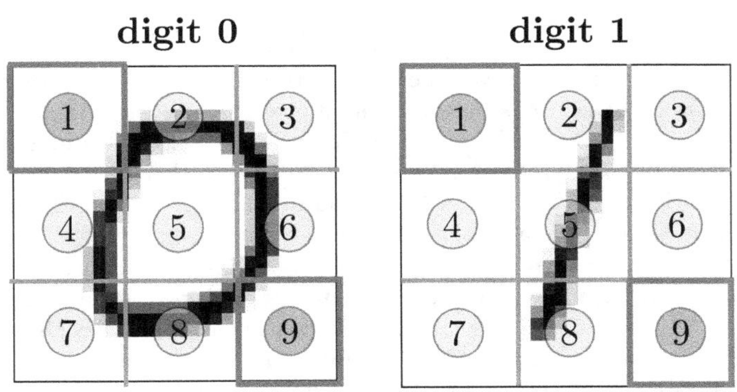

Figure 12.4: Each fraction of the image is observed by a different agent. Agents 1 and 9, highlighted in blue and purple, respectively, correspond to the least informed agents.

As we can see in Figure 12.4, different agents will observe data with different levels of informativeness, e.g., agents 1 and 9 will dispose of little or no information, within their attributed image patch, to distinguish digits 0 and 1. To overcome this lack of local information, the agents are allowed to cooperate by interacting over a network. Specifically, they are linked according to the strong undirected graph in Figure 12.5 (all nodes have a self-loop, not shown in the figure), and equipped with a combination matrix generated using the uniform-averaging rule — see Table 4.1.

In the training phase, each agent is trained independently over a balanced set of 200 labeled images, using an MLP (see Example 12.2) with activation function $\sigma_a = \tanh$ and $L = 2$ layers. The first layer has $n_1 = 64$ nodes. The second layer has $n_2 = 1$ node, conforming with the binary classification problem.

To minimize the empirical risk from (12.45), we use a mini-batch stochastic gradient algorithm over multiple epochs (also called runs) [155]. Specifically, we consider a batch size equal to 10, a learning rate equal to 0.001, and 30 epochs. At each iteration of the algorithm, the samples belonging to the batch are randomly selected.

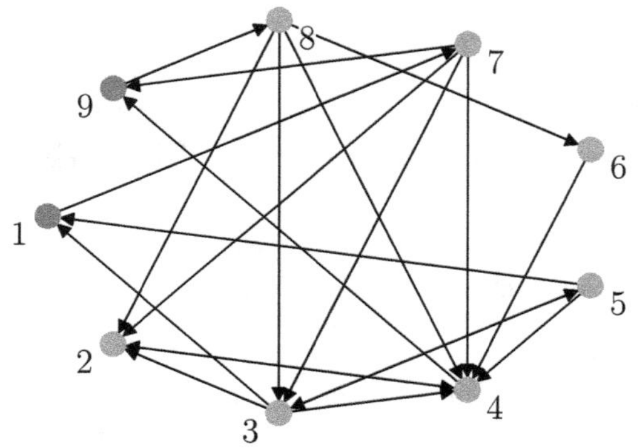

Figure 12.5: Network topology used in Example 12.3. The graph is undirected and all agents are assumed to have a self-loop (not shown in the figure).

The evolution over the training epochs of the empirical risk for each agent is shown in the left panel of Figure 12.6. We repeated the process over 5 training sessions. Each risk curve shown in the figure is obtained as an average over the training sessions. As expected, we see that classifiers 1 and 9 exhibit the least reliable training performance; their empirical risks are indeed higher than the risks of the other agents, and also exhibit a higher variability across the epochs. This could be problematic if these agents were to solve the classification problem on their own, but we will see that their individual poor classification performance is mitigated when collaborating within the network.

At the end of the training phase, each agent k is equipped with a learned decision statistic, in the de-biased form — see (12.53). Then, in the prediction phase, the agents observe unlabeled images over time. The nature of the images changes every 1000 time instants. Specifically, the agents observe images representing digit 0 for $t \in [1, 1000]$, and digit 1 for $t \in [1001, 2000]$. Then, from instant $t = 2001$ the images switch back to digits 0, and so on.

We implement the social learning strategy (12.11) in its adaptive version, i.e., with nonzero adaptation parameter δ. Specifically, we set $\delta = 0.01$. In the right panel of Figure 12.6, we display the evolution over time of the log belief ratio of agent 9, $\log \frac{\mu_{9,t}(+1)}{\mu_{9,t}(-1)}$. According to the MAP criterion, each agent k chooses the hypothesis that maximizes its belief, which is tantamount to saying that agent k opts for $\theta = +1$ (i.e., digit 1) or $\theta = -1$ (i.e., digit 0) depending on whether the log belief ratio stays above or below 0 (the dashed line in the right panel of Figure 12.6). The instantaneous decision of each agent k at time t can be represented as

$$\theta_{k,t}^{\mathsf{SML}} = \text{sign}\left(\log \frac{\mu_{k,t}(+1)}{\mu_{k,t}(-1)}\right). \tag{12.64}$$

We see how, despite the limited information available during training, agent 1 is able to clearly distinguish digits 0 and 1.

Example 12.4 (Comparison with AdaBoost). We compare the performance of the social machine learning strategy (12.11) with a classic strategy to aggregate multiple

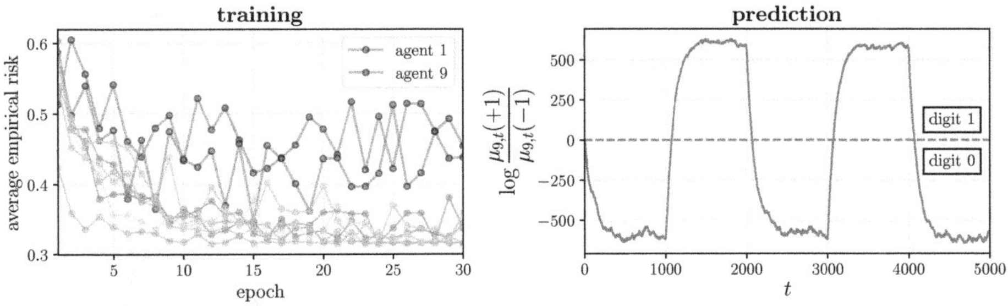

Figure 12.6: Training and prediction phases of the SML strategy, under the setting described in Example 12.3. (*Left*) Evolution over the training epochs of the empirical risk of all agents. Each curve is obtained by averaging the risk over 5 training sessions. The risks corresponding to agents 1 and 9 are highlighted in blue and purple, respectively. (*Right*) Evolution during the prediction phase of the log belief ratio $\log \frac{\mu_{k,t}(+1)}{\mu_{k,t}(-1)}$, for agent $k = 9$, obtained by running the SML strategy (12.11) in its adaptive version, with adaptation parameter $\delta = 0.01$. The observed images represent digit 0 within interval $[1, 1000]$, then the digit changes every 1000 time instants.

classifiers known as AdaBoost [73, 155]. In the AdaBoost strategy, the agents are trained sequentially in a series. The training of one agent is performed by taking into account the performance estimated for the previous agents in the series. This is done to motivate the current agent to pay particular attention to the samples for which the previous agents perform worse. After training, each agent k is endowed with a decision statistic $h_k^{\sf boost}(x)$ and a weight $w_k^{\sf boost}$ representing the accuracy of its classification performance over the training set. Then, during the prediction phase, each agent makes an individual decision with the learned decision statistic. These local decisions, scaled by the aforementioned weights, are then aggregated in a centralized manner — see [73, 155] for details on the implementation of the AdaBoost strategy. In order to perform a fair comparison with the SML strategy considered in the previous example, we apply the AdaBoost strategy by considering that the decision structures used by the agents are MLPs with the same architecture described before.

During the prediction phase, each agent k at time t observes the unlabeled data $x_{k,t}$ and computes a decision

$$\theta_{k,t}^{\sf boost} = \text{sign}\left(h_k^{\sf boost}(x_{k,t})\right). \tag{12.65}$$

The collective decision at time t, denoted by $\theta_t^{\sf boost}$, is performed by using the boosting weights determined during training, according to the fusion rule

$$\theta_t^{\sf boost} = \text{sign}\left(\sum_{k=1}^{K} w_k^{\sf boost}\, \theta_{k,t}^{\sf boost}\right). \tag{12.66}$$

Note that computing $\theta_t^{\sf boost}$ requires centralized information, i.e., knowledge of the instantaneous decisions of all agents. We compare this centralized boosting decision with the decision of agent 1 from the SML strategy, whose log belief ratio was seen in the right panel of Figure 12.6.

In Figure 12.7 we compare the SML and Adaboost strategies, under the same setting used in Example 12.3. We see that the SML strategy makes wrong decisions only during short periods after state transitions occur, whereas the AdaBoost strategy makes

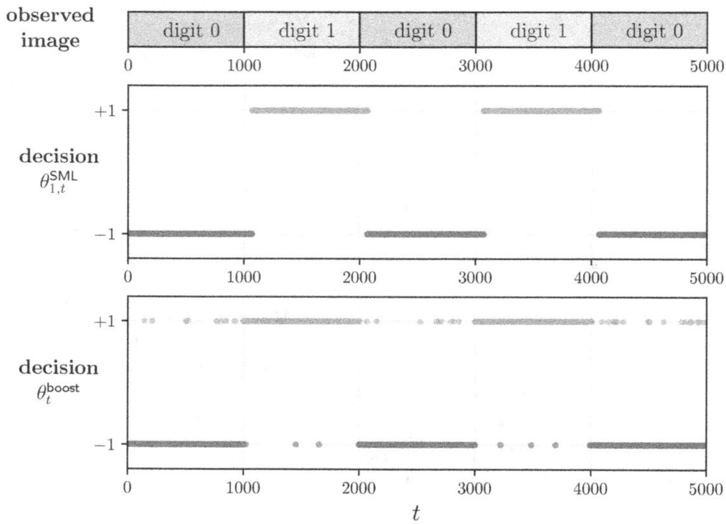

Figure 12.7: Comparison between SML and Adaboost as described in Example 12.4. (*Top*) Sequence of digits occurring during the prediction phase. The observed images represent digit 0 within interval [1,1000], then the digit changes every 1000 time instants. (*Center*) Decision of agent 1 when using the adaptive SML strategy. (*Bottom*) AdaBoost decision.

mistakes throughout the prediction phase. The improvement achieved with the SML strategy is examined and explained in the next example.

Example 12.5 (SML performance). In this example we examine the SML performance under two prediction scenarios: a *stationary* scenario over 20 time instants, where the true underlying digit is 0 throughout the prediction horizon; and a *nonstationary* scenario over 40 time instants, where the true underlying digit is initially 0 and switches to digit 1 at instant $t = 20$. For the stationary setting we implement the SML strategy in the nonadaptive version that corresponds to the algorithm in listing (12.11) with $\delta = 0$. For the nonstationary setting we consider instead the adaptive version that corresponds to using $0 < \delta < 1$ in the same listing (in this example, we set $\delta = 0.1$). In both cases we also implement the AdaBoost strategy. For all the considered strategies, the preliminary training phase is implemented as described in the previous examples, with the following two differences: For each agent, the number of labeled images in its training set is 40, and the number of first-layer nodes in its MLP is $n_1 = 10$.

In Figure 12.8 we display the error probabilities, estimated from 1000 Monte Carlo runs, achieved by the AdaBoost strategy and by agent 1 under the SML strategies. Specifically, the left panel refers to the stationary setting, where we see that the SML strategy in the nonadaptive version quickly surpasses AdaBoost and attains a significantly improved accuracy over time. The right panel focuses on the nonstationary setting, where we see that the SML strategy in the adaptive version successfully adapts its predictive behavior in view of the change in the underlying class of digits, surpassing the performance of the AdaBoost strategy after a relatively short adaptation time.

The improved performance attained by the SML strategy can be explained as follows.

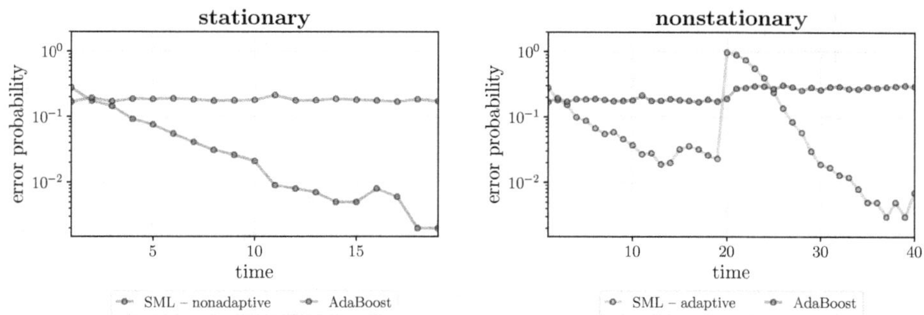

Figure 12.8: Evolution over time of the error probabilities, estimated from 1000 Monte Carlo runs, for SML (agent 1) and AdaBoost (centralized decision), as described in Example 12.5. (*Left*) Nonadaptive case: SML algorithm (12.11) run with $\delta = 0$. Here the true state corresponds to digit 0. (*Right*) Adaptive case: SML algorithm (12.11) run with adaptation parameter $\delta = 0.1$. Here, the true state corresponds to digit 0 until instant $t = 19$, and to digit 1 afterwards.

As was repeatedly observed, social learning strategies introduce a combination *over time*, where streaming data are continually incorporated into the beliefs; and a combination *over the network*, where each agent aggregates locally the information received from its neighbors. In contrast, AdaBoost does not perform any kind of combination over time (since it does not aggregate information sequentially) or over the network (since the solution is centralized) and therefore there is no adaptation time associated with its behavior. Note that the fact that AdaBoost makes instantaneous decisions without aggregating information over time results in an error probability that does not change over time if the underlying hypothesis does not change. For this reason, in the stationary case represented in the left panel of Figure 12.8, the performance is constant over time, whereas in the nonstationary case represented in the right panel, the error probability drifts when the true hypothesis changes, i.e., at time 20. Remarkably, in both the stationary and nonstationary scenarios, AdaBoost is significantly outperformed by the SML strategy as time elapses. This is because the SML strategy benefits from integrating information over time. Moreover, we observe that a small delay is present in the SML strategy, at the beginning of the learning process or right after a change. The delay at $t = 0$ is related to belief aggregation over space, i.e., to the time necessary for the agents to converge to a coordinated solution. The delay after the hypothesis change is the adaptation time characterized by Corollary 10.1. While also including a transient related to the network, the adaptation time is mainly determined by the number of iterations necessary to delete the memory accumulated from the observations before the change.

As a concluding example, we now show that the SML strategy can also be employed successfully to solve classification problems with more than two hypotheses. Referring back to the general classification setting of Section 12.3, in the H-ary case with $H > 2$ we assume that the agents run the social learning algorithm from listing (12.11), with the $(H-1)$-dimensional decision statistic $\widehat{\boldsymbol{h}}_k(x)$ estimated during the training phase by minimizing the empirical risk from (12.34) over a given function family $\mathcal{H}_k$, namely,

$$\widehat{\boldsymbol{h}}_k = \boldsymbol{h}_k^o = \underset{h_k \in \mathcal{H}_k}{\arg\min}\, \widehat{\boldsymbol{R}}_k(h_k). \tag{12.67}$$

Example 12.6 (Multiclass MNIST). Let us consider a similar setup to the one presented in Example 12.3, except that now we take into account all classes contained in the MNIST dataset, that is, $H = 10$ classes representing the digits $0, 1, 2, \ldots, 9$. Specifically, class $\theta = 1$ corresponds to "digit 0," class $\theta = 2$ corresponds to "digit 1," and so on. We consider the same network shown in Figure 12.5, where each agent sees a patch of the handwritten image according to Figure 12.4.

In the training phase, each agent is trained independently over a balanced set of 1000 labeled images (100 images per digit), using an MLP (see Example 12.2) with activation function $\sigma_a = \tanh$ and $L = 2$ layers. The first layer has $n_1 = 64$ nodes. The second layer has $n_2 = 9$ nodes, conforming with a classification problem with 10 hypotheses.

Minimization of the empirical risk from (12.34) is performed with a mini-batch stochastic gradient algorithm with multiple epochs, and with randomly selected batch samples [155]. Specifically, the batch size is equal to 10, the learning rate is equal to 0.001, and the algorithm is run over 30 epochs. The top panel of Figure 12.9 shows the evolution over the training epochs of the empirical training risk of each agent, averaged over 5 training sessions.

In the prediction phase, the agents observe unlabeled images over time. The nature of the images changes every 100 samples. The agents observe images representing digit 0 for $t \in [1, 100]$, digit 1 for $t \in [101, 200]$, and so on. We implement the SML strategy based on the social learning algorithm from listing (12.11), in its adaptive version with the choice $\delta = 0.1$. The bottom panel of Figure 12.9 displays the time evolution of the beliefs of agent 1. We see that the algorithm is able to learn well under the considered nonstationary scenario, exhibiting remarkable adaptation properties. In fact, all the belief mass is placed on the true hypothesis that changes dynamically over time.

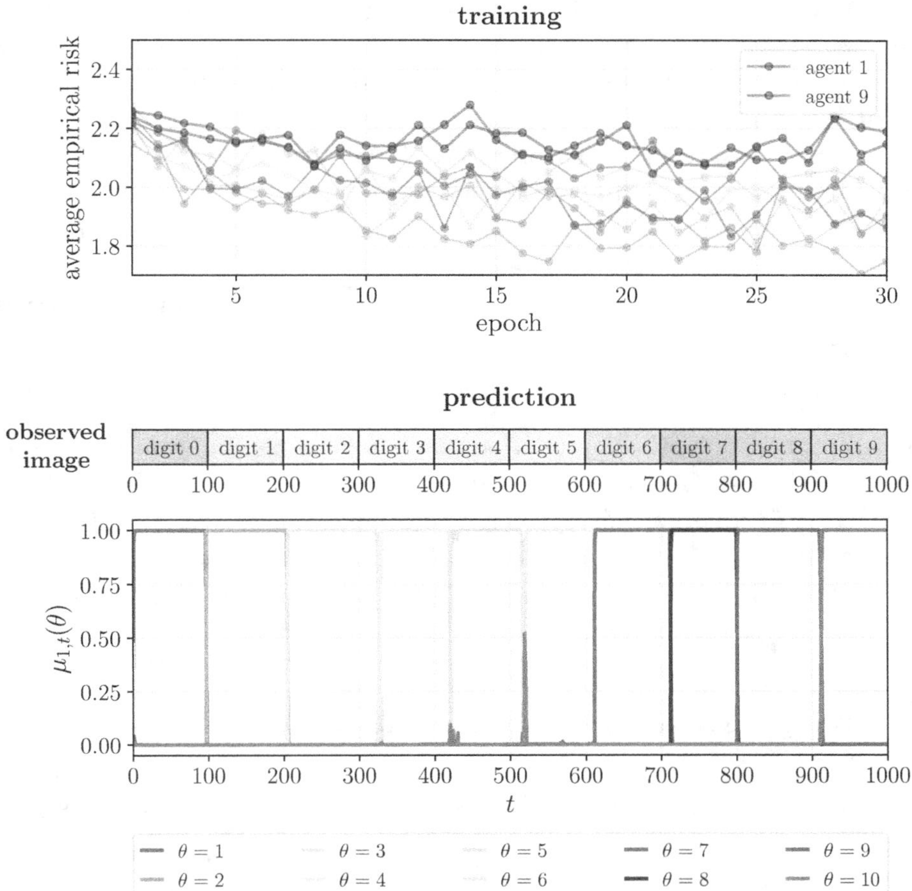

Figure 12.9: Adaptive social machine learning strategy (with adaptation parameter $\delta = 0.1$) operating under the setting in Example 12.6, with digits $0, 1, \ldots, 9$. (*Top*) Evolution over the training epochs of the empirical risk for all agents, averaged over 5 training sessions. The risks corresponding to agents 1 and 9, the least informed agents, are highlighted in blue and purple, respectively. (*Bottom*) Belief evolution over time for agent 1. The bar displayed at the top shows the evolution of the true state, which changes every 100 time instants. Specifically, the agents observe images representing digit 0 (corresponding to hypothesis $\theta = 1$) for $t \in [1, 100]$, then digit 1 (corresponding to hypothesis $\theta = 2$) for $t \in [101, 200]$, and so on.

12.A Appendix: Notation for Binary Decision Problems

The appendices at the end of this chapter are devoted to the proof of Theorems 12.1 and 12.2. We start by focusing on Theorem 12.1, where we claim the consistency of the social machine learning strategy for the binary case ($H = 2$). We recall that in this case the set of hypotheses is chosen for convenience as $\Theta = \{+1, -1\}$, and the decision statistic of any agent k is scalar, with $h_k(x) = h_k(x; +1)$ — see (12.38).

The first step to prove Theorem 12.1 is to specialize to the case $H = 2$ the condition for consistency that was formulated in Lemma 12.2 for an arbitrary number H of hypotheses. It is convenient to introduce some notation to carry out the analysis.

We stack the individual agent functions h_k into a vector-valued function h, namely,

$$h \triangleq [h_1, h_2, \ldots, h_K] : \mathcal{X}_1 \times \mathcal{X}_2 \times \cdots \times \mathcal{X}_K \mapsto \mathbb{R}^K. \tag{12.68}$$

Recall that $\mathcal{H}_k$ denotes the function family from which h_k can be selected. The function family to which h belongs, resulting from (12.68), will be denoted by $\mathcal{H}$. In other words, when we write $h \in \mathcal{H}$, we mean that $h_k \in \mathcal{H}_k$ for each k.

In the next definition we introduce a compact notation to describe some useful averaging operators.

Definition 12.3 (Averaging operators applied to a decision statistic h_k). Given a function h_k defined on a space $\mathcal{X}_k$:

$$h_k : \mathcal{X}_k \mapsto \mathbb{R}, \tag{12.69}$$

the expected values (assumed to be finite) of $h_k(\boldsymbol{x})$ computed under the likelihood models corresponding to $\vartheta^o = +1$ and $\vartheta^o = -1$ are denoted, respectively, by

$$\eta_k^+(h_k) \triangleq \mathbb{E}_{\ell_{k,+1}} h_k(\boldsymbol{x}), \qquad \eta_k^-(h_k) \triangleq \mathbb{E}_{\ell_{k,-1}} h_k(\boldsymbol{x}), \tag{12.70}$$

where the subscripts on $\mathbb{E}$ emphasize that the expectation is computed assuming that $\boldsymbol{x}$ is distributed according to $\ell_{k,+1}$ or $\ell_{k,-1}$.
It is also convenient to introduce the statistical and empirical means computed over the training set. Since the classes in the training set are balanced in view of (12.3), the statistical mean of $h_k(\widehat{\boldsymbol{x}}_{k,n})$ computed over the training set can be evaluated as

$$\eta_k(h_k) \triangleq \mathbb{E} h_k(\widehat{\boldsymbol{x}}_{k,n}) = \frac{1}{2}\Big(\eta_k^+(h_k) + \eta_k^-(h_k)\Big), \tag{12.71}$$

where the expectation is computed with respect to the random quantity $\widehat{\boldsymbol{x}}_{k,n}$, i.e., with respect to the distribution of the features in the training set. The *empirical* mean over the training set is instead given by

$$\widehat{\boldsymbol{\eta}}_k(h_k) \triangleq \frac{1}{E_k}\sum_{n=1}^{E_k} h_k(\widehat{\boldsymbol{x}}_{k,n}). \tag{12.72}$$

Finally, we define the network counterparts of the above quantities, namely,

$$\eta^+(h) \triangleq \sum_{k=1}^{K} v_k\, \eta_k^+(h_k), \qquad \eta^-(h) \triangleq \sum_{k=1}^{K} v_k\, \eta_k^-(h_k) \tag{12.73}$$

and

$$\eta(h) \triangleq \sum_{k=1}^{K} v_k \eta_k(h_k), \qquad \widehat{\boldsymbol{\eta}}(h) \triangleq \sum_{k=1}^{K} v_k\, \widehat{\boldsymbol{\eta}}_k(h_k), \tag{12.74}$$

where, as usual, v_k denotes the kth entry of the Perron vector of the combination matrix A.

The notation introduced in Definition 12.3 allows us to write the consistency condition (12.18), specialized to the binary case, as

$$\eta^+(h) > 0, \qquad \eta^-(h) < 0. \tag{12.75}$$

In particular, in Theorem 12.1 we are interested in establishing the consistency of the *de-biased* decision statistic introduced in Definition 12.2. Recall that to construct this statistic, we first compute an intermediate function by minimizing the empirical risk in (12.34):

$$\boldsymbol{h}_k^o = \arg\min_{h_k \in \mathcal{H}_k} \widehat{\boldsymbol{R}}_k(h_k), \tag{12.76}$$

and subsequently shift it by subtracting its empirical average $\widehat{\boldsymbol{\eta}}_k(\boldsymbol{h}_k^o)$ (see definition (12.72)) to produce the final de-biased statistic

$$\widetilde{\boldsymbol{h}}_k(x) = \boldsymbol{h}_k^o(x) - \widehat{\boldsymbol{\eta}}_k(\boldsymbol{h}_k^o). \tag{12.77}$$

If we now apply the consistency condition (12.75) to the function $\widetilde{\boldsymbol{h}} = [\widetilde{\boldsymbol{h}}_1, \widetilde{\boldsymbol{h}}_2, \ldots, \widetilde{\boldsymbol{h}}_K]$, we obtain

$$\eta^+\left(\widetilde{\boldsymbol{h}}\right) > 0, \qquad \eta^-\left(\widetilde{\boldsymbol{h}}\right) < 0, \tag{12.78}$$

which, using (12.77), corresponds to

$$\eta^+(\boldsymbol{h}^o) - \widehat{\boldsymbol{\eta}}(\boldsymbol{h}^o) > 0, \qquad \eta^-(\boldsymbol{h}^o) - \widehat{\boldsymbol{\eta}}(\boldsymbol{h}^o) < 0, \tag{12.79}$$

resulting in the following alternative expression for the probability of consistent learning in (12.37):

$$P_c = \mathbb{P}\left[\eta^+(\boldsymbol{h}^o) > \widehat{\boldsymbol{\eta}}(\boldsymbol{h}^o),\ \eta^-(\boldsymbol{h}^o) < \widehat{\boldsymbol{\eta}}(\boldsymbol{h}^o)\right]. \tag{12.80}$$

We note in passing that the averaging operator η is deterministic, while its argument is random since the function $\boldsymbol{h}^o$ results from an optimization procedure performed over the (random) training set. In comparison, the averaging operator $\widehat{\boldsymbol{\eta}}$ is random since it depends on the training set, and is also applied to the random argument $\boldsymbol{h}^o$.

Before concluding this section, it is useful to examine the rationale behind the de-biasing operation. Consider first the optimized decision statistic $\boldsymbol{h}^o$ *without de-biasing*. For this statistic, the consistency conditions in (12.79) amount to

$$\eta^+(\boldsymbol{h}^o) > 0, \qquad \eta^-(\boldsymbol{h}^o) < 0. \tag{12.81}$$

In other words, we require $\eta^+(\boldsymbol{h}^o)$ to be positive and $\eta^-(\boldsymbol{h}^o)$ to be negative in order to attain consistent learning. However, it might happen that the estimated statistics are biased toward one class, for example, we might have $\eta^-(\boldsymbol{h}^o) > 0$. In this case it would be reasonable to expect that we could still make reliable decisions provided that $\eta^+(\boldsymbol{h}^o) > \eta^-(\boldsymbol{h}^o)$. This is in fact possible by using the de-biased decision statistic from (12.77).

To see why, observe that for sufficiently large training set sizes the empirical and true means are close to each other, namely,

$$\widehat{\boldsymbol{\eta}}(\boldsymbol{h}^o) \approx \frac{1}{2}\left(\eta^+(\boldsymbol{h}^o) + \eta^-(\boldsymbol{h}^o)\right). \tag{12.82}$$

Under this approximation, the two conditions in (12.79) become

$$\eta^+(\boldsymbol{h}^o) > \eta^-(\boldsymbol{h}^o). \tag{12.83}$$

We see from (12.83) that, due to de-biasing, consistent learning is satisfied by requiring that the expectation under $\vartheta^o = +1$ is greater than the expectation under $\vartheta^o = -1$, regardless of the sign of the individual terms $\eta^+(\boldsymbol{h}^o)$ and $\eta^-(\boldsymbol{h}^o)$. Therefore, consistent learning becomes possible even when the trained decision statistics are biased, for example, when $\eta^+(\boldsymbol{h}^o) > \eta^-(\boldsymbol{h}^o) > 0$.

12.B Appendix: Bounds for Consistent Learning

In the next two sections, we establish two results useful to prove Theorem 12.1. First, in Lemma 12.3 we obtain a lower bound on the probability of consistent learning, which is composed of the two probability terms appearing on the RHS of (12.85). The first term depends on the distance between the empirical and true means, whereas the second term depends on the risk function. Both terms can be bounded by using the uniform laws of large numbers established in Lemma 12.4, which characterize the discrepancy between the empirical and true means/risks. The combination of the results from Lemmas 12.3 and 12.4 is exploited in Section 12.C to prove Theorem 12.1.

12.B.1 Probability of Consistent Learning

Lemma 12.3 (Bound for the probability of consistent learning). Let Assumptions 5.1, 12.1, and 12.2 be satisfied. Let $R_k(h_k)$ be the exact risk of agent k associated with the decision statistic h_k, as defined in (12.44), and introduce the *network* risk associated with the vector-valued function h defined in (12.68):

$$R(h) \triangleq \sum_{k=1}^{K} v_k\, R_k(h_k). \tag{12.84}$$

Then, for any $y > 0$, the probability of consistent learning in (12.80) obeys the following lower bound:

$$P_c \geq 1 - \mathbb{P}\Big[\big|\widehat{\eta}(\boldsymbol{h}^o) - \eta(\boldsymbol{h}^o)\big| > y\Big] - \mathbb{P}\Big[R(\boldsymbol{h}^o) \geq \log\big(1 + e^{-y}\big)\Big]. \tag{12.85}$$

Proof. To begin with, we introduce two events that will be useful in the proof:

$$\mathcal{A} \triangleq \left\{\big|\widehat{\eta}(\boldsymbol{h}^o) - \eta(\boldsymbol{h}^o)\big| \geq \frac{1}{2}\Big(\eta^+(\boldsymbol{h}^o) - \eta^-(\boldsymbol{h}^o)\Big)\right\}, \tag{12.86}$$

$$\mathcal{B} \triangleq \left\{\frac{1}{2}\Big(\eta^+(\boldsymbol{h}^o) - \eta^-(\boldsymbol{h}^o)\Big) > y\right\}. \tag{12.87}$$

The following chain of equalities holds, where the notation $\mathcal{E}^{\mathrm{c}}$ denotes the complement of event $\mathcal{E}$:

$$\begin{aligned}
1 - P_c &\overset{(a)}{=} 1 - \mathbb{P}\left[\{\eta^+(\boldsymbol{h}^o) > \widehat{\eta}(\boldsymbol{h}^o)\} \cap \{\eta^-(\boldsymbol{h}^o) < \widehat{\eta}(\boldsymbol{h}^o)\}\right] \\
&= \mathbb{P}\left[\Big\{ \{\eta^+(\boldsymbol{h}^o) > \widehat{\eta}(\boldsymbol{h}^o)\} \cap \{\eta^-(\boldsymbol{h}^o) < \widehat{\eta}(\boldsymbol{h}^o)\} \Big\}^{\mathrm{c}}\right] \\
&\overset{(b)}{=} \mathbb{P}\Big[\{\eta^+(\boldsymbol{h}^o) \leq \widehat{\eta}(\boldsymbol{h}^o)\} \cup \{\eta^-(\boldsymbol{h}^o) \geq \widehat{\eta}(\boldsymbol{h}^o)\} \Big] \\
&= \mathbb{P}\Big[\{\eta^+(\boldsymbol{h}^o) - \eta(\boldsymbol{h}^o) \leq \widehat{\eta}(\boldsymbol{h}^o) - \eta(\boldsymbol{h}^o)\} \\
&\qquad\qquad \cup \{\eta^-(\boldsymbol{h}^o) - \eta(\boldsymbol{h}^o) \geq \widehat{\eta}(\boldsymbol{h}^o) - \eta(\boldsymbol{h}^o)\}\Big] \\
&\overset{(c)}{=} \mathbb{P}\bigg[\Big\{\widehat{\eta}(\boldsymbol{h}^o) - \eta(\boldsymbol{h}^o) \geq \frac{1}{2}\Big(\eta^+(\boldsymbol{h}^o) - \eta^-(\boldsymbol{h}^o)\Big)\Big\} \\
&\qquad\qquad \cup \Big\{\widehat{\eta}(\boldsymbol{h}^o) - \eta(\boldsymbol{h}^o) \leq -\frac{1}{2}\Big(\eta^+(\boldsymbol{h}^o) - \eta^-(\boldsymbol{h}^o)\Big)\Big\}\bigg]
\end{aligned}$$

$$\stackrel{\text{(d)}}{=} \mathbb{P}[\mathcal{A}] \stackrel{\text{(e)}}{=} \mathbb{P}[\mathcal{A} \cap \mathcal{B}] + \mathbb{P}[\mathcal{A} \cap \mathcal{B}^{c}], \tag{12.88}$$

where (a) follows from (12.80); (b) holds because, in view of De Morgan's law [21], the complement of the intersection of two sets is the union of the complements of the sets; (c) follows from the identities

$$\eta^{+}(\boldsymbol{h}^{o}) - \eta(\boldsymbol{h}^{o}) = \frac{1}{2}\Big(\eta^{+}(\boldsymbol{h}^{o}) - \eta^{-}(\boldsymbol{h}^{o})\Big), \tag{12.89}$$

$$\eta^{-}(\boldsymbol{h}^{o}) - \eta(\boldsymbol{h}^{o}) = -\frac{1}{2}\Big(\eta^{+}(\boldsymbol{h}^{o}) - \eta^{-}(\boldsymbol{h}^{o})\Big), \tag{12.90}$$

which are obtained by using (12.71), (12.73), and (12.74); in step (d) we apply the definition of absolute value and the definition of $\mathcal{A}$ from (12.86); and in (e) we introduce the event $\mathcal{B}$ defined in (12.87) and apply the law of total probability. We now focus on the two probabilities obtained after step (e) in (12.88). Regarding the first probability, from (12.86) and (12.87) we obtain the relation

$$\mathcal{A} \cap \mathcal{B} \implies \Big\{\Big|\eta(\boldsymbol{h}^{o}) - \widehat{\boldsymbol{\eta}}(\boldsymbol{h}^{o})\Big| > y\Big\}, \tag{12.91}$$

which implies

$$\mathbb{P}[\mathcal{A} \cap \mathcal{B}] \le \mathbb{P}\Big[\,|\eta(\boldsymbol{h}^{o}) - \widehat{\boldsymbol{\eta}}(\boldsymbol{h}^{o})| > y\Big]. \tag{12.92}$$

For the second probability on the RHS of (12.88), recalling that the probability of the intersection of two events cannot be larger than the probability of any of the individual events, and using the definition of $\mathcal{B}$ from (12.87), we can write

$$\mathbb{P}[\mathcal{A} \cap \mathcal{B}^{c}] \le \mathbb{P}[\mathcal{B}^{c}] = \mathbb{P}\left[\frac{1}{2}\Big(\eta^{+}(\boldsymbol{h}^{o}) - \eta^{-}(\boldsymbol{h}^{o})\Big) \le y\right]. \tag{12.93}$$

Using (12.92) and (12.93) in (12.88), we get

$$1 - P_{c} \le \mathbb{P}\Big[\,|\eta(\boldsymbol{h}^{o}) - \widehat{\boldsymbol{\eta}}(\boldsymbol{h}^{o})| > y\Big] + \mathbb{P}\left[\frac{1}{2}\Big(\eta^{+}(\boldsymbol{h}^{o}) - \eta^{-}(\boldsymbol{h}^{o})\Big) \le y\right]. \tag{12.94}$$

To complete the proof of the lemma, we need to focus on the second term on the RHS of (12.94). Consider the network risk in (12.84), applied to the vector-valued function h defined by (12.68):

$$\begin{aligned}
R(h) &= \sum_{k=1}^{K} v_k\, R_k(h_k) \\
&= \sum_{k=1}^{K} v_k\, \mathbb{E}\log\left(1 + \exp\left\{-\widehat{\boldsymbol{\theta}}_{k,n}\, h_k(\widehat{\boldsymbol{x}}_{k,n})\right\}\right) \\
&\stackrel{\text{(a)}}{\ge} \sum_{k=1}^{K} v_k\, \log\left(1 + \exp\left\{-\mathbb{E}\left[\widehat{\boldsymbol{\theta}}_{k,n}\, h_k(\widehat{\boldsymbol{x}}_{k,n})\right]\right\}\right) \\
&\stackrel{\text{(b)}}{\ge} \log\left(1 + \exp\left\{-\sum_{k=1}^{K} v_k\, \mathbb{E}\left[\widehat{\boldsymbol{\theta}}_{k,n}\, h_k(\widehat{\boldsymbol{x}}_{k,n})\right]\right\}\right) \\
&\stackrel{\text{(c)}}{=} \log\left(1 + \exp\left\{-\frac{1}{2}\Big(\eta^{+}(h) - \eta^{-}(h)\Big)\right\}\right),
\end{aligned} \tag{12.95}$$

where in (a) and (b) we apply Jensen's inequality (Theorem C.5) to the convex function $\log(1+e^x)$, specifically, with respect to the expectation operator $\mathbb{E}$ in inequality (a), and with respect to the convex combination with weights $\{v_k\}$ in inequality (b) — see (C.10). In (c), we compute the expectation by using the assumption of uniform priors during training and the definitions in (12.73) and (12.73).

Exploiting (12.95) we obtain the following implication:

$$\frac{1}{2}\Big(\eta^+(h) - \eta^-(h)\Big) \le y \implies R(h) \ge \log\left(1 + e^{-y}\right), \tag{12.96}$$

which, when applied to the optimized functions $\boldsymbol{h}_k^o$, implies the following bound:

$$\mathbb{P}\left[\frac{1}{2}\Big(\eta^+(\boldsymbol{h}^o) - \eta^-(\boldsymbol{h}^o)\Big) \le y\right] \le \mathbb{P}\Big[R(\boldsymbol{h}^o) \ge \log\left(1 + e^{-y}\right)\Big]. \tag{12.97}$$

Using (12.97) in (12.94) yields the bound in (12.85), which completes the proof of the lemma.

■

12.B.2 Uniform Laws of Large Numbers

In this section we establish two concentration bounds to quantify the proximity between the true and empirical risks, as well as the true and empirical means. Regarding the risks, we consider the following general form, which includes the binary cross-entropy as a special case:

$$R_k(h_k) = \mathbb{E}\,\mathscr{Q}\left(\widehat{\boldsymbol{\theta}}_{k,n}\, h_k(\widehat{\boldsymbol{x}}_{k,n})\right), \tag{12.98a}$$

$$\widehat{\boldsymbol{R}}_k(h_k) = \frac{1}{E_k}\sum_{n=1}^{E_k} \mathscr{Q}\left(\widehat{\boldsymbol{\theta}}_{k,n}\, h_k(\widehat{\boldsymbol{x}}_{k,n})\right), \tag{12.98b}$$

where $\mathscr{Q} : \mathbb{R} \mapsto \mathbb{R}$ is an $\mathscr{L}$-Lipschitz loss function. The network true and empirical risk functions will be, respectively,

$$R(h) = \sum_{k=1}^{K} v_k\, R_k(h_k), \tag{12.99a}$$

$$\widehat{\boldsymbol{R}}(h) = \sum_{k=1}^{K} v_k\, \widehat{\boldsymbol{R}}_k(h_k). \tag{12.99b}$$

The following theorem characterizes the deviations between the empirical and true risks, and the deviations between the empirical and true means. In particular, the theorem provides bounds on the probability that these deviations exceed some threshold.

Lemma 12.4 (Uniform laws of large numbers). Let Assumptions 5.1, 12.1, and 12.2 be satisfied. Assume that the loss function $\mathcal{Q} : \mathbb{R} \mapsto \mathbb{R}$ is $\mathcal{L}$−Lipschitz and that, for $k = 1, 2, \ldots, K$, the decision statistic $h_k : \mathcal{X}_k \mapsto \mathbb{R}$ belongs to a family $\mathcal{H}_k$ of bounded functions:

$$|h_k(x)| \leq h_{k,\mathsf{max}} \quad \forall x \in \mathcal{X}_k, \quad \text{with } 0 < h_{k,\mathsf{max}} < \infty. \tag{12.100}$$

Let ρ_k be the Rademacher complexity (see Definition G.1) associated with the family $\mathcal{H}_k$, and let ρ_{net} be the *network* Rademacher complexity defined in (12.49). Denote by h the vector-valued function defined in (12.68) and by $\mathcal{H}$ the resulting family to which h belongs. Then, we have the following two results:

$$\mathbb{P}\left[\sup_{h\in\mathcal{H}} \left|\widehat{\boldsymbol{R}}(h) - R(h)\right| \geq y\right] \leq \exp\left\{-\frac{E_{\mathsf{max}}\left(y - 2\mathcal{L}\,\rho_{\mathsf{net}}\right)^2}{2h_{\mathsf{net}}^2\mathcal{L}^2}\right\} \tag{12.101}$$

for all $y > 2\mathcal{L}\,\rho_{\mathsf{net}}$, and

$$\mathbb{P}\left[\sup_{h\in\mathcal{H}} |\widehat{\boldsymbol{\eta}}(h) - \eta(h)| \geq y\right] \leq \exp\left\{-\frac{E_{\mathsf{max}}\left(y - 2\rho_{\mathsf{net}}\right)^2}{2h_{\mathsf{net}}^2}\right\} \tag{12.102}$$

for all $y > 2\rho_{\mathsf{net}}$, where

$$E_{\mathsf{max}} \triangleq \max_{k\in\{1,2,\ldots,K\}} E_k \tag{12.103}$$

and h_{net} is defined in (12.58).

Proof. We develop the proofs of (12.101) and (12.102) separately.

Bound (12.101). Consider the difference between the network empirical and true risks,

$$\widehat{\boldsymbol{R}}(h) - R(h) = \chi(h) - \sum_{k=1}^{K}\frac{v_k}{E_k}\sum_{n=1}^{E_k}\mathcal{Q}\left(\widehat{\boldsymbol{\theta}}_{k,n}\, h_k(\widehat{\boldsymbol{x}}_{k,n})\right), \tag{12.104}$$

where we introduced the auxiliary functional

$$\chi(h) = \sum_{k=1}^{K} v_k\, \mathbb{E}\,\mathcal{Q}\left(\widehat{\boldsymbol{\theta}}_{k,n}\, h_k(\widehat{\boldsymbol{x}}_{k,n})\right). \tag{12.105}$$

Our focus is on the supremum of the absolute risk deviation taken over the function family $\mathcal{H}$, namely, on

$$\sup_{h\in\mathcal{H}}\left|\widehat{\boldsymbol{R}}(h) - R(h)\right| = \sup_{h\in\mathcal{H}}\left|\chi(h) - \sum_{k=1}^{K}\frac{v_k}{E_k}\sum_{n=1}^{E_k}\mathcal{Q}\left(\widehat{\boldsymbol{\theta}}_{k,n}\, h_k(\widehat{\boldsymbol{x}}_{k,n})\right)\right|. \tag{12.106}$$

In order to bound the probability that this maximum deviation exceeds some threshold, we will call upon McDiarmid's inequality — see Theorem C.4. To this end, it is necessary to choose the random vectors $\boldsymbol{z}_n$ and the function g mentioned in Theorem C.4. The vectors $\boldsymbol{z}_n$ are constructed as follows. First, we stack the features $\widehat{\boldsymbol{x}}_{k,n}$ and the labels $\widehat{\boldsymbol{\theta}}_{k,n}$ from across the agents $k = 1, 2, \ldots, K$ into the vectors

$$\widehat{\boldsymbol{x}}_n \triangleq \operatorname{col}\left\{\widehat{\boldsymbol{x}}_{k,n}\right\}_{k=1}^{K}, \qquad \widehat{\boldsymbol{\theta}}_n \triangleq \operatorname{col}\left\{\widehat{\boldsymbol{\theta}}_{k,n}\right\}_{k=1}^{K}, \tag{12.107}$$

where $n = 1, 2, \ldots, E_{\max}$ and where, we recall, $\text{col}\{x_k\}_{k=1}^{K}$ denotes the $K \times 1$ vector obtained by stacking into a single column its vector entries. Observe that the agents are allowed to have training sets with different sizes E_k. Therefore, for a given n, some agents might have a number of training samples $E_k < n$. In this case, the vector $\widehat{\boldsymbol{x}}_n$ appearing in (12.107) would not contain the features from these agents. The same argument applies to $\widehat{\boldsymbol{\theta}}_n$.

To apply McDiarmid's inequality, we form the vector $\boldsymbol{z}_n$ by stacking the vectors $\widehat{\boldsymbol{x}}_n$ and $\widehat{\boldsymbol{\theta}}_n$ into a single vector, namely,

$$\boldsymbol{z}_n \triangleq \text{col}\left\{\widehat{\boldsymbol{x}}_n, \widehat{\boldsymbol{\theta}}_n\right\}. \tag{12.108}$$

Then, we define the function

$$g(\boldsymbol{z}_1, \boldsymbol{z}_2, \ldots, \boldsymbol{z}_{E_{\max}}) \triangleq \sup_{h \in \mathcal{H}} \left| \chi(h) - \sum_{k=1}^{K} \frac{v_k}{E_k} \sum_{n=1}^{E_k} \mathscr{Q}\left(\widehat{\boldsymbol{\theta}}_{k,n}\, h_k(\widehat{\boldsymbol{x}}_{k,n})\right)\right|. \tag{12.109}$$

Note that, in view of (12.106), we have the following identity:

$$g(\boldsymbol{z}_1, \boldsymbol{z}_2, \ldots, \boldsymbol{z}_{E_{\max}}) = \sup_{h \in \mathcal{H}} \left| \widehat{\boldsymbol{R}}(h) - R(h) \right|. \tag{12.110}$$

Accordingly, if we apply McDiarmid's inequality to the function g we obtain a bound on the probability that the empirical risk deviates from the true risk (uniformly for all functions h in the family $\mathcal{H}$). However, to apply McDiarmid's inequality we need to verify that the chosen function g meets the bounded-difference condition (C.5). To this end, we must consider all sequences

$$\{z_1, z_2, \ldots, z_i, \ldots, z_{E_{\max}}\} \qquad \text{and} \qquad \{z_1, z_2, \ldots, \breve{z}_i, \ldots, z_{E_{\max}}\} \tag{12.111}$$

that differ only in their respective ith vectors,

$$z_i = \text{col}\left\{\widehat{x}_i, \widehat{\theta}_i\right\} \quad \text{and} \quad \breve{z}_i = \text{col}\left\{\breve{x}_i, \breve{\theta}_i\right\}. \tag{12.112}$$

Applying (12.109) to the second sequence in (12.111) we have

$$\begin{aligned}
&g(z_1, z_2, \ldots, \breve{z}_i, \ldots, z_{E_{\max}}) \\
&= \sup_{h \in \mathcal{H}} \left| \chi(h) - \sum_{k=1}^{K} \frac{v_k}{E_k} \left[\mathscr{Q}\left(\breve{\theta}_{k,i}\, h_k(\breve{x}_{k,i})\right) + \sum_{\substack{n=1 \\ n \neq i}}^{E_k} \mathscr{Q}\left(\widehat{\theta}_{k,n}\, h_k(\widehat{x}_{k,n})\right)\right]\right| \\
&= \sup_{h \in \mathcal{H}} \left| \chi(h) - \sum_{k=1}^{K} \frac{v_k}{E_k} \sum_{n=1}^{E_k} \mathscr{Q}\left(\widehat{\theta}_{k,n}\, h_k(\widehat{x}_{k,n})\right) \right. \\
&\left. + \sum_{k=1}^{K} \frac{v_k}{E_k} \left[\mathscr{Q}\left(\widehat{\theta}_{k,i}\, h_k(\widehat{x}_{k,i})\right) - \mathscr{Q}\left(\breve{\theta}_{k,i}\, h_k(\breve{x}_{k,i})\right)\right] \mathbb{I}[i \le E_k] \right|,
\end{aligned} \tag{12.113}$$

where $\mathbb{I}$ denotes, as usual, the indicator function, which appears since the difference between $(\widehat{x}_{k,i}, \widehat{\theta}_{k,i})$ and $(\breve{x}_{k,i}, \breve{\theta}_{k,i})$ is present only if $i \le E_k$, because agent k has E_k samples in its training set.

By introducing the functionals

$$S(h) \triangleq \chi(h) - \sum_{k=1}^{K} \frac{v_k}{E_k} \sum_{n=1}^{E_k} \mathscr{Q}\left(\widehat{\theta}_{k,n}\, h_k(\widehat{x}_{k,n})\right), \tag{12.114}$$

$$T(h) \triangleq \sum_{k=1}^{K} \frac{v_k}{E_k} \left[\mathscr{Q}\left(\widehat{\theta}_{k,i}\, h_k(\widehat{x}_{k,i})\right) - \mathscr{Q}\left(\breve{\theta}_{k,i}\, h_k(\breve{x}_{k,i})\right) \right] \mathbb{I}[i \le E_k], \tag{12.115}$$

we see that (12.109) and (12.113) can be written as

$$g(z_1, z_2, \dots, z_i, \dots, z_{E_{\max}}) = \sup_{h \in \mathcal{H}} |S(h)|, \tag{12.116}$$

$$g(z_1, z_2, \dots, \breve{z}_i, \dots, z_{E_{\max}}) = \sup_{h \in \mathcal{H}} |S(h) + T(h)|. \tag{12.117}$$

Applying Lemma 12.5 to the functionals defined in (12.114) and (12.115), from (12.116) and (12.117) we obtain

$$\begin{aligned}
&\Big| g(z_1, z_2, \dots, z_i, \dots, z_{E_{\max}}) - g(z_1, z_2, \dots, \breve{z}_i, \dots, z_{E_{\max}}) \Big| \le \sup_{h \in \mathcal{H}} |T(h)| \\
&= \sup_{h \in \mathcal{H}} \left| \sum_{k=1}^{K} \frac{v_k}{E_k} \left[\mathscr{Q}\left(\widehat{\theta}_{k,i}\, h_k(\widehat{x}_{k,i})\right) - \mathscr{Q}\left(\breve{\theta}_{k,i}\, h_k(\breve{x}_{k,i})\right) \right] \mathbb{I}[i \le E_k] \right| \\
&\overset{(a)}{\le} \sum_{k=1}^{K} \frac{v_k}{E_k} \sup_{h_k \in \mathcal{H}_k} \left| \mathscr{Q}\left(\widehat{\theta}_{k,i}\, h_k(\widehat{x}_{k,i})\right) - \mathscr{Q}\left(\breve{\theta}_{k,i}\, h_k(\breve{x}_{k,i})\right) \right| \\
&\overset{(b)}{\le} \mathscr{L} \sum_{k=1}^{K} \frac{v_k}{E_k} \sup_{h_k \in \mathcal{H}_k} \left| \widehat{\theta}_{k,i}\, h_k(\widehat{x}_{k,i}) - \breve{\theta}_{k,i}\, h_k(\breve{x}_{k,i}) \right| \\
&\overset{(c)}{\le} 2\mathscr{L} \sum_{k=1}^{K} \frac{v_k\, h_{k,\max}}{E_k} \overset{(d)}{=} \frac{2\, h_{\text{net}} \mathscr{L}}{E_{\max}},
\end{aligned} \tag{12.118}$$

where (a) follows from the subadditivity of the supremum and the fact that the indicator function is bounded by 1; (b) follows from the Lipschitz property of $\mathscr{Q}$; (c) follows from the boundedness assumption $|h_k(x)| \le h_{k,\max}$ and the fact that the labels are equal to ± 1; and (d) follows from the fact that $e_k = E_{\max}/E_k$ and from the definition of h_{net} in (12.58). The final bound resulting from (12.118) shows that the function g defined by (12.109) satisfies the bounded difference condition (C.5) with the choice $c_i = 2\, h_{\text{net}} \mathscr{L}/E_{\max}$. It is therefore legitimate to apply McDiarmid's inequality. Specifically, applying (C.8a) and further recalling (12.106), we get

$$\mathbb{P}\left[\sup_{h \in \mathcal{H}} \left| \widehat{\boldsymbol{R}}(h) - R(h) \right| - \mathbb{E} \sup_{h \in \mathcal{H}} \left| \widehat{\boldsymbol{R}}(h) - R(h) \right| \ge a \right] \le \exp\left\{ -\frac{a^2 E_{\max}}{2\, h_{\text{net}}^2 \mathscr{L}^2} \right\}, \tag{12.119}$$

holding for any $a > 0$. On the other hand, Lemma 12.7 allows us to bound the expected value appearing on the LHS of (12.119) as follows:

$$\mathbb{E} \sup_{h \in \mathcal{H}} \left| \widehat{\boldsymbol{R}}(h) - R(h) \right| \le 2\, \mathscr{L} \rho_{\text{net}}, \tag{12.120}$$

with ρ_{net} being the network Rademacher complexity from (12.49). In view of (12.120) we can write

$$\begin{aligned} & \sup_{h\in\mathcal{H}} \left|\widehat{\boldsymbol{R}}(h) - R(h)\right| \geq y \\ & \quad \Longrightarrow \sup_{h\in\mathcal{H}} \left|\widehat{\boldsymbol{R}}(h) - R(h)\right| - \mathbb{E}\sup_{h\in\mathcal{H}} \left|\widehat{\boldsymbol{R}}(h) - R(h)\right| \geq y - 2\mathscr{L}\rho_{\text{net}}, \end{aligned} \tag{12.121}$$

which implies that

$$\begin{aligned} & \mathbb{P}\left[\sup_{h\in\mathcal{H}} \left|\widehat{\boldsymbol{R}}(h) - R(h)\right| \geq y\right] \\ & \leq \mathbb{P}\left[\sup_{h\in\mathcal{H}} \left|\widehat{\boldsymbol{R}}(h) - R(h)\right| - \mathbb{E}\sup_{h\in\mathcal{H}} \left|\widehat{\boldsymbol{R}}(h) - R(h)\right| \geq y - 2\mathscr{L}\rho_{\text{net}}\right]. \end{aligned} \tag{12.122}$$

Considering a value $y > 2\mathscr{L}\rho_{\text{net}}$ and setting $a = y - 2\mathscr{L}\rho_{\text{net}}$ in (12.119), from (12.122) we obtain

$$\mathbb{P}\left[\sup_{h\in\mathcal{H}} \left|\widehat{\boldsymbol{R}}(h) - R(h)\right| \geq y\right] \leq \exp\left\{-\frac{(y - 2\mathscr{L}\rho_{\text{net}})^2 E_{\max}}{2\,h_{\text{net}}^2\mathscr{L}^2}\right\}, \tag{12.123}$$

and the proof of (12.101) is complete.

Bound **(12.102).** The proof of (12.102) is similar to the proof of (12.101), and will be presented in a concise manner. We start by using McDiarmid's inequality (Theorem C.4) with $\boldsymbol{z}_n = \widehat{\boldsymbol{x}}_n$ and with the following choice of the function g:

$$g(\boldsymbol{z}_1, \boldsymbol{z}_2, \ldots, \boldsymbol{z}_{E_{\max}}) = \sup_{h\in\mathcal{H}} \left|\chi(h) - \sum_{k=1}^{K} \frac{v_k}{E_k} \sum_{n=1}^{E_k} h_k(\widehat{\boldsymbol{x}}_{k,n})\right|, \tag{12.124}$$

where the auxiliary functional $\chi(h)$ is now defined as

$$\chi(h) = \sum_{k=1}^{K} v_k\, \mathbb{E} h_k\left(\widehat{\boldsymbol{x}}_{k,n}\right). \tag{12.125}$$

We follow similar steps to those used to prove (12.101), which result in the following bound:

$$\mathbb{P}\left[\sup_{h\in\mathcal{H}} \left|\widehat{\boldsymbol{\eta}}(h) - \eta(h)\right| - \mathbb{E}\sup_{h\in\mathcal{H}} \left|\widehat{\boldsymbol{\eta}}(h) - \eta(h)\right| \geq a\right] \leq \exp\left\{-\frac{a^2 E_{\max}}{2\,h_{\text{net}}^2}\right\}, \tag{12.126}$$

holding for any $a > 0$. We use again Lemma 12.7 to bound the expected value appearing on the LHS of (12.126). Specifically, examining the claim of Lemma 12.7, we see that the function g defined by (12.124) corresponds to the particular setup where the loss function is the identity function (and, hence, $\mathscr{L} = 1$) and the labels are deterministically equal to 1; with these choices, Eq. (12.167) gives

$$\mathbb{E}\sup_{h\in\mathcal{H}} \left|\widehat{\boldsymbol{\eta}}(h) - \eta(h)\right| \leq 2\rho_{\text{net}}. \tag{12.127}$$

Using this bound in (12.126) and setting $a = y - 2\rho_{\text{net}}$ (for $y > 2\rho_{\text{net}}$) yields (12.102).

■

12.C Appendix: Proof of Theorem 12.1

Proof. From Lemma 12.3 we obtain the lower bound in (12.85) for the probability of consistent learning. Next, we need to examine each of the terms on the RHS of (12.85).

Regarding the first term, by taking the supremum over the family of functions and then applying the uniform bound provided by (12.102) we can write

$$\begin{aligned}\mathbb{P}\Big[\big|\widehat{\boldsymbol{\eta}}(\boldsymbol{h}^o)-\eta(\boldsymbol{h}^o)\big|>y\Big] &\le \mathbb{P}\left[\sup_{h\in\mathcal{H}}\big|\widehat{\boldsymbol{\eta}}(h)-\eta(h)\big|\ge y\right]\\ &\le \exp\left\{\frac{-E_{\max}\,(y-2\rho_{\mathsf{net}})^2}{2\,h_{\mathsf{net}}^2}\right\}\\ &= \exp\left\{\frac{-2\,E_{\max}\,(y/2-\rho_{\mathsf{net}})^2}{h_{\mathsf{net}}^2}\right\}\end{aligned} \tag{12.128}$$

for any y such that

$$\frac{y}{2}>\rho_{\mathsf{net}}. \tag{12.129}$$

Next, we examine the second term on the RHS of (12.85). We call upon Lemma 12.6. In particular the functionals $S(h)$ and $T(h)$ and the related quantities $h_S^\star$ and $T^\star$ appearing in that lemma are chosen as follows. The functional $S(h)$ is chosen as the network risk $R(h)$, whereas the functional $T(h)$ is chosen as the network empirical risk $\widehat{\boldsymbol{R}}(h)$. Since we have[6]

$$\boldsymbol{h}^o=\arg\min_{h\in\mathcal{H}}\widehat{\boldsymbol{R}}(h),\qquad \mathsf{R}_{\mathsf{net}}^o=\inf_{h\in\mathcal{H}}R(h), \tag{12.130}$$

the minimizer $h_S^\star$ becomes $\boldsymbol{h}^o$ and $T^\star$ becomes the network target risk $\mathsf{R}_{\mathsf{net}}^o$. With these choices, from (12.164) we obtain

$$R(\boldsymbol{h}^o)-\mathsf{R}_{\mathsf{net}}^o\le 2\sup_{h\in\mathcal{H}}\left|\widehat{\boldsymbol{R}}(h)-R(h)\right|. \tag{12.131}$$

Choose now the parameter $y>0$ in the range of values that satisfy the following inequality:

$$\log\left(1+e^{-y}\right)>\mathsf{R}_{\mathsf{net}}^o. \tag{12.132}$$

These values certainly exist since $\mathsf{R}_{\mathsf{net}}^o<\log 2$ by assumption. In view of (12.131) we can write

$$\begin{aligned}&\mathbb{P}\left[R(\boldsymbol{h}^o)\ge\log\left(1+e^{-y}\right)\right]\\ &=\mathbb{P}\left[R(\boldsymbol{h}^o)-\mathsf{R}_{\mathsf{net}}^o\ge\log\left(1+e^{-y}\right)-\mathsf{R}_{\mathsf{net}}^o\right]\\ &\le\mathbb{P}\left[\sup_{h\in\mathcal{H}}\left|\widehat{\boldsymbol{R}}(h)-R(h)\right|\ge\frac{\log\left(1+e^{-y}\right)-\mathsf{R}_{\mathsf{net}}^o}{2}\right].\end{aligned} \tag{12.133}$$

The last probability can be upper bounded by using (12.101). Specifically, the threshold value y in (12.101) is replaced by the value $(\log(1+e^{-y})-\mathsf{R}_{\mathsf{net}}^o)/2$ and the loss function is chosen as $\mathscr{Q}(z)=\log(1+e^z)$ (yielding a Lipschitz constant $\mathscr{L}=1$), which corresponds

[6]Equation (12.130) follows by observing that *i)* the network risks in (12.99a) and (12.99b) are linear combinations, with positive weights, of the individual agent risks; and *ii)* the vector-valued function h is composed of the individual agent functions h_k.

to the binary cross-entropy risk adopted in our framework. With these choices, from (12.101) we get

$$\begin{aligned}\mathbb{P}\left[\sup_{h\in\mathcal{H}}\left|\widehat{\boldsymbol{R}}(h)-R(h)\right|\geq\frac{1}{2}\Big(\log\left(1+e^{-y}\right)-\mathsf{R}^{o}_{\mathsf{net}}\Big)\right]\\\leq\exp\left\{-\frac{E_{\mathsf{max}}}{2\,h^2_{\mathsf{net}}}\left(\frac{\log\left(1+e^{-y}\right)-\mathsf{R}^{o}_{\mathsf{net}}}{2}-2\rho_{\mathsf{net}}\right)^2\right\}\\=\exp\left\{-\frac{2\,E_{\mathsf{max}}}{h^2_{\mathsf{net}}}\left(\frac{\log\left(1+e^{-y}\right)-\mathsf{R}^{o}_{\mathsf{net}}}{4}-\rho_{\mathsf{net}}\right)^2\right\},\end{aligned}\tag{12.134}$$

where the inequality holds for any y such that

$$\frac{\log\left(1+e^{-y}\right)-\mathsf{R}^{o}_{\mathsf{net}}}{4}>\rho_{\mathsf{net}}.\tag{12.135}$$

By introducing the auxiliary functions

$$e_1(y)\triangleq\left(\frac{y}{2}-\rho_{\mathsf{net}}\right)^2,\qquad e_2(y)\triangleq\left(\frac{\log(1+e^{-y})-\mathsf{R}^{o}_{\mathsf{net}}}{4}-\rho_{\mathsf{net}}\right)^2\tag{12.136}$$

and using (12.128) and (12.134) in (12.85), we obtain the following bound on the probability of consistent learning:

$$\begin{aligned}P_c&\geq1-\exp\left\{\frac{-2\,E_{\mathsf{max}}\,e_1(y)}{h^2_{\mathsf{net}}}\right\}-\exp\left\{\frac{-2\,E_{\mathsf{max}}\,e_2(y)}{h^2_{\mathsf{net}}}\right\}\\&\geq1-2\exp\left\{\frac{-2\,E_{\mathsf{max}}\,\min\{e_1(y),e_2(y)\}}{h^2_{\mathsf{net}}}\right\}.\end{aligned}\tag{12.137}$$

To find the tightest bound, we can maximize the quantity $\min\{e_1(y),e_2(y)\}$ over the parameter y, under constraints (12.129) and (12.135). To this end, observe that under these constraints the function $e_1(y)$ is an increasing function of y, while $e_2(y)$ is a decreasing function of y. Accordingly, if there exists a value $y^\star$ that satisfies the equality

$$e_1(y^\star)=e_2(y^\star)\tag{12.138}$$

and meets constraints (12.129) and (12.135), then the maximum of $\min\{e_1(y),e_2(y)\}$ computed under these constraints will be equal to $e_1(y^\star)=e_2(y^\star)$. We now show that such a solution $y^\star$ exists.

Since the terms within brackets appearing in both $e_1(y)$ and $e_2(y)$ from (12.136) are constrained to be positive, Eq. (12.138) corresponds to the equation

$$y^\star=\frac{\log(1+e^{-y^\star})-\mathsf{R}^{o}_{\mathsf{net}}}{2}\tag{12.139}$$

solved under constraints (12.129) and (12.135). Equation (12.139) can be written as

$$e^{\mathsf{R}^{o}_{\mathsf{net}}}\,e^{3y^\star}-e^{y^\star}-1=0.\tag{12.140}$$

Therefore, if we set $e^{y^\star}=z$, we must solve the third-order equation

$$e^{\mathsf{R}^{o}_{\mathsf{net}}}\,z^3-z-1=0,\tag{12.141}$$

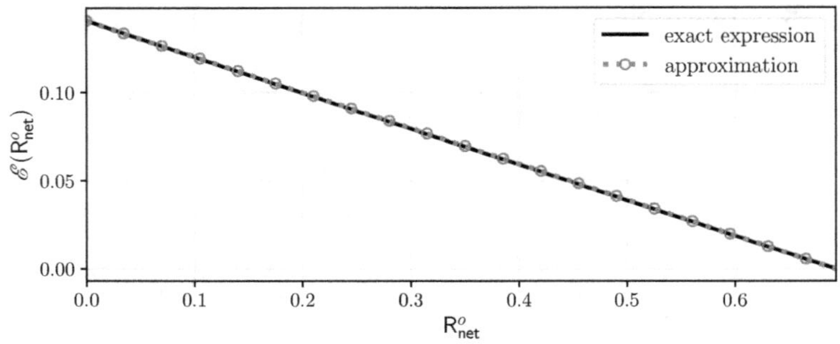

Figure 12.10: Comparison between the exact expression in (12.142) and the approximation in (12.148).

whose unique real-valued solution $z^\star$ is available in closed form as

$$z^\star = \frac{2 \times 3^{1/3} + 2^{1/3} e^{-\mathsf{R}^o_{\mathsf{net}}} [\mathsf{f}(\mathsf{R}^o_{\mathsf{net}})]^{2/3}}{6^{2/3} [\mathsf{f}(\mathsf{R}^o_{\mathsf{net}})]^{1/3}}, \tag{12.142}$$

where

$$\mathsf{f}(\mathsf{R}^o_{\mathsf{net}}) = 9e^{2\mathsf{R}^o_{\mathsf{net}}} + \sqrt{3e^{3\mathsf{R}^o_{\mathsf{net}}}(-4 + 27e^{\mathsf{R}^o_{\mathsf{net}}})}. \tag{12.143}$$

Recalling that $e^{y^\star} = z$, we have

$$y^\star = \log \frac{2 \times 3^{1/3} + 2^{1/3} e^{-\mathsf{R}^o_{\mathsf{net}}} [\mathsf{f}(\mathsf{R}^o_{\mathsf{net}})]^{2/3}}{6^{2/3} [\mathsf{f}(\mathsf{R}^o_{\mathsf{net}})]^{1/3}}. \tag{12.144}$$

It can be verified that $y^\star > 0$ within the range $\mathsf{R}^o_{\mathsf{net}} \in [0, \log 2]$. As a result, the inequality

$$\rho_{\mathsf{net}} < \frac{y^\star}{2} \tag{12.145}$$

is meaningful, in the sense that there exists a range of values of the network Rademacher complexity ρ_{net} that satisfy (12.145). When (12.145) holds, $y^\star$ meets constraint (12.129). Moreover, in view of (12.139), constraint (12.135) is also satisfied under (12.145). Finally, by defining

$$\mathscr{E}(\mathsf{R}^o_{\mathsf{net}}) \triangleq \frac{y^\star}{2} = \frac{1}{2} \log \frac{2 \times 3^{1/3} + 2^{1/3} e^{-\mathsf{R}^o_{\mathsf{net}}} [\mathsf{f}(\mathsf{R}^o_{\mathsf{net}})]^{2/3}}{6^{2/3} [\mathsf{f}(\mathsf{R}^o_{\mathsf{net}})]^{1/3}} \tag{12.146}$$

and evaluating (12.137) with $e_1(y^\star) = e_2(y^\star)$, we obtain the desired bound:

$$P_c \geq 1 - 2\exp\left\{ -\frac{2E_{\mathsf{max}}}{h^2_{\mathsf{net}}} \left(\mathscr{E}(\mathsf{R}^o_{\mathsf{net}}) - \rho_{\mathsf{net}} \right)^2 \right\}, \tag{12.147}$$

holding for $\rho_{\mathsf{net}} < \mathscr{E}(\mathsf{R}^o_{\mathsf{net}})$.

■

A good approximation for the function $\mathscr{E}(\mathsf{R}^o)$ is the linear fit

$$\mathscr{E}(\mathsf{R}^o_{\mathsf{net}}) \approx \mathscr{E}(0) \left(1 - \frac{\mathsf{R}^o_{\mathsf{net}}}{\log 2} \right), \tag{12.148}$$

where

$$\mathscr{E}(0) = 0.1406 \tag{12.149}$$

is computed from (12.146). Figure 12.10 shows the function $\mathscr{E}(\mathsf{R}^o_{\mathsf{net}})$ from (12.146), along with the linear fit from (12.148), for $\mathsf{R}^o_{\mathsf{net}} \in [0, \log 2]$. We see that the approximation is excellent.

12.D Appendix: Proof of Theorem 12.2

Proof. The proof relies on the bound in (12.57). Recall that this bound was obtained under the condition $\rho_{\mathsf{net}} < \mathscr{E}(\mathsf{R}^o_{\mathsf{net}})$. Since by assumption we have $\rho_k \le C_k/\sqrt{E_k}$, from the definition of ρ_{net} in (12.49) we have

$$\rho_{\mathsf{net}} \le \frac{C_{\mathsf{net}}}{\sqrt{E_{\mathsf{max}}}}, \tag{12.150}$$

where C_{net} was defined in (12.62). Accordingly, the condition $\rho_{\mathsf{net}} < \mathscr{E}(\mathsf{R}^o_{\mathsf{net}})$ is certainly verified if

$$\frac{C_{\mathsf{net}}}{\sqrt{E_{\mathsf{max}}}} < \mathscr{E}(\mathsf{R}^o_{\mathsf{net}}). \tag{12.151}$$

If condition (12.151) is satisfied, we can apply (12.57) and write

$$\begin{aligned} P_c &\ge 1 - 2\exp\left\{ -2\, E_{\mathsf{max}} \left(\frac{\mathscr{E}(\mathsf{R}^o_{\mathsf{net}}) - \rho_{\mathsf{net}}}{h_{\mathsf{net}}} \right)^2 \right\} \\ &\ge 1 - 2\exp\left\{ -\frac{2E_{\mathsf{max}}}{h^2_{\mathsf{net}}} \left(\mathscr{E}(\mathsf{R}^o_{\mathsf{net}}) - \frac{C_{\mathsf{net}}}{\sqrt{E_{\mathsf{max}}}} \right)^2 \right\}, \end{aligned} \tag{12.152}$$

where the last inequality follows from (12.150) and (12.151).

According to the claim of the theorem, we want to guarantee a minimum probability $1-\varepsilon$ of consistent learning, i.e.,

$$P_c \ge 1 - \varepsilon. \tag{12.153}$$

This condition is guaranteed if we impose that the last lower bound in (12.152) is not smaller than $1-\varepsilon$, which amounts to requiring

$$2\exp\left\{ -\frac{2E_{\mathsf{max}}}{h^2_{\mathsf{net}}} \left(\mathscr{E}(\mathsf{R}^o_{\mathsf{net}}) - \frac{C_{\mathsf{net}}}{\sqrt{E_{\mathsf{max}}}} \right)^2 \right\} \le \varepsilon, \tag{12.154}$$

or

$$\left(\sqrt{E_{\mathsf{max}}}\, \mathscr{E}(\mathsf{R}^o_{\mathsf{net}}) - C_{\mathsf{net}} \right)^2 \ge \frac{h^2_{\mathsf{net}}}{2} \log\left(\frac{2}{\varepsilon}\right). \tag{12.155}$$

In the range prescribed by constraint (12.151), the last inequality is satisfied when

$$\sqrt{E_{\mathsf{max}}}\, \mathscr{E}(\mathsf{R}^o_{\mathsf{net}}) \ge C_{\mathsf{net}} + \sqrt{\frac{h^2_{\mathsf{net}}}{2} \log\left(\frac{2}{\varepsilon}\right)}, \tag{12.156}$$

and the final result of the theorem is established by squaring both sides of (12.156). ∎

12.E Appendix: Auxiliary Results

The next three lemmas are used in the proofs of Theorem 12.1 and Lemma 12.4.

Lemma 12.5 (Difference of suprema). Assume that $S(h)$ and $T(h)$ are functionals of a function h belonging to a family $\mathcal{H}$, and consider the following quantities:

$$s_1 = \sup_{h\in\mathcal{H}} |S(h)|, \quad s_2 = \sup_{h\in\mathcal{H}} |S(h) + T(h)|. \tag{12.157}$$

Then

$$|s_1 - s_2| \leq \sup_{h\in\mathcal{H}} |T(h)| \tag{12.158}$$

Proof. The proof is split into two cases.

Case $s_2 \geq s_1$.

$$\begin{aligned} |s_1 - s_2| &= s_2 - s_1 \\ &= \sup_{h\in\mathcal{H}} |S(h) + T(h)| - \sup_{h\in\mathcal{H}} |S(h)| \\ &\leq \sup_{h\in\mathcal{H}} |S(h)| + \sup_{h\in\mathcal{H}} |T(h)| - \sup_{h\in\mathcal{H}} |S(h)| \\ &= \sup_{h\in\mathcal{H}} |T(h)|, \end{aligned} \tag{12.159}$$

where the inequality follows from the triangle inequality and the subadditivity of the supremum.

Case $s_2 < s_1$.

$$\begin{aligned} |s_1 - s_2| &= s_1 - s_2 \\ &= \sup_{h\in\mathcal{H}} |S(h)| - \sup_{h\in\mathcal{H}} |S(h) + T(h)| \\ &= \sup_{h\in\mathcal{H}} (|S(h)| - s_2) \\ &\overset{(a)}{\leq} \sup_{h\in\mathcal{H}} (|S(h)| - |S(h) + T(h)|) \\ &\leq \sup_{h\in\mathcal{H}} \Big| \, |S(h)| - |S(h) + T(h)| \, \Big| \\ &\overset{(b)}{\leq} \sup_{h\in\mathcal{H}} |S(h) - S(h) - T(h)| \\ &= \sup_{h\in\mathcal{H}} |T(h)|, \end{aligned} \tag{12.160}$$

where (a) follows because, from the definition of s_2, we have $-s_2 \leq -|S(h) + T(h)|$ for all h, and (b) follows from the reverse triangle inequality, i.e.,

$$|a - b| \geq \Big| \, |a| - |b| \, \Big|. \tag{12.161}$$

Grouping (12.159) and (12.160), we obtain the desired result in (12.158). ∎

Lemma 12.6 (Useful bound for wrong minimizers). Assume that $S(h)$ and $T(h)$ are functionals of a function h belonging to a family $\mathcal{H}$. Let

$$h_S^\star = \arg\min_{h\in\mathcal{H}} S(h) \tag{12.162}$$

be the minimizer of $S(h)$ and let

$$T^\star = \inf_{h\in\mathcal{H}} T(h) \tag{12.163}$$

be the infimum of $T(h)$. Then, the error $T(h_S^\star) - T^\star$, between the functional T evaluated at the minimizer of S and the infimum $T^\star$, can be related to the error between $S(h)$ and $T(h)$ through the following upper bound:

$$T(h_S^\star) - T^\star \le 2\sup_{h\in\mathcal{H}} |S(h) - T(h)|\,. \tag{12.164}$$

Proof. We have the following chain of equalities and inequalities:

$$\begin{aligned}
T(h_S^\star) - T^\star &= T(h_S^\star) - \inf_{h\in\mathcal{H}} T(h)\\
&= T(h_S^\star) - S(h_S^\star) + S(h_S^\star) - \inf_{h\in\mathcal{H}} T(h)\\
&= T(h_S^\star) - S(h_S^\star) + \sup_{h\in\mathcal{H}}\Big(S(h_S^\star) - T(h)\Big)\\
&\stackrel{\text{(a)}}{\le} T(h_S^\star) - S(h_S^\star) + \sup_{h\in\mathcal{H}}\Big(S(h) - T(h)\Big)\\
&\le 2\sup_{h\in\mathcal{H}} |S(h) - T(h)|\,,
\end{aligned} \tag{12.165}$$

where (a) follows from the fact that $S(h_S^\star) \le S(h)$ for all $h \in \mathcal{H}$ in view of (12.162), and the proof is complete. ∎

Lemma 12.7 (Useful bound for Lipschitz-continuous loss functions). Let Assumptions 5.1, 12.1, and 12.2 be satisfied. Let $h_k : \mathcal{X}_k \mapsto \mathbb{R}$ be a function belonging to a family $\mathcal{H}_k$. Denote by h the vector-valued function defined in (12.68) and by $\mathcal{H}$ the resulting family to which h belongs. Let also $\mathscr{Q} : \mathbb{R} \mapsto \mathbb{R}$ be an $\mathscr{L}$–Lipschitz function, and introduce the functional

$$\chi_k(h_k) \triangleq \mathbb{E}\mathscr{Q}\left(\widehat{\boldsymbol{\theta}}_{k,n}\, h_k(\widehat{\boldsymbol{x}}_{k,n})\right). \tag{12.166}$$

Then

$$\mathbb{E}\sup_{h\in\mathcal{H}}\left|\sum_{k=1}^{K} v_k\left[\chi_k(h_k)-\frac{1}{E_k}\sum_{n=1}^{E_k}\mathscr{Q}\left(\widehat{\boldsymbol{\theta}}_{k,n}\,h_k(\widehat{\boldsymbol{x}}_{k,n})\right)\right]\right|\le 2\mathscr{L}\,\rho_{\text{net}},\qquad(12.167)$$

where v_k is the kth entry of the Perron vector and ρ_{net} is the network Rademacher complexity defined by (12.49).

Proof. From the triangle inequality and the subadditivity of the supremum we can write

$$\begin{aligned}&\mathbb{E}\sup_{h\in\mathcal{H}}\left|\sum_{k=1}^{K} v_k\left[\chi_k(h_k)-\frac{1}{E_k}\sum_{n=1}^{E_k}\mathscr{Q}\left(\widehat{\boldsymbol{\theta}}_{k,n}\,h_k(\widehat{\boldsymbol{x}}_{k,n})\right)\right]\right|\\ &\le\sum_{k=1}^{K} v_k\,\mathbb{E}\sup_{h_k\in\mathcal{H}_k}\left|\chi_k(h_k)-\frac{1}{E_k}\sum_{n=1}^{E_k}\mathscr{Q}\left(\widehat{\boldsymbol{\theta}}_{k,n}\,h_k(\widehat{\boldsymbol{x}}_{k,n})\right)\right|.\end{aligned}\qquad(12.168)$$

Let us focus on the expected values appearing in the summation on the RHS of (12.168). From definition (12.166), owing to the identical distribution across n, we have the equality

$$\chi_k(h_k)=\frac{1}{E_k}\sum_{n=1}^{E_k}\mathbb{E}\,\mathscr{Q}\left(\widehat{\boldsymbol{\theta}}_{k,n}\,h_k(\widehat{\boldsymbol{x}}_{k,n})\right),\qquad(12.169)$$

which allows us to write

$$\begin{aligned}&\mathbb{E}\sup_{h_k\in\mathcal{H}_k}\left|\chi_k(h_k)-\frac{1}{E_k}\sum_{n=1}^{E_k}\mathscr{Q}\left(\widehat{\boldsymbol{\theta}}_{k,n}\,h_k(\widehat{\boldsymbol{x}}_{k,n})\right)\right|\\ &=\mathbb{E}_{x,\theta}\sup_{h_k\in\mathcal{H}_k}\left|\frac{1}{E_k}\sum_{n=1}^{E_k}\mathbb{E}_{x',\theta'}\mathscr{Q}\left(\widehat{\boldsymbol{\theta}}'_{k,n}\,h_k(\widehat{\boldsymbol{x}}'_{k,n})\right)-\frac{1}{E_k}\sum_{n=1}^{E_k}\mathscr{Q}\left(\widehat{\boldsymbol{\theta}}_{k,n}\,h_k(\widehat{\boldsymbol{x}}_{k,n})\right)\right|,\end{aligned}\qquad(12.170)$$

where we introduced a *fictitious* training set

$$\mathcal{T}'_k\triangleq\left\{\widehat{\boldsymbol{x}}'_{k,n},\widehat{\boldsymbol{\theta}}'_{k,n}\right\}_{n=1}^{E_k}\overset{\text{d}}{=}\mathcal{T}_k,\qquad(12.171)$$

with the equality meaning that $\mathcal{T}'_k$ shares the same distribution as the original training set $\mathcal{T}_k$ defined by (12.2). Moreover, we assume that $\mathcal{T}_k$ and $\mathcal{T}'_k$ are statistically independent.

From (12.170) we can also write

$$\begin{aligned}&\mathbb{E}\sup_{h_k\in\mathcal{H}_k}\left|\chi_k(h_k)-\frac{1}{E_k}\sum_{n=1}^{E_k}\mathscr{Q}\left(\widehat{\boldsymbol{\theta}}_{k,n}\,h_k(\widehat{\boldsymbol{x}}_{k,n})\right)\right|\\ &=\mathbb{E}_{x,\theta}\sup_{h_k\in\mathcal{H}_k}\left|\frac{1}{E_k}\sum_{n=1}^{E_k}\mathbb{E}_{x',\theta'}\left[\mathscr{Q}\left(\widehat{\boldsymbol{\theta}}'_{k,n}\,h_k(\widehat{\boldsymbol{x}}'_{k,n})\right)-\mathscr{Q}\left(\widehat{\boldsymbol{\theta}}_{k,n}\,h_k(\widehat{\boldsymbol{x}}_{k,n})\right)\right]\right|\\ &\le\mathbb{E}\sup_{h_k\in\mathcal{H}_k}\left|\frac{1}{E_k}\sum_{n=1}^{E_k}\left[\mathscr{Q}\left(\widehat{\boldsymbol{\theta}}'_{k,n}\,h_k(\widehat{\boldsymbol{x}}'_{k,n})\right)-\mathscr{Q}\left(\widehat{\boldsymbol{\theta}}_{k,n}\,h_k(\widehat{\boldsymbol{x}}_{k,n})\right)\right]\right|.\end{aligned}\qquad(12.172)$$

Note that, when necessary, we have used the subscripts x, θ and x', θ' to distinguish which random quantities the expectation is taken over. The inequality in (12.172) holds since the absolute value of the expectation is upper bounded by the expectation of the absolute value, and the supremum of the expectation is upper bounded by the expectation of the supremum.

To complete the proof, inspired by the arguments used in [13, 30], we develop the following symmetrization procedure. Let us focus on the last term in (12.172). We have the identity

$$\begin{aligned}
&\mathbb{E} \sup_{h_k \in \mathcal{H}_k} \left| \frac{1}{E_k} \sum_{n=1}^{E_k} (+1) \times \left[\mathscr{Q}\left(\widehat{\boldsymbol{\theta}}'_{k,n}\, h_k(\widehat{\boldsymbol{x}}'_{k,n})\right) - \mathscr{Q}\left(\widehat{\boldsymbol{\theta}}_{k,n}\, h_k(\widehat{\boldsymbol{x}}_{k,n})\right) \right] \right| \\
&= \mathbb{E} \sup_{h_k \in \mathcal{H}_k} \left| \frac{1}{E_k} \sum_{n=1}^{E_k} (-1) \times \left[\mathscr{Q}\left(\widehat{\boldsymbol{\theta}}'_{k,n}\, h_k(\widehat{\boldsymbol{x}}'_{k,n})\right) - \mathscr{Q}\left(\widehat{\boldsymbol{\theta}}_{k,n}\, h_k(\widehat{\boldsymbol{x}}_{k,n})\right) \right] \right|, \qquad (12.173)
\end{aligned}$$

which follows from the fact that the training sets $\mathcal{T}_k$ and $\mathcal{T}'_k$ are iid and, hence, exchanging them is immaterial. Consider now a sequence of iid Rademacher random variables $\boldsymbol{r}_n$ (i.e., binary variables taking on values ± 1 with equal probability). Furthermore, assume that the sequences $\{\boldsymbol{r}_n\}_{n=1}^{E_k}$, $\mathcal{T}_k$, and $\mathcal{T}'_k$ are mutually independent. In view of (12.173), the last term in (12.172) can also be written as

$$\mathbb{E} \sup_{h_k \in \mathcal{H}_k} \left| \frac{1}{E_k} \sum_{n=1}^{E_k} \boldsymbol{r}_n \left[\mathscr{Q}\left(\widehat{\boldsymbol{\theta}}'_{k,n}\, h_k(\widehat{\boldsymbol{x}}'_{k,n})\right) - \mathscr{Q}\left(\widehat{\boldsymbol{\theta}}_{k,n}\, h_k(\widehat{\boldsymbol{x}}_{k,n})\right) \right] \right|, \qquad (12.174)$$

where the expectation is taken over all involved random variables, including the Rademacher variables $\boldsymbol{r}_n$.

The quantity appearing in (12.174) can be bounded as follows:

$$\begin{aligned}
&\mathbb{E} \sup_{h_k \in \mathcal{H}_k} \left| \frac{1}{E_k} \sum_{n=1}^{E_k} \boldsymbol{r}_n \left[\mathscr{Q}\left(\widehat{\boldsymbol{\theta}}'_{k,n}\, h_k(\widehat{\boldsymbol{x}}'_{k,n})\right) - \mathscr{Q}\left(\widehat{\boldsymbol{\theta}}_{k,n}\, h_k(\widehat{\boldsymbol{x}}_{k,n})\right) \right] \right| \\
&\stackrel{\text{(a)}}{=} \mathbb{E} \sup_{h_k \in \mathcal{H}_k} \left| \frac{1}{E_k} \sum_{n=1}^{E_k} \boldsymbol{r}_n \left[\widetilde{\mathscr{Q}}\left(\widehat{\boldsymbol{\theta}}'_{k,n}\, h_k(\widehat{\boldsymbol{x}}'_{k,n})\right) - \widetilde{\mathscr{Q}}\left(\widehat{\boldsymbol{\theta}}_{k,n}\, h_k(\widehat{\boldsymbol{x}}_{k,n})\right) \right] \right| \\
&\stackrel{\text{(b)}}{\leq} \mathbb{E} \sup_{h_k \in \mathcal{H}_k} \left\{ \left| \frac{1}{E_k} \sum_{n=1}^{E_k} \boldsymbol{r}_n \widetilde{\mathscr{Q}}\left(\widehat{\boldsymbol{\theta}}'_{k,n}\, h_k(\widehat{\boldsymbol{x}}'_{k,n})\right) \right| + \left| \frac{1}{E_k} \sum_{n=1}^{E_k} \boldsymbol{r}_n \widetilde{\mathscr{Q}}\left(\widehat{\boldsymbol{\theta}}_{k,n}\, h_k(\widehat{\boldsymbol{x}}_{k,n})\right) \right| \right\} \\
&\stackrel{\text{(c)}}{=} 2\mathbb{E} \sup_{h_k \in \mathcal{H}_k} \left| \frac{1}{E_k} \sum_{n=1}^{E_k} \boldsymbol{r}_n \widetilde{\mathscr{Q}}\left(\widehat{\boldsymbol{\theta}}_{k,n}\, h_k(\widehat{\boldsymbol{x}}_{k,n})\right) \right|. \qquad (12.175)
\end{aligned}$$

In (a) we introduced the shifted function

$$\widetilde{\mathscr{Q}}(z) \triangleq \mathscr{Q}(z) - \mathscr{Q}(0). \qquad (12.176)$$

We note that $\widetilde{\mathscr{Q}}(0) = 0$ and that $\widetilde{\mathscr{Q}}(z)$ is $\mathscr{L}$-Lipschitz since so is $\mathscr{Q}(z)$ and since Lipschitz continuity is shift-invariant — see (G.5). Step (b) applies the triangle inequality, while (c) follows from the subadditivity of the supremum and the fact that $\mathcal{T}_k$ and $\mathcal{T}'_k$ are identically distributed.

We now want to upper bound the last term in (12.175) by exploiting the Lipschitz property of $\widetilde{\mathscr{Q}}(z)$ associated with the contraction principle of the Rademacher complexity (Lemma G.1). Specifically, for $n = 1, 2, \ldots, E_k$, we consider the samples

$$\boldsymbol{\xi}_n \triangleq \text{col}\left\{\widehat{\boldsymbol{x}}_{k,n}, \widehat{\boldsymbol{\theta}}_{k,n}\right\} \tag{12.177}$$

and the function family $\mathcal{G}$ composed of the functions

$$g(\boldsymbol{\xi}_n) = \widehat{\boldsymbol{\theta}}_{k,n}\, h_k(\widehat{\boldsymbol{x}}_{k,n}), \tag{12.178}$$

where h_k spans the family $\mathcal{H}_k$. Applying Lemma G.1 with these choices, we obtain

$$\begin{aligned}\mathbb{E} \sup_{h_k \in \mathcal{H}_k} \left| \frac{1}{E_k} \sum_{n=1}^{E_k} \boldsymbol{r}_n \widetilde{\mathscr{Q}}\left(\widehat{\boldsymbol{\theta}}_{k,n}\, h_k(\widehat{\boldsymbol{x}}_{k,n})\right)\right| &= \mathbb{E} \sup_{h_k \in \mathcal{H}_k} \left| \frac{1}{E_k} \sum_{n=1}^{E_k} \boldsymbol{r}_n \widetilde{\mathscr{Q}}(g(\boldsymbol{\xi}_n))\right| \\ \le \mathscr{L}\, \mathbb{E} \sup_{h_k \in \mathcal{H}_k} \left| \frac{1}{E_k} \sum_{n=1}^{E_k} \boldsymbol{r}_n\, g(\boldsymbol{\xi}_n)\right| &= \mathscr{L}\, \mathbb{E} \sup_{h_k \in \mathcal{H}_k} \left| \frac{1}{E_k} \sum_{n=1}^{E_k} \boldsymbol{r}_n\, \widehat{\boldsymbol{\theta}}_{k,n}\, h_k(\widehat{\boldsymbol{x}}_{k,n})\right|. \end{aligned} \tag{12.179}$$

It can be readily verified that the random variables $\boldsymbol{r}_n \widehat{\boldsymbol{\theta}}_{k,n}\, h_k(\widehat{\boldsymbol{x}}_{k,n})$ and $\boldsymbol{r}_n\, h_k(\widehat{\boldsymbol{x}}_{k,n})$ share the same distribution since $\widehat{\boldsymbol{\theta}}_{k,n}$ assumes values ± 1 with equal probability and $\boldsymbol{r}_n$ and $-\boldsymbol{r}_n$ are equally distributed and independent of the pairs $(\widehat{\boldsymbol{x}}_{k,n}, \widehat{\boldsymbol{\theta}}_{k,n})$. Since we also have independence across n, the equality in distribution holding for the individual n extends to the whole sequences. Therefore, we can write

$$\begin{aligned} &2\mathscr{L}\, \mathbb{E} \sup_{h_k \in \mathcal{H}_k} \left| \frac{1}{E_k} \sum_{n=1}^{E_k} \boldsymbol{r}_n\, \widehat{\boldsymbol{\theta}}_{k,n}\, h_k(\widehat{\boldsymbol{x}}_{k,n})\right| \\ &\quad = 2\mathscr{L}\, \mathbb{E} \sup_{h_k \in \mathcal{H}_k} \left| \frac{1}{E_k} \sum_{n=1}^{E_k} \boldsymbol{r}_n\, h_k(\widehat{\boldsymbol{x}}_{k,n})\right| = 2\mathscr{L} \rho_k, \end{aligned} \tag{12.180}$$

where the last equality follows from the definition of the Rademacher complexity (Definition G.1). Substituting (12.180) into (12.179) and using the resulting bound in (12.175) yields

$$\mathbb{E} \sup_{h_k \in \mathcal{H}_k} \left| \frac{1}{E_k} \sum_{n=1}^{E_k} \boldsymbol{r}_n \left[\mathscr{Q}\left(\widehat{\boldsymbol{\theta}}'_{k,n}\, h_k(\widehat{\boldsymbol{x}}'_{k,n})\right) - \mathscr{Q}\left(\widehat{\boldsymbol{\theta}}_{k,n}\, h_k(\widehat{\boldsymbol{x}}_{k,n})\right)\right]\right| \le 2\mathscr{L}\rho_k. \tag{12.181}$$

Recalling that the LHS of (12.181) is equal to the RHS of (12.172), we conclude that

$$\mathbb{E} \sup_{h_k \in \mathcal{H}_k} \left| \chi_k(h_k) - \frac{1}{E_k} \sum_{n=1}^{E_k} \mathscr{Q}\left(\widehat{\boldsymbol{\theta}}_{k,n}\, h_k(\widehat{\boldsymbol{x}}_{k,n})\right)\right| \le 2\mathscr{L}\rho_k. \tag{12.182}$$

Combining this result with (12.168) and recalling the definition of ρ_{net} from (12.49), we obtain (12.167), which is the claim of the lemma.

■

Chapter 13

Extensions and Conclusions

In this concluding chapter we give an overview of some recent advances on social learning, which are the subject of ongoing investigations.

13.1 Non-Bayesian Updates

As explained in Chapter 3, in *non-Bayesian* social learning the agents form their beliefs by iterating the following procedure. During the self-learning step, each agent performs an individual Bayesian update and then shares it with its neighbors. During the combination step, each agent blends the received beliefs according to some pooling rule. Even if the local updates are Bayesian, the overall learning scheme is *non-Bayesian*, since it does not amount to computing the overall posterior distribution given the data from all agents.

Let us focus on the objective evidence model described in Section 5.3, where a true underlying hypothesis ϑ^o exists for all agents. In Chapters 5 and 7 we showed that, despite being non-Bayesian, traditional social learning schemes learn well, in the sense that the full belief mass is placed on the true hypothesis as the number of observations grows. This implies in particular that the probability of choosing the true hypothesis, e.g., by selecting the maximum entry of the belief vector, converges to 1 or, equivalently, that the error probability vanishes as $t \to \infty$. However, we do not know whether traditional social learning schemes reach the best attainable performance, or how they compare with benchmark schemes. One benchmark scheme to assess the goodness of a social learning strategy is a *centralized* Bayesian construction that has access to all agents' data and computes the Bayesian posterior over them. Some fundamental questions arise. *How much does non-Bayesian learning lose with respect to the centralized Bayesian scheme? Can we modify traditional non-Bayesian schemes to attain improved performance?*

One meaningful performance index to compare decision-making schemes is the error probability, which is unfortunately too difficult to compute for general statistical models. However, in Chapter 6 we characterized the decay rate to 0 of the error probability for social learning with geometric averaging. In particular, we proved that, under suitable conditions, the error probabilities of all agents vanish exponentially fast and we illustrated a procedure to evaluate the error exponent that rules this decay. In this section we will use the error exponent as a performance index to compare different schemes.

The analysis presented in this section stems from the work started in [23], where the NB^2 (non-Bayesian learning with non-Bayesian updates) strategy is introduced. This strategy uses the following update (recall (2.90)):

$$\psi_{k,t}(\theta) \propto \mu_{k,t-1}(\theta)\ell_k^{\gamma_k}(x_{k,t}|\theta), \quad \gamma_k > 0, \tag{13.1}$$

which departs from the Bayesian update used in traditional social learning (except for $\gamma_k = 1$). We will discuss the relevance of this modified update in the rest of this section. In particular, we will examine the performance of the centralized Bayesian scheme for the case where the data are independent across the agents, and for the case where the agents are partitioned into clusters with data highly dependent within the same cluster. Then, we will compare the centralized Bayesian scheme against traditional social learning and the NB^2 strategy.

13.1.1 Performance Results

In the following analysis, we assume that the conditions used in Theorem 6.3 are verified. Moreover, we stick to the objective evidence model in Section 5.3, where the observations collected by each agent k are distributed according to $\ell_k(x|\vartheta^o)$, for a true underlying hypothesis $\vartheta^o \in \Theta$ common to all agents.

Centralized Bayesian scheme. Let

$$\boldsymbol{x}_{\mathsf{cen},t} \triangleq \mathrm{col}\{\boldsymbol{x}_{1,t}, \boldsymbol{x}_{2,t}, \ldots, \boldsymbol{x}_{K,t}\} \tag{13.2}$$

be the vector collecting all agents' data at time t. Let us further denote by $\boldsymbol{\mu}_{\mathsf{cen},t}$ the belief vector of the centralized system at time t, and by

$\ell(x|\theta)$ (no subscript k here) the likelihood for the global data $\boldsymbol{x}_{\mathsf{cen},t}$. Given the true hypothesis ϑ^o, the data are assumed to be drawn from the joint statistical model $\ell(x|\vartheta^o)$. Note that the specific form of $\ell(x|\theta)$ depends on the statistical dependence across the agents. We will write this likelihood explicitly for some cases examined next. However, for the results in this section to hold, we do not need to assume any specific form for it.

Since for the centralized scheme we consider the *joint* model ℓ_θ rather than the marginal models $\ell_{k,\theta}$, the finiteness of the KL divergences assumed in (5.37) must be rephrased in terms of this joint model. In other words, we assume that for all pairs (θ, θ'),

$$D(\ell_\theta || \ell_{\theta'}) < \infty. \tag{13.3}$$

By exploiting independence and identical distribution over time, we have that the Bayesian posterior at time t is given by

$$\boldsymbol{\mu}_{\mathsf{cen},t}(\theta) \propto \mu_{\mathsf{cen},0}(\theta) \prod_{\tau=1}^{t} \ell(\boldsymbol{x}_{\mathsf{cen},\tau}|\theta). \tag{13.4}$$

where we assume that $\mu_{\mathsf{cen},0}(\theta) > 0$ for all $\theta \in \Theta$. From (13.4) we compute the centralized log belief ratio

$$\boldsymbol{\beta}_{\mathsf{cen},t}(\theta) \triangleq \log \frac{\boldsymbol{\mu}_{\mathsf{cen},t}(\vartheta^o)}{\boldsymbol{\mu}_{\mathsf{cen},t}(\theta)} = \log \frac{\mu_{\mathsf{cen},0}(\vartheta^o)}{\mu_{\mathsf{cen},0}(\theta)} + \sum_{\tau=1}^{t} \boldsymbol{\lambda}_{\mathsf{cen},\tau}(\theta), \tag{13.5}$$

where

$$\boldsymbol{\lambda}_{\mathsf{cen},\tau}(\theta) \triangleq \log \frac{\ell(\boldsymbol{x}_{\mathsf{cen},\tau}|\vartheta^o)}{\ell(\boldsymbol{x}_{\mathsf{cen},\tau}|\theta)}. \tag{13.6}$$

To quantify the performance, we focus on the large deviation analysis and evaluate the pertinent error exponent. We know from Chapter 6 that we need to examine the asymptotic behavior of the log belief ratio divided by t,

$$\bar{\boldsymbol{\beta}}_{\mathsf{cen},t}(\theta) \triangleq \frac{1}{t}\boldsymbol{\beta}_{\mathsf{cen},t}(\theta) = \frac{1}{t} \log \frac{\mu_{\mathsf{cen},0}(\vartheta^o)}{\mu_{\mathsf{cen},0}(\theta)} + \frac{1}{t}\sum_{\tau=1}^{t} \boldsymbol{\lambda}_{\mathsf{cen},\tau}(\theta). \tag{13.7}$$

It can be verified that the vanishing term that depends on the initial state is immaterial to the evaluation of the error exponent.[1] Therefore, neglecting this term, the RHS of (13.7) is an empirical average of iid variables, which is the traditional case addressed by Theorem E.1. Applying this theorem, we

[1]One easy way to show that the term depending on the initial state is immaterial is to call upon Theorem E.2.

conclude in particular that the error probability $\mathbb{P}[\boldsymbol{\beta}_{\mathsf{cen},t}(\theta) \leq 0]$ vanishes exponentially fast as $t \to \infty$, and that the error exponent ruling this convergence is computed as follows. First, we introduce the LMGF of $\boldsymbol{\lambda}_{\mathsf{cen},t}(\theta)$,

$$\Lambda_{\mathsf{cen}}(s;\theta) \triangleq \log \mathbb{E} \exp \left\{ s\, \boldsymbol{\lambda}_{\mathsf{cen},t}(\theta) \right\}. \tag{13.8}$$

Then, we introduce the Fenchel-Legendre transform of $\Lambda_{\mathsf{cen}}(s;\theta)$,

$$\Lambda^*_{\mathsf{cen}}(y;\theta) = \sup_{s\in\mathbb{R}} \Big(sy - \Lambda_{\mathsf{cen}}(s;\theta) \Big). \tag{13.9}$$

Finally, the error exponent corresponds to this Fenchel-Legendre transform evaluated at the decision threshold $y = 0$:

$$\mathscr{E}_{\mathsf{cen}}(\theta, \vartheta^o) \triangleq \Lambda^*_{\mathsf{cen}}(0;\theta) = -\inf_{s\in\mathbb{R}} \Lambda_{\mathsf{cen}}(s;\theta). \tag{13.10}$$

Note that in these calculations there is an implicit dependence on the true hypothesis ϑ^o (since expectations are computed relative to the true distribution defined by ϑ^o). This dependence is now made explicit by the notation $\mathscr{E}_{\mathsf{cen}}(\theta, \vartheta^o)$.

Using (13.6) in (13.8), we can further represent the LMGF of the log likelihood ratio as

$$\Lambda_{\mathsf{cen}}(s;\theta) = \log \mathbb{E} \left[\left(\frac{\ell(\boldsymbol{x}_{\mathsf{cen},t}|\vartheta^o)}{\ell(\boldsymbol{x}_{\mathsf{cen},t}|\theta)} \right)^s \right], \tag{13.11}$$

where we recall that the expectation is computed under the true hypothesis ϑ^o. Substituting (13.11) into (13.10), the error exponent can be rewritten as

$$\mathscr{E}_{\mathsf{cen}}(\theta, \vartheta^o) = \Lambda^*_{\mathsf{cen}}(0;\theta) = -\inf_{s\in\mathbb{R}} \log \mathbb{E} \left[\left(\frac{\ell(\boldsymbol{x}_{\mathsf{cen},t}|\vartheta^o)}{\ell(\boldsymbol{x}_{\mathsf{cen},t}|\theta)} \right)^s \right], \tag{13.12}$$

which, as was seen in Example 6.4, is referred to as the Chernoff information between $\ell(x|\vartheta^o)$ and $\ell(x|\theta)$ [59, 60]. The exponent $\mathscr{E}_{\mathsf{cen}}(\theta, \vartheta^o)$ characterizes the decay rate, as $t \to \infty$, of the probability of choosing θ in place of ϑ^o:

$$\mathbb{P}[\boldsymbol{\beta}_{\mathsf{cen},t}(\theta) \leq 0] \doteq e^{-\mathscr{E}_{\mathsf{cen}}(\theta,\vartheta^o)\, t}. \tag{13.13}$$

The probability of error given ϑ^o is dominated by the worst-case (i.e., minimum) Chernoff information across the hypotheses $\theta \neq \vartheta^o$:

$$\mathbb{P}\left[\vartheta^o \neq \arg\max_{\theta\in\Theta} \boldsymbol{\mu}_{\mathsf{cen},t}(\theta) \right] \doteq e^{-\min\limits_{\theta\neq\vartheta^o} \mathscr{E}_{\mathsf{cen}}(\theta,\vartheta^o)\, t}. \tag{13.14}$$

Finally, the *total* error probability, (i.e., the probability averaged over all true hypotheses ϑ^o by using the prior $\mu_{\mathsf{cen},0}$), is ruled by the minimum Chernoff information across *all pairs of hypotheses*:

$$\sum_{\vartheta^o \in \Theta} \mu_{\mathsf{cen},0}(\vartheta^o)\, \mathbb{P}\left[\vartheta^o \neq \arg\max_{\theta \in \Theta} \boldsymbol{\mu}_{\mathsf{cen},t}(\theta)\right] \doteq e^{-\min\limits_{\vartheta^o} \min\limits_{\theta \neq \vartheta^o} \mathscr{E}_{\mathsf{cen}}(\theta,\vartheta^o)\, t}. \quad (13.15)$$

It is well known that the MAP rule (which in this case amounts to choosing the hypothesis that maximizes the centralized Bayesian posterior) attains the best (i.e., the smallest) total error probability. In view of (13.15), this implies that the best (i.e., the largest) attainable error exponent for the total error probability is given by the minimum Chernoff information, and is also attained with the centralized Bayesian posterior [107].

Traditional social learning. We have seen in Chapter 6 that the performance of traditional social learning with geometric averaging can be characterized, for large t, in terms of the *network* random variables

$$\boldsymbol{\lambda}_{\mathsf{net},t}(\theta) = \sum_{k=1}^{K} v_k \boldsymbol{\lambda}_{k,t}(\theta). \quad (13.16)$$

More specifically, the large deviation performance obtained in Theorem 6.3 reveals that the error exponent is given by

$$\mathscr{E}_{\mathsf{net}}(\theta, \vartheta^o) \triangleq -\inf_{s \in \mathbb{R}} \Lambda_{\mathsf{net}}(s;\theta), \quad (13.17)$$

where

$$\Lambda_{\mathsf{net}}(s;\theta) = \log \mathbb{E} \exp\left\{s\, \boldsymbol{\lambda}_{\mathsf{net},t}(\theta)\right\} \quad (13.18)$$

is the LMGF of $\boldsymbol{\lambda}_{\mathsf{net},t}$. Therefore, the random variable $\boldsymbol{\lambda}_{\mathsf{net},t}(\theta)$ is all we need to evaluate the error exponent $\mathscr{E}_{\mathsf{net}}(\theta, \vartheta^o)$. Likewise, for the centralized Bayesian scheme we need the random variable $\boldsymbol{\lambda}_{\mathsf{cen},t}(\theta)$ defined by (13.6). The variable $\boldsymbol{\lambda}_{\mathsf{cen},t}(\theta)$ is a log likelihood ratio pertaining to the optimal centralized Bayesian scheme. Therefore, its specific form depends on the joint distribution of the data across the agents. In comparison, we see from (13.16) that $\boldsymbol{\lambda}_{\mathsf{net},t}(\theta)$ is a linear combination of the log likelihood ratios of the individual agents. This is a direct consequence of the fact that the combination step in listing (3.16) is a geometric-averaging rule, amounting to a linear combination in the log domain. Furthermore, the linear combination in (13.16) is weighted by the entries of the Perron vector. In other words, the network topology plays a role in the learning behavior. In contrast, the optimal Bayesian scheme being centralized, there is no topology influence on it.

It is now legitimate to ask what is the *performance loss* introduced by the constrained structure of $\boldsymbol{\lambda}_{\mathsf{net},t}(\theta)$, and if there are alternative schemes to suitably modify this structure. To answer, we start by introducing the NB^2 strategy from [23].

NB² strategy. In the context of distributed Bayesian filtering, it has been observed that modifying the Bayesian update by raising the likelihoods to some constant power (equal for all agents) can reduce the error in tracking the centralized Bayesian posterior [92, 99]. Starting from this observation, in [23] a social learning scheme with *non-Bayesian updates* is introduced. This non-Bayesian learning scheme with non-Bayesian updates (NB^2) is summarized in listing (13.19). The combination step is the geometric-averaging rule. The fundamental modification lies in the update step, where the likelihood of each agent k is raised to some positive constant γ_k. Differently from what was proposed in [92, 99], the constant γ_k is allowed to be agent-dependent, a property that will be shown to be critical in the sequel.

NB^2: Social learning with non-Bayesian updates

start from the prior belief vectors $\mu_{k,0}$ for $k = 1, 2, \ldots, K$
choose the update parameters $\gamma_k > 0$ for $k = 1, 2, \ldots, K$
for $t = 1, 2, \ldots$
  for $k = 1, 2, \ldots, K$
    agent k observes $x_{k,t}$
    for $\theta = 1, 2, \ldots, H$

$$\psi_{k,t}(\theta) = \frac{\mu_{k,t-1}(\theta)\ell_k^{\gamma_k}(x_{k,t}|\theta)}{\sum_{\theta' \in \Theta} \mu_{k,t-1}(\theta')\ell_k^{\gamma_k}(x_{k,t}|\theta')} \qquad \text{(self-learning)}$$

    end
  end

  for $k = 1, 2, \ldots, K$
    for $\theta = 1, 2, \ldots, H$

$$\mu_{k,t}(\theta) = \frac{\prod_{j \in \mathcal{N}_k} [\psi_{j,t}(\theta)]^{a_{jk}}}{\sum_{\theta' \in \Theta} \prod_{j \in \mathcal{N}_k} [\psi_{j,t}(\theta')]^{a_{jk}}} \qquad \text{(cooperation)}$$

    end
  end
end

(13.19)

Unfolding the recursion arising from the algorithm described in listing (13.19), it is readily seen that the only modification with respect to the recursion obtained from traditional social learning is that the log likelihood ratios $\boldsymbol{\lambda}_{j,t}$ in (6.27) are now multiplied by the update constants γ_j. This implies that in the NB^2 case the relevant network variable is

$$\boldsymbol{\lambda}_{\mathsf{NB}^2,t}(\theta) \triangleq \sum_{k=1}^{K} v_k \gamma_k \boldsymbol{\lambda}_{k,t}(\theta) \tag{13.20}$$

and that the error exponents can be evaluated by computing the LMGF

$$\Lambda_{\mathsf{NB}^2}(s;\theta) \triangleq \log \mathbb{E} \exp\left\{ s\, \boldsymbol{\lambda}_{\mathsf{NB}^2,t}(\theta)\right\} \tag{13.21}$$

and then, for the NB^2 strategy, the error exponent relative to hypotheses θ and ϑ^o is

$$\mathscr{E}_{\mathsf{NB}^2}(\theta,\vartheta^o) \triangleq -\inf_{s\in\mathbb{R}} \Lambda_{\mathsf{NB}^2}(s;\theta). \tag{13.22}$$

13.1.2 Independent Agents

In this section we examine the case where the data are independent across the agents. Accordingly, the log likelihood ratio of the optimal centralized Bayesian scheme becomes

$$\boldsymbol{\lambda}_{\mathsf{cen},t}(\theta) = \sum_{k=1}^{K} \boldsymbol{\lambda}_{k,t}(\theta). \tag{13.23}$$

We start by comparing traditional social learning against the centralized Bayesian scheme.

Consider first what happens when the combination matrix is doubly stochastic, which implies that $v_k = 1/K$, yielding, in view of (13.20),

$$\boldsymbol{\lambda}_{\mathsf{net},t}(\theta) = \frac{1}{K}\sum_{k=1}^{K} \boldsymbol{\lambda}_{k,t}(\theta). \tag{13.24}$$

Comparing (13.24) against (13.23), we see that, for doubly stochastic matrices, $\boldsymbol{\lambda}_{\mathsf{net},t}(\theta)$ and $\boldsymbol{\lambda}_{\mathsf{cen},t}(\theta)$ differ by a scaling constant K, which suggests that in this case the decentralized and centralized schemes should behave similarly in terms of decision performance. We now show that this is the case by examining the error exponents. In view of (13.24) and (13.23), the LMGFs of $\boldsymbol{\lambda}_{\mathsf{net},t}(\theta)$ and $\boldsymbol{\lambda}_{\mathsf{cen},t}(\theta)$ are related as follows:

$$\begin{aligned}\Lambda_{\mathsf{net}}(s;\theta) &= \log \mathbb{E} \exp\left\{ s\, \boldsymbol{\lambda}_{\mathsf{net},t}(\theta)\right\}\\ &= \log \mathbb{E} \exp\left\{ s\, \boldsymbol{\lambda}_{\mathsf{cen},t}(\theta)/K\right\} = \Lambda_{\mathsf{cen}}(s/K;\theta).\end{aligned} \tag{13.25}$$

Using (13.25) along with (13.17) and (13.10), we have

$$\begin{aligned}\mathscr{E}_{\mathsf{net}}(\theta,\vartheta^o) &= -\inf_{s\in\mathbb{R}} \Lambda_{\mathsf{net}}(s;\theta) = -\inf_{s\in\mathbb{R}} \Lambda_{\mathsf{cen}}(s/K;\theta)\\ &= -\inf_{s\in\mathbb{R}} \Lambda_{\mathsf{cen}}(s;\theta) = \mathscr{E}_{\mathsf{cen}}(\theta,\vartheta^o),\end{aligned} \tag{13.26}$$

which shows that, when the data are statistically independent across the agents, traditional social learning with a doubly stochastic combination matrix achieves the same error exponents, for all pairs (θ, ϑ^o) with $\theta \neq \vartheta^o$, as the centralized Bayesian scheme.

Consider next the setting where the combination matrix is *not* doubly stochastic (recall that it must be left stochastic). This setting plays an important role in several applications, especially over *directed* graphs, where it can be difficult to construct a doubly stochastic combination matrix. When the matrix is not doubly stochastic, the Perron vector v cannot have uniform entries.[2] Therefore, the equivalence between $\boldsymbol{\lambda}_{\mathsf{net},t}(\theta)$ and $\boldsymbol{\lambda}_{\mathsf{cen},t}(\theta)$ is lost, and, hence, the equivalence between the distributed and centralized scheme is lost. This happens because an agent with a higher Perron vector entry gives more credit to its own likelihood with respect to agents with lower entries. However, giving uneven degree of importance to the likelihoods is not supported from a statistical viewpoint, since the optimal Bayesian scheme would assign equal weights to the likelihoods — see (13.23). Notably, this lack of optimality was already proved for the case of *adaptive* social learning [94, 95].

We now show how the NB^2 strategy from listing (13.19) can overcome the limitations of traditional social learning thanks to the insertion of non-Bayesian updates into the social learning loop. In fact, the update parameters $\{\gamma_k\}$ can be used to compensate for the unequal assignment of importance across the agents' likelihoods. Specifically, the choice

$$\gamma_k = \frac{1}{v_k} \tag{13.28}$$

leads to $\boldsymbol{\lambda}_{\mathsf{NB}^2,t}(\theta) = \boldsymbol{\lambda}_{\mathsf{cen},t}(\theta)$, and, hence, the NB^2 strategy achieves the same error exponent as the centralized Bayesian scheme, even with left stochastic combination matrices. Note that the Perron vector need not

[2]The columns of A add up to 1 since A is left stochastic by assumption. If $v_k = 1/K$ for all k, from (4.5) we have

$$A\frac{\mathbb{1}}{K} = \frac{\mathbb{1}}{K} \iff A\mathbb{1} = \mathbb{1}, \tag{13.27}$$

which implies that the rows of A also add up to 1. Therefore, if v has uniform entries, A must be doubly stochastic.

be known beforehand by the agents. It can be estimated by means of a standard distributed consensus protocol [58], as explained in [23].

Example 13.1 (Independent agents). Consider the following social learning problem with statistically independent data across the agents. The network topology is displayed in the left panel of Figure 13.1. It is a strong undirected graph (all nodes have a self-loop, not shown in the figure). On top of it, we construct a left stochastic combination matrix through the uniform-averaging rule — see Table 4.1. Regarding the agents' data, consider the following family of Laplace probability density functions with three different means and unit scale parameter, namely,

$$g_n(x) = \frac{1}{2}e^{-|x-0.1\,n|}, \qquad n = 1, 2, 3. \tag{13.29}$$

The distributions of the agents are chosen from among these Laplace densities, in the specific way reported in Table 13.1. We observe that this assignment results in a globally identifiable problem for any choice of ϑ^o.

Table 13.1: Identifiability setup for the learning problem in Example 13.1.

Agent k	**Likelihood model: $\ell_k(x\|\theta)$**		
	$\theta = 1$	$\theta = 2$	$\theta = 3$
$1-3$	$g_1(x)$	$g_1(x)$	$g_3(x)$
$4-6$	$g_1(x)$	$g_3(x)$	$g_3(x)$
$7-10$	$g_1(x)$	$g_2(x)$	$g_1(x)$

We run all algorithms by assuming that the initial beliefs are uniform. Accordingly, the overall error probability (13.15), corresponding to the centralized system, becomes

$$\bar{p}_t = \frac{1}{H}\sum_{\vartheta^o\in\Theta}\mathbb{P}_{\vartheta^o}\left[\vartheta^o \neq \arg\max_{\theta\in\Theta}\boldsymbol{\mu}_{\mathsf{cen},t}(\theta)\right]. \tag{13.30}$$

Likewise, for the distributed strategies (i.e., the NB^2 strategy and traditional social learning), to obtain a compact performance descriptor we further average the error probabilities of all agents. The overall error probability $\bar{p}_t$ in this case is defined as

$$\bar{p}_t = \frac{1}{KH}\sum_{k=1}^{K}\sum_{\vartheta^o\in\Theta}\mathbb{P}_{\vartheta^o}\left[\vartheta^o \neq \arg\max_{\theta\in\Theta}\boldsymbol{\mu}_{k,t}(\theta)\right]. \tag{13.31}$$

Figure 13.1 shows the evolution over time of this average error probability for: *i)* the NB^2 strategy from listing (13.19) with the update parameters $\{\gamma_k\}$ chosen according to (13.28); *ii)* traditional social learning (SL) from listing (3.16); and *iii)* the centralized Bayesian posterior in (13.4). We see that the NB^2 strategy outperforms traditional social learning, and attains the same error exponent as the centralized Bayesian posterior.

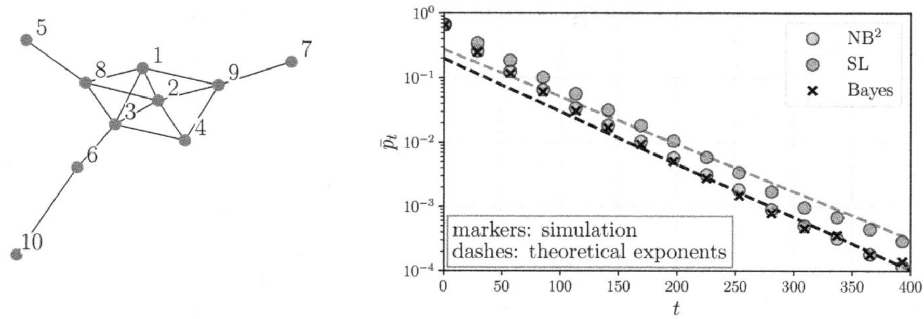

Figure 13.1: (*Left*) Network topology used in Example 13.1. The graph is undirected and all agents are assumed to have a self-loop (not shown in the figure). (*Right*) Average error probability (Eqs. (13.30) and (13.31)) as a function of time, for the independent data case and a left stochastic matrix. We compare: *i)* the NB^2 strategy from listing (13.19) with the update parameters $\{\gamma_k\}$ chosen according to (13.28); *ii)* traditional social learning (SL) from listing (3.16); and *iii)* the centralized Bayesian posterior in (13.4).

13.1.3 Clusters of Highly Dependent Agents

Interestingly, the NB^2 strategy can achieve the optimal Bayesian exponent and outperform traditional social learning even in some scenarios with high statistical dependence across the agents. This is shown in [23] for the limiting case where the network is divided into clusters wherein agents have the same data (i.e., they have maximal statistical dependence). Let us now examine this case in greater detail.

Specifically, the network is partitioned into M clusters or groups, denoted by $\mathcal{G}_1, \mathcal{G}_2, \ldots, \mathcal{G}_M$. The cluster to which agent k belongs will be denoted by $\mathcal{C}_k$. For example, if we have 3 agents and $M = 2$ clusters, with

$$\mathcal{G}_1 = \{1, 2\}, \qquad \mathcal{G}_2 = \{3\}, \tag{13.32}$$

then the clusters to which the individual agents belong are

$$\mathcal{C}_1 = \mathcal{G}_1, \qquad \mathcal{C}_2 = \mathcal{G}_1, \qquad \mathcal{C}_3 = \mathcal{G}_2. \tag{13.33}$$

We assume that the agents within the same cluster observe the same data, i.e., if j and k belong to the same cluster,

$$\boldsymbol{x}_{j,t} = \boldsymbol{x}_{k,t}. \tag{13.34}$$

In this scenario, traditional non-Bayesian social learning ends up counting multiple times the same data arising from agents belonging to the same cluster. In contrast, a centralized Bayesian posterior taking into account the statistical dependence would discard all redundant data and compute the product only between the remaining likelihoods. In other words, the centralized problem can be reformulated as an equivalent problem with only M independent data samples at each time t. Assume that, given the mth cluster, the centralized system picks from this cluster only the data from a single agent in the cluster, denoted by j_m. Accordingly, the optimal log likelihood ratio would be the sum of the log likelihood ratios for these independent data samples, namely,

$$\boldsymbol{\lambda}_{\mathsf{cen},t} = \sum_{m=1}^{M} \boldsymbol{\lambda}_{j_m,t}. \tag{13.35}$$

Note that we can also represent the log likelihood ratio in (13.35) by including the log likelihood ratios from all agents, by writing

$$\boldsymbol{\lambda}_{\mathsf{cen},t} = \sum_{k=1}^{K} \frac{1}{|\mathcal{C}_k|} \boldsymbol{\lambda}_{k,t}. \tag{13.36}$$

In fact, in (13.36) the log likelihood ratio of each agent k is divided by the cardinality of the cluster to which agent k belongs. Since the log likelihood ratios corresponding to the same cluster are identical, the representation in (13.36) is equivalent to including a single log likelihood ratio per cluster.

Consider now the NB^2 strategy. In order to achieve the same exponent as the centralized Bayesian scheme, we need to match (13.20) with (13.36). This is easily obtained by setting

$$\gamma_k = \frac{1}{v_k |\mathcal{C}_k|}. \tag{13.37}$$

For example, if one cluster is made of 2 agents, say j and k, we have $|\mathcal{C}_j| = |\mathcal{C}_k| = 2$, and rule (13.37) (apart from compensating for the Perron vector entries, as explained in the previous section) discounts the log likelihood by a factor $1/2$ to split its contribution equally between the two agents in the cluster. In contrast, traditional non-Bayesian social learning neglects the dependence and simply treats the data in the cluster as if they were independent. As a consequence, in traditional non-Bayesian social learning the data in the cluster are given more relevance than what they would deserve according to the optimal Bayesian processing.

Example 13.2 (Clusters of highly dependent agents). Consider the same network topology used in Example 13.1. To explore the potential benefits of the NB^2 strategy with dependent data, for the experiments in Figure 13.2 we consider the following setup.

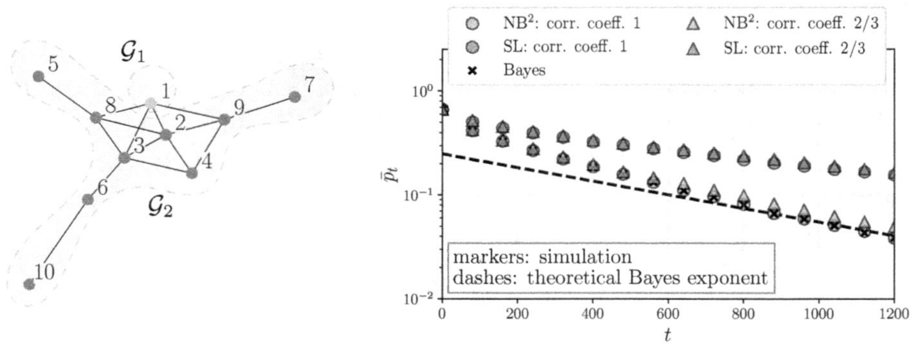

Figure 13.2: (*Left*) Network topology used in Example 13.2. The shaded areas represent the clusters of agents. The graph is undirected and all agents are assumed to have a self-loop (not shown in the figure). (*Right*) Average error probability (Eqs. (13.30) and (13.31)) as a function of time, for the highly dependent data case and a doubly stochastic matrix. We compare: *i)* the NB^2 strategy from listing (13.19) with the update parameters $\{\gamma_k\}$ chosen according to (13.37); *ii)* traditional social learning (SL) from listing (3.16); and *iii)* the centralized Bayesian posterior in (13.4), computed by assuming perfect correlation among the data within the same cluster.

The data samples of agent 1 originate from a unit-scale Laplace distribution with mean equal to 0.1; the data samples of all other agents originate from a unit-scale Laplace distribution with mean equal to 0.05, and these agents (i.e., from 2 to 10) form a cluster with dependent data. The two groups of agents, $\mathcal{G}_1 = \{1\}$ and $\mathcal{G}_2 = \{2, 3, \dots, 10\}$, are highlighted in the network topology depicted in the left panel of Figure 13.2. For the dependence enforced within $\mathcal{G}_2$, we consider the following scenarios: the limiting case where all data within the cluster are the same (corresponding to a Pearson correlation coefficient equal to 1); the more practical case where we first generate the same data samples for all the agents within $\mathcal{G}_2$, and then add to these samples independent Gaussian variables with zero mean and unit variance. In this way, the observations of agents $2, 3, \dots, 10$ are highly correlated but not equal (specifically, they feature a Pearson correlation coefficient equal to 2/3).

In this example we want to emphasize the role of the dependence among the agents, rather than of the asymmetries arising from unequal Perron vector entries. Therefore, we choose a doubly stochastic combination matrix (specifically, a Metropolis matrix — see Table 4.1), which implies that the Perron vector has uniform entries, i.e., $v_k = 1/K$ for $k = 1, 2, \dots, K$. According to (13.37), the update constants for the NB^2 strategy are set as

$$\gamma_k = \frac{K}{|\mathcal{C}_k|}. \tag{13.38}$$

We remember that this design choice has been obtained for a model where the data within the same cluster are *exactly the same*. Observe that this model holds for the considered case with Pearson correlation coefficient equal to 1, while it does not hold for the case with Pearson correlation coefficient equal to 2/3. Remarkably, in Figure 13.2 we see that the NB^2 strategy significantly outperforms traditional social learning for both the considered values of the correlation coefficient, i.e., also in the more practical case where the data within the same cluster are different.

In Figure 13.2 we also show the performance of the centralized Bayesian posterior that assumes perfect correlation among the data within the same cluster. Remarkably, the error exponent attained by the NB^2 strategy, for both the considered values of the correlation coefficient, is close to the error exponent attained by this Bayesian posterior.

13.1.4 More General Update Rules

The update rule (13.1) was already obtained in Chapter 2 — see (2.90). In particular, we showed there that, from an information-theoretic viewpoint, this modified rule arises when one modifies the free energy by weighting the KL divergence term by $1/\gamma_k$. From a stochastic-optimization viewpoint, the parameter γ_k plays the role of the step-size of a stochastic mirror descent algorithm. Rule (13.1) is not the only possibility for deriving posterior beliefs based on specific constraints [174]. As also discussed in the last paragraph of Chapter 2, different update rules would arise by considering variations of the cost functions in (2.61), (2.70), or (2.72), by scaling the individual terms with different weights, so as to unbalance the relative importance of past information (encoded in the prior) and fresh data (encoded in the likelihood). We have seen other instances of this general approach in Chapter 8 when we introduced the *adaptive* update rule

$$\psi_{k,t}(\theta) \propto \mu_{k,t-1}^{1-\delta}(\theta)\ell_k(x_{k,t}|\theta). \tag{13.39}$$

Comparing (13.39) with (13.1) we see that in the adaptive rule (13.39) the belief, rather than the likelihood, is raised to some power. Note also that in the adaptive case we did not consider an agent-dependent parameter δ_k, even if this choice is possible. The reason why an agent-independent parameter δ works is that, to infuse adaptation, we do not need to differentiate among the agents. What we need to do is to reduce the importance of past data in comparison with new data. To this end, we reduced the importance of the previous-lag belief with respect to the likelihood. In contrast, as we explained in the previous section, in the NB^2 update (13.1) what matters is to assign different degrees of importance to the likelihoods of different agents, for example, to compensate for different Perron entries or to avoid redundancy in the case of dependent data. That is why in (13.1) we consider an agent-dependent parameter γ_k.

Another interesting aspect concerns the intrinsic non-Bayesian nature of the update in (13.1). As we showed in Section 8.2.3, Eq. (13.39) can be interpreted as a Bayesian update with respect to a flattened belief $\widehat{\mu}_{k,t-1}(\theta) \propto \mu_{k,t-1}^{1-\delta}(\theta)$. In contrast, Eq. (13.1) cannot be interpreted as a Bayesian update, since if we try to normalize $\ell_k^{\gamma_k}(x_{k,t}|\theta)$ to get a probability (mass or density) function, we get a normalization constant that depends on θ, and we lose the Bayesian-update structure.

The above discussion suggests that we can also consider a more general update rule in the form

$$\psi_{k,t}(\theta) \propto \mu_{k,t-1}^{1-\delta}(\theta)\ell_k^{\gamma_k}(x_{k,t}|\theta), \tag{13.40}$$

where we raise to suitable powers both the belief and the likelihood. In this way, we can design social learning algorithms that are at the same time adaptive (thanks to the adaptation parameter δ) and able to control the discrepancies among the agents so as to optimize the performance (thanks to the update constants $\{\gamma_k\}$).

13.1.5 Bayesian or Non-Bayesian?

The local Bayesian update employed in traditional social learning is motivated by observing that, when given a prior $\mu_{k,t-1}$ and a new observation $x_{k,t}$, an agent k acting rationally would build the updated belief $\mu_{k,t}$ via Bayes' rule. Since agent k has all the necessary knowledge to perform such update locally, then it makes sense to assume that the local update step is Bayesian. Then, traditional social learning becomes *globally* non-Bayesian due to the combination step, which aggregates the marginal agents' likelihoods without implementing Bayes' rule globally, i.e., at the network level.

We have shown that in some useful cases the NB^2 strategy attains the same asymptotic performance as the optimal centralized Bayesian scheme, while traditional social learning does not. Intriguingly, the NB^2 strategy achieves this improvement by adding a further non-Bayesian layer, since the update step is no longer Bayesian. Therefore, we obtain a curious result: Two non-Bayesian steps lead to a Bayesian behavior! This might be regarded as an instance of the double-negation case where *two negatives cancel each other out.*

One explanation for this behavior is that the local Bayesian update is a greedy choice, since it is optimal only locally. In making this greedy choice, the agent is not considering that this local step is one part of a global learning process, which involves cooperation with other agents. Therefore, it is legitimate that a rational agent modifies its local behavior to reach improved global performance.

In practice, establishing whether in a social learning algorithm the local updates must be Bayesian or non-Bayesian is an open question.

From a behavioral viewpoint, social learning models attempt to mimic the behavior of real-world social groups. Useful theories developed in social and cognitive sciences support the thesis that agents act individually in a Bayesian manner. It would be interesting to explore whether these behavioral theories can incorporate models that depart from the assumption of local Bayesian updates. In this perspective, a non-Bayesian update scheme like the one implemented by the NB^2 strategy can be interpreted as a more powerful notion of rationality. The agents are cognizant of belonging to a social system and accordingly modify their updates (from Bayesian to non-Bayesian) to optimize the *social*, rather than the individual, performance. However, from an experimental viewpoint, it is not known whether the non-Bayesian updates would match well the effective cognitive mechanism observed in real-world groups.

From an engineering design perspective, we have shown that a social learning algorithm using the non-Bayesian update rule (13.1) can lead to superior performance in some scenarios. For this to be true, each agent k should incorporate (through the parameter γ_k) into its own update some information regarding the social aspects. For example, the Perron vector that is related to the graph of social interactions, or the statistical dependence across the agents. However, when this knowledge regarding the distributed network features is not available, or in scenarios different from the ones considered in [23], the NB^2 strategy could be outperformed by a scheme with Bayesian updates.

We could sum up by saying that a local Bayesian update is a more general-purpose rule motivated by assuming local rationality of the agents, which does not need any information regarding the distributed network setting. In comparison, the NB^2 strategy is a more focused strategy that can outperform traditional social learning in some cases, by incorporating ad-hoc information regarding the distributed scenario.

13.2 Censored Beliefs

The ASL strategy introduced in Chapter 8 enables adaptation in social learning by modifying the Bayesian update rule (3.10a) into the adaptive update rule (8.6). We next illustrate another possibility to enable adaptation in social learning. Referring back to the social learning scheme (3.10a)–(3.10b), as done for the ASL strategy, we continue to use geometric averaging for the cooperation step (3.10b) and focus instead on the modification of the self-learning step (3.10a), namely, on the computation of the intermediate belief vector $\psi_{k,t}$.

The basic idea is to avoid that the agents become too "extreme" in their convictions. To this end, we should avoid that the beliefs about the discarded hypotheses become too small. Therefore, while we want *i)* an intermediate belief vector $\psi_{k,t}$ close to the Bayesian update $\mu^{\mathsf{Bu}}_{k,t}$, we also want *ii)* that the entries of $\psi_{k,t}$ do not fall below some minimum value $\psi_{\mathsf{min}} > 0$. In this way, these entries will remain bounded away from 0. These two requirements can be translated into the following optimization problem:

$$\psi_{k,t} = \arg\min_{p\in\Delta_H} D\left(p||\mu^{\mathsf{Bu}}_{k,t}\right), \quad \text{subject to } p(\theta) \geq \psi_{\mathsf{min}} > 0 \quad \forall\theta\in\Theta. \tag{13.41}$$

Note that the minimum admissible belief must fulfill the condition

$$\psi_{\mathsf{min}} \leq \frac{1}{H}, \tag{13.42}$$

otherwise the vector p would have all entries larger than $1/H$. Then the sum of its entries will exceed 1, and p could not be a probability vector. Preliminarily, we observe that problem (8.1) is feasible under (13.42) because there exists at least one feasible point, namely, the uniform solution $p(\theta) = 1/H$ for all $\theta \in \Theta$.

We now show how to solve (13.41) by using the Karush-Kuhn-Tucker (KKT) conditions [33, 155]. In order to state these conditions, it is first necessary to rewrite (13.41) as a convex problem in standard form. To this end, we introduce the following notation:

$$J(p) \triangleq D(p||\mu^{\mathsf{Bu}}_{k,t}), \qquad p\in\mathbb{R}^H_+, \tag{13.43}$$

$$\mathsf{f}(p) \triangleq \sum_{\theta\in\Theta} p(\theta) - 1, \tag{13.44}$$

$$\mathsf{g}_\theta(p) \triangleq \psi_{\mathsf{min}} - p(\theta), \quad \theta\in\Theta, \tag{13.45}$$

which allows us to rewrite the problem in (13.41) as

$$\psi_{k,t} = \arg\min_{p\in\mathbb{R}^H_+} J(p), \quad \text{subject to } \mathsf{f}(p) = 0 \text{ and } \mathsf{g}_\theta(p) \leq 0 \ \ \forall\theta\in\Theta, \tag{13.46}$$

where $\mathbb{R}_+$ is the set of positive real numbers. Note that the cost function is strictly convex in its argument p, the equality constraint is affine, and the inequality constraints are convex. Therefore, we have a convex optimization problem, which is feasible since we showed that there exists at least one feasible point.

We continue by introducing the Lagrangian

$$L(p,\zeta,\nu) \triangleq J(p) + \nu\,\mathsf{f}(p) + \sum_{\theta\in\Theta} \zeta_\theta\,\mathsf{g}_\theta(p), \tag{13.47}$$

where $\nu \in \mathbb{R}$ is the Lagrange multiplier associated with the equality constraint, and $\zeta = [\zeta_\theta]$ is the vector collecting the nonnegative Lagrange multipliers associated with the inequality constraints.

Under differentiability and convexity, it is known that a point p is a solution to (13.41) if, and only if, it fulfills the KKT conditions, which are the following [33, 155]:

$$\mathsf{f}(p) = 0, \tag{13.48}$$

$$\mathsf{g}_\theta(p) \le 0, \tag{13.49}$$

$$\zeta_\theta \ge 0, \tag{13.50}$$

$$\zeta_\theta\,\mathsf{g}_\theta(p) = 0, \tag{13.51}$$

$$\nabla_p L(p,\zeta,\nu) = 0, \tag{13.52}$$

where the conditions relative to θ are intended to hold for all $\theta \in \Theta$, and where ∇_p denotes the gradient computed with respect to p.

Let us start by evaluating the θth entry of the gradient in (13.52), which, by exploiting (13.43)–(13.45), can be evaluated as

$$\begin{aligned}\frac{\partial L(p,\zeta,\nu)}{\partial p(\theta)} &= \frac{\partial J(p)}{\partial p(\theta)} + \nu\frac{\partial \mathsf{f}(p)}{\partial p(\theta)} + \zeta_\theta\frac{\partial \mathsf{g}_\theta(p)}{\partial p(\theta)}\\ &= 1 + \nu + \log\frac{p(\theta)}{\mu^{\mathsf{Bu}}_{k,t}(\theta)} - \zeta_\theta.\end{aligned} \tag{13.53}$$

Imposing condition (13.52), we find that the sought-after solution $\psi_{k,t}(\theta) = p(\theta)$ must satisfy

$$\psi_{k,t}(\theta) = \chi\,\mu^{\mathsf{Bu}}_{k,t}(\theta)\,e^{\zeta_\theta}, \tag{13.54}$$

where we introduced the scaling constant $\chi = e^{-(1+\nu)}$. Accounting for (13.51), we conclude that

$$\psi_{k,t}(\theta) = \begin{cases}\psi_{\mathsf{min}} & \text{if } \zeta_\theta > 0,\\ \chi\,\mu^{\mathsf{Bu}}_{k,t}(\theta) & \text{if } \zeta_\theta = 0,\end{cases} \tag{13.55}$$

which, defining the set $\mathcal{S} \triangleq \{\theta : \zeta_\theta > 0\}$, can be rewritten as

$$\psi_{k,t}(\theta) = \begin{cases}\psi_{\mathsf{min}} & \text{if } \theta \in \mathcal{S},\\ \chi\,\mu^{\mathsf{Bu}}_{k,t}(\theta) & \text{if } \theta \in \mathcal{S}^{\mathsf{c}},\end{cases} \tag{13.56}$$

where $\mathcal{S}^c$ denotes the complement of $\mathcal{S}$. Imposing the equality constraint (13.48), we have

$$1 = \sum_{\theta\in\Theta} \psi_{k,t}(\theta) = |\mathcal{S}|\psi_{\min} + \chi \sum_{\theta\in\mathcal{S}^c} \mu^{\mathsf{Bu}}_{k,t}(\theta) \implies \chi = \frac{1-|\mathcal{S}|\psi_{\min}}{\sum\limits_{\theta\in\mathcal{S}^c} \mu^{\mathsf{Bu}}_{k,t}(\theta)}. \quad (13.57)$$

The solution in (13.56) is not yet determined since we have not specified how to determine the set $\mathcal{S}$. Moreover, since we must guarantee that $\psi_{k,t}(\theta) \geq \psi_{\min}$ for all θ, in view of (13.56), when $\theta \in \mathcal{S}^c$ we must impose the condition

$$\chi\, \mu^{\mathsf{Bu}}_{k,t}(\theta) \geq \psi_{\min}, \quad (13.58)$$

which depends on the constant χ and, hence, is also affected by $\mathcal{S}$.

The set $\mathcal{S}$ can be obtained by implementing the straightforward algorithmic procedure shown in listing (13.59).

Adaptive update with censored beliefs

initialize $\psi_{k,t} = \mu^{\mathsf{Bu}}_{k,t}$, $\quad \mathcal{S} = \{\theta : \psi_{k,t}(\theta) \leq \psi_{\min}\}$

while $\psi_{k,t}(\theta) < \psi_{\min}$ for some θ

  for each $\theta \in \Theta$

$$\chi = \frac{1-|\mathcal{S}|\psi_{\min}}{\sum\limits_{\theta\in\mathcal{S}^c} \mu^{\mathsf{Bu}}_{k,t}(\theta)} \quad (13.59)$$

$$\psi_{k,t}(\theta) = \begin{cases} \psi_{\min} & \text{if } \theta \in \mathcal{S} \\ \chi\, \mu^{\mathsf{Bu}}_{k,t}(\theta) & \text{if } \theta \in \mathcal{S}^c \end{cases}$$

  end

  $\mathcal{S} = \mathcal{S} \bigcup \{\theta \in \mathcal{S}^c : \psi_{k,t}(\theta) \leq \psi_{\min}\}$

end

In the algorithm (see also Figure 13.3 for a graphical illustration), we start by collecting into a set $\mathcal{S}$ the hypotheses θ for which the initial beliefs $\mu^{\mathsf{Bu}}_{k,t}(\theta)$ are smaller than or equal to $\psi_{\min}$. Then we replace with $\psi_{\min}$ all the entries that are smaller than $\psi_{\min}$. The additional mass necessary to fill the gap between these entries and $\psi_{\min}$ is taken from the beliefs $\mu^{\mathsf{Bu}}_{k,t}(\theta)$ that exceed $\psi_{\min}$, which correspond to $\theta \in \mathcal{S}^c$. Specifically, this mass is redistributed by applying (13.57) and setting $\psi_{k,t}(\theta) = \chi\, \mu^{\mathsf{Bu}}_{k,t}(\theta)$ for $\theta \in \mathcal{S}^c$. If the resulting vector $\psi_{k,t}$ continues to violate the inequality constraints, we update the set $\mathcal{S}$, saturate the entries smaller than $\psi_{\min}$,

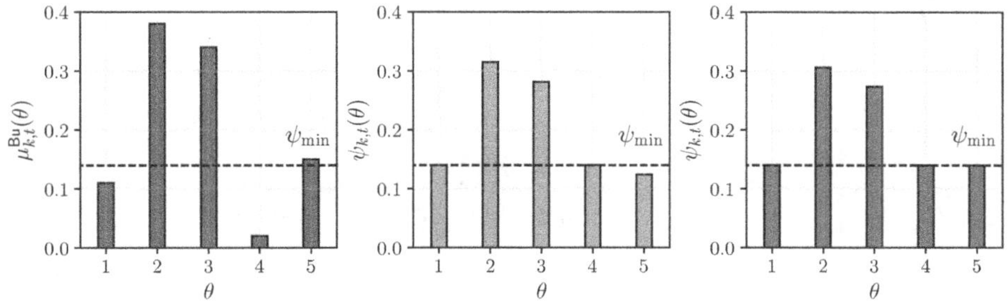

Figure 13.3: An example illustrating the algorithm in listing (13.59). (*Left*) Bayesian update $\mu^{\mathsf{Bu}}_{k,t}$. (*Center/Right*) Beliefs computed by the algorithm in the two iterations necessary to converge.

and apply again (13.57). The procedure is repeated until the vector $\psi_{k,t}$ is feasible, i.e., until it fulfills the constraints $\psi_{k,t}(\theta) \geq \psi_{\mathsf{min}}$ for all $\theta \in \Theta$. Note that the algorithm can perform at most $H-1$ iterations and that it must necessarily find an admissible solution. To see why, observe that at each iteration the algorithm adds at least one new entry equal to ψ_{min}. Accordingly, if the algorithm has run for $H-1$ iterations, the current belief has $H-1$ entries equal to ψ_{min}. However, since at each iteration the algorithm produces a valid pmf by construction, and since $\psi_{\mathsf{min}} \leq 1/H$, we have two cases. If $\psi_{\mathsf{min}} = 1/H$, to obtain a pmf the remaining entry must be equal to $1/H$. This means that the updated $\mathcal{S}^{\mathsf{c}}$ is empty, and the algorithm terminates with a uniform belief. If instead $\psi_{\mathsf{min}} < 1/H$, the remaining entry cannot be smaller than ψ_{min} (otherwise we would not have a pmf) and, hence, the algorithm terminates. We conclude that the algorithm finds always an admissible solution in at most $H-1$ iterations.

In summary, we arrive at an algorithmic procedure that implements a belief update with censored beliefs. As a result of keeping the belief-vector entries away from zero, we infuse the resulting social learning algorithm with adaptation capabilities. To understand why, it is useful to consider the simplified setting with a single agent (we accordingly drop the subscript k in the following). Consider the sequential Bayesian updates seen in (2.21):

$$\boldsymbol{\mu}_t(\theta) \propto \boldsymbol{\mu}_{t-1}(\theta)\ell(\boldsymbol{x}_t|\theta), \tag{13.60}$$

which lead to the relation

$$\log \frac{\boldsymbol{\mu}_t(\theta)}{\boldsymbol{\mu}_t(\theta')} = \log \frac{\boldsymbol{\mu}_{t-1}(\theta)}{\boldsymbol{\mu}_{t-1}(\theta')} + \log \frac{\ell(\boldsymbol{x}_t|\theta)}{\ell(\boldsymbol{x}_t|\theta')}$$
$$= \log \frac{\mu_0(\theta)}{\mu_0(\theta')} + \sum_{\tau=1}^{t} \log \frac{\ell(\boldsymbol{x}_\tau|\theta)}{\ell(\boldsymbol{x}_\tau|\theta')}. \tag{13.61}$$

As shown in Lemma 2.2, under correct likelihood models, the belief about the true hypothesis tends to 1 as time elapses. This result follows from the fact that the log belief ratio between the true and a wrong hypothesis diverges asymptotically (as shown in (2.42)). For instance, if the true state is θ, the log belief ratio in (13.61) diverges to ∞.

Assume now that the true state is θ until a given time instant $T \gg 1$, and then the true state changes from θ to θ'. To examine the learning behavior under this new condition, it is useful to write the following relation for $t > T$:

$$\log \frac{\boldsymbol{\mu}_t(\theta)}{\boldsymbol{\mu}_t(\theta')} = \log \frac{\boldsymbol{\mu}_T(\theta)}{\boldsymbol{\mu}_T(\theta')} + \sum_{\tau=T+1}^{t} \log \frac{\ell(\boldsymbol{x}_\tau|\theta)}{\ell(\boldsymbol{x}_\tau|\theta')}. \tag{13.62}$$

If T is large, we have $\mu_T(\theta) \approx 1$ (hence, $\mu_T(\theta') \approx 0$). Thus, the first term on the RHS of (13.62) is large. As observed, this is a desired behavior to learn the hypothesis θ that is in force until T. On the other hand, under the new true hypothesis θ', the log belief ratio in (13.62) should invert its trend and diverge to $-\infty$, thus resulting in a belief that is correctly maximized at θ'. However, it is virtually impossible to achieve correct learning within a short time period, due to the almost infinite initial condition at $t = T$. To mitigate this effect, we can *censor* the beliefs in such a way that they remain sufficiently away from zero, which would guarantee the boundedness of the first term on the RHS of (13.62).

The shape of the resulting intermediate beliefs in (13.56) has an interesting interpretation. Depending on the values of the traditional Bayesian update $\mu_{k,t}^{\mathsf{Bu}}$, some entries of $\psi_{k,t}$ are censored and set to the minimum admissible credibility ψ_{min}. The other entries of $\psi_{k,t}$ are a *scaled version of the traditional Bayesian update.* In a nutshell, we could say that $\psi_{k,t}$ follows as much as possible the shape of the Bayesian update, while remaining compatible with the minimum credibility constraint. It is useful to make some analogies and distinctions with respect to the adaptive social learning (ASL) strategy introduced in Chapter 8.

- We have already seen that the ASL strategy leads to a linear combination of log likelihood ratios — see, e.g., (9.1). In contrast, when applied in a social learning context, the censored-belief approach entails repeated nonlinear steps related to the censoring operation. While it is still true that, with the geometric combination rule, the log ratios $\log \frac{\psi_{j,t}(\vartheta^\star)}{\psi_{j,t}(\theta)}$ are linearly combined, this combination takes place after a censoring operation that destroys the overall linear structure of the recursion.

- The censoring strategy implemented by (13.56) is reminiscent of the philosophy underlying traditional *change detection* or *quickest detection* procedures, such as Page's CUSUM test [14, 141, 163, 176]. For a single-agent, binary change-detection problem, it is well known that censoring the decision statistic is beneficial from the quickest detection viewpoint. In particular, in the single-agent binary detection case with true distribution corresponding either to $\ell(x|1)$ or $\ell(x|2)$, and with data generated from a hypothesis that suddenly varies at some instant, Page's rule optimizes in a suitable mathematical sense the trade-off between false alarms and time to detect the change [133, 142]. It would be interesting to explore whether this trade-off is also optimized by the belief-censoring strategy adopted in the social learning setting. It would also be interesting to compare the ASL strategy against the strategy with censored beliefs. This type of comparison is akin to the classic comparison existing in the literature between EWMA and CUSUM control charts [114] for single-agent binary detection problems.

- Another aspect is the impact of censoring on the *value itself of the belief*, rather than on the decision made by the agents. The belief value plays an important role in opinion formation and contains more information than the decision. It provides the degree of confidence assigned by the agents to each hypothesis, and, in particular, reveals how confident an agent is about a particular decision. For example, consider the case where sufficient evidence has been collected in support of the target hypothesis, so that the belief corresponding to the wrong hypotheses has reached the minimum value $\psi_{\min}$. This means that the belief corresponding to the target hypothesis has reached the maximum value tolerated by the censored strategy, which is $\psi_{\max} = 1 - (H-1)\psi_{\min}$. Assume now that the data distribution changes by providing more evidence in support of the currently chosen hypothesis. The censored strategy is not able to translate this increase of evidence into an increase of belief, since the maximum value $\psi_{\max}$ has already been reached. In contrast, the ASL strategy will reflect the change by fluctuating around a higher belief. This is because the ASL strategy inherently tracks an analog decision statistic, whereas the censored strategy can track only until censoring takes place.

- We observed in Chapter 5 that traditional social learning allows the belief about the target hypothesis to reach values astronomically close to 1, making the agents reluctant to change their mind in the presence of drifting conditions. One might ask whether such an endless improvement makes sense in practice and which mechanisms can be implemented to contrast it. In this connection, both the ASL and the censored strategies prevent the belief about the target hypothesis to approach indefinitely the value 1, albeit in two very different manners. For the ASL strategy, the belief "collapse" is avoided by preserving some degree of uncertainty in the beliefs, which keep on fluctuating randomly as time elapses. The censored strategy operates more "abruptly"; it forces the beliefs to stay always above a minimum credibility, which implies that the belief about the target hypothesis stays bounded away from 1.

- The comparison of different adaptation mechanisms is particularly interesting from a behavioral perspective, where numerous questions arise. For example, which adaptation mechanism would represent more faithfully the behavior observed in a certain distributed cognitive process? Are different categories of individuals represented by different adaptation mechanisms?

13.3 Learning the Social Graph

We have explained in the previous chapters that the network graph plays a critical role in the social learning performance. For example, we saw in Chapter 5 how diverse learning modes are observed over connected or weak graphs. In social learning applications, there is however another important learning problem that we should consider, namely, the inverse problem of learning the underlying graph structure from the observation of the beliefs. In other words, instead of focusing on *what* the agents learn through their social learning algorithm (which is the goal of the *direct* learning problem), we can focus on the *dual* problem that deals with *how* the nodes learn (i.e., on discovering the hidden interconnections that drive the social learning process). Figure 13.4 summarizes this scenario. In the direct problem, we start from a topology, run the social learning algorithm, and examine its performance (e.g., the convergence of the agents' beliefs or the probability of misclassifying the target hypothesis) and the dependence of this performance on the graph. Conversely, in the dual problem we collect the streams of beliefs $\psi_{k,t}$ publicly exchanged by the agents, and focus on discovering the underlying graph.

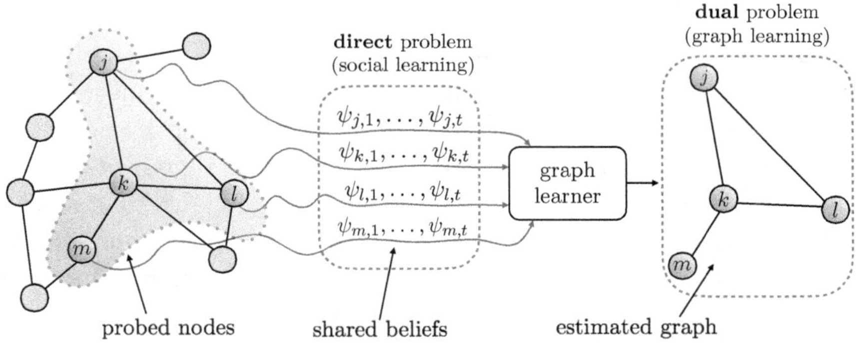

Figure 13.4: Illustration of the social graph learning problem. The agents of a network run a social learning algorithm to construct their beliefs about some hypotheses of interest (the *direct* learning problem). The network graph influences the way each agent shares its opinions with its neighbors. An inferential engine can probe the subset of agents $\{j, k, l, m\}$ and collect the pertinent beliefs shared over the network. Based on these beliefs, the goal of the *dual* learning problem is to estimate the subgraph of connections between nodes j, k, l, m.

The general problem of learning a graph topology from measurements collected at the nodes arises across several disciplines. It is therefore not surprising that this problem is referred to in multiple ways, including: graph learning, topology inference, network tomography, graph reconstruction, graph estimation. The graph learning problem can provide answers to many interesting questions. For example, by observing the evolution of the nodes' signals, can one establish which nodes are sharing information with each other? How is privacy reflected in these interactions? Can one discover which nodes have a magnified influence on the overall network behavior? Numerous applications would benefit from such answers: tracing the relationships between the users in a social network to capture the opinion formation mechanism or to locate the source of fake news [115, 118,

140, 160]; discovering clandestine information flows over the Internet [116, 168]; learning the synchronized cognitive behavior of a school of fish evading predators [51, 138]; investigating the interaction between structural and functional connectivity in the brain [112]; characterizing the evolution of urban traffic [61]; discovering hidden relationships in financial data [86]; estimating gene regulatory networks from gene expression data [75].

Owing to several forms of physical limitations, the difficulty in addressing the graph learning problem is compounded by the fact that usually only a limited fraction of the nodes can be probed. This is especially true over large-scale networks. A second important question arises: Despite the presence of *latent, unobserved* nodes, can partial observations still be sufficient to discover the graph linking the probed nodes? For example, in probing signals from the brain, only certain regions of the brain are examined. Likewise, in probing signal flows over a social network with millions of members, only a limited number of observations may be available. Under this regime, which is referred to as the *partial observability* regime, graph learning becomes more complicated than usual, since the signals collected at the probed nodes are subject to the influence (through information diffusion) of the unobserved nodes. Interestingly, recent works show that, despite the influence of the latent nodes, the graph linking the probed nodes can be estimated faithfully under suitable conditions [47, 121, 122, 124, 150].

There exist some useful survey articles related to the topic of graph learning [63, 81, 117, 121]. Since the focus of this book is on social learning, it is useful to mention two recent works that deal with the graph learning problem in the context of the social learning models that we have introduced and examined in the previous chapters.

The first work is [118], which addresses the topology inference problem for a social learning algorithm ruled by a weak graph — see Section 4.5. Specifically, given the beliefs collected by probing a node belonging to a receiving network, the goal is to discover the global influence that any sending network might have exerted on that node. This problem is also referred to as *macroscopic topology learning* in [118]. This global influence is measured by the sum of the limiting combination weights from all nodes in a given sending network to the node under consideration. The limiting combination weights automatically embody the effect of multi-hop paths from the agents in the sending networks to the agents in the receiving networks — see (4.55). The following strong interplay between social and topology learning is discovered in [118]: Given H hypotheses and S sending networks, for macroscopic topology learning to be feasible it is necessary that the number of sending networks is smaller than or equal to the number of hypotheses, i.e., that $S \leq H$. This is only a necessary condition. For example, by exploiting the theory of Euclidean distance matrices, it is shown in [118] that if the data within the sending networks follow a very structured Gaussian model (see details in [118]), macroscopic topology inference is feasible only for $S = 2$. However, and remarkably, one fundamental result from [118] is that macroscopic topology inference becomes feasible whenever $S \leq H$ (that is, the condition $S \leq H$ becomes also sufficient) provided that a certain degree of diversity exists in the statistical models of the sending networks.

The second work that we would like to mention is [160], which addresses the problem of estimating, from the beliefs publicly exchanged by the agents, the entire combination matrix underlying the social learning algorithm. A graph learning algorithm is proposed, whose analytical characterization reveals that it is possible: *i)* to estimate faithfully the combination matrix, which allows to learn the underlying topology and quantify the pairwise influences between agents; *ii)* to identify the influence that each individual agent has on the objective of truth learning and accordingly quantify its degree of informativeness, further allowing to identify the *influencers* and the *influenced agents*. The proposed algorithm works under nonstationary environments where either the true state of nature or the graph topology are allowed to drift over time. The operation of the algorithm is illustrated by applying it to different subnetworks of Twitter users to identify the most influential users by using the text contained in their public tweets.

Appendices

Appendix A

Convex Functions

In this book we use some properties of convex functions. For this reason, we collect in this appendix minimal elements about convex functions that are directly applied in our proofs. For a broader treatment see, e.g., [33, 155]. Here and in the forthcoming appendices, most classic results are stated without proof; for these results, we provide references where the interested reader can find the necessary technical background and derivations.

We start with the definition of convex sets.

Definition A.1 (Convex sets). A set $\mathcal{S} \subseteq \mathbb{R}^d$ is convex when, for all pairs of distinct points $x, y \in \mathcal{S}$ and all $p \in (0, 1)$, the point $z = px + (1 - p)y$ belongs to $\mathcal{S}$.

Geometrically, a convex function is ∪-shaped: Given two points x and y, the line segment (*chord*) connecting $f(x)$ to $f(y)$ lies above or on the function evaluated along the line segment that connects x to y.

Definition A.2 (Convex functions). A function $f : \mathbb{R}^d \mapsto \mathbb{R}$ is convex when its domain $\mathrm{dom}(f)$ is convex and

$$f(px + (1 - p)y) \leq pf(x) + (1 - p)f(y) \quad \forall x, y \in \mathrm{dom}(f), \quad \forall p \in [0, 1]. \tag{A.1}$$

The function is called strictly convex if the inequality is strict whenever $x \neq y$ and $0 < p < 1$.

When the function is differentiable, convexity can be defined in terms of the gradient.

Lemma A.1 (First-order condition for convexity [33]). Let $f : \mathbb{R}^d \mapsto \mathbb{R}$ be defined in an open domain $\text{dom}(f)$ and differentiable on it. Then, convexity is equivalent to the condition

$$f(y) - f(x) \geq [\nabla f(x)]^\mathsf{T} (y - x) \quad \forall x, y \in \text{dom}(f), \tag{A.2}$$

and strict convexity holds when the inequality is strict for all $x \neq y$.

In particular, when $0 \in \text{dom}(f)$ and $f(0) = 0$, we have the following two relations that are useful in some of our proofs:

$$f(y) \geq [\nabla f(0)]^\mathsf{T} y \quad \forall y \in \text{dom}(f), \tag{A.3a}$$

$$[\nabla f(x)]^\mathsf{T} x \geq f(x) \quad \forall x \in \text{dom}(f). \tag{A.3b}$$

Equation (A.3a) is obtained by setting $x = 0$ in (A.2), whereas Eq. (A.3b) is obtained by setting $y = 0$.

When the function is twice differentiable, convexity can be directly expressed in terms of the Hessian matrix.

Lemma A.2 (Second-order condition for convexity [33]). Let $f : \mathbb{R}^d \mapsto \mathbb{R}$ be defined in an open domain $\text{dom}(f)$ and twice differentiable on it. Then, f is convex if, and only if, the Hessian matrix $\nabla^2 f(x)$ is positive semidefinite (i.e., all its eigenvalues are nonnegative) for all $x \in \text{dom}(f)$. Strict convexity holds when the matrix is positive definite (i.e., all its eigenvalues are positive).

Convexity is especially useful in optimization problems, because it provides information about the existence and/or uniqueness of minima. Consider an open domain $\text{dom}(f)$. Recall that, regardless of convexity, if x^o is a minimum (even local) of a differentiable function f, then we must have $\nabla f(x^o) = 0$. Convexity implies that annihilation of the gradient is also a sufficient condition for a minimum to exist. Moreover, for convex functions this minimum is always a global minimum, but there might be multiple global minima. For *strictly* convex functions, this is not possible, and when a minimum exists, it is the unique global minimum.

Appendix B

Entropy and KL Divergence

This appendix collects some information-theoretic quantities that are useful in our treatment. For more details see, for example, [52].

In the forthcoming definitions we adopt the following conventions (based on continuity arguments):

$$0\log\frac{1}{0} = 0\log 0 = 0\log\frac{0}{0} = 0, \qquad \log\frac{a}{0} = \infty, \qquad \log\frac{0}{a} = -\infty, \quad \text{(B.1)}$$

for $a > 0$. We start by presenting the definition of *Shannon's entropy* for discrete random variables, which quantifies the uncertainty associated with a probability mass function.

Definition B.1 (Shannon's entropy). Let $\boldsymbol{y}$ be a discrete random variable defined on a set $\mathcal{Y}$ and having pmf $p(y)$. The entropy of $\boldsymbol{y}$ or, equivalently, the entropy of the pmf $p(y)$, is

$$H(p) \triangleq \sum_{y\in\mathcal{Y}} p(y)\log\frac{1}{p(y)} = \mathbb{E}\log\frac{1}{p(\boldsymbol{y})}. \quad \text{(B.2)}$$

Next, we define the *conditional entropy* of a random variable $\boldsymbol{z}$ given another random variable $\boldsymbol{y}$, which quantifies the residual uncertainty contained in $\boldsymbol{z}$ once $\boldsymbol{y}$ is observed.

Definition B.2 (Conditional entropy). Let $\boldsymbol{y}$ and $\boldsymbol{z}$ be two discrete random variables defined on sets $\mathcal{Y}$ and $\mathcal{Z}$, respectively, and having joint pmf $p(y, z)$. The entropy of $\boldsymbol{z}$ given an observed value y is the entropy of the conditional pmf

$p(z|y)$,

$$\sum_{z \in \mathcal{Z}} p(z|y) \log \frac{1}{p(z|y)}. \tag{B.3}$$

The conditional entropy of $\boldsymbol{z}$ given $\boldsymbol{y}$ is the expectation of the above quantity taken over the distribution of $\boldsymbol{y}$,

$$H_{z|y}(p) \triangleq \sum_{y \in \mathcal{Y}} \sum_{z \in \mathcal{Z}} p(y, z) \log \frac{1}{p(z|y)} = \mathbb{E} \log \frac{1}{p(\boldsymbol{z}|\boldsymbol{y})}. \tag{B.4}$$

In the following definition we introduce the *cross-entropy* between a reference pmf $p(y)$ (which is assumed to be the actual pmf of the random variable $\boldsymbol{y}$) and another pmf $q(y)$.

Definition B.3 (Cross-entropy). Let $\boldsymbol{y}$ be a discrete random variable defined on a set $\mathcal{Y}$ and having pmf $p(y)$. Let $q(y)$ be another pmf defined on the same set. The cross-entropy between $p(y)$ and $q(y)$ is

$$H(p, q) \triangleq \sum_{y \in \mathcal{Y}} p(y) \log \frac{1}{q(y)} = \mathbb{E} \log \frac{1}{q(\boldsymbol{y})}. \tag{B.5}$$

We continue by defining a quantity related to the cross-entropy, called *Kullback-Leibler* (KL) *divergence* or *relative entropy*, which is useful to measure the dissimilarity between a reference pmf $p(y)$ and another pmf $q(y)$.

Definition B.4 (KL divergence). Let $\boldsymbol{y}$ be a discrete random variable defined on a set $\mathcal{Y}$ and having pmf $p(y)$. Let $q(y)$ be another pmf defined on the same set. The KL divergence between $p(y)$ and $q(y)$ is

$$D(p||q) \triangleq \sum_{y \in \mathcal{Y}} p(y) \log \frac{p(y)}{q(y)} = \mathbb{E} \log \frac{p(\boldsymbol{y})}{q(\boldsymbol{y})}. \tag{B.6}$$

Since $\log(p(y)/q(y)) = \log p(y) - \log q(y)$, by using Definitions B.1 and B.3 we get the following representation:

$$D(p||q) = H(p, q) - H(p). \tag{B.7}$$

Moreover, the KL divergence is always nonnegative, and is equal to 0 if, and only if, $p = q$.

The KL divergence can also be introduced for two pdfs $p(y)$ and $q(y)$ defined on $\mathbb{R}^d$:

$$D(p||q) = \mathbb{E} \log \frac{p(\boldsymbol{y})}{q(\boldsymbol{y})} = \int_{\mathbb{R}^d} p(y) \log \frac{p(y)}{q(y)} dy, \tag{B.8}$$

and it can be extended to more general probability measures by appealing to the concept of the Radon-Nikodym derivative [54].

Similarly to what was done in Definition B.2, we can introduce the *conditional cross-entropy* as follows.

Definition B.5 (Conditional cross-entropy). Let $\boldsymbol{y}$ and $\boldsymbol{z}$ be two discrete random variables defined on sets $\mathcal{Y}$ and $\mathcal{Z}$, respectively, and having joint pmf $p(y, z)$. Let $q(z|y)$ be a conditional pmf defined for $y \in \mathcal{Y}$ and $z \in \mathcal{Z}$. Given an observed value y, the cross-entropy between the conditional pmfs $p(z|y)$ and $q(z|y)$ is

$$\sum_{z \in \mathcal{Z}} p(z|y) \log \frac{1}{q(z|y)}. \tag{B.9}$$

The conditional cross-entropy $H_{z|y}(p, q)$ is the expectation of the above quantity taken over the distribution of $\boldsymbol{y}$,

$$H_{z|y}(p, q) \triangleq \sum_{y \in \mathcal{Y}} \sum_{z \in \mathcal{Z}} p(y, z) \log \frac{1}{q(z|y)} = \mathbb{E} \log \frac{1}{q(\boldsymbol{z}|\boldsymbol{y})}. \tag{B.10}$$

Likewise, we define the *conditional* KL *divergence* as follows.

Definition B.6 (Conditional KL divergence). Let $\boldsymbol{y}$ and $\boldsymbol{z}$ be two discrete random variables defined on sets $\mathcal{Y}$ and $\mathcal{Z}$, respectively, and having joint pmf $p(y, z)$. Let $q(z|y)$ be a conditional pmf defined for $y \in \mathcal{Y}$ and $z \in \mathcal{Z}$. Given an observed value y, the KL divergence between the conditional pmfs $p(z|y)$ and $q(z|y)$ is

$$\sum_{z \in \mathcal{Z}} p(z|y) \log \frac{p(z|y)}{q(z|y)}. \tag{B.11}$$

The conditional KL divergence $D_{z|y}(p||q)$ is the expectation of the above quantity taken over the distribution of $\boldsymbol{y}$,

$$D_{z|y}(p||q) \triangleq \sum_{y \in \mathcal{Y}} \sum_{z \in \mathcal{Z}} p(y, z) \log \frac{p(z|y)}{q(z|y)} = \mathbb{E} \log \frac{p(\boldsymbol{z}|\boldsymbol{y})}{q(\boldsymbol{z}|\boldsymbol{y})}. \tag{B.12}$$

Since $\log(p(z|y)/q(z|y)) = \log p(z|y) - \log q(z|y)$, by using Definitions B.2 and B.5 we get the following representation:

$$D_{z|y}(p||q) = H_{z|y}(p, q) - H_{z|y}(p). \tag{B.13}$$

Appendix C

Probabilistic Inequalities

We start by introducing three famous inequalities that are employed to obtain concentration bounds. These bounds are useful to estimate the probability that a random variable exceeds some threshold or deviates from expected behavior.

The first inequality relates the probability that a nonnegative random variable exceeds a threshold to the expectation of the random variable.

Theorem C.1 (Markov's inequality [21, 85]). If z is a nonnegative random variable and $a > 0$, then

$$\mathbb{P}[z \geq a] \leq \frac{\mathbb{E}z}{a}. \tag{C.1}$$

The second inequality can be obtained as a corollary of Markov's inequality, and establishes a connection between the variance of a random variable and the probability that the random variable deviates from its expectation.

Theorem C.2 (Chebyshev's inequality [85]). If z is a random variable and $a > 0$, then

$$\mathbb{P}[|z - \mathbb{E}z| \geq a] \leq \frac{\mathbb{E}\left[(z - \mathbb{E}z)^2\right]}{a^2}. \tag{C.2}$$

The third inequality can also be obtained from Markov's inequality, and provides an exponential bound on the probability that a random variable exceeds a threshold.

Theorem C.3 (Chernoff's bound [31]). If $\boldsymbol{z}$ is a random variable, $a \in \mathbb{R}$, and $s \geq 0$, then

$$\mathbb{P}[\boldsymbol{z} \geq a] \leq \frac{\mathbb{E}e^{s\boldsymbol{z}}}{e^{sa}}. \tag{C.3}$$

Another concentration bound that is useful in our treatment is the *independent bounded differences inequality*, which is also known as *McDiarmid's inequality* [125, 155].

Theorem C.4 (McDiarmid's inequality [125, 155]). Consider N independent random vectors

$$\boldsymbol{z}_1 \in \mathcal{Z}_1, \boldsymbol{z}_2 \in \mathcal{Z}_2, \ldots, \boldsymbol{z}_N \in \mathcal{Z}_N, \tag{C.4}$$

and a function $g : \mathcal{Z}_1 \times \mathcal{Z}_2 \times \cdots \times \mathcal{Z}_N \mapsto \mathbb{R}$ that satisfies, for $i = 1, 2, \ldots, N$, the condition

$$|g(z_1, z_2, \ldots, z_i, \ldots, z_N) - g(z_1, z_2, \ldots, \check{z}_i, \ldots, z_N)| \leq c_i \tag{C.5}$$

for some constants c_i and for all sequences

$$\{z_1, z_2, \ldots, z_i, \ldots, z_N\} \qquad \text{and} \qquad \{z_1, z_2, \ldots, \check{z}_i, \ldots, z_N\} \tag{C.6}$$

that differ only in their respective ith vectors. Then, letting

$$c = \sum_{i=1}^{N} c_i^2, \tag{C.7}$$

for all $a > 0$ we have the following concentration bounds:

$$\mathbb{P}\Big[g(\boldsymbol{z}_1, \boldsymbol{z}_2, \ldots, \boldsymbol{z}_N) - \mathbb{E}g(\boldsymbol{z}_1, \boldsymbol{z}_2, \ldots, \boldsymbol{z}_N) \geq a\Big] \leq e^{-2a^2/c}, \tag{C.8a}$$

$$\mathbb{P}\Big[g(\boldsymbol{z}_1, \boldsymbol{z}_2, \ldots, \boldsymbol{z}_N) - \mathbb{E}g(\boldsymbol{z}_1, \boldsymbol{z}_2, \ldots, \boldsymbol{z}_N) \leq -a\Big] \leq e^{-2a^2/c}. \tag{C.8b}$$

The next statement introduces a fundamental inequality concerning the interplay between random variables and convex functions.

Theorem C.5 (Jensen's inequality [7]). Let $g(z)$ be a convex function from I to $\mathbb{R}$, where $I \subseteq \mathbb{R}$ is an open interval. Let $\boldsymbol{z}$ be a finite-mean random variable such that $\mathbb{P}[\boldsymbol{z} \in I] = 1$. Then

$$g\big(\mathbb{E}\boldsymbol{z}\big) \leq \mathbb{E}g(\boldsymbol{z}). \tag{C.9}$$

Theorem C.5 has the following immediate implication. Assume that $g(z)$ is a convex function from an open interval I to $\mathbb{R}$. Given a collection of points $\{z_n\}_{n=1}^N$ belonging to I, and a collection of convex weights (i.e., nonnegative scalars that add up to 1) $\{a_n\}_{n=1}^N$, then we have

$$g\left(\sum_{n=1}^{N} a_n\, z_n\right) \leq \sum_{n=1}^{N} a_n\, g(z_n). \tag{C.10}$$

This result follows readily from Theorem C.5 because the sequence of weights can be interpreted as a probability vector.

The next inequality is an essential tool in the theory of measure and integration. We state it directly in the probabilistic form used in our treatment. This form relates the joint moment of two random variables to their individual moments of specific orders.

Theorem C.6 (Hölder's inequality [145]). Given two random variables $\boldsymbol{z}_1$ and $\boldsymbol{z}_2$, for any $r_1, r_2 \in [1, \infty)$ with $1/r_1 + 1/r_2 = 1$,

$$\mathbb{E}|\boldsymbol{z}_1\boldsymbol{z}_2| \leq \left(\mathbb{E}|\boldsymbol{z}_1|^{r_1}\right)^{\frac{1}{r_1}} \left(\mathbb{E}|\boldsymbol{z}_2|^{r_2}\right)^{\frac{1}{r_2}}. \tag{C.11}$$

We remark that, for $r_1 = r_2 = 2$, Hölder's inequality coincides with the Cauchy-Schwarz inequality for expected values.

We conclude the appendix with a useful inequality that relates the KL divergence to the so-called *total variation* distance, which is defined as follows.

Definition C.1 (Total variation distance). The total variation distance between two pmfs $p(z)$ and $q(z)$ defined on a set $\mathcal{Z}$ is

$$D_{\mathsf{TV}}(p, q) = \frac{1}{2} \sum_{z \in \mathcal{Z}} |p(z) - q(z)|. \tag{C.12}$$

Likewise, the total variation distance between two pdfs $p(z)$ and $q(z)$ defined on $\mathbb{R}^d$ is

$$D_{\mathsf{TV}}(p, q) = \frac{1}{2} \int_{\mathbb{R}^d} |p(z) - q(z)| dz. \tag{C.13}$$

Note that the total variation distance is basically an L_1 distance [145] (but for a constant factor). Technically, there exists a more general definition of total variation between probability measures, but the definition provided here is sufficient for our purposes.

The inequality connecting the KL divergence to the total variation distance is the following.

Theorem C.7 (Pinsker's inequality [139, 155, 165]). Given two pmfs or pdfs p and q,

$$D(p||q) \geq \frac{1}{2} D_{\mathsf{TV}}^2(p, q). \tag{C.14}$$

Actually, Pinsker proved the inequality with a constant larger than 1/2 [139]. The inequality in the form (C.14) was derived independently by Csiszár, Kemperman, and Kullback — see [165].

Appendix D

Stochastic Convergence

In this appendix we focus on the convergence of infinite sequences of random vectors,

$$\boldsymbol{z}_1, \boldsymbol{z}_2, \ldots, \qquad \text{with } \boldsymbol{z}_n \in \mathbb{R}^d \quad \forall n \in \mathbb{N}. \tag{D.1}$$

To denote the sequence, we will compactly write $\{\boldsymbol{z}_n\}$, implicitly implying that $n \in \mathbb{N}$.

D.1 Types of Stochastic Convergence

Consider a sequence $\{\boldsymbol{z}_n\}$ of random vectors in $\mathbb{R}^d$, defined on a common probability space identified by the triple $(\Omega, \mathscr{F}, \mathbb{P})$. We recall that, in this triple, Ω denotes the sample space that collects the possible outcomes of the random experiment under consideration. The realization of the sequence associated with a particular outcome ω is denoted by

$$\boldsymbol{z}_1(\omega), \boldsymbol{z}_2(\omega), \ldots \tag{D.2}$$

We also recall that the symbol $\mathscr{F}$ denotes a σ-field of subsets of Ω, called *events*, and $\mathbb{P}$ is a probability measure defined on $\mathscr{F}$.

There are different types of *stochastic* convergence. The simplest type is the following immediate generalization of the standard concept of the limit of a *deterministic* sequence: We require that, for *every* $\omega \in \Omega$, the sequence in (D.2) converges to a limiting value $\boldsymbol{z}(\omega)$. In this case, we would say that the sequence $\{\boldsymbol{z}_n\}$ converges *surely* to the random vector $\boldsymbol{z}$.

However, in probability theory, the strong requirement of convergence for *every* outcome is more conveniently replaced by requiring convergence for almost all realizations, i.e., possibly excluding a set of outcomes occurring with probability 0. This leads to the concept of *almost-sure* convergence, a.k.a. *convergence with probability* 1.

Definition D.1 (Almost-sure convergence). Let $\{z_n\}$ be a sequence of random vectors in $\mathbb{R}^d$. We say that $\{z_n\}$ converges almost surely (or with probability 1) to a random vector z when

$$\mathbb{P}\left[\lim_{n\to\infty} z_n = z\right] = 1. \tag{D.3}$$

That is, when the set of outcomes

$$\left\{\omega \in \Omega : \lim_{n\to\infty} z_n(\omega) = z(\omega)\right\} \tag{D.4}$$

has probability 1. Our notation for almost-sure convergence is

$$z_n \xrightarrow[n\to\infty]{\text{a.s.}} z. \tag{D.5}$$

A weaker notion of convergence is convergence *in probability.* In this case, instead of requiring that z_n converges to z for almost all realizations, we require that *the probability* that z_n deviates from z vanishes as $n \to \infty$. In the following, the symbol $\|\cdot\|$ denotes the Euclidean norm.

Definition D.2 (Convergence in probability). Let $\{z_n\}$ be a sequence of random vectors in $\mathbb{R}^d$. We say that $\{z_n\}$ converges in probability to a random vector $z \in \mathbb{R}^d$ when, for all $\varepsilon > 0$, we have

$$\lim_{n\to\infty} \mathbb{P}\left[\|z_n - z\| > \varepsilon\right] = 0. \tag{D.6}$$

Our notation for convergence in probability is

$$z_n \xrightarrow[n\to\infty]{\text{P}} z. \tag{D.7}$$

Another useful notion is convergence *in the mean.* Here we require that the expected value of the absolute deviation of z_n from z vanishes. In particular, we can consider convergence in the rth mean by taking the rth absolute moment. Before introducing this type of convergence, it is necessary to define the L_r norm of a vector. Given a vector $x = [x_j] \in \mathbb{R}^d$, its L_r norm is

$$\|x\|_r \triangleq \left(\sum_{j=1}^{d} |x_j|^r\right)^{1/r}. \tag{D.8}$$

Definition D.3 (Convergence in the rth mean). Let $\{z_n\}$ be a sequence of random vectors in $\mathbb{R}^d$ and let $r > 0$. We say that $\{z_n\}$ converges in the rth mean (or in the L_r norm) to a random vector z when

$$\lim_{n\to\infty} \mathbb{E}\left[\|z_n - z\|_r^r\right] = 0. \tag{D.9}$$

Our notation for convergence in the rth mean is

$$z_n \xrightarrow[n\to\infty]{L_r} z. \tag{D.10}$$

It is possible to show that, for $r \geq 1$, convergence in the rth mean is implied by convergence with respect to higher-order moments, i.e.,

$$z_n \xrightarrow[n\to\infty]{L_s} z \quad \Longrightarrow \quad z_n \xrightarrow[n\to\infty]{L_r} z \tag{D.11}$$

for all $s > r$.

Finally, we introduce the notion of convergence *in distribution*, a.k.a. *weak* convergence or convergence *in law*. Here we require that the cumulative distribution function (cdf) of z_n converges to some limiting cdf. In the following, $\mathrm{int}(\mathcal{S})$ and $\mathrm{cl}(\mathcal{S})$ denote the interior and the closure of a set $\mathcal{S}$, respectively, and $\partial\mathcal{S} = \mathrm{cl}(\mathcal{S})\backslash\mathrm{int}(\mathcal{S})$ denotes the boundary of $\mathcal{S}$.

Definition D.4 (Convergence in distribution). Let $\{z_n\}$ be a sequence of random vectors in $\mathbb{R}^d$ and z a random vector in $\mathbb{R}^d$. Let $F_n(z)$ and $F(z)$ be the cdfs of the random vectors z_n and z, respectively. We say that the sequence $\{z_n\}$ converges in distribution (or weakly, or in law) to z when

$$\lim_{n\to\infty} F_n(z) = F(z) \tag{D.12}$$

for any z that is a continuity point of $F(z)$. Our notation for convergence in distribution is

$$z_n \xrightarrow[n\to\infty]{\mathrm{d}} z. \tag{D.13}$$

An equivalent definition that is useful in our treatment is the following. The sequence $\{z_n\}$ converges in distribution to z when

$$\lim_{n\to\infty} \mathbb{P}[z_n \in \mathcal{S}] = \mathbb{P}[z \in \mathcal{S}] \tag{D.14}$$

for all sets $\mathcal{S} \subset \mathbb{R}^d$ that satisfy the condition

$$\mathbb{P}[z \in \partial\mathcal{S}] = 0. \tag{D.15}$$

In other words, convergence in distribution takes place when the probability that z_n belongs to $\mathcal{S}$ converges to the probability that the limiting variable z belongs to $\mathcal{S}$, for all sets $\mathcal{S}$ that have a boundary where z lies with zero probability.

Note that for convergence with probability 1, in probability, or in the rth mean, we need to compare the actual values of the random vectors $\boldsymbol{z}_n$ and $\boldsymbol{z}$. For example, for convergence in probability we have to evaluate the deviation $\|\boldsymbol{z}_n - \boldsymbol{z}\|$. This implies that these variables must be defined on the same probability space. In contrast, convergence in distribution requires that the cdf of $\boldsymbol{z}_n$ converges to the cdf of $\boldsymbol{z}$, without requiring a direct comparison in terms of random vectors. As a result, for convergence in distribution, the random vectors $\boldsymbol{z}_n$ and $\boldsymbol{z}$, as well as the random vectors in the sequence $\{\boldsymbol{z}_n\}$, need not be defined on the same probability space.

There exist some useful connections between the different types of stochastic convergence. Applying the pertinent definitions, it is readily verified that almost-sure convergence implies convergence in probability, which in turn implies convergence in distribution:

$$\boldsymbol{z}_n \xrightarrow[n\to\infty]{\text{a.s.}} \boldsymbol{z} \quad \Longrightarrow \quad \boldsymbol{z}_n \xrightarrow[n\to\infty]{\text{p}} \boldsymbol{z} \quad \Longrightarrow \quad \boldsymbol{z}_n \xrightarrow[n\to\infty]{\text{d}} \boldsymbol{z}. \tag{D.16}$$

Likewise, convergence in the rth mean is stronger than convergence in probability, namely,

$$\boldsymbol{z}_n \xrightarrow[n\to\infty]{L_r} \boldsymbol{z} \quad \Longrightarrow \quad \boldsymbol{z}_n \xrightarrow[n\to\infty]{\text{p}} \boldsymbol{z}, \tag{D.17}$$

as can be verified by applying Markov's inequality (Theorem C.1). Note that there is no implication involving almost-sure convergence and convergence in the rth mean. In fact, in general, almost-sure convergence does not imply, nor is implied by convergence in the rth mean. Additional conditions (like uniform integrability) are needed to establish a link between these two types of convergence — see, e.g., [159].

D.2 Fundamental Asymptotic Results

This section collects some classic results that we call upon during our treatment.

We start with a theorem that is very useful when one needs to prove convergence in distribution. The theorem establishes a strong link between convergence in distribution and the behavior of the characteristic function.

Theorem D.1 (Lévy-Cramér continuity theorem [159, Thm. 1.9]). Let $\{\boldsymbol{z}_n\}$ be a sequence of random vectors in $\mathbb{R}^d$, and consider the associated sequence of characteristic functions

$$\varphi_n(s) \triangleq \mathbb{E}e^{\iota s^\mathsf{T} z_n}, \qquad s \in \mathbb{R}^d, \tag{D.18}$$

where $\iota = \sqrt{-1}$ is the imaginary unit. Then, the sequence $\{z_n\}$ converges in distribution to a random vector z if, and only if, the sequence of characteristic functions $\{\varphi_n(s)\}_{n\in\mathbb{N}}$ converges to the characteristic function of z, namely, to

$$\varphi(s) \triangleq \mathbb{E}e^{\iota s^\mathsf{T} z}, \qquad s \in \mathbb{R}^d. \tag{D.19}$$

In other words, we have the following double implication:

$$z_n \xrightarrow[n\to\infty]{\mathrm{d}} z \iff \lim_{n\to\infty} \varphi_n(s) = \varphi(s) \;\; \forall s \in \mathbb{R}^d. \tag{D.20}$$

One useful consequence of the Lévy-Cramér continuity theorem is the following result, which is often referred to as the Cramér-Wold device. The result establishes that convergence in distribution of a sequence of random vectors is equivalent to convergence in distribution of *all* linear combinations of the entries of the vectors.

Theorem D.2 (Cramér-Wold device [159, Thm. 1.9]). Let $\{z_n\}$ be a sequence of random vectors in $\mathbb{R}^d$. Then, the sequence $\{z_n\}$ converges in distribution to a random vector $z \in \mathbb{R}^d$ if, and only if,

$$c^\mathsf{T} z_n \xrightarrow[n\to\infty]{\mathrm{d}} c^\mathsf{T} z \;\text{ for all deterministic vectors } c \in \mathbb{R}^d. \tag{D.21}$$

The next theorem examines what happens, in terms of convergence, when the random vectors in a sequence are transformed by a continuous mapping. Loosely speaking, the theorem ascertains that the limit of the mapping is equal to the mapping of the limit for three types of convergence, namely, almost-sure convergence, convergence in probability, and convergence in distribution.

Theorem D.3 (Continuous mapping theorem [65, Thm. 3.2.10]). Let $\{z_n\}$ be a sequence of random vectors in $\mathbb{R}^d$ and z a random vector in $\mathbb{R}^d$. Let

$$g : \mathbb{R}^d \mapsto \mathbb{R}^m \tag{D.22}$$

be a mapping continuous at every point of a set $\mathcal{A}$ such that $\mathbb{P}[z \in \mathcal{A}] = 1$. Then we have the following implications:

i)

$$z_n \xrightarrow[n\to\infty]{\text{a.s.}} z \implies g(z_n) \xrightarrow[n\to\infty]{\text{a.s.}} g(z). \quad \text{(D.23)}$$

ii)

$$z_n \xrightarrow[n\to\infty]{\text{p}} z \implies g(z_n) \xrightarrow[n\to\infty]{\text{p}} g(z). \quad \text{(D.24)}$$

iii)

$$z_n \xrightarrow[n\to\infty]{\text{d}} z \implies g(z_n) \xrightarrow[n\to\infty]{\text{d}} g(z). \quad \text{(D.25)}$$

In our proofs, we often exploit the following properties of stochastic convergence.

Lemma D.1 (Useful properties of stochastic convergence [166]). Let $\{\boldsymbol{y}_n\}$ and $\{\boldsymbol{z}_n\}$ be sequences of random vectors in $\mathbb{R}^d$.

P1) If $\boldsymbol{y}_n$ converges in probability to $\boldsymbol{y}$ and $\boldsymbol{z}_n$ converges in probability to $\boldsymbol{z}$, then

$$\boldsymbol{y}_n + \boldsymbol{z}_n \xrightarrow[n\to\infty]{\text{p}} \boldsymbol{y} + \boldsymbol{z}. \quad \text{(D.26)}$$

P2) $\boldsymbol{z}_n$ converges in probability to $\boldsymbol{z}$ if, and only if, all its entries converge in probability to the corresponding entries of $\boldsymbol{z}$.

P3) If $\boldsymbol{y}_n$ converges in distribution to a deterministic constant c, then $\boldsymbol{y}_n$ converges in probability to c.

The next lemma shows a sufficient condition for the product of random sequences to vanish in probability. This lemma is perhaps less known, and for this reason, it is stated with proof.

Lemma D.2 (Product of random sequences). Let $\boldsymbol{z}_n = \boldsymbol{w}_n \boldsymbol{y}_n$, where $\{\boldsymbol{w}_n\}$ and $\{\boldsymbol{y}_n\}$ are two sequences of nonnegative random variables satisfying the following conditions:

i)

$$\boldsymbol{w}_n \xrightarrow[n\to\infty]{\text{p}} 0. \quad \text{(D.27)}$$

ii) For sufficiently large values $y > 0$,

$$\mathbb{P}[\boldsymbol{y}_n > y] \leq g(y), \quad \text{(D.28)}$$

where $g(y)$ is a nonnegative function such that

$$\lim_{y\to\infty} g(y) = 0. \quad \text{(D.29)}$$

Then,

$$\boldsymbol{z}_n \xrightarrow[n\to\infty]{\mathrm{P}} 0. \tag{D.30}$$

Proof. For any two positive values y and z,

$$\left\{\boldsymbol{w}_n \leq z/y\right\} \bigcap \left\{\boldsymbol{y}_n \leq y\right\} \implies \left\{\boldsymbol{w}_n \boldsymbol{y}_n \leq z\right\}. \tag{D.31}$$

Now, for any two events $\mathcal{A}$ and $\mathcal{B}$, the statement $\mathcal{A} \implies \mathcal{B}$ is equivalent to the statement $\mathcal{B}^{\mathrm{c}} \implies \mathcal{A}^{\mathrm{c}}$ (the notation $\mathcal{A}^{\mathrm{c}}$ denotes the complement of $\mathcal{A}$). Therefore, Eq. (D.31) is equivalent to

$$\left\{\boldsymbol{w}_n \boldsymbol{y}_n \leq z\right\}^{\mathrm{c}} \implies \left(\left\{\boldsymbol{w}_n \leq z/y\right\} \bigcap \left\{\boldsymbol{y}_n \leq y\right\}\right)^{\mathrm{c}}, \tag{D.32}$$

which is in turn equivalent to

$$\left\{\boldsymbol{w}_n \boldsymbol{y}_n > z\right\} \implies \left\{\boldsymbol{w}_n > z/y\right\} \bigcup \left\{\boldsymbol{y}_n > y\right\}, \tag{D.33}$$

because, in view of De Morgan's law [21], the complement of the intersection of two sets is the union of the complements of the sets. Moreover, since the condition $\mathcal{A} \implies \mathcal{B}$ implies that $\mathbb{P}[\mathcal{A}] \leq \mathbb{P}[\mathcal{B}]$, from (D.33) and using the union bound, we conclude that

$$\mathbb{P}[\boldsymbol{z}_n > z] \leq \mathbb{P}[\boldsymbol{w}_n > z/y] + \mathbb{P}[\boldsymbol{y}_n > y]. \tag{D.34}$$

Now, let us fix a value $\varepsilon > 0$ and choose y sufficiently large to use the upper bound in (D.28) and to ensure that $g(y) < \varepsilon/2$ (the latter choice is possible thanks to (D.29)). Therefore, for such a y, Eq. (D.34) implies

$$\mathbb{P}[\boldsymbol{z}_n > z] \leq \mathbb{P}[\boldsymbol{w}_n > z/y] + g(y) < \mathbb{P}[\boldsymbol{w}_n > z/y] + \varepsilon/2. \tag{D.35}$$

On the other hand, since by assumption $\boldsymbol{w}_n$ converges to 0 in probability, for given values of y and z there exists a sufficiently large n_0 such that, for all $n \geq n_0$, the quantity $\mathbb{P}[\boldsymbol{w}_n > z/y]$ is also strictly upper bounded by $\varepsilon/2$. Using this result in (D.35) we conclude that

$$\mathbb{P}[\boldsymbol{z}_n > z] < \varepsilon \quad \forall n \geq n_0, \tag{D.36}$$

which means that the limit of $\mathbb{P}[\boldsymbol{z}_n > z]$ is 0 for all choices of $z > 0$. Since $\boldsymbol{z}_n$ is nonnegative, this conclusion corresponds to the statement that $\boldsymbol{z}_n$ converges to 0 in probability, and the claim of the lemma is proved.

■

The next theorem is a classic result that is useful to examine the convergence in distribution of sums or products of two random sequences, one converging in distribution, the other converging to a deterministic value in probability.

Theorem D.4 (Slutsky's theorem [159, Thm. 1.11]). Let $\{\boldsymbol{z}_n\}$ be a sequence of random variables converging in distribution to a random variable $\boldsymbol{z}$. Let $\{\boldsymbol{c}_n\}$ be another sequence of random variables converging in probability to a *deterministic* value c. Then

$$\boldsymbol{z}_n + \boldsymbol{c}_n \xrightarrow[n\to\infty]{\mathrm{d}} \boldsymbol{z} + c, \tag{D.37}$$

$$\boldsymbol{c}_n\, \boldsymbol{z}_n \xrightarrow[n\to\infty]{\mathrm{d}} c\, \boldsymbol{z}. \tag{D.38}$$

Note that with Slutsky's theorem we are able to draw useful conclusions about sums or products of random variables *without* knowing the joint characterization of these random variables. We also remark that Slutsky's theorem can be generalized to handle random vectors [166]. In particular, if $\{\boldsymbol{z}_n\}$ and $\{\boldsymbol{c}_n\}$ are sequences of random vectors in $\mathbb{R}^d$, then

$$\boldsymbol{z}_n \xrightarrow[n\to\infty]{\mathrm{d}} \boldsymbol{z} \quad \text{and} \quad \boldsymbol{c}_n \xrightarrow[n\to\infty]{\mathrm{p}} c \quad \Longrightarrow \quad \boldsymbol{z}_n + \boldsymbol{c}_n \xrightarrow[n\to\infty]{\mathrm{d}} \boldsymbol{z} + c. \tag{D.39}$$

Example D.1 (Stochastic convergence with continuous parameter). When dealing with adaptive social learning in Chapters 8, 9, and 10, we are faced with questions related to stochastic convergence when a certain *continuous* parameter δ approaches zero. We can easily restate the definitions of stochastic convergence in terms of a continuous parameter δ, as opposed to a discrete index n. We explain in this example how the theorems listed before can be adapted to the continuous-parameter case — see [145, p. 213] for more discussion on how to move from *countable* sequences to *continuous* parameters. The adjustment is achieved by resorting to the sequential property of the limit, which states that, given a deterministic function $f(\delta)$, we have [144, Thm. 4.2]

$$\lim_{\delta\to 0} f(\delta) = \ell \tag{D.40}$$

if, and only if,

$$\lim_{n\to\infty} f(\delta_n) = \ell \text{ for all sequences } \{\delta_n\} \text{ such that } \delta_n \neq 0 \text{ and } \lim_{n\to\infty} \delta_n = 0. \tag{D.41}$$

If desired, the limiting point 0 can be replaced by any arbitrary value δ_0, also $\pm\infty$. Assume now that we want to apply, for example, Theorem D.4 to a family of random variables $\{\boldsymbol{z}_\delta\}$ indexed by the continuous parameter δ. The hypotheses of Slutsky's theorem for a continuous parameter become

$$\boldsymbol{z}_\delta \xrightarrow[\delta\to 0]{\mathrm{d}} \boldsymbol{z}, \qquad \boldsymbol{c}_\delta \xrightarrow[\delta\to 0]{\mathrm{p}} c. \tag{D.42}$$

Let us denote by $F_\delta(z)$ and $F(z)$ the cdfs of $\boldsymbol{z}_\delta$ and $\boldsymbol{z}$, respectively. According to Definition D.4, the convergence of $\boldsymbol{z}_\delta$ to $\boldsymbol{z}$ in (D.42) means that $\lim_{\delta\to 0} F_\delta(z) = F(z)$ for all continuity points of $F(z)$. Using the sequential property of the limit, (specifically, considering that (D.40) implies (D.41)), we deduce that, for any sequence $\{\delta_n\}$ such that $\delta_n \neq 0$ and $\lim_{n\to\infty} \delta_n = 0$, the extracted sequence of functions $F_{\delta_n}(z)$ converges to $F(z)$ (for all continuity points of $F(z)$), which means that

$$\boldsymbol{z}_{\delta_n} \xrightarrow[n\to\infty]{\mathrm{d}} \boldsymbol{z}. \tag{D.43}$$

A similar argument applies to the convergence of $\boldsymbol{c}_\delta$ to c in (D.42). In summary, if we extract the following n−indexed sequences $\{\boldsymbol{z}_n\}$ and $\{\boldsymbol{c}_n\}$ from $\{\boldsymbol{z}_\delta\}$ and $\{\boldsymbol{c}_\delta\}$, respectively:

$$\boldsymbol{z}_n \triangleq \boldsymbol{z}_{\delta_n}, \qquad \boldsymbol{c}_n \triangleq \boldsymbol{c}_{\delta_n}, \tag{D.44}$$

we can write

$$\boldsymbol{z}_n \xrightarrow[n\to\infty]{\mathrm{d}} \boldsymbol{z}, \qquad \boldsymbol{c}_n \xrightarrow[n\to\infty]{\mathrm{p}} c, \tag{D.45}$$

which means that the sequences $\{\boldsymbol{z}_n\}$ and $\{\boldsymbol{c}_n\}$ fulfill the hypotheses of Theorem D.4. This implies that (D.37) and (D.38) hold for any sequence $\{\delta_n\}$ such that $\delta_n \neq 0$ and $\lim_{n\to\infty} \delta_n = 0$. Applying again the sequential property of the limit (this time in the reverse direction, i.e., from (D.41) to (D.40)), we deduce that (D.37) and (D.38) hold when we replace $\boldsymbol{z}_n$ with $\boldsymbol{z}_\delta$ and $\boldsymbol{c}_n$ with $\boldsymbol{c}_\delta$, and let $\delta \to 0$, which is the continuous version of Slutsky's theorem that we wanted to obtain.

We conclude this section with two fundamental results in measure theory and probability theory, which concern the interchange of limits and expectations. The first result is Fatou's lemma, which establishes a useful inequality relating the limit inferior and the expectation for a sequence of random variables.

Theorem D.5 (Fatou's lemma [65, Thm. 1.6.5]). If $\{\boldsymbol{z}_n\}$ is a sequence of nonnegative random variables, then

$$\liminf_{n\to\infty} \mathbb{E}\boldsymbol{z}_n \geq \mathbb{E}\left[\liminf_{n\to\infty} \boldsymbol{z}_n\right]. \tag{D.46}$$

If all random variables in the sequence are bounded by a random variable with finite mean, the conclusion from Fatou's lemma can be strengthened in two directions, namely, by allowing for random variables of arbitrary sign, and by establishing an exact convergence.

Theorem D.6 (Dominated convergence theorem [65, Thm. 1.6.7]). Let $\{\boldsymbol{z}_n\}$ be a sequence of random variables, and $\boldsymbol{y}$ a random variable with finite mean such that (almost surely) $|\boldsymbol{z}_n| \leq \boldsymbol{y}$ for all n. If $\boldsymbol{z}_n$ converges almost surely to a random variable $\boldsymbol{z}$, then

$$\lim_{n\to\infty} \mathbb{E}\boldsymbol{z}_n = \mathbb{E}\boldsymbol{z}, \tag{D.47}$$

i.e., the order of the limit and expectation operations can be interchanged.

D.3 Convergence of Sums and Recursions

This section collects some results on the stochastic convergence of sequences that arise from sums or recursions involving random variables or vectors. The first result is the famous law of large numbers. In particular, we focus on its strong version, known as the strong law of large numbers. This law establishes that, for a sequence of iid random variables with finite statistical mean, the empirical average (i.e., the arithmetic mean) converges *almost surely* to the statistical mean.

Theorem D.7 (Strong law of large numbers [35, Thm. 3.30]). Let $\{\boldsymbol{y}_n\}$ be a sequence of iid random variables with finite mean $\bar{y} = \mathbb{E}\boldsymbol{y}_n$. Then, the sequence of empirical averages

$$\boldsymbol{z}_n = \frac{1}{n}\sum_{i=1}^{n} \boldsymbol{y}_i \tag{D.48}$$

converges almost surely to $\bar{y}$:

$$\boldsymbol{z}_n \xrightarrow[n\to\infty]{\text{a.s.}} \bar{y}. \tag{D.49}$$

The next theorem is another pillar of asymptotic statistics, whose centrality is highlighted by the name itself: the central limit theorem (CLT). The theorem establishes that the *distribution* of a sum of n iid random vectors, centered by subtracting the mean and divided by $\sqrt{n}$, converges to a zero-mean multivariate Gaussian distribution with the covariance matrix of the individual vectors.

Theorem D.8 (Central limit theorem [35, Thm. 11.10]). Let $\{\boldsymbol{y}_n\}$ be a sequence of iid random vectors in $\mathbb{R}^d$, with finite mean $\bar{y} = \mathbb{E}\boldsymbol{y}_n$ and covariance matrix $\Sigma = \mathbb{E}\left[(\boldsymbol{y}_n - \bar{y})(\boldsymbol{y}_n - \bar{y})^\top\right]$ with finite entries. Let $\mathscr{G}(0, \Sigma)$ denote a random vector having a zero-mean multivariate Gaussian distribution with covariance matrix Σ. Then, the sequence $\{\boldsymbol{z}_n\}$ defined by

$$\boldsymbol{z}_n = \frac{1}{\sqrt{n}}\sum_{i=1}^{n} (\boldsymbol{y}_i - \bar{y}), \tag{D.50}$$

converges in distribution to a zero-mean Gaussian vector with covariance matrix Σ:

$$\boldsymbol{z}_n \xrightarrow[n\to\infty]{\text{d}} \mathscr{G}(0, \Sigma). \tag{D.51}$$

We also state (a particular case of) the Lindeberg-Feller central limit theorem [166], which handles independent but not identically distributed random vectors, and turns out to be useful to prove the asymptotic normality results in Chapters 6 and 9.

Theorem D.9 (CLT under the Lindeberg condition [166, Prop. 2.27]). Let $\{\boldsymbol{y}_n\}$ be a sequence of independent random vectors in $\mathbb{R}^d$, possessing the following three properties. First,

$$\mathbb{E}\boldsymbol{y}_n = 0. \tag{D.52}$$

Second,

$$\lim_{n\to\infty} \frac{1}{n}\sum_{i=1}^{n} \mathbb{E}\left[\boldsymbol{y}_i\boldsymbol{y}_i^\top\right] = \Sigma, \tag{D.53}$$

where Σ has finite entries. Third, the following condition (called the Lindeberg condition) is satisfied:

$$\lim_{n\to\infty} \frac{1}{n}\sum_{i=1}^{n} \mathbb{E}\left[\|\boldsymbol{y}_i\|^2\,\mathbb{I}\left[\|\boldsymbol{y}_i\|^2 > \varepsilon n\right]\right] = 0 \quad \forall\varepsilon > 0. \tag{D.54}$$

Then, the sequence $\{\boldsymbol{z}_n\}$ defined by

$$\boldsymbol{z}_n = \frac{1}{\sqrt{n}}\sum_{i=1}^{n} \boldsymbol{y}_i, \tag{D.55}$$

converges in distribution to a zero-mean Gaussian vector with covariance matrix Σ:

$$\boldsymbol{z}_n \xrightarrow[n\to\infty]{\mathrm{d}} \mathscr{G}(0,\Sigma). \tag{D.56}$$

We conclude the section with a lemma characterizing a vector recursion that is repeatedly encountered in our treatment.

Lemma D.3 (Useful vector recursion). Let $\{\boldsymbol{y}_n\}$ be a sequence of iid random vectors in $\mathbb{R}^d$ with finite mean $\bar{y}$. Consider the sequence $\{\boldsymbol{z}_n\}$ formed by the vectors defined through the following recursion, for $n \in \mathbb{N}$:

$$\boldsymbol{z}_n = A^\top(\boldsymbol{z}_{n-1} + \boldsymbol{y}_n), \tag{D.57}$$

where z_0 is an initial deterministic vector, and A is a $d \times d$ deterministic left stochastic matrix. Since all left stochastic matrices are Cesàro-summable (Theorem 4.4), there exists a limiting matrix $A^\bullet$ such that

$$\lim_{n\to\infty} \frac{1}{n}\sum_{i=1}^{n} A^i = A^\bullet. \tag{D.58}$$

Then

$$\frac{1}{n}\boldsymbol{z}_n \xrightarrow[n\to\infty]{\mathrm{a.s.}} (A^\bullet)^\top\,\bar{y}. \tag{D.59}$$

Proof. Unfolding the recursion in (D.57) we get

$$\boldsymbol{z}_n = (A^n)^\mathsf{T} z_0 + \sum_{i=1}^{n} (A^i)^\mathsf{T} \boldsymbol{y}_{n-i+1}. \tag{D.60}$$

Once scaled by $1/n$, the first term on the RHS vanishes as $n \to \infty$ in view of the properties of A. We focus on the second term. It is useful to rewrite the summation in (D.60) as follows:

$$\frac{1}{n}\sum_{i=1}^{n} (A^i)^\mathsf{T} \boldsymbol{y}_{n-i+1} = \frac{1}{n}\sum_{i=1}^{n} (A^i)^\mathsf{T} \bar{y} + \frac{1}{n}\sum_{i=1}^{n} (A^i)^\mathsf{T} \left(\boldsymbol{y}_{n-i+1} - \bar{y}\right). \tag{D.61}$$

The first term on the RHS converges to $(A^\bullet)^\mathsf{T}\,\bar{y}$ as $n \to \infty$ in view of (D.58). As a result, the claim of the lemma will be proved if we establish that the second term on the RHS of (D.61) vanishes almost surely as $n \to \infty$. To show that this is the case, observe that this term is a vector and consider its kth entry:

$$\frac{1}{n}\sum_{i=1}^{n}\sum_{j=1}^{d} [A^i]_{jk} \left(\boldsymbol{y}_{j,n-i+1} - \bar{y}_j\right), \tag{D.62}$$

where $[A^i]_{jk}$ denotes the (j,k) entry of A^i, while $\boldsymbol{y}_{j,n}$ and $\bar{y}_j$ denote the jth entries of $\boldsymbol{y}_n$ and $\bar{y}$, respectively. Interchanging the summations and rearranging the summands in the summation running over i, the quantity in (D.62) can be rewritten as

$$\sum_{j=1}^{d} \frac{1}{n}\sum_{i=1}^{n} [A^{n-i+1}]_{jk} \left(\boldsymbol{y}_{j,i} - \bar{y}_j\right). \tag{D.63}$$

Let us focus on a single term of the outer summation, i.e., a fixed value of j. By setting

$$b_{ni} \triangleq [A^{n-i+1}]_{jk}, \qquad \boldsymbol{\xi}_i \triangleq \boldsymbol{y}_{j,i} - \bar{y}_j, \tag{D.64}$$

the jth term of the outer summation in (D.63) becomes

$$\frac{1}{n}\sum_{i=1}^{n} b_{ni}\, \boldsymbol{\xi}_i, \tag{D.65}$$

where b_{ni} defines a triangular array[1] with nonnegative entries bounded by 1, and the random variables $\boldsymbol{\xi}_i$ are zero-mean and iid. Therefore, the random variable in (D.65) is a weighted sum of independent random variables, where the weights form a triangular

[1] Let $n \in \mathbb{N}$ and define, for each n, a sequence of real values y_{ni}, with $i = 1, 2, \ldots, n$. Then we say that these values form a triangular array. The term stems from the following visual representation:

$$\begin{array}{llll} n=1 & y_{11} & & \\ n=2 & y_{21}, & y_{22} & \\ n=3 & y_{31}, & y_{32} & y_{33} \\ \vdots & & & \end{array} \tag{D.66}$$

array with bounded entries. Under these conditions, a strong law of large numbers holds, specifically, we have that [45]

$$\frac{1}{n}\sum_{i=1}^{n} b_{ni}\,\boldsymbol{\xi}_i \xrightarrow[n\to\infty]{\text{a.s.}} 0, \tag{D.67}$$

which concludes the proof.

■

D.4 Martingales

In some proofs related to the convergence of social learning algorithms (see Chapters 7 and 11), we rely on the concept of *martingales*. Preliminarily, we introduce the definition of *filtrations*.

Definition D.5 (Filtrations). Let $(\Omega, \mathscr{F}, \mathbb{P})$ be a probability space. A filtration $\{\mathcal{F}_n\}$ is an increasing sequence of sub-σ-fields of $\mathscr{F}$:

$$\mathcal{F}_1 \subset \mathcal{F}_2 \subset \dots \tag{D.68}$$

We can now define martingales, submartingales, and supermartingales.

Definition D.6 (Martingales). Let $\{\boldsymbol{z}_n\}$ be a sequence of finite-mean random variables defined on a probability space $(\Omega, \mathscr{F}, \mathbb{P})$, and let $\{\mathcal{F}_n\}$ be a filtration according to Definition D.5. Then, $\{\boldsymbol{z}_n\}$ is said to be a martingale relative to $\{\mathcal{F}_n\}$ when, for all $n > 1$,

$$\mathbb{E}[\boldsymbol{z}_n|\mathcal{F}_{n-1}] = \boldsymbol{z}_{n-1} \qquad \text{almost surely.} \tag{D.69}$$

Likewise, it is said to be a submartingale when

$$\mathbb{E}[\boldsymbol{z}_n|\mathcal{F}_{n-1}] \geq \boldsymbol{z}_{n-1} \qquad \text{almost surely,} \tag{D.70}$$

and a supermartingale when

$$\mathbb{E}[\boldsymbol{z}_n|\mathcal{F}_{n-1}] \leq \boldsymbol{z}_{n-1} \qquad \text{almost surely.} \tag{D.71}$$

The next example helps illustrate what a martingale can be.

Example D.2 (Random walk). Consider a sequence $\{\boldsymbol{y}_n\}$ of iid zero-mean random variables, and define the so-called *random walk*

$$\boldsymbol{z}_n = \sum_{i=1}^{n} \boldsymbol{y}_i, \qquad n \in \mathbb{N}. \tag{D.72}$$

We now show that $\{z_n\}$ is a martingale.

Observe that we can write

$$z_n = y_n + \sum_{i=1}^{n-1} y_i \tag{D.73}$$

and consider the filtration over past variables,

$$\mathcal{F}_{n-1} = \sigma(y_1, y_2, \ldots, y_{n-1}), \tag{D.74}$$

where the notation signifies that $\mathcal{F}_{n-1}$ is the σ-field generated by the random variables $y_1, y_2, \ldots, y_{n-1}$. The conditional expectation of z_n given $\mathcal{F}_{n-1}$ is

$$\begin{aligned}\mathbb{E}[z_n|\mathcal{F}_{n-1}] &= \mathbb{E}[y_n|\mathcal{F}_{n-1}] + \mathbb{E}\left[\sum_{i=1}^{n-1} y_i|\mathcal{F}_{n-1}\right] \\ &= \underbrace{\mathbb{E}y_n}_{=0} + \sum_{i=1}^{n-1} y_i = z_{n-1},\end{aligned} \tag{D.75}$$

where in the first term on the RHS we remove the conditioning since y_n is independent from the past, whereas in the second term the summation is "frozen" given the filtration since it is a deterministic function of $y_1, y_2, \ldots, y_{n-1}$. Equation (D.75) (along with the fact that z_n has finite mean for all n) shows that $\{z_n\}$ is a martingale.

The next theorem is a fundamental convergence result in the theory of martingales.

Theorem D.10 (Martingale convergence theorem [35, Thm. 5.14]). Let $\{z_n\}$ be a submartingale satisfying the condition

$$\limsup_{n\to\infty} \mathbb{E}|z_n| < \infty. \tag{D.76}$$

Then, there exists a finite-mean random variable z such that

$$z_n \xrightarrow[n\to\infty]{\text{a.s.}} z. \tag{D.77}$$

Similar claims hold for martingales (since a martingale is also a submartingale) and supermartingales (since if z_n is a supermartingale, then $-z_n$ is a submartingale). The next corollary is useful in our treatment.

Corollary D.1 (Nonpositive submartingales). Let $\{z_n\}$ be a nonpositive submartingale, i.e., a submartingale satisfying (almost surely) the condition $z_n \leq 0$

for all n. Then

$$0 \geq \mathbb{E}\boldsymbol{z}_n \geq \mathbb{E}\boldsymbol{z}_{n-1} \geq \ldots \geq \mathbb{E}\boldsymbol{z}_1. \tag{D.78}$$

Since $\boldsymbol{z}_n$ is nonpositive, Eq. (D.78) implies condition (D.76), which, in view of Theorem D.10, implies the existence of a finite-mean random variable $\boldsymbol{z}$ such that

$$\boldsymbol{z}_n \xrightarrow[n\to\infty]{\text{a.s.}} \boldsymbol{z}. \tag{D.79}$$

Appendix E

Large Deviations

In this appendix we collect some fundamental results from the theory of *large deviations.* The main concepts of this theory are well described by considering the case of the empirical average of independent and identically distributed observations, and by examining the probability that this average deviates from the statistical mean of the observations. This problem is illustrated in the next section, which culminates with one of the earliest results on large deviations, known as Cramér's theorem [53]. Then, in Section E.2, we consider a more general setting by enunciating the *large deviation principle* and by addressing the case of dependent observations through the Gärtner-Ellis theorem [68, 78].

E.1 Empirical Averages

Consider a sequence $\{\boldsymbol{y}_n\}$ of iid random variables with mean $\bar{y} = \mathbb{E}\boldsymbol{y}_n$ and variance σ^2, and define the empirical average

$$\bar{\boldsymbol{y}}_n = \frac{1}{n}\sum_{i=1}^{n} \boldsymbol{y}_i. \tag{E.1}$$

The strong law of large numbers in Theorem D.7 establishes that $\bar{\boldsymbol{y}}_n$ converges almost surely to $\bar{y}$. Recalling that almost-sure convergence implies convergence in probability, in view of Definition D.2 the probability that $\bar{\boldsymbol{y}}_n$ deviates from $\bar{y}$ by any arbitrary amount vanishes as $n \to \infty$. This implies in particular that

$$\lim_{n\to\infty} \mathbb{P}\left[\bar{\boldsymbol{y}}_n \geq y\right] = 0 \quad \forall y > \bar{y}, \tag{E.2a}$$

$$\lim_{n\to\infty} \mathbb{P}\left[\bar{\boldsymbol{y}}_n \leq y\right] = 0 \quad \forall y < \bar{y}. \tag{E.2b}$$

The aim of a large deviation analysis it to characterize the rate at which these probabilities converge to 0.

Example E.1 (Gaussian variables). Assume that, for all $n \in \mathbb{N}$, the variable $\boldsymbol{y}_n$ is Gaussian, with mean $\bar{y}$ and variance σ^2. In this case, the empirical average (E.1) will also be Gaussian, with mean $\bar{y}$ and variance σ^2/n. Accordingly, the probabilities in (E.2a) and (E.2b) can be evaluated as

$$\mathbb{P}\left[\bar{\boldsymbol{y}}_n \geq y\right] = Q\left(\sqrt{n}\,\frac{y-\bar{y}}{\sigma}\right), \qquad \mathbb{P}\left[\bar{\boldsymbol{y}}_n \leq y\right] = Q\left(\sqrt{n}\,\frac{\bar{y}-y}{\sigma}\right), \tag{E.3}$$

where we use the Q-function

$$Q(y) \triangleq \frac{1}{\sqrt{2\pi}} \int_y^{\infty} \exp\left\{-\frac{x^2}{2}\right\} dx, \tag{E.4}$$

which is the complementary cumulative distribution function of a standard (i.e., with zero mean and unit variance) Gaussian variable. Applying to (E.3) the approximation

$$Q(y) \approx \frac{1}{\sqrt{2\pi}\,y} \exp\left\{-\frac{y^2}{2}\right\}, \qquad y > 0, \tag{E.5}$$

we get, for $y > \bar{y}$,

$$\begin{aligned}\mathbb{P}\left[\bar{\boldsymbol{y}}_n \geq y\right] &\approx \frac{\sigma\, n^{-1/2}}{\sqrt{2\pi}\,(y-\bar{y})} \exp\left\{-n\frac{(y-\bar{y})^2}{2\sigma^2}\right\} \\ &= \exp\left\{-n\frac{(y-\bar{y})^2}{2\sigma^2} - \frac{1}{2}\log n + \log\frac{\sigma}{\sqrt{2\pi}\,(y-\bar{y})}\right\}.\end{aligned} \tag{E.6}$$

We see that the probability of exceeding the mean decays exponentially with n, but for higher-order corrections, namely, the term varying logarithmically with n and the constant term. From (E.6) we can compute the leading exponent as

$$\lim_{n\to\infty} \frac{1}{n} \log \mathbb{P}\left[\bar{\boldsymbol{y}}_n \geq y\right] = -\frac{(y-\bar{y})^2}{2\sigma^2}, \qquad y > \bar{y}. \tag{E.7}$$

Using similar arguments we can also show that

$$\lim_{n\to\infty} \frac{1}{n} \log \mathbb{P}\left[\bar{\boldsymbol{y}}_n \leq y\right] = -\frac{(y-\bar{y})^2}{2\sigma^2}, \qquad y < \bar{y}. \tag{E.8}$$

We remark that (E.7) and (E.8) can be obtained rigorously, i.e., without resorting to the approximation of the Q-function. This can be done by using classic lower and upper bounds on the Q-function, or by using L'Hôpital's rule [144] to establish the limit

$$\lim_{n\to\infty} \frac{1}{n} \log Q\left(\sqrt{n}y\right) = -\frac{y^2}{2}, \qquad y \neq 0. \tag{E.9}$$

The exponential decay shown in Example E.1 is not observed only for Gaussian variables. In fact, under suitable regularity conditions on the tail of the probability distribution of $\boldsymbol{y}_n$ (which, as seen later, are expressed in terms of the moment generating function of $\boldsymbol{y}_n$), the probability of deviating from $\bar{y}$ would continue to vanish *exponentially* with the number of samples n. Focusing for brevity on the regime $y > \bar{y}$, this condition can be written generically as

$$-\lim_{n\to\infty} \frac{1}{n} \log \mathbb{P}\left[\bar{\boldsymbol{y}}_n \geq y\right] = I(y) \tag{E.10}$$

for some function $I(y)$ or, equivalently, as

$$\mathbb{P}\left[\bar{\boldsymbol{y}}_n \geq y\right] = e^{-n[I(y)+o(1)]}, \qquad y > \bar{y}, \tag{E.11}$$

where the notation $o(1)$ refers to a quantity that vanishes as $n \to \infty$ — see Table 1.1. Therefore, relation (E.11) means that the leading exponential decay is given by the linear term $nI(y)$. We will say in this case that the probability in (E.11) vanishes at *rate* n and with *rate function* $I(y)$. The quantity $n \times o(1)$ collects higher-order corrections to the leading term. Note that these corrections can diverge with n. For example, on the RHS of (E.6), we have

$$n \times o(1) = -\frac{1}{2}\log n + \log \frac{\sigma}{\sqrt{2\pi}\,(y-\bar{y})}. \tag{E.12}$$

The rate function $I(y)$ provides the main *exponent* ruling the exponential decay to 0 of the probability of exceeding the mean $\bar{y}$ by an amount $y - \bar{y}$. This exponent is a function of y. For instance, for the Gaussian case in Example E.1, Eq. (E.7) reveals that the rate function is

$$I(y) = \frac{(y-\bar{y})^2}{2\sigma^2} \qquad \text{[Gaussian case]}. \tag{E.13}$$

As expected, the larger the deviation from the mean is, the larger the exponent will become, and the faster the convergence to 0 will be. An alternative and common notation to represent (E.11) is to write [52]

$$\mathbb{P}\left[\bar{\boldsymbol{y}}_n \geq y\right] \doteq e^{-nI(y)}, \qquad y > \bar{y}, \tag{E.14}$$

which masks the higher-order corrections. We hasten to add that one cannot use the expression $e^{-nI(y)}$ to *approximate* the probability $\mathbb{P}\left[\bar{\boldsymbol{y}}_n \geq y\right]$. This is because a large deviation analysis provides only the leading exponent $I(y)$. For example, the two probabilities $e^{-nI(y)}$ and $100e^{-nI(y)}$ share the same exponent, but they differ by two orders of magnitude!

A famous theorem proved by Cramér [53] establishes the shape of the rate function $I(y)$. Before stating this theorem, it is necessary to introduce and characterize some relevant tools, which are illustrated in the next two sections.

E.1.1 Fenchel-Legendre Transform

Given a function

$$f : \mathbb{R} \mapsto (-\infty, \infty], \tag{E.15}$$

we introduce its *Fenchel-Legendre transform*, a.k.a. the *convex conjugate* of $f(y)$,

$$f^*(y) \triangleq \sup_{s\in\mathbb{R}} \Big(sy - f(s)\Big), \quad y \in \mathbb{R}. \tag{E.16}$$

Note that the function $f^*(y)$ can be equal to ∞ for some y.

It is immediate to verify that $f^*(y)$ is convex. Indeed, for any $\alpha \in (0,1)$ and any pair of points $y_1, y_2 \in \mathbb{R}$,

$$\begin{aligned} & f^*(\alpha y_1 + (1-\alpha)y_2) \\ &= \sup_{s\in\mathbb{R}} \Big(s(\alpha y_1 + (1-\alpha)y_2) - f(s)\Big) \\ &= \sup_{s\in\mathbb{R}} \Big(s(\alpha y_1 + (1-\alpha)y_2) - \alpha f(s) - (1-\alpha)f(s)\Big) \\ &\leq \alpha \sup_{s\in\mathbb{R}} \Big(sy_1 - f(s)\Big) + (1-\alpha) \sup_{s\in\mathbb{R}} \Big(sy_2 - f(s)\Big) \\ &= \alpha f^*(y_1) + (1-\alpha) f^*(y_2), \end{aligned} \tag{E.17}$$

where the inequality holds because the supremum of the sum of functions is upper bounded by the sum of the suprema of the functions. A second property of the Fenchel-Legendre transform is lower semicontinuity, which means that, for all $y_0 \in \mathbb{R}$,

$$\liminf_{y\to y_0} f^*(y) \geq f^*(y_0). \tag{E.18}$$

This is in fact true since we can write, for all $s \in \mathbb{R}$,

$$\liminf_{y\to y_0} f^*(y) \geq \liminf_{y\to y_0} \Big(sy - f(s)\Big) = sy_0 - f(s). \tag{E.19}$$

Accordingly, we have

$$\liminf_{y\to y_0} f^*(y) \geq \sup_{s\in\mathbb{R}} \Big(sy_0 - f(s)\Big) = f^*(y_0), \tag{E.20}$$

which shows that $f^*(y)$ is lower semicontinuous. Lower semicontinuity plays a role in the following, when we characterize the general shape that any rate function must have.

E.1.2 Generating Functions

The *moment generating function* (MGF) of a random variable $\boldsymbol{y}$ is defined as

$$M(s) \triangleq \mathbb{E}e^{s\boldsymbol{y}}, \qquad s \in \mathbb{R}. \tag{E.21}$$

The function $M(s)$ is allowed to be equal to ∞ for some s. The *effective domain* of $M(s)$ is

$$\mathcal{D}_M \triangleq \{s \in \mathbb{R} : M(s) < \infty\}. \tag{E.22}$$

We verify that $\mathcal{D}_M$ must be an interval. To this end, we recall that a subset $\mathcal{S}$ of the real line is an interval if, and only if, given two points s_1, s_2 in $\mathcal{S}$, any point $s \in (s_1, s_2)$ also belongs to $\mathcal{S}$. Observe that we have $0 \in \mathcal{D}_M$ since $M(0) = 1$. Moreover, if there exists a positive value s_0 such that $M(s_0) < \infty$, then it is readily seen that $M(s) < \infty$ in the interval $[0, s_0)$. Likewise, if there exists a negative value $-s_0$ such that $M(-s_0) < \infty$, then it is also seen that $M(s) < \infty$ for all $s \in (-s_0, 0]$. As a result, if for two points s_1 and s_2 we have $M(s_1) < \infty$ and $M(s_2) < \infty$, then $M(s) < \infty$ for all $s \in (s_1, s_2)$. We conclude that the effective domain $\mathcal{D}_M$ is an interval, which can be open or closed depending on the particular random variable. Note that we can also have $\mathcal{D}_M = \{0\}$, that is, $M(s) = \infty$ for all $s \neq 0$. This happens, e.g., for the Cauchy pdf

$$p(y) = \frac{1}{\pi\sigma} \frac{1}{1 + \left(\dfrac{y - m}{\sigma}\right)^2}, \tag{E.23}$$

defined for $m \in \mathbb{R}$ and $\sigma > 0$. Moreover, if $\mathcal{D}_M$ is a nondegenerate interval (i.e., if $\mathcal{D}_M \neq \{0\}$), from the properties of the exponential function it is possible to show that $M(s)$ is infinitely differentiable on $\mathrm{int}(\mathcal{D}_M)$ (the interior of $\mathcal{D}_M$), and that its derivatives can be computed by interchanging the differentiation and integration operators, which yields [21]

$$M^{(n)}(s) = \mathbb{E}\left[\boldsymbol{y}^n e^{s\boldsymbol{y}}\right], \tag{E.24}$$

where we denote by $M^{(n)}(s)$ the nth derivative of $M(s)$. Therefore, we see that if $M(s)$ is finite in a neighborhood of the origin $s = 0$, then we have the identity, for all $n \in \mathbb{N}$,

$$M^{(n)}(0) = \mathbb{E}\boldsymbol{y}^n, \tag{E.25}$$

which justifies the name "moment generating function."

Of particular interest for the theory of large deviations is the *logarithmic moment generating function* (LMGF), a.k.a. *cumulant generating function*,

$$\Lambda(s) \triangleq \log M(s). \tag{E.26}$$

One important property of $\Lambda(s)$ is convexity, which follows from Hölder's inequality — see Theorem C.6. In fact, for any $\alpha \in (0,1)$ and any pair of points $s_1, s_2 \in \mathbb{R}$, we can write

$$\begin{aligned}\Lambda(\alpha s_1 + (1-\alpha)s_2) &= \log \mathbb{E} e^{(\alpha s_1 + (1-\alpha)s_2)\boldsymbol{y}} \\ &= \log \mathbb{E}\left[\left(e^{s_1 \boldsymbol{y}}\right)^{\alpha}\left(e^{s_2 \boldsymbol{y}}\right)^{1-\alpha}\right] \\ &\leq \log\left(\left(\mathbb{E} e^{s_1 \boldsymbol{y}}\right)^{\alpha}\left(\mathbb{E} e^{s_2 \boldsymbol{y}}\right)^{1-\alpha}\right) \\ &= \alpha \log \mathbb{E} e^{s_1 \boldsymbol{y}} + (1-\alpha)\log \mathbb{E} e^{s_2 \boldsymbol{y}} \\ &= \alpha\Lambda(s_1) + (1-\alpha)\Lambda(s_2),\end{aligned} \tag{E.27}$$

where the inequality follows by making the following choices in (C.11):

$$\boldsymbol{z}_1 = \left(e^{s_1 \boldsymbol{y}}\right)^{\alpha}, \quad \boldsymbol{z}_2 = \left(e^{s_2 \boldsymbol{y}}\right)^{1-\alpha}, \quad r_1 = \frac{1}{\alpha}, \quad r_2 = \frac{1}{1-\alpha}, \tag{E.28}$$

Since $M(s) \neq 0$, we see that $-\infty < \Lambda(s) \leq \infty$, and is equal to ∞ when $M(s) = \infty$. Thus, the effective domain $\mathcal{D}_\Lambda$ of $\Lambda(s)$ is equal to the effective domain of $M(s)$. Moreover, from the rules of differentiation, when $M(s)$ is infinitely differentiable so is $\Lambda(s)$. Accordingly, when $\mathcal{D}_\Lambda$ is a nondegenerate interval, we can compute the first two derivatives of $\Lambda(s)$ for any s belonging to the interior of $\mathcal{D}_\Lambda$, namely,

$$\Lambda'(s) = \frac{d}{ds}\log M(s) = \frac{M'(s)}{M(s)} = \frac{\mathbb{E}[\boldsymbol{y}\, e^{s\boldsymbol{y}}]}{\mathbb{E} e^{s\boldsymbol{y}}} \tag{E.29}$$

and

$$\begin{aligned}\Lambda''(s) &= \frac{d}{ds}\frac{\mathbb{E}[\boldsymbol{y}\, e^{s\boldsymbol{y}}]}{M(s)} = \frac{\mathbb{E}[\boldsymbol{y}^2 e^{s\boldsymbol{y}}]M(s) - M'(s)\mathbb{E}[\boldsymbol{y}\, e^{s\boldsymbol{y}}]}{M^2(s)} \\ &= \mathbb{E}\left[\boldsymbol{y}^2 \frac{e^{s\boldsymbol{y}}}{M(s)}\right] - \left(\mathbb{E}\left[\boldsymbol{y}\frac{e^{s\boldsymbol{y}}}{M(s)}\right]\right)^2.\end{aligned} \tag{E.30}$$

We remark that $\Lambda(s)$ is called the cumulant generating function because, under the aforementioned assumption of finiteness in a neighborhood of the origin, the quantity $\Lambda^{(n)}(0)$, for $n \in \mathbb{N}$, is equal to the nth cumulant of $\boldsymbol{y}$. For example, by evaluating (E.29) and (E.30) at $s = 0$ we obtain the identities

$$\Lambda'(0) = \mathbb{E}\boldsymbol{y}, \qquad \Lambda''(0) = \mathbb{E}\boldsymbol{y}^2 - (\mathbb{E}\boldsymbol{y})^2, \tag{E.31}$$

which are consistent with the term "cumulant generating function," since the first cumulant is equal to the mean, while the second cumulant is equal to the variance.

It is now useful to introduce the concept of *exponential tilting*. Let $\mathbb{P}(dy)$ denote the probability measure associated with the random variable $\boldsymbol{y}$, and consider the exponentially tilted measure

$$\mathbb{P}_s(dy) = \frac{e^{sy}}{M(s)}\mathbb{P}(dy) = e^{sy-\Lambda(s)}\mathbb{P}(dy). \tag{E.32}$$

Note that $\mathbb{P}_s(dy)$ is a probability measure as well, since we have

$$\int_{\mathbb{R}} \mathbb{P}_s(dy) = \int_{\mathbb{R}} \frac{e^{sy}}{M(s)}\mathbb{P}(dy) = \frac{\mathbb{E}e^{sy}}{M(s)} = 1. \tag{E.33}$$

Expectation under the tilted distribution will be denoted by $\mathbb{E}_s$. Using this notation, Eqs. (E.29) and (E.30) become, respectively,

$$\Lambda'(s) = \mathbb{E}_s\boldsymbol{y} \tag{E.34}$$

and

$$\Lambda''(s) = \mathbb{E}_s\boldsymbol{y}^2 - (\mathbb{E}_s\boldsymbol{y})^2, \tag{E.35}$$

where the last difference is the variance of $\boldsymbol{y}$ computed under the tilted distribution. Note that when $\boldsymbol{y}$ is not deterministic, $\mathbb{P}(dy)$ does not place all the probability mass on a single value. We see from (E.32) that in this case, even $\mathbb{P}_s(dy)$ does not place all the probability mass on a single value. As a result, for nondeterministic variables, the variance computed under the tilted distribution is positive. In view of (E.35), this implies

$$\Lambda''(s) > 0 \quad \forall s \in \mathrm{int}(\mathcal{D}_\Lambda), \tag{E.36}$$

which means that $\Lambda(s)$ is *strictly* convex on $\mathrm{int}(\mathcal{D}_\Lambda)$ — see Lemma A.2.

E.1.3 Cramér's Theorem

We state next Cramér's theorem.

Theorem E.1 (Cramér's theorem [53] [60, Thm. I.4]). Let $\{\boldsymbol{y}_n\}$ be a sequence of iid random variables with LMGF $\Lambda(s)$ satisfying

$$\Lambda(s) < \infty \qquad \forall s \in \mathbb{R}. \tag{E.37}$$

Let

$$\bar{\boldsymbol{y}}_n = \frac{1}{n}\sum_{i=1}^{n} \boldsymbol{y}_i. \tag{E.38}$$

Then, for all $y > \bar{y}$,

$$\lim_{n\to\infty} \frac{1}{n} \log \mathbb{P}[\bar{\boldsymbol{y}}_n \geq y] = -\Lambda^*(y). \tag{E.39}$$

That is, the rate function is given by the Fenchel-Legendre transform of $\Lambda(s)$.

Cramér's theorem establishes that, under the regularity condition (E.37) for the LMGF, the probability that the empirical average deviates from the mean vanishes exponentially with the number of samples, and that the rate function ruling this decay is given by the Fenchel-Legendre transform of the LMGF. It is useful to remark that Cramér's theorem can be proved in a more general setting, for example, with reference to random vectors $\boldsymbol{y}_n \in \mathbb{R}^d$ and/or by relaxing condition (E.37) — see [59, 60] for a comprehensive treatment.

Example E.2 (Gaussian variables, revisited). Let us apply Cramér's theorem to the Gaussian case considered in Example E.1. The LMGF of a Gaussian variable with mean $\bar{y}$ and variance σ^2 is given by [159]

$$\Lambda(s) = s\bar{y} + \frac{\sigma^2 s^2}{2}. \tag{E.40}$$

Accordingly, the Fenchel-Legendre transform in (E.16) can be computed by maximizing over $s \in \mathbb{R}$ the function

$$J(s) = sy - s\bar{y} - \frac{\sigma^2 s^2}{2} = (y - \bar{y})s - \frac{\sigma^2 s^2}{2}. \tag{E.41}$$

Taking the derivative of $J(s)$ and equating it to 0 yields

$$\frac{d}{ds}J(s) = (y - \bar{y}) - \sigma^2 s = 0 \implies s = \frac{y - \bar{y}}{\sigma^2}. \tag{E.42}$$

Substituting into (E.41) we get

$$\Lambda^*(y) = \max_{s\in\mathbb{R}} J(s) = \frac{(y - \bar{y})^2}{2\sigma^2}. \tag{E.43}$$

In view of Cramér's theorem, $\Lambda^*(y)$ is the desired rate function, which is displayed in Figure E.1. Note that the result agrees with what we obtained in Example E.1 in expression (E.7) by direct evaluation of the probability $\mathbb{P}[\bar{\boldsymbol{y}}_n > y]$. However, a direct approach was possible in the Gaussian case since the distribution of the empirical average was known. In general, it is not possible to compute this distribution in closed form, and the theory of large deviations (here, more specifically, Cramér's theorem) allows us to compute the rate function from the knowledge of the LMGF.

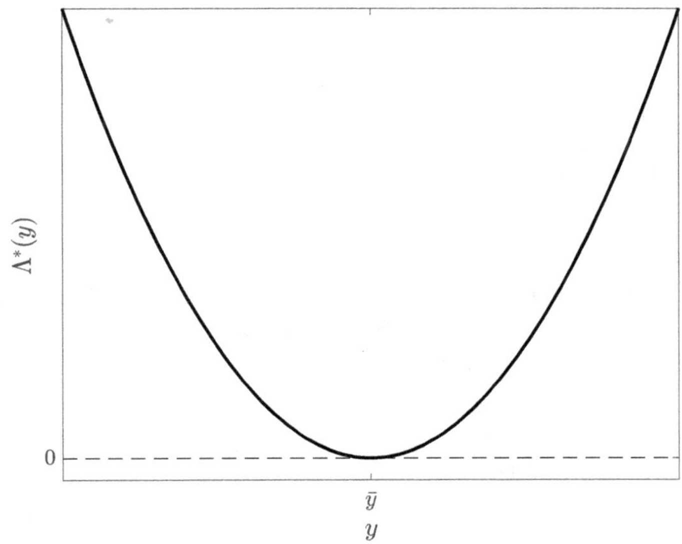

Figure E.1: Rate function for the Gaussian case considered in Example E.2.

Example E.3 (Bernoulli variables). We move away from the Gaussian case and consider an application of Cramér's theorem to Bernoulli random variables. Let

$$\mathbb{P}[\boldsymbol{y}_n = 1] = p = 1 - \mathbb{P}[\boldsymbol{y}_n = 0], \qquad 0 < p < 1. \tag{E.44}$$

The LMGF of a Bernoulli variable is immediately seen to be

$$\Lambda(s) = \log \mathbb{E} e^{s\boldsymbol{y}} = \log\big(pe^s + (1-p)\big). \tag{E.45}$$

The Fenchel-Legendre transform in (E.16) is then computed by taking the supremum over $s \in \mathbb{R}$ of the function

$$J(s) = sy - \log\big(pe^s + (1-p)\big). \tag{E.46}$$

Note that, for all $s > 0$,

$$\log\big(pe^s + (1-p)\big) < \log\big(pe^s + (1-p)e^s\big) = \log e^s = s. \tag{E.47}$$

Therefore, for $y > 1$ we have that

$$\Lambda^*(y) = \sup_{s\in\mathbb{R}} J(s) \geq \sup_{s>0} J(s) > \sup_{s>0} s(y-1) = \infty. \tag{E.48}$$

Likewise we can show that

$$\Lambda^*(y) = \infty \qquad \forall y < 0. \tag{E.49}$$

Note that this conclusion is convincing since the probability that the empirical average is below 0 or above 1 is zero, and, hence, we must have

$$\lim_{n\to\infty} \frac{1}{n} \log \mathbb{P}[\bar{\boldsymbol{y}}_n > y] = \lim_{n\to\infty} \frac{1}{n} \log 0 = -\infty \quad \forall y > 1, \tag{E.50a}$$

$$\lim_{n\to\infty} \frac{1}{n} \log \mathbb{P}[\bar{\boldsymbol{y}}_n < y] = \lim_{n\to\infty} \frac{1}{n} \log 0 = -\infty \quad \forall y < 0. \tag{E.50b}$$

Let us now evaluate the Fenchel-Legendre transform at $y = 1$. We have

$$\Lambda^*(1) = \sup_{s\in\mathbb{R}} \Big(s - \log\big(pe^s + (1-p)\big)\Big). \tag{E.51}$$

Observe that

$$\begin{aligned}\lim_{s\to\infty} \Big(s - \log\big(pe^s + (1-p)\big)\Big) &= \lim_{s\to\infty} \Big(\log e^s - \log\big(pe^s + (1-p)\big)\Big)\\ &= \lim_{s\to\infty} \log \frac{e^s}{pe^s + (1-p)} = \log\frac{1}{p}. \end{aligned}\tag{E.52}$$

Since, for $0 < p < 1$, the function $s - \log\big(pe^s + (1-p)\big)$ is strictly increasing in s (this can be readily verified, e.g., by showing that the first derivative of $s - \log\big(pe^s + (1-p)\big)$ is strictly positive for $0 < p < 1$), Eq. (E.52) yields

$$\Lambda^*(1) = \sup_{s\in\mathbb{R}} \Big(s - \log\big(pe^s + (1-p)\big)\Big) = \log\frac{1}{p}. \tag{E.53}$$

It can be shown similarly that

$$\Lambda^*(0) = \log\frac{1}{1-p}. \tag{E.54}$$

We finally focus on the interval $0 < y < 1$. Computing the derivative of $J(s)$ and equating it to 0, we get

$$\frac{d}{ds}J(s) = y - \frac{pe^s}{pe^s + (1-p)} = y - \frac{1}{1 + \dfrac{(1-p)}{p}e^{-s}} = 0, \tag{E.55}$$

which, after some straightforward algebra, yields the solution

$$s = \log\frac{(1-p)y}{(1-y)p}. \tag{E.56}$$

Substituting this solution into (E.46) we obtain

$$\begin{aligned}\Lambda^*(y) &= y\log\frac{(1-p)y}{(1-y)p} - \log\left(p\frac{(1-p)y}{(1-y)p} + (1-p)\right)\\ &= y\log\frac{(1-p)y}{(1-y)p} - \log\frac{1-p}{1-y}\\ &= y\log\frac{y}{p} + (1-y)\log\frac{1-y}{1-p}\\ &= D_b(y\|p),\end{aligned}\tag{E.57}$$

where $D_b(y\|p)$ is the shortcut for the binary KL divergence introduced in (6.98), which means that $D_b(y\|p)$ denotes the KL divergence between the pmfs $[y, 1-y]$ and $[p, 1-p]$. It follows that

$$\Lambda^*(y) = \begin{cases} D_b(y\|p) & \text{if } 0 \le y \le 1,\\ \infty & \text{otherwise,}\end{cases} \tag{E.58}$$

where, for $y = 0$ and $y = 1$, we mean that we compute the limits

$$\lim_{y\to 0+} D_b(y\|p) = \log\frac{1}{1-p}, \qquad \lim_{y\to 1-} D_b(y\|p) = \log\frac{1}{p}. \tag{E.59}$$

The rate function $\Lambda^*(y)$ is illustrated in Figure E.2.

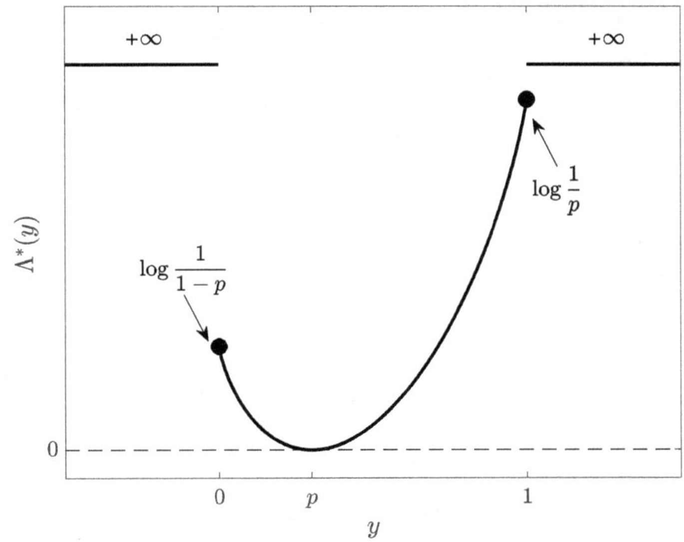

Figure E.2: Rate function for the Bernoulli case considered in Example E.3.

The rate functions found in the last two examples have some characteristic shape. They are nonnegative strictly convex functions inside the domain delimited by the infimum and supremum of the support of the random variables $\boldsymbol{y}_n$, and they are equal to ∞ outside this domain. Moreover, they are minimized at $y = \bar{y}$. Actually, these properties do not arise only in these two examples, but are typical of all rate functions appearing in Cramér's theorem.

Before illustrating the properties of the rate function, it is useful to introduce a formal definition for the support of a probability distribution.

Definition E.1 (Support of a probability distribution). The support of the probability distribution of a random variable $\boldsymbol{y}$, also referred to as support of $\boldsymbol{y}$ and denoted by supp_y, is the closure of the set that contains all points such that any neighborhood of these points has nonzero probability, formally,

$$\text{supp}_y \triangleq \text{cl}(\mathcal{S}), \tag{E.60}$$

where

$$\mathcal{S} = \Big\{ y_0 \in \mathbb{R} : \forall \varepsilon > 0,\, \mathbb{P}[\boldsymbol{y} \in (y_0 - \varepsilon, y_0 + \varepsilon)] > 0 \Big\}. \tag{E.61}$$

For example, from Definition E.1 we have that the support of a Gaussian random variable is $\mathbb{R}$, the support of a random variable uniform in $[0, 1]$ is the closed interval $[0, 1]$, and the support of a Bernoulli random variable is the discrete finite set $\{0, 1\}$.

The following lemma collects the relevant properties that identify the general shape of the rate function.

Lemma E.1 (Properties of the rate function). Assume that the random variable $\boldsymbol{y}$ is not deterministic (if it is deterministic, the rate function can be examined trivially), and that

$$\Lambda(s) = \log \mathbb{E} e^{s\boldsymbol{y}} < \infty \quad \forall s \in \mathbb{R}. \tag{E.62}$$

Introduce the expected value $\bar{y} = \mathbb{E}\boldsymbol{y}$ and the extremes of the support of the probability distribution associated with $\boldsymbol{y}$ (see Definition E.1),

$$y_{\mathsf{inf}} \triangleq \inf\left(\mathrm{supp}_{\boldsymbol{y}}\right), \qquad y_{\mathsf{sup}} \triangleq \sup\left(\mathrm{supp}_{\boldsymbol{y}}\right). \tag{E.63}$$

Consider then the Fenchel-Legendre transform

$$\Lambda^*(y) = \sup_{s\in\mathbb{R}} \Big(sy - \Lambda(s)\Big) \tag{E.64}$$

and define its effective domain as

$$\mathcal{D}_{\Lambda^*} = \{y \in \mathbb{R} : \Lambda^*(y) < \infty\}. \tag{E.65}$$

We have the following properties:

P1) **Nonnegativity.** $0 \leq \Lambda^*(y) \leq \infty$ for all $y \in \mathbb{R}$, and $\Lambda^*(\bar{y}) = 0$.

P2) **Alternative expressions:**

$$\Lambda^*(y) = \sup_{s\geq 0} \Big(sy - \Lambda(s)\Big) \quad \text{for } y \geq \bar{y}, \tag{E.66a}$$

$$\Lambda^*(y) = \sup_{s\leq 0} \Big(sy - \Lambda(s)\Big) \quad \text{for } y \leq \bar{y}. \tag{E.66b}$$

P3) **Interior of the effective domain:** $\mathrm{int}(\mathcal{D}_{\Lambda^*}) = (y_{\mathsf{inf}}, y_{\mathsf{sup}})$.

P4) **Smoothness and strict convexity.** $\Lambda^*(y)$ is infinitely differentiable and strictly convex on $(y_{\mathsf{inf}}, y_{\mathsf{sup}})$. Thus, in view of P1, $\Lambda^*(y)$ attains its unique minimum at $\bar{y}$.

P5) **Values at the boundary of the effective domain:**

$$\text{if } y_{\mathsf{inf}} = -\infty, \qquad \lim_{y\to y_{\mathsf{inf}}} \Lambda^*(y) = \infty, \tag{E.67a}$$

$$\text{if } y_{\mathsf{inf}} > -\infty, \qquad \Lambda^*(y_{\mathsf{inf}}) = -\log \mathbb{P}[\boldsymbol{y} = y_{\mathsf{inf}}], \tag{E.67b}$$

$$\text{if } y_{\mathsf{sup}} < \infty, \qquad \Lambda^*(y_{\mathsf{sup}}) = -\log \mathbb{P}[\boldsymbol{y} = y_{\mathsf{sup}}], \tag{E.67c}$$

$$\text{if } y_{\mathsf{sup}} = \infty, \qquad \lim_{y\to y_{\mathsf{sup}}} \Lambda^*(y) = \infty, \tag{E.67d}$$

where the expressions in the form $-\log p$ should be read as ∞ when $p = 0$.

Proof. It is convenient to prove the different properties separately.

Proof of P1. Since by definition $\Lambda(0) = \log \mathbb{E}[1] = 0$, we have

$$\Lambda^*(y) = \sup_{s\in\mathbb{R}}(sy - \Lambda(s)) \geq 0 \times y - \Lambda(0) = 0 \tag{E.68}$$

and, hence, $\Lambda^*(y)$ is nonnegative, and it can be equal to ∞ since the supremum in the definition of the Fenchel-Legendre transform can be infinite. Moreover, in view of the convexity of the exponential function, we can call upon Jensen's inequality (Theorem C.5) to obtain

$$\Lambda(s) = \log \mathbb{E} e^{s\boldsymbol{y}} \geq \mathbb{E} \log e^{s\boldsymbol{y}} = s\bar{y} \tag{E.69}$$

and, hence,

$$\Lambda^*(\bar{y}) = \sup_{s\in\mathbb{R}} \underbrace{\Big(s\bar{y} - \Lambda(s)\Big)}_{\leq 0 \text{ from (E.69)}} \leq 0, \tag{E.70}$$

which implies

$$\Lambda^*(\bar{y}) = 0, \tag{E.71}$$

since we know from (E.68) that $\Lambda^*(y)$ is nonnegative. Thus, P1 is proved.

Proof of P2. For all $y \geq \bar{y}$ and all $s < 0$ we can write

$$sy - \Lambda(s) \leq s\bar{y} - \Lambda(s) \leq \sup_{s\in\mathbb{R}} \Big(s\bar{y} - \Lambda(s)\Big) = \Lambda^*(\bar{y}) = 0. \tag{E.72}$$

Since from property P1 we know that the supremum of $sy - \Lambda(s)$ is nonnegative, the fact that $sy - \Lambda(s) \leq 0$ for all $s < 0$ implies (E.66a). Equation (E.66b) is obtained similarly.

Proof of P3. Let us introduce the function

$$J(s) = sy - \Lambda(s) \quad \Longrightarrow \quad \Lambda^*(y) = \sup_{s\in\mathbb{R}} J(s). \tag{E.73}$$

Since $\Lambda(s)$ is strictly convex and infinitely differentiable on $\mathbb{R}$, the function $J(s)$ is strictly concave and infinitely differentiable on $\mathbb{R}$, with

$$J'(s) = y - \Lambda'(s). \tag{E.74}$$

Accordingly, strict concavity of $J(s)$ implies that its supremum appearing in (E.73) will be in fact the unique maximum of $J(s)$ if the equation $J'(s) = 0$ has a solution, i.e., if

$$\Lambda'(s) = y \tag{E.75}$$

has a solution. Since we know from (E.36) that $\Lambda'(s)$ is strictly increasing, it makes sense to define the following limits:

$$\lim_{s\to-\infty} \Lambda'(s) = \Lambda'_{\mathsf{inf}}, \qquad \lim_{s\to\infty} \Lambda'(s) = \Lambda'_{\mathsf{sup}}, \tag{E.76}$$

and we conclude that Eq. (E.75) will have a unique solution $s(y)$ for each $y \in (\Lambda'_{\mathsf{inf}}, \Lambda'_{\mathsf{sup}})$. This solution is the maximizer of $J(s)$. Therefore, we have the identity $\sup_{s\in\mathbb{R}} J(s) = J\big(s(y)\big)$, which, when substituted into (E.73), yields

$$\Lambda^*(y) = s(y)\, y - \Lambda\big(s(y)\big) < \infty. \tag{E.77}$$

If $\Lambda'_{\mathsf{inf}} = -\infty$ and $\Lambda'_{\mathsf{sup}} = \infty$, Eq. (E.77) holds for all $y \in \mathbb{R}$, which means that in this case the effective domain of $\Lambda^*(y)$ is $\mathcal{D}_{\Lambda^*} = \mathbb{R}$.

Consider instead the case where $\Lambda'_{\mathsf{sup}} < \infty$. We now show that in this case $\Lambda^*(y)$ is infinite for $y > \Lambda'_{\mathsf{sup}}$. In fact, by applying the first-order condition for convexity from (A.2) to the strictly convex function $\Lambda(s)$ (exploiting the fact that $\Lambda(0) = 0$ and using in particular (A.3b)), we can write, for $s \neq 0$,

$$\Lambda'(s)s > \Lambda(s), \tag{E.78}$$

which implies

$$J(s) = sy - \Lambda(s) > s\Big(y - \Lambda'(s)\Big) \quad \forall s \neq 0. \tag{E.79}$$

When $y > \Lambda'_{\mathsf{sup}}$, the term on the RHS diverges to ∞ as $s \to \infty$, which means that in this case

$$\Lambda^*(y) = \sup_{s \in \mathbb{R}} J(s) = \infty. \tag{E.80}$$

Likewise, if $\Lambda'_{\mathsf{inf}} > -\infty$ and $y < \Lambda'_{\mathsf{inf}}$, the term on the RHS of (E.79) diverges to ∞ as $s \to -\infty$, implying that $\Lambda^*(y) = \infty$. We have thus shown that, when one or both boundary points Λ'_{inf} and Λ'_{sup} are finite, $\Lambda^*(y)$ will be equal to ∞ outside these boundaries. On the other hand, we have shown before that $\Lambda^*(y)$ is finite for any $y \in (\Lambda'_{\mathsf{inf}}, \Lambda'_{\mathsf{sup}})$, which implies that the interior of the effective domain is

$$\operatorname{int}(\mathcal{D}_{\Lambda^*}) = (\Lambda'_{\mathsf{inf}}, \Lambda'_{\mathsf{sup}}). \tag{E.81}$$

The behavior of $\Lambda^*(y)$ at the boundary points is still undetermined. We will address this point when proving property P6. To complete the proof of property P3, we need to show that Λ'_{inf} and Λ'_{sup} coincide with the extremes of the support of $\boldsymbol{y}$. We will focus on the right boundary Λ'_{sup}. The proof for the left boundary can be obtained similarly.[1] Consider a point y_0 such that $y_{\mathsf{inf}} < y_0 < y_{\mathsf{sup}}$, where y_{inf} and y_{sup} are, respectively, the infimum and the supremum of the support of $\boldsymbol{y}$, defined by (E.63). Using (E.34) we can write

$$\begin{aligned}\Lambda'(s) = \mathbb{E}_s\boldsymbol{y} &= y_0 + \mathbb{E}_s(\boldsymbol{y} - y_0)\\ &= y_0 + \mathbb{E}_s\Big[(\boldsymbol{y} - y_0)\mathbb{I}[\boldsymbol{y} < y_0]\Big] + \mathbb{E}_s\Big[(\boldsymbol{y} - y_0)\mathbb{I}[\boldsymbol{y} \geq y_0]\Big]\\ &\geq y_0 + \mathbb{E}_s\Big[(\boldsymbol{y} - y_0)\mathbb{I}[\boldsymbol{y} < y_0]\Big],\end{aligned} \tag{E.82}$$

where $\mathbb{E}_s$ denotes the expectation computed under the exponentially tilted measure defined by (E.32). We now show that the last term vanishes as $s \to \infty$. To this end, observing that this term is nonpositive, let us change its sign and make explicit the definition of the expectation under the tilted measure to obtain

$$\begin{aligned}\mathbb{E}_s\Big[(y_0 - \boldsymbol{y})\mathbb{I}[\boldsymbol{y} < y_0]\Big] &= \frac{\mathbb{E}\Big[(y_0 - \boldsymbol{y})e^{s\boldsymbol{y}}\,\mathbb{I}[\boldsymbol{y} < y_0]\Big]}{M(s)}\\ &= \mathbb{E}\Big[(y_0 - \boldsymbol{y})e^{s\boldsymbol{y} - \Lambda(s)}\mathbb{I}[\boldsymbol{y} < y_0]\Big]\\ &= e^{sy_0 - \Lambda(s)}\,\mathbb{E}\Big[(y_0 - \boldsymbol{y})e^{s(\boldsymbol{y} - y_0)}\mathbb{I}[\boldsymbol{y} < y_0]\Big],\end{aligned} \tag{E.83}$$

[1] It is actually not necessary to repeat the proof for the left boundary. In fact, if we consider the random variable $\boldsymbol{z} = -\boldsymbol{y}$, the LMGF of $\boldsymbol{z}$ is equal to $\Lambda(-s)$, and the roles of y_{sup} and y_{inf} are interchanged.

where we further used the definition $\Lambda(s) = \log M(s)$. Now, by exploiting Chernoff's bound (Theorem C.3), for all nonnegative s we have

$$\mathbb{P}[\boldsymbol{y} \geq y_0] \leq \frac{\mathbb{E}e^{s\boldsymbol{y}}}{e^{sy_0}} = M(s)e^{-sy_0} = e^{\Lambda(s)-sy_0}. \tag{E.84}$$

Accordingly, observing that $\mathbb{P}[\boldsymbol{y} \geq y_0] > 0$ since $y_{\mathsf{inf}} < y_0 < y_{\mathsf{sup}}$, and using (E.84) in (E.83), we obtain

$$\mathbb{E}_s\left[(y_0 - \boldsymbol{y})\mathbb{I}[\boldsymbol{y} \leq y_0]\right] \leq \frac{\mathbb{E}\left[(y_0 - \boldsymbol{y})e^{s(\boldsymbol{y}-y_0)}\mathbb{I}[\boldsymbol{y} < y_0]\right]}{\mathbb{P}[\boldsymbol{y} \geq y_0]}. \tag{E.85}$$

Note that

$$\lim_{s\to\infty} e^{s(\boldsymbol{y}-y_0)}\mathbb{I}[\boldsymbol{y} < y_0] = 0. \tag{E.86}$$

Since for all $s > 0$ we have the bound

$$e^{s(\boldsymbol{y}-y_0)}\mathbb{I}[\boldsymbol{y} < y_0] < 1, \tag{E.87}$$

and since the random variable $\boldsymbol{y}$ has finite mean, from the dominated convergence theorem (Theorem D.6) we conclude that

$$\lim_{s\to\infty} \mathbb{E}\left[(y_0 - \boldsymbol{y})e^{s(\boldsymbol{y}-y_0)}\mathbb{I}[\boldsymbol{y} < y_0]\right] = 0, \tag{E.88}$$

which, in view of (E.82) and (E.85), implies that

$$\liminf_{s\to\infty} \Lambda'(s) \geq y_0. \tag{E.89}$$

Since y_0 can be any point such that $y_{\mathsf{inf}} < y_0 < y_{\mathsf{sup}}$, if $y_{\mathsf{sup}} = \infty$ the value of y_0 can be chosen arbitrarily large, and Eq. (E.89) implies that the limit inferior of $\Lambda'(s)$ is equal to ∞, which in turn implies $\Lambda'_{\mathsf{sup}} = y_{\mathsf{sup}}$. If instead $y_{\mathsf{sup}} < \infty$, we can choose $y_0 = y_{\mathsf{sup}} - \varepsilon$ for a small $\varepsilon > 0$, and conclude from the arbitrariness of ε that

$$\liminf_{s\to\infty} \Lambda'(s) \geq y_{\mathsf{sup}}. \tag{E.90}$$

On the other hand, from the definition of y_{sup} we have

$$\Lambda'(s) = \frac{\mathbb{E}[\boldsymbol{y}\, e^{s\boldsymbol{y}}]}{M(s)} \leq \frac{\mathbb{E}e^{s\boldsymbol{y}}}{M(s)} y_{\mathsf{sup}} = y_{\mathsf{sup}}, \tag{E.91}$$

which, when combined with (E.90), gives

$$\lim_{s\to\infty} \Lambda'(s) = y_{\mathsf{sup}}, \tag{E.92}$$

and the proof of P3 is complete.

Proof of P4. In the proof of P3 we have established that, for all $y \in (y_{\mathsf{inf}}, y_{\mathsf{sup}})$, Eq. (E.75) implicitly defines a function $s(y)$ through the equation

$$\Lambda'\big(s(y)\big) = y. \tag{E.93}$$

The theorem about differentiation of the inverse function [144, p. 114] allows us to conclude that the derivative of the function $s(y)$ can be computed as

$$s'(y) = \frac{1}{\Lambda''\big(s(y)\big)} > 0 \tag{E.94}$$

and that $s(y)$ is infinitely differentiable on $(y_{\mathsf{inf}}, y_{\mathsf{sup}})$. Then, from (E.77) we can write

$$\frac{d}{dy}\Lambda^*(y) = s(y) + y\, s'(y) - \underbrace{\Lambda'\big(\, s(y) \big)}_{=y \text{ from (E.93)}}\, s'(y) = s(y), \tag{E.95}$$

and

$$\frac{d^2}{dy^2}\Lambda^*(y) = s'(y) \overset{\text{(E.94)}}{>} 0, \tag{E.96}$$

which shows that $\Lambda^*(y)$ is strictly convex on $(y_{\mathsf{inf}}, y_{\mathsf{sup}})$ in view of Lemma A.2. We already know from property P1 that the point $\bar{y}$ is a global minimizer for $\Lambda^*(y)$. From strict convexity, it is actually the unique minimizer, with $\Lambda^*(y)$ being strictly decreasing for $y < \bar{y}$ and strictly increasing for $y > \bar{y}$.

Proof of P5. We focus on y_{sup}, with the proof for y_{inf} being similar. We must accordingly establish (E.67c) and (E.67d). The fact that $\Lambda^*(y) \to \infty$ when $y_{\mathsf{sup}} = \infty$, i.e., relation (E.67d), is readily established by taking a point $y_0 > \bar{y}$ and applying the first-order condition for convexity (Lemma A.1), which yields

$$\Lambda^*(y) > \Lambda^*(y_0) + \frac{d}{dy}\Lambda^*(y)\Big|_{y=y_0} (y - y_0). \tag{E.97}$$

Observing that the function $\Lambda^*(y)$ is strictly increasing for $y > \bar{y}$, its derivative at y_0 is positive, and we conclude from (E.97) that $\Lambda^*(y) \to \infty$ as $y \to \infty$.

Consider next the case where y_{sup} is finite, corresponding to (E.67c). We must prove that

$$\Lambda^*(y_{\mathsf{sup}}) = -\log \mathbb{P}[\boldsymbol{y} = y_{\mathsf{sup}}], \tag{E.98}$$

which should be read as $\Lambda^*(y_{\mathsf{sup}}) = \infty$ if $\mathbb{P}[\boldsymbol{y} = y_{\mathsf{sup}}] = 0$. To prove (E.98), observe that

$$\begin{aligned} \Lambda^*(y_{\mathsf{sup}}) &= \sup_{s\in\mathbb{R}} \Big(s y_{\mathsf{sup}} - \Lambda(s) \Big) = \sup_{s\in\mathbb{R}} \Big(s y_{\mathsf{sup}} - \log \mathbb{E} e^{s\boldsymbol{y}} \Big) \\ &= \sup_{s\in\mathbb{R}} \Big(\log e^{s y_{\mathsf{sup}}} - \log \mathbb{E} e^{s\boldsymbol{y}} \Big) = \sup_{s\in\mathbb{R}} \Big(\log \frac{e^{s y_{\mathsf{sup}}}}{\mathbb{E} e^{s\boldsymbol{y}}} \Big) \\ &= \sup_{s\in\mathbb{R}} \Big(\log \frac{1}{\mathbb{E} e^{s(\boldsymbol{y} - y_{\mathsf{sup}})}} \Big) = -\inf_{s\in\mathbb{R}} \log \mathbb{E} e^{s(\boldsymbol{y} - y_{\mathsf{sup}})}, \end{aligned} \tag{E.99}$$

where the first two equalities follow from the definitions of the Fenchel-Legendre transform and the LMGF, respectively. The last equality follows from the properties of the infimum and supremum, whereas the remaining equalities result from straightforward algebraic manipulations. We now want to evaluate the infimum appearing in (E.99). Due to the monotonicity of the logarithm, it is sufficient to evaluate the infimum of $\mathbb{E} e^{s(\boldsymbol{y} - y_{\mathsf{sup}})}$. From the definition of the expected value, this quantity can be represented as

$$\mathbb{E} e^{s(\boldsymbol{y} - y_{\mathsf{sup}})} = \mathbb{P}[\boldsymbol{y} = y_{\mathsf{sup}}] + \mathbb{E}\left[e^{s(\boldsymbol{y} - y_{\mathsf{sup}})} \mathbb{I}\left[\boldsymbol{y} < y_{\mathsf{sup}}\right] \right]. \tag{E.100}$$

The first term on the RHS of (E.100) does not depend on s. The second term vanishes as $s \to \infty$, in view of the dominated convergence theorem (Theorem D.6). However, since this term is nonnegative,[2] the fact that it vanishes implies that its infimum is zero. Accordingly, from (E.100) we conclude that

$$\inf_{s\in\mathbb{R}} \mathbb{E} e^{s(\boldsymbol{y} - y_{\mathsf{sup}})} = \mathbb{P}[\boldsymbol{y} = y_{\mathsf{sup}}], \tag{E.101}$$

[2]Actually, it is strictly positive since $\boldsymbol{y} < y_{\mathsf{sup}}$ with nonzero probability. This follows from the fact that y_{sup} is the supremum of the support of $\boldsymbol{y}$ and $\boldsymbol{y}$ is not deterministic.

which, from the monotonicity of the logarithm, implies

$$\inf_{s\in\mathbb{R}} \log \mathbb{E}e^{s(\boldsymbol{y}-y_{\mathsf{sup}})} = \log \mathbb{P}[\boldsymbol{y} = y_{\mathsf{sup}}], \tag{E.102}$$

with the understanding that the expression is equal to $-\infty$ when $\mathbb{P}[\boldsymbol{y} = y_{\mathsf{sup}}] = 0$. Using (E.102) in (E.99) proves (E.98). This completes the proof of P5 and, hence, of the lemma. ■

The next lemma generalizes the previous one by characterizing the Fenchel-Legendre transform of a special function that is useful to characterize the large deviations of random sums examined later in Appendix F, and ultimately to characterize the error exponent for adaptive social learning in Chapter 9.

Lemma E.2 (Properties of a useful function). Let $\Lambda(s)$ be the LMGF of a nondeterministic random variable $\boldsymbol{y}$ (the conclusions for the deterministic case are trivial), with $\Lambda(s) < \infty$ for all $s \in \mathbb{R}$, and introduce the function

$$\phi(s) \triangleq \int_0^s \frac{\Lambda(\varsigma)}{\varsigma} d\varsigma. \tag{E.103}$$

Then we have the following property:

Q0) **Smoothness and strict convexity of $\phi(s)$.** The function $\phi(s)$ is infinitely differentiable on $\mathbb{R}$ and, for all $r \in \mathbb{N}$,

$$\phi^{(r)}(s) = \frac{1}{s^r} \int_0^s \Lambda^{(r)}(\varsigma)\varsigma^{r-1} d\varsigma, \tag{E.104}$$

where, for $s = 0$, the above equation should be read as

$$\phi^{(r)}(0) = \lim_{s\to 0} \frac{1}{s^r} \int_0^s \Lambda^{(r)}(\varsigma)\varsigma^{r-1} d\varsigma = \frac{\Lambda^{(r)}(0)}{r}. \tag{E.105}$$

In particular, we have that

$$\phi''(s) = \frac{\Lambda'(s)\, s - \Lambda(s)}{s^2} > 0 \qquad \forall s \in \mathbb{R}, \tag{E.106}$$

with $\phi''(0) = \lim_{s\to 0} \phi''(s) = \Lambda''(0)/2$.

Let us further introduce the expected value $\bar{y} = \mathbb{E}\boldsymbol{y}$ and the extremes of the support of the probability distribution associated with $\boldsymbol{y}$ (see Definition E.1),

$$y_{\mathsf{inf}} \triangleq \inf\left(\mathrm{supp}_y\right), \qquad y_{\mathsf{sup}} \triangleq \sup\left(\mathrm{supp}_y\right). \tag{E.107}$$

Consider then the Fenchel-Legendre transform

$$\phi^*(y) = \sup_{s\in\mathbb{R}} \left(sy - \phi(s)\right) \tag{E.108}$$

and define its effective domain as

$$\mathcal{D}_{\phi^*} = \{y \in \mathbb{R} : \phi^*(y) < \infty\}. \tag{E.109}$$

We have the following properties:

Q1) **Nonnegativity.** $0 \le \phi^*(y) \le \infty$ for all $y \in \mathbb{R}$, and $\phi^*(\bar{y}) = 0$.

Q2) **Alternative expressions:**

$$\phi^*(y) = \sup_{s \ge 0} \Big(sy - \phi(s) \Big) \quad \text{for } y \ge \bar{y}, \tag{E.110a}$$

$$\phi^*(y) = \sup_{s \le 0} \Big(sy - \phi(s) \Big) \quad \text{for } y \le \bar{y}. \tag{E.110b}$$

Q3) **Interior of the effective domain.** The interior of the effective domain of $\phi^*(y)$ is the open interval $\text{int}(\mathcal{D}_{\phi^*}) = (y_{\text{inf}}, y_{\text{sup}})$.

Q4) **Smoothness and strict convexity.** $\phi^*(y)$ is infinitely differentiable and strictly convex on $(y_{\text{inf}}, y_{\text{sup}})$. Thus, in view of Q1, $\phi^*(y)$ attains its unique minimum at $\bar{y}$.

Q5) **Values at the boundary of the effective domain.** If $y_{\text{sup}} < \infty$, then $\phi^*(y_{\text{sup}}) = \infty$, and, likewise, if $y_{\text{inf}} > -\infty$, then $\phi^*(y_{\text{inf}}) = \infty$. Thus, the effective domain is the open interval $(y_{\text{inf}}, y_{\text{sup}})$, i.e., we have $\mathcal{D}_{\phi^*} = \text{int}(\mathcal{D}_{\phi^*})$. Moreover,

$$\lim_{y \to y_{\text{inf}}^+} \phi^*(y) = \infty, \qquad \lim_{y \to y_{\text{sup}}^-} \phi^*(y) = \infty. \tag{E.111}$$

A typical shape of $\phi^*(y)$ is illustrated in Figure E.3.

Proof. It is convenient to prove the different properties separately.

Proof of Q0. The function $\phi(s)$ defined in (E.103) shares some properties with the LMGF $\Lambda(s)$, and for this reason the proof will be similar to that of Lemma E.1. Preliminarily, we observe that the function $\Lambda(s)/s$ is continuous over the entire real axis, once we evaluate its value at $s = 0$ using

$$\lim_{s \to 0} \frac{\Lambda(s)}{s} = \Lambda'(0) = \bar{y}, \tag{E.112}$$

where the equality follows from the first relation in (E.31).

Let us start by establishing (E.104) for the case $s \neq 0$; we proceed by induction. Relation (E.104) trivially holds for $r = 1$. We show that if (E.104) holds for r, then it must hold for $r + 1$. Indeed,

$$\phi^{(r+1)}(s) = \frac{d}{ds}\phi^{(r)}(s) = \frac{d}{ds}\left(\frac{1}{s^r} \int_0^s \Lambda^{(r)}(\varsigma)\varsigma^{r-1} d\varsigma \right). \tag{E.113}$$

Applying the rule of integration by parts we have

$$\frac{1}{s^r} \int_0^s \Lambda^{(r)}(\varsigma)\varsigma^{r-1} d\varsigma = \frac{\Lambda^{(r)}(s)}{r} - \frac{1}{r s^r} \int_0^s \Lambda^{(r+1)}(\varsigma)\varsigma^r d\varsigma. \tag{E.114}$$

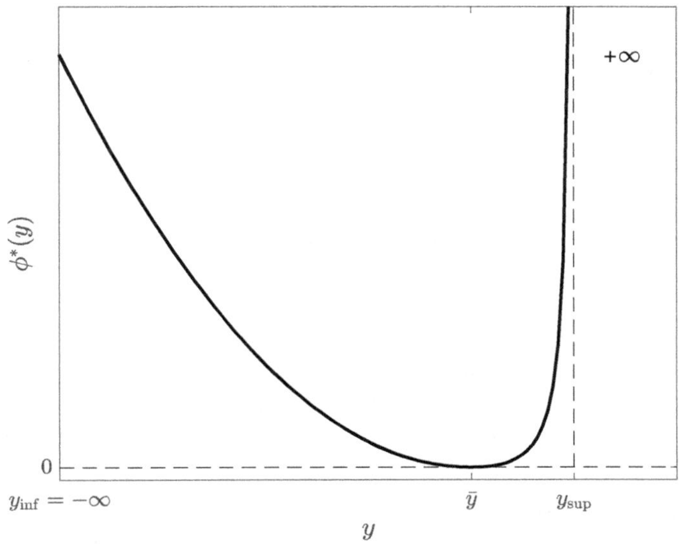

Figure E.3: Typical shape of the function $\phi^*(y)$ defined in Lemma E.2.

Differentiating the above expression yields

$$\phi^{(r+1)}(s) = \frac{\Lambda^{(r+1)}(s)}{r} - \frac{\Lambda^{(r+1)}(s)}{r} + \frac{1}{s^{r+1}} \int_0^s \Lambda^{(r+1)}(\varsigma)\varsigma^r d\varsigma, \tag{E.115}$$

which corresponds to (E.104) for $r+1$. This completes the proof by induction, and the identity in Eq. (E.104) is proved for the case $s \neq 0$. To get (E.105), observe that when the limit of $\phi^{(r)}(s)$ as $s \to 0$ exists and is finite, then $\phi^{(r)}(0)$ exists and is equal to this limit.[3] By applying L'Hôpital's rule to (E.104), we obtain

$$\phi^{(r)}(0) = \lim_{s\to 0} \phi^{(r)}(s) = \lim_{s\to 0} \frac{1}{s^r} \int_0^s \Lambda^{(r)}(\varsigma)\varsigma^{r-1} d\varsigma = \frac{\Lambda^{(r)}(0)}{r}. \tag{E.117}$$

Specializing (E.104) to the case $r = 2$ we get (E.106), where the inequality for $s \neq 0$ follows from (E.78), whereas for $s = 0$ we have $\phi''(0) = \lim_{s\to 0} \phi''(s) = \Lambda''(0)/2$, which is positive because $\Lambda''(0)$ is the variance of the nondeterministic random variable $\boldsymbol{y}$ — see (E.31).

The proof of Q0 is complete. We focus next on the regularity properties of the Fenchel-Legendre transform $\phi^*(y)$.

[3]To see that this property holds, let g be a function defined on an interval $I_\varepsilon = (-\varepsilon, \varepsilon)$, for some $\varepsilon > 0$, and differentiable on $I_\varepsilon \backslash \{0\}$. Assume that $\lim_{x\to 0} g'(x) = l$. We want to show that $g'(0)$ exists and is equal to l. From the mean-value theorem [144, Thm. 5.9] we have $g'(\bar{x}) = (g(x) - g(0))/x$ for some $\bar{x} \in (0, x)$. Moreover, from the squeeze theorem [144, Thm. 3.19] it follows that $\bar{x} \to 0$ as $x \to 0$. Then we can write

$$g'(0) = \lim_{x\to\infty} \frac{g(x) - g(0)}{x} = \lim_{\bar{x}\to\infty} g'(\bar{x}) = l, \tag{E.116}$$

where the first equality is the definition of the derivative of g in 0.

Proof of Q1. Property Q1 is proved similarly to property P1 in Lemma E.1. First, we note that $\phi(0) = 0$, and, hence, Eq. (E.68) can be obtained with $\phi^*(y)$ and $\phi(s)$ in place of $\Lambda(s)$ and $\Lambda^*(y)$, respectively. In other words, we have $0 \le \phi(y) \le \infty$ for all $y \in \mathbb{R}$. Second, using (E.69) we can write, for all $s > 0$,

$$\phi(s) = \int_0^s \frac{\Lambda(\varsigma)}{\varsigma} d\varsigma \ge \int_0^s \bar{y}\, d\varsigma = s\bar{y}. \tag{E.118}$$

Likewise, for all $s < 0$ it holds that (observe that for negative s we have the identity $s = -|s|$)

$$\phi(s) = \int_0^{-|s|} \frac{\Lambda(\varsigma)}{\varsigma} d\varsigma = -\int_{-|s|}^0 \frac{\Lambda(\varsigma)}{\varsigma} d\varsigma \ge -\int_{-|s|}^0 \bar{y}\, d\varsigma = s\bar{y}. \tag{E.119}$$

Therefore, we established the inequality $\phi(s) \ge s\bar{y}$ for all $s \in \mathbb{R}$, namely, we proved (E.69) with $\phi(s)$ in place of $\Lambda(s)$. Now, combining the two results: *i)* $0 \le \phi(y) \le \infty$ for all $y \in \mathbb{R}$; and *ii)* $\phi(s) \ge s\bar{y}$ for all $s \in \mathbb{R}$, property Q1 follows from the same arguments used in the proof of property P1.

Proof of Q2. Q2 is obtained from Q1 in the same manner as P2 is obtained from P1 in Lemma E.1.

Proof of Q3. We can follow the same argument used to establish property P3 in Lemma E.1. To prove P3 we relied on the strict convexity of $\Lambda(s)$, and observed that $\mathrm{int}(\mathcal{D}_{\Lambda^*})$ is the open interval with boundaries given by the limiting values of $\Lambda'(s)$ as $s \to \pm\infty$. Since $\phi(s)$ is also strictly convex, the same argument applies and we conclude that

$$\mathrm{int}(\mathcal{D}_{\phi^*}) = \left(\lim_{s\to-\infty} \phi'(s), \lim_{s\to\infty} \phi'(s) \right). \tag{E.120}$$

Recalling that

$$\phi'(s) = \frac{\Lambda(s)}{s}, \tag{E.121}$$

we want to show that

$$\lim_{s\to-\infty} \frac{\Lambda(s)}{s} = y_{\mathsf{inf}}, \qquad \lim_{s\to\infty} \frac{\Lambda(s)}{s} = y_{\mathsf{sup}}. \tag{E.122}$$

Actually, we now prove the limit relative to the right boundary y_{sup}, with the proof for the left boundary y_{inf} being similar.

Let us consider a point y_0 such that $y_{\mathsf{inf}} < y_0 < y_{\mathsf{sup}}$, where y_{inf} and y_{sup} are, respectively, the infimum and the supremum of the support of $\boldsymbol{y}$, defined by (E.107). Using (E.84) we can write, for $s > 0$,

$$\frac{\Lambda(s)}{s} \ge y_0 + \frac{\log \mathbb{P}[\boldsymbol{y} \ge y_0]}{s}. \tag{E.123}$$

We remark that $\mathbb{P}[\boldsymbol{y} \ge y_0] > 0$ since $y_{\mathsf{inf}} < y_0 < y_{\mathsf{sup}}$. From (E.123) we get

$$\liminf_{s\to\infty} \frac{\Lambda(s)}{s} \ge y_0. \tag{E.124}$$

If $y_{\mathsf{sup}} = \infty$ the result is proved due to the arbitrariness of y_0. If $y_{\mathsf{sup}} < \infty$, we can choose $y_0 = y_{\mathsf{sup}} - \varepsilon$, and conclude that the limit inferior in (E.124) is equal to y_{sup}. The fact that the corresponding limit superior is equal to y_{sup} follows by observing that for all $s > 0$

the quantity $\Lambda(s)/s$ is upper bounded by y_{sup} since $\Lambda(s) = \log \mathbb{E} e^{s\boldsymbol{y}} \leq \log \mathbb{E} e^{s y_{\mathsf{sup}}} = s\, y_{\mathsf{sup}}$. We have in fact established that

$$\liminf_{s\to\infty} \frac{\Lambda(s)}{s} = \limsup_{s\to\infty} \frac{\Lambda(s)}{s} = y_{\mathsf{sup}} = \lim_{s\to\infty} \frac{\Lambda(s)}{s}, \tag{E.125}$$

as desired.

Proof of Q4. To prove P4 in Lemma E.1 we relied on the smoothness and the strict convexity of $\Lambda(s)$. We can therefore prove Q4 as done for P4, by exploiting the smoothness and the strict convexity of $\phi(s)$ established in property Q0 of the present lemma.

Proof of Q5. Finally, we characterize the behavior of $\phi^*(y)$ at the boundaries of $\mathrm{int}(\mathcal{D}_{\phi^*})$. We focus again on the right boundary y_{sup}, with the proof for y_{inf} being similar. When $y_{\mathsf{sup}} = \infty$, it suffices to notice that the rate function $\phi^*(y)$ is strictly convex on $\mathrm{int}(\mathcal{D}_{\phi^*})$ and is strictly increasing for $y > \bar{y}$ (see Figure E.3) to conclude that the rate function diverges to ∞ as $y \to y_{\mathsf{sup}}$. Technically, this conclusion can be obtained from the first-order condition for convexity as done for $\Lambda^*(y)$ in (E.97).

We next examine the case $y_{\mathsf{sup}} < \infty$. Consider a point $y_0 < y_{\mathsf{sup}}$ and let $q = \mathbb{P}[\boldsymbol{y} \geq y_0]$. We can write, for all $s > 0$,

$$\begin{aligned}\Lambda(s) &= \log\Big(\mathbb{E}\Big[e^{s\boldsymbol{y}}\mathbb{I}[\boldsymbol{y} < y_0]\Big] + \mathbb{E}\Big[e^{s\boldsymbol{y}}\mathbb{I}[\boldsymbol{y} \geq y_0]\Big]\Big)\\ &\leq \log\Big((1-q)e^{s y_0} + q e^{s y_{\mathsf{sup}}}\Big)\\ &= s y_{\mathsf{sup}} + \log\big((1-q)e^{-s(y_{\mathsf{sup}} - y_0)} + q\big).\end{aligned} \tag{E.126}$$

Since $y_0 < y_{\mathsf{sup}}$,

$$\lim_{s\to\infty} e^{-s(y_{\mathsf{sup}} - y_0)} = 0. \tag{E.127}$$

This implies that, for any $\varepsilon > 0$, there exists $s_0 > 0$ such that

$$(1-q)e^{-s(y_{\mathsf{sup}} - y_0)} \leq \varepsilon q \quad \forall s \geq s_0. \tag{E.128}$$

Combining (E.126) and (E.128), we obtain

$$\Lambda(s) \leq s y_{\mathsf{sup}} + \log\big((1+\varepsilon)q\big) \quad \forall s \geq s_0, \tag{E.129}$$

which, when used in (E.103), gives

$$\begin{aligned}\phi(s) &= \int_0^s \frac{\Lambda(\varsigma)}{\varsigma} d\varsigma = \int_0^{s_0} \frac{\Lambda(\varsigma)}{\varsigma} d\varsigma + \int_{s_0}^s \frac{\Lambda(\varsigma)}{\varsigma} d\varsigma\\ &\leq \phi(s_0) + y_{\mathsf{sup}}(s - s_0) + \int_{s_0}^s \frac{\log\big((1+\varepsilon)q\big)}{\varsigma} d\varsigma\\ &= \phi(s_0) + y_{\mathsf{sup}}(s - s_0) + \log\big((1+\varepsilon)q\big)\, \log\frac{s}{s_0}.\end{aligned} \tag{E.130}$$

Performing straightforward algebraic manipulations, we conclude that

$$s y_{\mathsf{sup}} - \phi(s) \geq -\phi(s_0) + s_0\, y_{\mathsf{sup}} + \log\frac{1}{(1+\varepsilon)q}\, \log\frac{s}{s_0} \tag{E.131}$$

for all $s \geq s_0$. Using this result in (E.108) yields

$$\phi^*(y_{\mathsf{sup}}) \geq \sup_{s \geq s_0} [s y_{\mathsf{sup}} - \phi(s)] \geq -\phi(s_0) + s_0\, y_{\mathsf{sup}} + \log \frac{1}{(1+\varepsilon)q} \sup_{s \geq s_0} \log \frac{s}{s_0} = \infty, \tag{E.132}$$

where we assume ε small enough to ensure that $(1+\varepsilon)q < 1$. Finally, in view of (E.108) we can write, for a generic $s \in \mathbb{R}$,

$$\liminf_{y \to y_{\mathsf{sup}}} \phi^*(y) \geq \liminf_{y \to y_{\mathsf{sup}}} [sy - \phi(s)] = s y_{\mathsf{sup}} - \phi(s), \tag{E.133}$$

which, due to the arbitrariness of s, also implies that

$$\liminf_{y \to y_{\mathsf{sup}}} \phi^*(y) \geq \sup_{s \in \mathbb{R}} [s y_{\mathsf{sup}} - \phi(s)] = \infty, \tag{E.134}$$

where the equality follows from (E.132). We conclude that $\phi^*(y) \to \infty$ as $y \to y_{\mathsf{sup}}$, and the proof is complete.

■

Example E.4 (Why does the Fenchel-Legendre transform appear in large deviation analysis?). Consider the same setting used in Cramér's theorem and focus on the probability $\mathbb{P}[\bar{\boldsymbol{y}}_n \geq y]$ for $y > \bar{y}$. By applying Chernoff's bound (Theorem C.3) to the empirical average, for all $s \geq 0$ we can write

$$\mathbb{P}[\bar{\boldsymbol{y}}_n \geq y] \leq \frac{\mathbb{E} e^{n s \bar{\boldsymbol{y}}_n}}{e^{n s y}}. \tag{E.135}$$

On the other hand, we have the identities

$$\mathbb{E} e^{n s \bar{\boldsymbol{y}}_n} \overset{(a)}{=} \mathbb{E}\left[\prod_{i=1}^{n} e^{s \boldsymbol{y}_i}\right] \overset{(b)}{=} \prod_{i=1}^{n} \mathbb{E} e^{s \boldsymbol{y}_i} \overset{(c)}{=} M^n(s) = e^{n \log M(s)} = e^{n \Lambda(s)}, \tag{E.136}$$

where $M(s)$ and $\Lambda(s)$ denote the MGF and the LMGF of $\boldsymbol{y}_i$, respectively. In step (a) we apply the relation $n\, \bar{\boldsymbol{y}}_n = \sum_{i=1}^n \boldsymbol{y}_i$; step (b) holds because the random variables $\boldsymbol{y}_i$ are independent; and step (c) holds because they are identically distributed. Using (E.136) in (E.135) we get

$$\mathbb{P}[\bar{\boldsymbol{y}}_n \geq y] \leq \frac{e^{n\Lambda(s)}}{e^{n s y}} = e^{-n\left(sy - \Lambda(s)\right)}. \tag{E.137}$$

On the other hand, since s is an arbitrary nonnegative value, we can also write

$$\mathbb{P}[\bar{\boldsymbol{y}}_n \geq y] \leq \inf_{s \geq 0} e^{-n\left(sy - \Lambda(s)\right)} = e^{-n \sup_{s \geq 0} \left(sy - \Lambda(s)\right)} = e^{-n \Lambda^*(y)}, \tag{E.138}$$

where $\Lambda^*(y)$ is the Fenchel-Legendre transform of $\Lambda(s)$ — see (E.66a). We remark that, in the evaluation of the Fenchel-Legendre transform, the supremum is taken only for $s \geq 0$, which is legitimate in view or (E.110a), since we are considering the case $y > \bar{y}$. Equation (E.138) highlights the relevance of the Fenchel-Legendre transform $\Lambda^*(y)$ in evaluating the deviation of the empirical average from the statistical average. Note that this result does not prove Cramér's theorem, since it establishes only an upper bound. The derivation of a lower bound that allows to establish Cramér's theorem is provided in [59, 60].

Example E.5 (Why cannot the CLT be used to evaluate large deviations?). Just like Cramér's theorem, the central limit theorem (Theorem D.8) is a useful concentration result that sharpens the law of large numbers by providing information about the deviation of the empirical average from the statistical average. In terms of the empirical average

$$\bar{\boldsymbol{y}}_n = \frac{1}{n}\sum_{i=1}^{n} \boldsymbol{y}_i \tag{E.139}$$

of iid random variables $\boldsymbol{y}_i$ with mean $\bar{y}$ and variance σ^2, the convergence in distribution in (D.51) corresponds to the statement

$$\sqrt{n}\,(\bar{\boldsymbol{y}}_n - \bar{y}) \xrightarrow[n\to\infty]{\mathrm{d}} \mathscr{G}(0,\sigma^2). \tag{E.140}$$

Applying the definition of convergence in distribution seen in (D.12), we conclude from (E.140) that, for all $\gamma \in \mathbb{R}$,

$$\lim_{n\to\infty} \mathbb{P}\left[\sqrt{n}\,(\bar{\boldsymbol{y}}_n - \bar{y}) \geq \gamma\right] = Q(\gamma/\sigma). \tag{E.141}$$

By straightforward manipulations, Eq. (E.141) becomes

$$\lim_{n\to\infty} \mathbb{P}\left[\bar{\boldsymbol{y}}_n \geq \bar{y} + \frac{\gamma}{\sqrt{n}}\right] = Q(\gamma/\sigma). \tag{E.142}$$

Consider, for example, a value $\gamma > 0$. Equation (E.142) reveals that, for large n, the probability that $\bar{\boldsymbol{y}}_n$ deviates from $\bar{y}$ by a small positive amount $\gamma/\sqrt{n}$ is close to $Q(\gamma/\sigma)$. Now, recalling (E.2a), we can write

$$\lim_{n\to\infty} \mathbb{P}\left[\bar{\boldsymbol{y}}_n \geq y\right] = \lim_{n\to\infty} \mathbb{P}\left[\bar{\boldsymbol{y}}_n \geq \bar{y} + (y - \bar{y})\right] = 0 \quad \forall y > \bar{y}. \tag{E.143}$$

It is useful to compare (E.142) against (E.143). In (E.142), the deviation from the mean $\bar{y}$ is quantified by the term $\gamma/\sqrt{n}$, which vanishes as $n \to \infty$. A vanishing deviation of this form is usually referred to as *moderate* or *normal* [59, 60]. In contrast, in (E.143) the deviation $(y - \bar{y})$ is called a *large deviation*, since it does not converge to 0 as $n \to \infty$. We see from (E.143) that the probability of a large deviation converges to 0. In contrast, Eq. (E.142) reveals that the probability of a moderate deviation converges to a *nonzero* value $Q(\gamma/\sigma)$.

This behavior can be explained as follows. As n increases, the law of large numbers asserts that the distribution of the empirical average will be concentrated on $\bar{y}$. The CLT reveals that this distribution assumes a Gaussian shape, with a variance converging to 0 as $1/n$. This is the reason why variations on the order of $1/\sqrt{n}$ correspond to a constant limiting probability. For example, if we focus on the body of the distribution of $\bar{\boldsymbol{y}}_n$ containing approximately 95% of the probability mass, we will need to consider a range around the mean corresponding to twice the standard deviation, $2\sigma/\sqrt{n}$, yielding, for large n,

$$\mathbb{P}\left[|\bar{\boldsymbol{y}}_n - \bar{y}| \leq 2\frac{\sigma}{\sqrt{n}}\right] \approx 1 - 2\,Q(2) \approx 0.9545. \tag{E.144}$$

This situation is illustrated in Figure E.4, where we display four histograms of the empirical average $\bar{\boldsymbol{y}}_n$ of iid Bernoulli random variables with success probability $p = 0.1$.

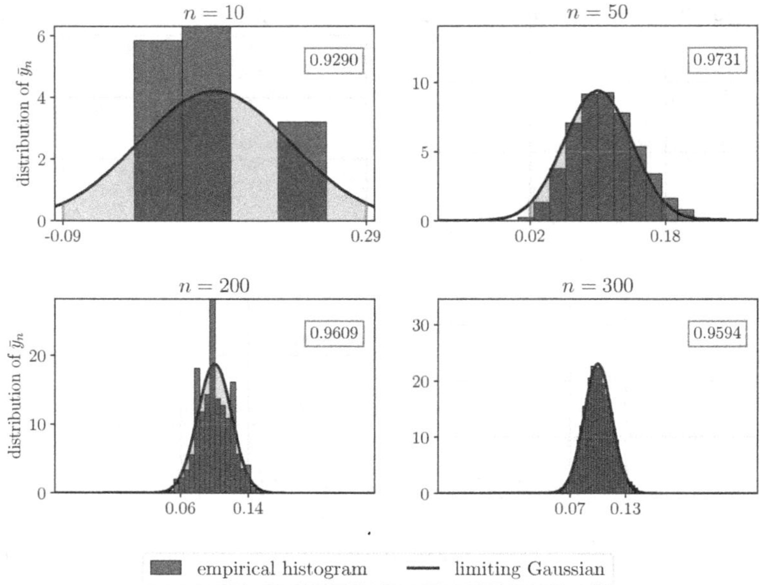

Figure E.4: Histograms computed from 10^6 independent realizations of the empirical average $\bar{\boldsymbol{y}}_n$ of iid Bernoulli random variables with success probability $p = 0.1$. The four panels correspond to $n = 10, 50, 200, 300$. The numerical value inside each panel is the *empirical* probability of belonging to the interval in (E.145). The pink area in the plots refers to this interval.

The four panels correspond to $n = 10, 50, 200, 300$. We see that the histograms approach a Gaussian shape as n increases. The area highlighted in pink corresponds to the range

$$\left[\bar{y} - 2\frac{\sigma}{\sqrt{n}}, \bar{y} + 2\frac{\sigma}{\sqrt{n}}\right] \tag{E.145}$$

and the numerical value in each panel is the *empirical* probability of belonging to such interval. We see that, as n increases, this probability approaches the value $1 - 2\,Q(2) \approx 0.9545$ predicted by the CLT. The main message of this analysis is that, as n increases, the region that contains most of the probability mass shrinks, becoming progressively concentrated on the mean. This is evident in Figure E.4, where the histograms approach a Dirac-δ shape as n increases. For this reason, if we consider intervals that shrink appropriately as $n \to \infty$, the probability of belonging to these intervals converge to a constant *nonzero* value, and the CLT is able to predict well this value through a Gaussian approximation. In other words, the effective *body* of the distribution is well approximated by a Gaussian distribution.

Consider now the large deviation perspective, where we are interested in evaluating the probability that the empirical average exceeds a threshold $y \neq \bar{y}$ that is *constant* with n — see (E.11). For example, let $y = 0.2$. Examining again the panels in Figure E.4, we see that, as $n \to \infty$, the value 0.2 becomes progressively farther from the body of the distribution, and this explains why, as predicted by Eq. (E.143), the probability that the empirical average exceeds this value converges to 0 as $n \to \infty$. In other words, the large deviation regime focuses on the *tails* of the distribution of $\bar{\boldsymbol{y}}_n$, rather than on the body.

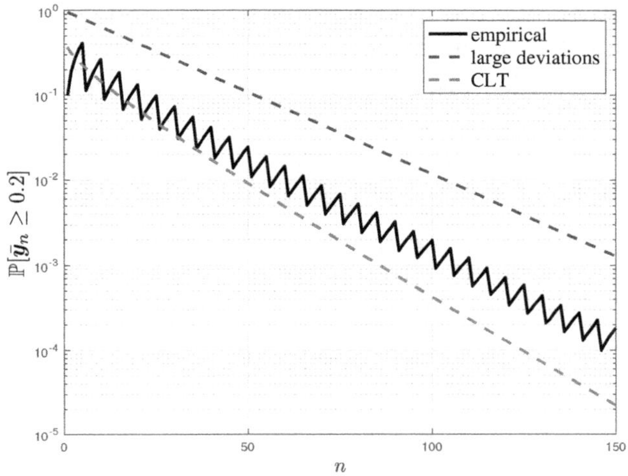

Figure E.5: Probability that the empirical average of iid Bernoulli variables with success probability $p = 0.1$ exceeds the value $y = 0.2$, displayed as a function of the number of samples n. The empirical probability (black) is estimated from 10^6 Monte Carlo runs. The red curve shows the Gaussian approximation in (E.146). The blue curve corresponds to the function $e^{-n\Lambda^*(0.2)}$, where $\Lambda^*(y)$ is the rate function given by (E.58).

In particular, under some regularity conditions on the tails, the probability of observing a large deviation vanishes at an exponential rate, and is accordingly characterized by evaluating the exponent that rules this decay — see (E.39). Note that, since the CLT is asymptotically exact for moderate (i.e., not large) deviations, it is not possible to use the CLT to evaluate the exponent. To give a concrete example, in Figure E.5 we consider the probability $\mathbb{P}[\bar{\boldsymbol{y}}_n \geq y]$, for $y = 0.2$ and several values of n, with reference to the same example considered before, i.e., iid Bernoulli variables with success probability $p = 0.1$. The black curve is obtained empirically by means of 10^6 Monte Carlo runs. The red curve corresponds to the Gaussian approximation

$$\mathbb{P}[\bar{\boldsymbol{y}}_n \geq y] \approx Q\left(\frac{y - \bar{y}}{\sigma/\sqrt{n}}\right), \tag{E.146}$$

where $\bar{y} = p$ and $\sigma^2 = p(1-p)$. We see that the CLT approximation fails to reproduce the correct behavior. Regarding the large deviation analysis, we depict in blue the curve $e^{-n\Lambda^*(0.2)}$, where $\Lambda^*(y)$ is the rate function given by (E.58). As already remarked, this curve is not intended to be an approximation for $\mathbb{P}[\bar{\boldsymbol{y}}_n \geq y]$, but only to represent the exponential decay. We see that this curve captures well the exponent exhibited by the black curve.

E.1.4 Probability of Belonging to Arbitrary Sets

Cramér's theorem focuses on the probability that the empirical average stays above or below the statistical average, and reveals that the exponential decay of this probability is well characterized through the rate function. It is therefore legitimate to ask whether the rate function can be useful to characterize the probability of belonging to arbitrary sets. To examine this question, it is useful to start with the following case:

$$\mathbb{P}[\bar{\boldsymbol{y}}_n \in \mathcal{S}], \qquad \mathcal{S} = (-\infty, y'] \cup [y'', \infty), \qquad y' < \bar{y} < y''. \tag{E.147}$$

We can write

$$\mathbb{P}[\bar{\boldsymbol{y}}_n \in \mathcal{S}] \geq \mathbb{P}[\bar{\boldsymbol{y}}_n \leq y'], \tag{E.148a}$$
$$\mathbb{P}[\bar{\boldsymbol{y}}_n \in \mathcal{S}] \geq \mathbb{P}[\bar{\boldsymbol{y}}_n \geq y''], \tag{E.148b}$$
$$\mathbb{P}[\bar{\boldsymbol{y}}_n \in \mathcal{S}] \leq \mathbb{P}[\bar{\boldsymbol{y}}_n \leq y'] + \mathbb{P}[\bar{\boldsymbol{y}}_n \geq y''], \tag{E.148c}$$

where the lower bounds hold since the probability of the union of events is not smaller than the probability of the individual events, while the upper bound follows from the union bound. From Cramér's theorem (Theorem E.1) we know that

$$\mathbb{P}[\bar{\boldsymbol{y}}_n \leq y'] \doteq e^{-n\Lambda^*(y')}, \qquad \mathbb{P}[\bar{\boldsymbol{y}}_n \geq y''] \doteq e^{-n\Lambda^*(y'')}. \tag{E.149}$$

Using these relations in (E.148a)–(E.148c), it is immediate to conclude that

$$\mathbb{P}[\bar{\boldsymbol{y}}_n \in \mathcal{S}] \doteq e^{-n\min\{\Lambda^*(y'),\Lambda^*(y'')\}}. \tag{E.150}$$

In this case we see that the exponent is not given by the rate function itself, but by the minimum value between $\Lambda^*(y')$ and $\Lambda^*(y'')$. From the general properties of the rate function (see Lemma E.1), this minimum is in fact the infimum of $\Lambda^*(y)$ over the set $\mathcal{S}$ — see Figure E.6 for a visual illustration. Note that the infimum corresponds to the smallest exponent. This behavior has the following useful interpretation.

On one hand, we know that the probability $\mathbb{P}[\bar{\boldsymbol{y}}_n \in \mathcal{S}]$ converges to 0 as $n \to \infty$. This means that $\{\bar{\boldsymbol{y}}_n \in \mathcal{S}\}$ is a *rare event* for large n. On the other hand, we know that $\mathbb{P}[\bar{\boldsymbol{y}}_n \in \mathcal{S}]$ vanishes at an exponential rate. The smaller the exponent, the rarer the event will be. We have seen that the exponent is determined by the infimum of the rate function over the set $\mathcal{S}$. This principle is nicely summarized in [60] by saying that "*any large deviation is done in the least unlikely of all the unlikely ways!*" It is now legitimate to ask whether this principle can be extended to sets $\mathcal{S}$ more general than the one in (E.147). We answer this question in the next section.

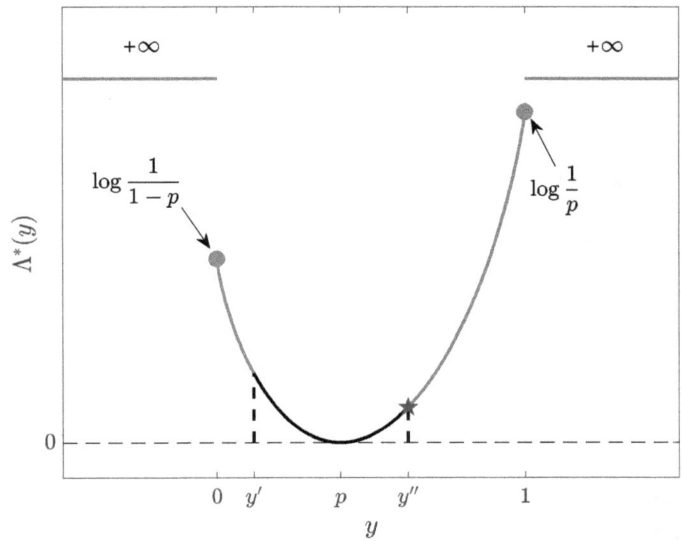

Figure E.6: Illustration of how to compute the exponent for the probability $\mathbb{P}[\bar{y}_n \in \mathcal{S}]$ in (E.147). The rate function $\Lambda^*(y)$ corresponds to Example E.3, and is given by (E.58). The part of the curve corresponding to $y \in \mathcal{S}$ is highlighted in red. The blue star denotes the value of the exponent, obtained as the infimum of $\Lambda^*(y)$ over $\mathcal{S}$.

E.2 Large Deviation Principle

The analysis in the previous sections focused only on the deviation of the empirical average of iid random variables from their statistical average. The theory of large deviations, however, is more general in several respects. First, it applies to more general families of random variables or vectors, and is not limited to empirical averages of iid samples. Second, the theory covers the probability of belonging to arbitrary sets.

The core concept is the *large deviation principle* (LDP). Although this principle can be formulated with reference to random vectors in $\mathbb{R}^d$ or general topological spaces, it is sufficient for our purposes to consider the simplest case of random variables. We recall that $\text{int}(\mathcal{S})$ and $\text{cl}(\mathcal{S})$ denote the interior and the closure of a set $\mathcal{S}$, respectively. Moreover, the infimum over an empty set is taken as ∞.

Definition E.2 (Large deviation principle). A family of random variables $\{\boldsymbol{y}_\varepsilon\}$ indexed by a (possibly continuous) parameter ε is said to satisfy the LDP with rate $1/\varepsilon$ and with rate function $I(y)$ when, for all sets $\mathcal{S} \subseteq \mathbb{R}$,

$$-\inf_{y\in\mathrm{int}(\mathcal{S})} I(y) \leq \liminf_{\varepsilon\to 0} \varepsilon \log \mathbb{P}[\boldsymbol{y}_\varepsilon \in \mathcal{S}] \leq \limsup_{\varepsilon\to 0} \varepsilon \log \mathbb{P}[\boldsymbol{y}_\varepsilon \in \mathcal{S}] \leq -\inf_{y\in\mathrm{cl}(\mathcal{S})} I(y), \tag{E.151}$$

where the function $I : \mathbb{R} \mapsto [0,\infty]$ must be lower semicontinuous, which means that for any $z \in [0,\infty)$, the level set

$$\{y \in \mathbb{R} : I(y) \leq z\} \tag{E.152}$$

is a closed subset of $\mathbb{R}$ or, equivalently, that

$$\liminf_{y\to y_0} I(y) \geq I(y_0) \quad \forall y_0 \in \mathbb{R}. \tag{E.153}$$

Examining the LDP definition, one might wonder about the need for the lower semicontinuity restriction on the family of rate functions. Likewise, one might ask why the LDP is defined in terms of the limit inferior (resp., superior) relative to the interior (resp., the closure) of $\mathcal{S}$, rather than simply by a limit relative to $\mathcal{S}$. Regarding lower semicontinuity, this property is useful to guarantee the uniqueness of the rate function — see [60, Thm. III.8] or [59, Lemma 4.1.4].

Regarding the second question, observe that whenever

$$\inf_{y\in\mathrm{int}(\mathcal{S})} I(y) = \inf_{y\in\mathrm{cl}(\mathcal{S})} I(y), \tag{E.154}$$

relation (E.151) implies the existence of the limit

$$\lim_{\varepsilon\to 0} \varepsilon \log \mathbb{P}[\boldsymbol{y}_\varepsilon \in \mathcal{S}] = -\inf_{y\in\mathcal{S}} I(y). \tag{E.155}$$

Sets fulfilling (E.154) are called I-continuity sets. There is no doubt that (E.155) is more direct and easier to deal with than (E.151). However, requiring (E.155) for *all* sets $\mathcal{S}$ can be too restrictive, and would exclude important classes of random variables.

To see why, assume that $\boldsymbol{y}_\varepsilon$ is a family of continuous random variables, and consider first the singleton $\mathcal{S} = \{y_0\}$. Then, since $\mathbb{P}[\boldsymbol{y}_\varepsilon \in \mathcal{S}] = \mathbb{P}[\boldsymbol{y}_\varepsilon = y_0] = 0$, by applying (E.155) we would get

$$\lim_{\varepsilon\to 0} \varepsilon \log \mathbb{P}[\boldsymbol{y}_\varepsilon = y_0] = -\infty = -I(y_0) \quad \forall y_0 \in \mathbb{R}. \tag{E.156}$$

Let us now consider another choice for the set $\mathcal{S}$, namely, $\mathcal{S} = \mathbb{R}$. Since $\mathbb{P}[\boldsymbol{y}_\varepsilon \in \mathbb{R}] = 1$, by applying (E.155) we would conclude that

$$\lim_{\varepsilon\to 0} \varepsilon \log \mathbb{P}[\boldsymbol{y}_\varepsilon \in \mathbb{R}] = 0 = -\inf_{y\in\mathbb{R}} I(y), \tag{E.157}$$

which is not compatible with (E.157). This reveals that, if the LDP were formulated by requiring (E.155) for all sets, continuous random variables would be excluded.

We are now ready to state the famous Gärtner-Ellis theorem, which generalizes Cramér's theorem to deal with sequences of random variables more general than empirical averages of iid samples.

Theorem E.2 (Gärtner-Ellis theorem [68, 78] [60, Thm. V.6] [59, Thm. 2.3.6]). Let $\{\boldsymbol{y}_\varepsilon\}$ be a family of random variables indexed by a (possibly continuous) parameter ε, and let

$$\Lambda_\varepsilon(s) = \log \mathbb{E} e^{s\boldsymbol{y}_\varepsilon} \tag{E.158}$$

be the LMGF of $\boldsymbol{y}_\varepsilon$. If, for all $s \in \mathbb{R}$,

$$\lim_{\varepsilon \to 0} \varepsilon \, \Lambda_\varepsilon(s/\varepsilon) = \Lambda(s) < \infty, \tag{E.159}$$

with $\Lambda(s)$ being differentiable on $\mathbb{R}$, then the family of random variables $\{\boldsymbol{y}_\varepsilon\}$ satisfies the LDP with rate $1/\varepsilon$ and with rate function $I(y) = \Lambda^*(y)$. Furthermore, $\Lambda^*(y)$ has compact level sets.[4]

To become familiar with the statement of the Gärtner-Ellis theorem, it is instructive to apply it to the empirical average of iid random variables, and verify whether we recover Cramér's theorem.

Example E.6 (Cramér's theorem from the Gärtner-Ellis theorem). Consider the setting of Cramér's theorem (Theorem E.1), where we have a sequence $\{\boldsymbol{y}_n\}$ of iid random variables, whose LMGF $\Lambda(s)$ satisfies the condition $\Lambda(s) < \infty$ for all $s \in \mathbb{R}$. Consider then the empirical average

$$\bar{\boldsymbol{y}}_n = \frac{1}{n} \sum_{i=1}^{n} \boldsymbol{y}_i, \tag{E.160}$$

whose LMGF can be computed in terms of the LMGF $\Lambda(s)$ of the individual variable $\boldsymbol{y}_n$ by using (E.136), which yields

$$\log \mathbb{E} e^{s\bar{\boldsymbol{y}}_n} = n\Lambda(s/n). \tag{E.161}$$

Performing the change of variable $\varepsilon = 1/n$ and denoting by $\Lambda_\varepsilon(s)$ the LMGF of $\bar{\boldsymbol{y}}_n$, we can write

$$\Lambda_\varepsilon(s) = \frac{1}{\varepsilon} \Lambda(\varepsilon \, s), \tag{E.162}$$

which in turn implies

$$\varepsilon \, \Lambda_\varepsilon(s/\varepsilon) = \Lambda(s). \tag{E.163}$$

[4]Some authors [59] use the term "good rate function" when the rate function has compact level sets. Other authors [60] embody directly the "goodness" of the rate functions in their definitions, i.e., they require that all rate functions must be "good" by definition.

Accordingly, condition (E.159) is verified with the limiting LMGF equal to the LMGF of the individual variable $\boldsymbol{y}_n$. Calling upon the Gärtner-Ellis theorem we conclude that the LDP is satisfied with rate $1/\varepsilon = n$ and with rate function $I(y) = \Lambda^*(y)$.

Now, let $\bar{y} < z < y_{\mathsf{sup}}$ and consider the set

$$\mathcal{S} = [z, \infty). \tag{E.164}$$

Property P4 in Lemma E.1 ensures that

$$\inf_{y \in \mathrm{int}(\mathcal{S})} \Lambda^*(y) = \inf_{y \in \mathrm{cl}(\mathcal{S})} \Lambda^*(y) = \Lambda^*(z), \tag{E.165}$$

i.e., $\mathcal{S}$ is a Λ^*-continuity set (see (E.154)) and the infima in (E.165) are in fact a minimum equal to $\Lambda^*(z)$. Thus, in this case the LDP from (E.151) corresponds to the statement

$$\lim_{n \to \infty} \frac{1}{n} \log \mathbb{P}[\bar{\boldsymbol{y}}_n \geq z] = -\Lambda^*(z), \tag{E.166}$$

which in turn corresponds to the claim of Cramér's theorem.

Reasoning in a similar manner it is possible to verify that when $y_{\mathsf{sup}} < \infty$, the set $[y_{\mathsf{sup}}, \infty)$ is *not* a Λ^*-continuity set (see, e.g., Figure E.2). This particular case, which is covered by Cramér's theorem, is not directly obtained from the statement of the Gärtner-Ellis theorem.

Owing to its generality, the Gärtner-Ellis theorem is a powerful tool that allows to cover numerous cases of practical relevance. In our treatment, it is used to characterize the error probability associated with traditional (Chapter 6) and adaptive (Chapter 9) social learning.

Appendix F

Random Sums and Series

This appendix focuses on the stochastic convergence of random sums. Section F.1 contains some classic results on convergent random series. Sections F.2 and F.3 focus instead on certain random sums that are relevant to the adaptive social learning paradigm considered in Chapters 8, 9, and 10; the results presented in these sections are either novel or borrowed from [25, 119, 120].

F.1 Convergent Random Series

We adopt the following standard terminology. Given a sequence of numbers $\{y_n\}$ and the partial sum $\sum_{i=1}^{n} y_i$, when the sequence of partial sums converges as $n \to \infty$ we say that the series $\sum_{i=1}^{\infty} y_i$ is convergent. Moreover, we say that the series $\sum_{i=1}^{\infty} y_i$ is *absolutely* convergent when the series $\sum_{i=1}^{\infty} |y_i|$ is convergent. Note that absolute convergence implies convergence.

The next lemma provides a sufficient condition for the almost-sure convergence of random series.

Lemma F.1 (Convergence of random series [113, Lemma 3.6′]). Let $\{\boldsymbol{y}_n\}$ be a sequence of independent random variables. If the series of expected values of $|\boldsymbol{y}_n|$ is convergent, i.e., if

$$\sum_{i=1}^{\infty} \mathbb{E}|\boldsymbol{y}_i| < \infty, \tag{F.1}$$

then $\sum_{i=1}^{\infty} \boldsymbol{y}_i$ is almost surely an absolutely convergent series.

We remark that the limiting value $\sum_{i=1}^{\infty} \boldsymbol{y}_i$ is in general *random*. For the characterization of a random series employed to characterize adaptive social learning in Chapter 9 (see footnote 2 in that chapter) it is useful to know that, if the summands $\boldsymbol{y}_i$ are not deterministic from some index onward, then the series $\sum_{i=1}^{\infty} \boldsymbol{y}_i$ will be a continuous[1] random variable. This property is guaranteed by the following result.

Lemma F.2 (Continuous nature of random series [111, Thm. XIII]). Let $\{\boldsymbol{y}_n\}$ be a sequence of independent random variables satisfying the condition

$$\sum_{i=1}^{\infty} \mathbb{E}|\boldsymbol{y}_i| < \infty. \tag{F.2}$$

If the random variables in the sequence are *not* deterministic from some index n_0 onward, then the series $\sum_{i=1}^{\infty} \boldsymbol{y}_i$ is a continuous random variable.

F.2 Random Sums Relevant to Adaptive Social Learning

In this section we derive some properties for a particular random sum that arises in the study of the adaptive social learning algorithms in Chapters 8, 9, and 10.

Definition F.1 (Useful random sums). Let $\{\boldsymbol{y}_n\}$ be a sequence of iid random variables with finite mean $\bar{y}$. Let also $0 < \delta < 1$ and consider the following partial sums:

$$\boldsymbol{z}_n(\delta) \triangleq \delta \sum_{i=1}^{n} (1-\delta)^{i-1} \alpha_i \boldsymbol{y}_i, \tag{F.3}$$

where $0 \le \alpha_i \le 1$, with α_i converging exponentially to some value $\alpha > 0$ and obeying the following upper bound for all $i \in \mathbb{N}$:

$$|\alpha_i - \alpha| \le \kappa \xi^i, \tag{F.4}$$

for some constants $\kappa > 0$ and $\xi \in (0, 1)$.

We will first study the convergence, as $n \to \infty$, of the partial sums in (F.3), and then characterize their behavior in the limit as $\delta \to 0$. To facilitate the presentation, the relevant properties are collected in a series of lemmas.

[1]We recall that a continuous random variable does not necessarily have a pdf, since it could be *continuous but singular* [21]. In the context of random series, one notable example that can exhibit this behavior is given by Bernoulli convolutions, which are sums of iid Bernoulli variables. For some values of the success probability, it has been shown that these sums converge to continuous but singular limiting variables [69].

Lemma F.3 (Useful random series). Consider the setting described in Definition F.1. Then, *irrespective of condition* (F.4), the partial sums in (F.3) converge almost surely as $n \to \infty$, namely, we can define the following random variable

$$\boldsymbol{z}(\delta) \triangleq \delta \sum_{i=1}^{\infty} (1-\delta)^{i-1} \alpha_i \boldsymbol{y}_i. \tag{F.5}$$

In particular, the series on the RHS is almost-surely *absolutely* convergent.

Proof. We start by establishing the convergence of the series of expectations

$$\delta \sum_{i=1}^{\infty} (1-\delta)^{i-1} \alpha_i \, \mathbb{E}|\boldsymbol{y}_i| = \bar{y}^{\mathsf{abs}} \, \delta \sum_{i=1}^{\infty} (1-\delta)^{i-1} \alpha_i, \tag{F.6}$$

where $\bar{y}^{\mathsf{abs}} = \mathbb{E}|\boldsymbol{y}_i|$. Note that $\bar{y}^{\mathsf{abs}} < \infty$ because the random variables $\boldsymbol{y}_i$ have finite mean. The series on the RHS of (F.6) is convergent for the following reason. Since $0 \le \alpha_i \le 1$ in view of Definition F.1, we have

$$\begin{aligned} \sum_{i=1}^{n} (1-\delta)^{i-1} \alpha_i &\le \sum_{i=1}^{n} (1-\delta)^{i-1} \le \sum_{i=1}^{\infty} (1-\delta)^{i-1} \\ &= \frac{1}{1-(1-\delta)} = \frac{1}{\delta} < \infty, \end{aligned} \tag{F.7}$$

where in the first equality we compute the known value of the geometric series. Observe that the partial sums

$$\sum_{i=1}^{n} (1-\delta)^{i-1} \alpha_i \tag{F.8}$$

consist of nonnegative terms; this implies that these sums form a monotone (nondecreasing) sequence. As a result, their limit

$$\lim_{n\to\infty} \sum_{i=1}^{n} (1-\delta)^{i-1} \alpha_i = \sum_{i=1}^{\infty} (1-\delta)^{i-1} \alpha_i \tag{F.9}$$

exists. In principle, the limit can be equal to ∞. However, this is not the case since, by letting $n \to \infty$ on the LHS of (F.7), we get

$$\sum_{i=1}^{\infty} (1-\delta)^{i-1} \alpha_i < \infty. \tag{F.10}$$

This proves that the series of expectations in (F.6) converges.

In view of Lemma F.1, convergence of the series of expectations in (F.6) implies that the random series $\boldsymbol{z}(\delta)$ in (F.5) is almost-surely absolutely convergent, and the lemma is proved. ∎

Lemma F.4 (First moment). Consider the setting described in Definition F.1, and let

$$\boldsymbol{z}(\delta) \triangleq \delta \sum_{i=1}^{\infty} (1-\delta)^{i-1} \alpha_i \boldsymbol{y}_i. \tag{F.11}$$

Then,

$$\mathbb{E}\boldsymbol{z}(\delta) = \bar{y}\,\delta \sum_{i=1}^{\infty} (1-\delta)^{i-1} \alpha_i = \alpha\,\bar{y} + O(\delta), \tag{F.12}$$

where $O(\delta)$ is a quantity such that the ratio $O(\delta)/\delta$ remains bounded as $\delta \to 0$ — see Table 1.1.

Proof. Since, in view of (F.6) and (F.7), the series

$$\delta \sum_{i=1}^{\infty} (1-\delta)^{i-1} \alpha_i\, \mathbb{E}|\boldsymbol{y}_i| \tag{F.13}$$

is convergent, so is the series

$$\sum_{i=1}^{\infty} (1-\delta)^{i-1} \alpha_i\, \mathbb{E}\boldsymbol{y}_i = \bar{y} \sum_{i=1}^{\infty} (1-\delta)^{i-1} \alpha_i. \tag{F.14}$$

The equality is obtained by using the definition $\bar{y} = \mathbb{E}\boldsymbol{y}_i$. On the other hand, from the triangle inequality we have the following upper bound for the random sum $\boldsymbol{z}_n(\delta)$ in (F.3):

$$|\boldsymbol{z}_n(\delta)| \leq \delta \sum_{i=1}^{n} (1-\delta)^{i-1} \alpha_i |\boldsymbol{y}_i| \leq \underbrace{\delta \sum_{i=1}^{\infty} (1-\delta)^{i-1} \alpha_i |\boldsymbol{y}_i|}_{\boldsymbol{z}^{\mathsf{abs}}(\delta)}, \tag{F.15}$$

where the convergence of the series defined by $\boldsymbol{z}^{\mathsf{abs}}(\delta)$ is guaranteed by Lemma F.3 (because the series $\boldsymbol{z}(\delta)$ converges *absolutely*). In view of Beppo Levi's monotone convergence theorem [65, Thm. 1.5.7], the expectation of the almost-sure limit $\boldsymbol{z}^{\mathsf{abs}}(\delta)$ is equal to the series of expectations, namely,

$$\mathbb{E}\boldsymbol{z}^{\mathsf{abs}}(\delta) = \delta \sum_{i=1}^{\infty} (1-\delta)^{i-1} \alpha_i\, \mathbb{E}|\boldsymbol{y}_i| < \infty, \tag{F.16}$$

where the inequality holds because of (F.7). From (F.15) and (F.16) we conclude that $|\boldsymbol{z}_n(\delta)|$ is upper bounded, for all n, by a random variable with finite mean. Therefore, the dominated convergence theorem (Theorem D.6) implies that the expectation of the almost-sure limit $\boldsymbol{z}(\delta)$ is equal to the convergent series of expectations (F.14), and the first equality in (F.12) follows. For the second equality, observe that

$$\delta \sum_{i=1}^{n} (1-\delta)^{i-1} \alpha_i = \delta \sum_{i=1}^{n} (1-\delta)^{i-1} (\alpha_i - \alpha) + \alpha\,\delta \sum_{i=1}^{n} (1-\delta)^{i-1}. \tag{F.17}$$

Reasoning as done to establish the convergence of (F.7), we can see that both partial sums on the RHS converge and, hence, we can write

$$\delta \sum_{i=1}^{\infty}(1-\delta)^{i-1}\alpha_i = \delta \sum_{i=1}^{\infty}(1-\delta)^{i-1}(\alpha_i-\alpha) + \alpha\, \underbrace{\delta \sum_{i=1}^{\infty}(1-\delta)^{i-1}}_{=1}. \tag{F.18}$$

In view of (F.4), the absolute value of the first summation on the RHS of (F.18) is dominated by

$$\kappa\,\delta \sum_{i=1}^{\infty}\xi^i(1-\delta)^{i-1} = \kappa\,\xi\,\delta \sum_{i=1}^{\infty}\big[\xi(1-\delta)\big]^{i-1} = \frac{\kappa\,\xi\,\delta}{1-\xi(1-\delta)} = O(\delta). \tag{F.19}$$

We conclude from (F.14), (F.18), and (F.19) that the second equality in (F.12) holds. ■

Lemma F.5 (Weak law of small δ). Consider the setting described in Definition F.1, and let

$$\boldsymbol{z}(\delta) \triangleq \delta \sum_{i=1}^{\infty}(1-\delta)^{i-1}\alpha_i \boldsymbol{y}_i. \tag{F.20}$$

Then, the series $\boldsymbol{z}(\delta)$ converges to $\alpha\,\bar{y}$ in probability as $\delta \to 0$, namely, for all $\varepsilon > 0$,

$$\lim_{\delta\to 0} \mathbb{P}\Big[|\boldsymbol{z}(\delta) - \alpha\,\bar{y}| > \varepsilon\Big] = 0. \tag{F.21}$$

Proof. If we assume finiteness of the second moment of $\boldsymbol{z}(\delta)$, the proof of this lemma is immediately obtained as an application of Chebyshev's inequality — see Theorem C.2. However, finiteness of the second moment is not required, and thus we proceed with a more technical proof that works without that assumption. Let

$$\zeta_i = \delta(1-\delta)^{i-1}\alpha_i \tag{F.22}$$

and consider the following centered variables:

$$\widetilde{\boldsymbol{z}}(\delta) = \boldsymbol{z}(\delta) - \mathbb{E}\,\boldsymbol{z}(\delta), \qquad \widetilde{\boldsymbol{y}}_i = \boldsymbol{y}_i - \mathbb{E}\boldsymbol{y}_i. \tag{F.23}$$

In view of Lemmas F.3 and F.4, the centered partial sums

$$\boldsymbol{z}_n(\delta) - \mathbb{E}\,\boldsymbol{z}_n(\delta) = \sum_{i=1}^{n}\zeta_i\,\widetilde{\boldsymbol{y}}_i \tag{F.24}$$

converge almost surely (hence, in distribution) to $\widetilde{\boldsymbol{z}}(\delta)$ as $n \to \infty$. From the Lévy-Cramér continuity theorem (Theorem D.1) the characteristic function of $\boldsymbol{z}_n - \mathbb{E}\,\boldsymbol{z}_n(\delta)$ must converge to the characteristic function of $\widetilde{\boldsymbol{z}}(\delta)$. Exploiting (F.24) and the fact that the random variables $\boldsymbol{y}_i$ are iid, the latter characteristic function can be represented as

$$\varphi_{\widetilde{z}}(s) = \prod_{i=1}^{\infty}\varphi_{\widetilde{y}}(\zeta_i s), \tag{F.25}$$

where $\varphi_{\widetilde{y}}(s)$ denotes the characteristic function of the centered variable $\widetilde{\boldsymbol{y}}_i$.

The claim of the lemma is that $\boldsymbol{z}(\delta) - \alpha\,\bar{y}$ converges in probability to 0 as $\delta \to 0$. In view of (F.12) and property P1 from Lemma D.1, it is enough to prove that $\widetilde{\boldsymbol{z}}(\delta)$ in (F.23) converges in probability to 0 as $\delta \to 0$. Furthermore, property P3 from Lemma D.1 allows us to conclude that it suffices to establish that $\widetilde{\boldsymbol{z}}(\delta)$ converges to 0 *in distribution*. To this end, we resort again to the Lévy-Cramér continuity theorem (and use it this time to examine the convergence for $\delta \to 0$). Since the characteristic function of a deterministic variable equal to 0 is identically 1 for all $s \in \mathbb{R}$, we need to show that $\varphi_{\widetilde{z}}(s)$ converges to 1 as $\delta \to 0$. Using (F.25) we can write[2]

$$|\varphi_{\widetilde{z}}(s) - 1| \le \sum_{i=1}^{\infty} |\varphi_{\widetilde{y}}(\zeta_i s) - 1|. \tag{F.27}$$

Consider a positive s (the proof for $s < 0$ is similar, whereas for $s = 0$ it is trivial). Since the random variables $\widetilde{\boldsymbol{y}}_i$ have finite mean, it is known that the first derivative of the characteristic function, $\varphi'_{\widetilde{y}}(s)$, is a continuous function [70], and from the mean-value theorem [144, Thm. 5.9] we can write (since in particular $\mathbb{E}\widetilde{\boldsymbol{y}}_i = 0$)

$$\varphi_{\widetilde{y}}(\zeta_i s) = 1 + \zeta_i s\, \varphi'_{\widetilde{y}}(s_m) \text{ for some } s_m \in (0, \zeta_i s). \tag{F.28}$$

Accordingly, we have

$$|\varphi_{\widetilde{y}}(\zeta_i s) - 1| = \zeta_i s\, |\varphi'_{\widetilde{y}}(s_m)| \le \zeta_i s \max_{\varsigma \in [0,\delta s]} |\varphi'_{\widetilde{y}}(\varsigma)|, \tag{F.29}$$

where the inequality holds because $s_m \in (0, \zeta_i s)$ and $\zeta_i \le \delta$ — see (F.22). Applying (F.29) to (F.27) we get

$$|\varphi_{\widetilde{z}}(s) - 1| \le s \max_{\varsigma \in [0,\delta s]} |\varphi'_{\widetilde{y}}(\varsigma)| \underbrace{\sum_{i=1}^{\infty} \zeta_i}_{\le 1}. \tag{F.30}$$

On the other hand, since $\varphi'_{\widetilde{y}}(0) = \mathbb{E}\widetilde{\boldsymbol{y}}_i = 0$, from the continuity of $\varphi'_{\widetilde{y}}(s)$ it follows that

$$\lim_{\delta \to 0} \max_{\varsigma \in [0,\delta s]} |\varphi'_{\widetilde{y}}(\varsigma)| = 0, \tag{F.31}$$

which, in view of (F.30), proves that $\varphi_{\widetilde{z}}(s)$ converges to 1 as $\delta \to 0$. From the Lévy-Cramér continuity theorem, this implies that $\boldsymbol{z}(\delta)$ converges to $\mathbb{E}\boldsymbol{z}(\delta)$ in distribution as $\delta \to 0$, and the proof is complete.

■

F.3 Vector Case for Network Behavior

In this section we extend Definition F.1 to examine the case where $\boldsymbol{y}_n$ is a vector. Actually, when we examine networks of agents in Chapters 8,

[2]The following inequality is known for complex numbers x_i, y_i, with $|x_i| \le 1$ and $|y_i| \le 1$ [70]:

$$\left| \prod_{i=1}^{n} x_i - \prod_{i=1}^{n} y_i \right| \le \sum_{i=1}^{n} |x_i - y_i|. \tag{F.26}$$

9, and 10, we always deal with this vector case (with the vector entries referring to distinct agents). In fact, the results proved for the scalar case in the previous sections can be readily applied to the vector case. However, for the results that we are going to prove in this section, an analysis specialized to the vector setting is necessary to account for the possible statistical dependence across the agents.

Definition F.2 (Random sums with random vectors). Let $\{\boldsymbol{y}_n\}$ be a sequence of iid random vectors in $\mathbb{R}^K$, whose entries have finite mean, and define

$$\boldsymbol{y}_n = [\boldsymbol{y}_{1,n}, \boldsymbol{y}_{2,n}, \ldots, \boldsymbol{y}_{K,n}], \qquad \bar{y} = \mathbb{E}\boldsymbol{y}_n. \tag{F.32}$$

Let also $0 < \delta < 1$, and consider the following (scalar) partial sums:

$$\boldsymbol{z}_n(\delta) = \delta \sum_{i=1}^{n} (1-\delta)^{i-1} \alpha_i^\mathsf{T} \boldsymbol{y}_i, \tag{F.33}$$

where

$$\alpha_n = [\alpha_{1,n}, \alpha_{2,n}, \ldots, \alpha_{K,n}] \tag{F.34}$$

is a deterministic vector with nonnegative entries bounded by 1, i.e.,

$$0 \le \alpha_{k,n} \le 1 \quad \text{for } k = 1, 2, \ldots, K \text{ and for all } n \in \mathbb{N}. \tag{F.35}$$

Moreover, assume that α_n converges to some vector

$$\alpha = [\alpha_1, \alpha_2, \ldots, \alpha_K] \tag{F.36}$$

with the following exponential law:

$$|\alpha_{k,n} - \alpha_k| \le \kappa\, \xi^n \tag{F.37}$$

for all k and n, and for some constants $\kappa > 0$ and $\xi \in (0,1)$. To avoid trivial cases, we assume that α has at least one nonzero entry.

Note that Lemma F.3 implies the convergence as $n \to \infty$ of the partial sums in (F.33), allowing us to define the series

$$\boldsymbol{z}(\delta) = \delta \sum_{i=1}^{\infty} (1-\delta)^{i-1} \alpha_i^\mathsf{T} \boldsymbol{y}_i. \tag{F.38}$$

Our main goal is to characterize the asymptotic behavior of $\boldsymbol{z}(\delta)$ as $\delta \to 0$. In particular, we will show that this behavior is ruled by the average variable

$$\boldsymbol{y}_{\mathsf{ave},n} \triangleq \alpha^\mathsf{T} \boldsymbol{y}_n = \sum_{k=1}^{K} \alpha_k \boldsymbol{y}_{k,n}. \tag{F.39}$$

For later use, we also introduce the mean and variance of this average variable, namely,

$$\bar{y}_{\text{ave}} \triangleq \mathbb{E}\boldsymbol{y}_{\text{ave},n} = \mathbb{E}\left[\alpha^{\mathsf{T}}\boldsymbol{y}_n\right] = \alpha^{\mathsf{T}}\bar{y} \tag{F.40}$$

and

$$\sigma^2_{\text{ave}} \triangleq \text{VAR}\left[\boldsymbol{y}_{\text{ave},n}\right] = \text{VAR}\left[\alpha^{\mathsf{T}}\boldsymbol{y}_n\right]. \tag{F.41}$$

Next, we examine the behavior of the second moment of $\boldsymbol{z}(\delta)$.

Lemma F.6 (Second moment). Consider the setting described in Definition F.2, and assume that $\boldsymbol{y}_n$ has finite second moment. Then, the variance of $\boldsymbol{z}(\delta)$ in (F.38) is

$$\text{VAR}[\boldsymbol{z}(\delta)] = \delta^2 \sum_{i=1}^{\infty} (1-\delta)^{2(i-1)}\,\text{VAR}[\alpha_i^{\mathsf{T}}\boldsymbol{y}_i] = \frac{1}{2}\sigma^2_{\text{ave}}\,\delta + O(\delta^2). \tag{F.42}$$

Proof. Let us introduce, for $i \in \mathbb{N}$, the centered, zero-mean vectors

$$\widetilde{\boldsymbol{y}}_i = \boldsymbol{y}_i - \bar{y} \tag{F.43}$$

and their expected squared norm

$$\sigma_y^2 = \mathbb{E}\|\widetilde{\boldsymbol{y}}_i\|^2. \tag{F.44}$$

Observe that, in view of the Cauchy-Schwarz inequality, we have

$$\text{VAR}\left[\alpha_i^{\mathsf{T}}\boldsymbol{y}_i\right] = \mathbb{E}\left[\left(\alpha_i^{\mathsf{T}}\widetilde{\boldsymbol{y}}_i\right)^2\right] \leq \|\alpha_i\|^2\,\mathbb{E}\|\widetilde{\boldsymbol{y}}_i\|^2 \leq K\sigma_y^2, \tag{F.45}$$

where in the last inequality we use definition (F.44) and the fact that the K entries of α_i are all bounded by 1. Since the scalar variables $\alpha_i^{\mathsf{T}}\boldsymbol{y}_i$ are independent, from the definition of $\boldsymbol{z}_n(\delta)$ in (F.33) we have

$$\begin{aligned}\text{VAR}[\boldsymbol{z}_n(\delta)] &= \delta^2 \sum_{i=1}^{n} (1-\delta)^{2(i-1)}\,\text{VAR}\left[\alpha_i^{\mathsf{T}}\boldsymbol{y}_i\right] \\ &\leq K\sigma_y^2\delta^2 \sum_{i=1}^{n}(1-\delta)^{2(i-1)},\end{aligned} \tag{F.46}$$

where the inequality follows from (F.45). Note that both partial sums in (F.46) admit limits since they consist of nonnegative terms. In particular, the limit of the partial sum on the RHS is finite since it is given by a convergent geometric series, and we can write

$$\begin{aligned}\lim_{n\to\infty}\text{VAR}[\boldsymbol{z}_n(\delta)] &= \delta^2 \sum_{i=1}^{\infty} (1-\delta)^{2(i-1)}\,\text{VAR}\left[\alpha_i^{\mathsf{T}}\boldsymbol{y}_i\right] \\ &\leq K\sigma_y^2\delta^2 \sum_{i=1}^{\infty}(1-\delta)^{2(i-1)} < \infty.\end{aligned} \tag{F.47}$$

Consider now the centered and squared variables

$$\left(\boldsymbol{z}_n(\delta) - \mathbb{E}\boldsymbol{z}_n(\delta)\right)^2 = \delta^2 \left(\sum_{i=1}^{n}(1-\delta)^{i-1}\alpha_i^{\mathsf{T}}\widetilde{\boldsymbol{y}}_i\right)^2. \tag{F.48}$$

In view of Lemmas F.3 and F.4, the quantity on the LHS converges almost surely, as $n \to \infty$, to

$$\left(\boldsymbol{z}(\delta) - \mathbb{E}\boldsymbol{z}(\delta)\right)^2. \tag{F.49}$$

Therefore, we can apply Fatou's lemma (Theorem D.5) to the variables $(\boldsymbol{z}_n(\delta) - \mathbb{E}\boldsymbol{z}_n(\delta))^2$, yielding

$$\lim_{n\to\infty} \mathsf{VAR}[\boldsymbol{z}_n(\delta)] \geq \mathsf{VAR}[\boldsymbol{z}(\delta)]. \tag{F.50}$$

In view of (F.47), this implies that the limiting variable $\boldsymbol{z}(\delta)$ has finite variance. But since the limiting variable $\boldsymbol{z}(\delta)$ can be written as

$$\boldsymbol{z}(\delta) = \boldsymbol{z}_n(\delta) + \delta \sum_{i=n+1}^{\infty} (1-\delta)^{i-1}\alpha_i^{\mathsf{T}}\boldsymbol{y}_i, \tag{F.51}$$

with the two quantities on the RHS being statistically independent, for any n the variance of $\boldsymbol{z}(\delta)$ cannot be smaller than the variance of $\boldsymbol{z}_n(\delta)$, implying that

$$\mathsf{VAR}[\boldsymbol{z}(\delta)] \geq \lim_{n\to\infty} \mathsf{VAR}[\boldsymbol{z}_n(\delta)]. \tag{F.52}$$

Combining (F.50) with (F.52) we see that the variance of the almost-sure limit $\boldsymbol{z}(\delta)$ is equal to the convergent series of variances, which is the first equality in (F.42).

In order to prove the second equality in (F.42) we write

$$\mathsf{VAR}\left[\boldsymbol{z}(\delta)\right] = \delta^2 \sum_{i=1}^{\infty}(1-\delta)^{2(i-1)}\, \mathbb{E}\left[\left(\alpha_i^{\mathsf{T}}\widetilde{\boldsymbol{y}}_i\right)^2\right]. \tag{F.53}$$

The expected values appearing in the last summation can be manipulated as follows:

$$\begin{aligned}
\mathbb{E}\left[\left(\alpha_i^{\mathsf{T}}\widetilde{\boldsymbol{y}}_i\right)^2\right] &= \mathbb{E}\left[\left(\alpha^{\mathsf{T}}\widetilde{\boldsymbol{y}}_i + (\alpha_i-\alpha)^{\mathsf{T}}\widetilde{\boldsymbol{y}}_i\right)^2\right] \\
&= \mathbb{E}\left[\left(\alpha^{\mathsf{T}}\widetilde{\boldsymbol{y}}_i\right)^2\right] + \mathbb{E}\left[\left((\alpha_i-\alpha)^{\mathsf{T}}\widetilde{\boldsymbol{y}}_i\right)^2\right] + 2\,\mathbb{E}\left[(\alpha_i-\alpha)^{\mathsf{T}}\widetilde{\boldsymbol{y}}_i\alpha^{\mathsf{T}}\widetilde{\boldsymbol{y}}_i\right] \\
&= \sigma^2_{\mathsf{ave}} + \mathbb{E}\left[\left((\alpha_i-\alpha)^{\mathsf{T}}\widetilde{\boldsymbol{y}}_i\right)^2\right] + 2\mathbb{E}\left[(\alpha_i-\alpha)^{\mathsf{T}}\widetilde{\boldsymbol{y}}_i\alpha^{\mathsf{T}}\widetilde{\boldsymbol{y}}_i\right],
\end{aligned} \tag{F.54}$$

where, in the last step, we used (F.41) and (F.43). Substituting (F.54) into (F.53) we obtain

$$\begin{aligned}
\mathsf{VAR}\left[\boldsymbol{z}(\delta)\right] &= \sigma^2_{\mathsf{ave}}\,\delta^2 \sum_{i=1}^{\infty}(1-\delta)^{2(i-1)} \\
&\quad + \delta^2 \sum_{i=1}^{\infty}(1-\delta)^{2(i-1)}\,\mathbb{E}\left[\left((\alpha_i-\alpha)^{\mathsf{T}}\widetilde{\boldsymbol{y}}_i\right)^2\right] \\
&\quad + 2\delta^2 \sum_{i=1}^{\infty}(1-\delta)^{2(i-1)}\,\mathbb{E}\left[(\alpha_i-\alpha)^{\mathsf{T}}\widetilde{\boldsymbol{y}}_i\alpha^{\mathsf{T}}\widetilde{\boldsymbol{y}}_i\right].
\end{aligned} \tag{F.55}$$

By evaluating the geometric series, the first term in (F.55) is equal to

$$\sigma_{\text{ave}}^2 \frac{\delta}{2-\delta} = \frac{\sigma_{\text{ave}}^2 \delta}{2} + \sigma_{\text{ave}}^2 \delta \left(\frac{1}{2-\delta} - \frac{1}{2}\right) = \frac{\sigma_{\text{ave}}^2 \delta}{2} + \frac{\sigma_{\text{ave}}^2 \delta^2}{2(2-\delta)} \tag{F.56}$$

and we see that the last term is $O(\delta^2)$. Therefore, the proof will be complete if we show that the last two terms on the RHS of (F.55) are $O(\delta^2)$.

To this end, we apply the Cauchy-Schwarz inequality to obtain the bound

$$\mathbb{E}\left[\left((\alpha_i - \alpha)^\mathsf{T} \widetilde{\boldsymbol{y}}_i\right)^2\right] \leq \|\alpha_i - \alpha\|^2 \, \mathbb{E}\left[\|\boldsymbol{y}_i\|^2\right] \leq K \kappa^2 \xi^{2i} \sigma_y^2, \tag{F.57}$$

where the second inequality follows from (F.37) and (F.44). Likewise, by applying the Cauchy-Schwarz inequality for expected values (see Theorem C.6 for $r_1 = r_2 = 2$), we can write

$$\begin{aligned}
&\left|\mathbb{E}\left[(\alpha_i - \alpha)^\mathsf{T} \widetilde{\boldsymbol{y}}_i \, \alpha^\mathsf{T} \widetilde{\boldsymbol{y}}_i\right]\right| \\
&\leq \Bigg(\underbrace{\mathbb{E}\left[\left((\alpha_i - \alpha)^\mathsf{T} \widetilde{\boldsymbol{y}}_i\right)^2\right]}_{\leq K \kappa^2 \xi^{2i} \sigma_y^2 \text{ from (F.57)}} \underbrace{\mathbb{E}\left[\left(\alpha^\mathsf{T} \widetilde{\boldsymbol{y}}_i\right)^2\right]}_{=\sigma_{\text{ave}}^2}\Bigg)^{1/2} \leq \sqrt{K}\, \sigma_y \, \sigma_{\text{ave}} \, \kappa \, \xi^i.
\end{aligned} \tag{F.58}$$

Using the bounds (F.57) and (F.58), in the second and third term on the RHS of (F.55), respectively, and computing the pertinent geometric series, it is readily seen that both these terms are $O(\delta^2)$. This proves the second equality in (F.42), thus completing the proof.

■

The next lemma establishes that, when properly shifted and scaled, $\boldsymbol{z}(\delta)$ is asymptotically normal as $\delta \to 0$.

Lemma F.7 (Asymptotic normality). Consider the setting described in Definition F.2, and assume that $\boldsymbol{y}_n$ has finite second moment. Recall from (F.40) and (F.41) that $\bar{y}_{\text{ave}}$ and σ_{ave}^2 denote the mean and variance, respectively, of the average variable $\boldsymbol{y}_{\text{ave},n}$ defined by (F.39). Consider the following shifted and scaled version of the random variable $\boldsymbol{z}(\delta)$ in (F.38):

$$\frac{\boldsymbol{z}(\delta) - \bar{y}_{\text{ave}}}{\sqrt{\delta}}, \tag{F.59}$$

where we remark that $\boldsymbol{z}(\delta)$ is shifted by subtracting the mean $\bar{y}_{\text{ave}}$ of the average variable $\boldsymbol{y}_{\text{ave},n}$. Then, $(\boldsymbol{z}(\delta) - \bar{y}_{\text{ave}})/\sqrt{\delta}$ converges in distribution to a zero-mean Gaussian variable with variance equal to half the variance σ_{ave}^2 of the average variable $\boldsymbol{y}_{\text{ave},n}$:

$$\frac{\boldsymbol{z}(\delta) - \bar{y}_{\text{ave}}}{\sqrt{\delta}} \xrightarrow[\delta \to 0]{\text{d}} \mathscr{G}\left(0, \frac{1}{2}\sigma_{\text{ave}}^2\right). \tag{F.60}$$

Proof. The claim in (F.60) is equivalent to stating that the random variable

$$\frac{\boldsymbol{z}(\delta) - \bar{y}_{\mathsf{ave}}}{\sqrt{\delta\sigma^2_{\mathsf{ave}}/2}} \tag{F.61}$$

converges in distribution to a standard Gaussian variable. Now, note that we can write

$$\frac{\boldsymbol{z}(\delta) - \bar{y}_{\mathsf{ave}}}{\sqrt{\delta\sigma^2_{\mathsf{ave}}/2}} = \frac{\boldsymbol{z}(\delta) - \mathbb{E}\boldsymbol{z}(\delta)}{\sqrt{\delta\sigma^2_{\mathsf{ave}}/2}} + \frac{\mathbb{E}\boldsymbol{z}(\delta) - \bar{y}_{\mathsf{ave}}}{\sqrt{\delta\sigma^2_{\mathsf{ave}}/2}}. \tag{F.62}$$

Since the second term in (F.62) converges to 0 in view of (F.12),[3] from Slutsky's theorem (see Theorem D.4) it suffices to show that the random variable

$$\frac{\boldsymbol{z}(\delta) - \mathbb{E}\boldsymbol{z}(\delta)}{\sqrt{\delta\sigma^2_{\mathsf{ave}}/2}} \tag{F.64}$$

converges in distribution to a standard Gaussian variable. To this end, we start by introducing, with a slight abuse of notation with respect to (F.22) and (F.23), the quantities

$$\zeta_i = \frac{\delta(1-\delta)^{i-1}}{\sqrt{\delta/2}} = \sqrt{2\delta}(1-\delta)^{i-1} \tag{F.65}$$

and

$$\widetilde{\boldsymbol{z}}(\delta) = \frac{\boldsymbol{z}(\delta) - \mathbb{E}[\boldsymbol{z}(\delta)]}{\sqrt{\delta\sigma^2_{\mathsf{ave}}/2}}, \qquad \widetilde{\boldsymbol{y}}_i = \frac{\boldsymbol{y}_i - \bar{y}}{\sigma_{\mathsf{ave}}}. \tag{F.66}$$

It is also useful to introduce the following centered and scaled version of the average variable $\boldsymbol{y}_{\mathsf{ave},i}$ defined by (F.39):

$$\widetilde{\boldsymbol{y}}_{\mathsf{ave},i} \triangleq \frac{\boldsymbol{y}_{\mathsf{ave},i} - \bar{y}_{\mathsf{ave}}}{\sigma_{\mathsf{ave}}} = \alpha^\top \widetilde{\boldsymbol{y}}_i. \tag{F.67}$$

Our aim is to establish that $\widetilde{\boldsymbol{z}}(\delta)$ converges in distribution to a standard Gaussian variable. In view of the Lévy-Cramér continuity theorem (Theorem D.1) this claim is equivalent to the convergence, as $\delta \to 0$, of the characteristic function of $\widetilde{\boldsymbol{z}}(\delta)$ to the characteristic function $e^{-s^2/2}$ (which is known to be the characteristic function of a standard Gaussian variable). From (F.38), (F.65), and (F.66) we see that

$$\widetilde{\boldsymbol{z}}(\delta) = \sum_{i=1}^{\infty} \zeta_i\, \alpha_i^\top \widetilde{\boldsymbol{y}}_i. \tag{F.68}$$

Reasoning as done to compute (F.25), the characteristic function of $\widetilde{\boldsymbol{z}}(\delta)$ in (F.68) is given by

$$\varphi_{\widetilde{z}}(s) = \prod_{i=1}^{\infty} \mathbb{E}\exp\left\{\iota s\, \zeta_i \alpha_i^\top \widetilde{\boldsymbol{y}}_i\right\}, \tag{F.69}$$

[3]We remark that Eq. (F.12) applies to the series $\boldsymbol{z}(\delta)$ defined by (F.11), where α_i and $\boldsymbol{y}_i$ are scalars. For the series $\boldsymbol{z}(\delta)$ in (F.38), where α_i and $\boldsymbol{y}_i$ are instead $K \times 1$ vectors, Eq. (F.12) becomes

$$\mathbb{E}\boldsymbol{z}(\delta) = \delta\sum_{i=1}^{\infty}(1-\delta)^{i-1}\alpha_i^\top \bar{y} = \alpha^\top \bar{y} + O(\delta) = \bar{y}_{\mathsf{ave}} + O(\delta). \tag{F.63}$$

This follows by noticing that $\alpha_i^\top \boldsymbol{y}_i = \sum_{k=1}^{K} \alpha_{k,i}\boldsymbol{y}_{k,i}$, and then applying (F.12), separately for each k, to the series (F.11) with $\alpha_{k,i}\boldsymbol{y}_{k,i}$ in place of $\alpha_i \boldsymbol{y}_i$.

where we recall that $\iota = \sqrt{-1}$ is the imaginary unit. Using the triangle inequality for complex numbers we can write

$$\left|\varphi_{\tilde{z}}(s) - e^{-\frac{s^2}{2}}\right| \le \left|\varphi_{\tilde{z}}(s) - \prod_{i=1}^{\infty} \varphi_{\mathsf{ave}}(\zeta_i s)\right| + \left|\prod_{i=1}^{\infty} \varphi_{\mathsf{ave}}(\zeta_i s) - e^{-\frac{s^2}{2}}\right|, \tag{F.70}$$

where

$$\varphi_{\mathsf{ave}}(s) \triangleq \mathbb{E} \exp\left\{\iota s \, \widetilde{\boldsymbol{y}}_{\mathsf{ave},i}\right\} \tag{F.71}$$

is the characteristic function of the random variable $\widetilde{\boldsymbol{y}}_{\mathsf{ave},i}$ defined by (F.67). Let us focus on the first term appearing on the RHS of (F.70). Since characteristic functions have magnitude not greater than 1, in view of (F.26) we can write

$$\begin{aligned}
\left|\varphi_{\tilde{z}}(s) - \prod_{i=1}^{\infty} \varphi_{\mathsf{ave}}(\zeta_i s)\right| &= \left|\prod_{i=1}^{\infty} \mathbb{E}\exp\left\{\iota s\, \zeta_i\, \alpha_i^{\mathsf{T}} \widetilde{\boldsymbol{y}}_i\right\} - \prod_{i=1}^{\infty} \mathbb{E}\exp\left\{\iota s\, \zeta_i\, \alpha^{\mathsf{T}} \widetilde{\boldsymbol{y}}_i\right\}\right| \\
&\le \sum_{i=1}^{\infty} \left|\mathbb{E}\left[\exp\left\{\iota s\, \zeta_i\, \alpha_i^{\mathsf{T}} \widetilde{\boldsymbol{y}}_i\right\} - \exp\left\{\iota s\, \zeta_i\, \alpha^{\mathsf{T}} \widetilde{\boldsymbol{y}}_i\right\}\right]\right| \\
&\le \sum_{i=1}^{\infty} \mathbb{E}\left|\exp\left\{\iota s\, \zeta_i\, \alpha_i^{\mathsf{T}} \widetilde{\boldsymbol{y}}_i\right\} - \exp\left\{\iota s\, \zeta_i\, \alpha^{\mathsf{T}} \widetilde{\boldsymbol{y}}_i\right\}\right| \\
&= \sum_{i=1}^{\infty} \mathbb{E}\left|1 - \exp\left\{\iota s\, \zeta_i\, (\alpha_i - \alpha)^{\mathsf{T}} \widetilde{\boldsymbol{y}}_i\right\}\right| \\
&\overset{(a)}{\le} |s| \sum_{i=1}^{\infty} \zeta_i\, \mathbb{E}\left|(\alpha_i - \alpha)^{\mathsf{T}} \widetilde{\boldsymbol{y}}_i\right| \\
&= |s| \sum_{i=1}^{\infty} \zeta_i\, \mathbb{E}\left|\sum_{k=1}^{K} (\alpha_{k,i} - \alpha_k)\, \widetilde{\boldsymbol{y}}_{k,i}\right| \\
&\le |s| \sum_{i=1}^{\infty} \zeta_i\, \mathbb{E}\left[\sum_{k=1}^{K} \underbrace{|\alpha_{k,i} - \alpha_k|}_{\le \kappa\, \xi^i \text{ from (F.37)}} \times |\widetilde{\boldsymbol{y}}_{k,i}|\right] \\
&\le |s| \sum_{i=1}^{\infty} \kappa\, \zeta_i\, \xi^i \underbrace{\mathbb{E}\|\widetilde{\boldsymbol{y}}_i\|_1}_{m_y} = \kappa\, m_y |s| \sum_{i=1}^{\infty} \zeta_i\, \xi^i,
\end{aligned} \tag{F.72}$$

where in step (a) we used the inequality $|1 - e^{\iota x}| \le |x|$ [65, Lemma 3.3.19], and where $\|\cdot\|_1$ denotes the L_1 norm. Using the definition of ζ_i from (F.65) and evaluating the geometric series, we see that the last summation in (F.72) converges to 0 as $\delta \to 0$. This proves that the first term on the RHS of (F.70) converges to 0. To complete the proof, we need to show that the second term on the RHS of (F.70) converges to 0, namely, that

$$\lim_{\delta \to 0} \prod_{i=1}^{\infty} \varphi_{\mathsf{ave}}(\zeta_i s) = e^{-\frac{s^2}{2}}. \tag{F.73}$$

To this end, we resort again to the triangle inequality for complex numbers and write

$$\left|\prod_{i=1}^{\infty} \varphi_{\mathsf{ave}}(\zeta_i s) - e^{-\frac{s^2}{2}}\right| \le \left|\prod_{i=1}^{\infty} \varphi_{\mathsf{ave}}(\zeta_i s) - e^{-\frac{1}{2}\sum_{i=1}^{\infty} \zeta_i^2 s^2}\right| + \left|e^{-\frac{1}{2}\sum_{i=1}^{\infty} \zeta_i^2 s^2} - e^{-\frac{s^2}{2}}\right|. \tag{F.74}$$

The second term on the RHS of (F.74) converges to 0 because

$$\lim_{\delta \to 0} \sum_{i=1}^{\infty} \zeta_i^2 = 1, \tag{F.75}$$

as can be seen by exploiting (F.65) and evaluating the pertinent geometric series. Let us now focus on the first term on the RHS of (F.74). We have the following chain of relations:

$$\begin{aligned}\left|\prod_{i=1}^{\infty} \varphi_{\mathsf{ave}}(\zeta_i s) - e^{-\frac{1}{2}\sum_{i=1}^{\infty} \zeta_i^2 s^2}\right| &= \left|\prod_{i=1}^{\infty} \varphi_{\mathsf{ave}}(\zeta_i s) - \prod_{i=1}^{\infty} e^{-\frac{\zeta_i^2 s^2}{2}}\right| \\ &\le \sum_{i=1}^{\infty} \left|\varphi_{\mathsf{ave}}(\zeta_i s) - e^{-\frac{\zeta_i^2 s^2}{2}}\right| \\ &\overset{(\text{F.26})}{\le} \sum_{i=1}^{\infty} \left|\varphi_{\mathsf{ave}}(\zeta_i s) - 1 + \frac{\zeta_i^2 s^2}{2}\right| \\ &\quad + \sum_{i=1}^{\infty} \left|e^{-\frac{\zeta_i^2 s^2}{2}} - 1 + \frac{\zeta_i^2 s^2}{2}\right|,\end{aligned} \tag{F.76}$$

where in the last step we applied the triangle inequality. Now, the second term on the RHS of (F.76) converges to 0 since for any positive number x we have $|e^{-x} - 1 + x| \le x^2/2$, which implies

$$\sum_{i=1}^{\infty} \left|e^{-\frac{\zeta_i^2 s^2}{2}} - 1 + \frac{\zeta_i^2 s^2}{2}\right| \le \frac{s^4}{8} \sum_{i=1}^{\infty} \zeta_i^4 \tag{F.77}$$

and it is immediate to show that (see the proof in [119]):

$$\lim_{\delta \to 0} \sum_{i=1}^{\infty} \zeta_i^4 = 0. \tag{F.78}$$

Next, we establish that also the first term on the RHS of (F.76) vanishes. To this end, consider the following identity:

$$\varphi_{\mathsf{ave}}(\zeta_i s) - 1 + \frac{\zeta_i^2 s^2}{2} = \mathbb{E}\left[e^{\iota\, \widetilde{\boldsymbol{y}}_{\mathsf{ave},i}\, \zeta_i s} - 1 - \iota\, \widetilde{\boldsymbol{y}}_{\mathsf{ave},i}\, \zeta_i s + \frac{1}{2}\, \widetilde{\boldsymbol{y}}_{\mathsf{ave},i}^2\, \zeta_i^2 s^2\right], \tag{F.79}$$

which holds because of the definition of $\varphi_{\mathsf{ave}}(s)$ in (F.71) and because $\widetilde{\boldsymbol{y}}_{\mathsf{ave},i}$ has zero mean and unit variance. Focusing on the argument of the expectation in (F.79) and using [65][Lemma 3.3.19] we can write, for an arbitrarily small $\varepsilon > 0$,

$$\begin{aligned}
&\left| e^{\iota\, \widetilde{\boldsymbol{y}}_{\mathsf{ave},i}\, \zeta_i s} - 1 - \iota\, \widetilde{\boldsymbol{y}}_{\mathsf{ave},i}\, \zeta_i s + \frac{1}{2}\, \widetilde{\boldsymbol{y}}^2_{\mathsf{ave},i}\, \zeta_i^2 s^2 \right| \\
&\le \mathbb{I}\Big[\left|\widetilde{\boldsymbol{y}}_{\mathsf{ave},i}\right| \zeta_i \le \varepsilon \Big] \frac{\left|\widetilde{\boldsymbol{y}}_{\mathsf{ave},i}\, \zeta_i s\right|^3}{6} + \mathbb{I}\Big[|\widetilde{\boldsymbol{y}}_{\mathsf{ave},i}|\, \zeta_i > \varepsilon \Big] \left(\widetilde{\boldsymbol{y}}_{\mathsf{ave},i}\, \zeta_i s\right)^2 \\
&\le \varepsilon\, \widetilde{\boldsymbol{y}}^2_{\mathsf{ave},i}\, \zeta_i^2\, \frac{|s|^3}{6} + \widetilde{\boldsymbol{y}}^2_{\mathsf{ave},i}\, \mathbb{I}\Big[|\widetilde{\boldsymbol{y}}_{\mathsf{ave},i}| \zeta_i > \varepsilon \Big]\, \zeta_i^2 s^2 \\
&\le \varepsilon\, \widetilde{\boldsymbol{y}}^2_{\mathsf{ave},i}\, \zeta_i^2\, \frac{|s|^3}{6} + \widetilde{\boldsymbol{y}}^2_{\mathsf{ave},i}\, \mathbb{I}\Big[|\widetilde{\boldsymbol{y}}_{\mathsf{ave},i}| > \varepsilon/\sqrt{2\delta} \Big]\, \zeta_i^2 s^2,
\end{aligned} \tag{F.80}$$

where the last inequality follows because $\zeta_i \le \sqrt{2\delta}$ — see (F.65). Computing the magnitude of both sides of (F.79), recalling that the magnitude of the expectation is upper bounded by the expectation of the magnitude, and using (F.80), we find that

$$\left| \varphi_{\mathsf{ave}}(\zeta_i s) - 1 + \frac{\zeta_i^2 s^2}{2} \right| \le \zeta_i^2 \left(\varepsilon \frac{|s|^3}{6} + s^2 g(\delta) \right), \tag{F.81}$$

where we define

$$g(\delta) \triangleq \mathbb{E}\left[\widetilde{\boldsymbol{y}}^2_{\mathsf{ave},i}\, \mathbb{I}\left[\widetilde{\boldsymbol{y}}^2_{\mathsf{ave},i} > \varepsilon/\sqrt{2\delta} \right] \right]. \tag{F.82}$$

The function $g(\delta)$ does not depend on i since the random variables $\widetilde{\boldsymbol{y}}_{\mathsf{ave},i}$ are identically distributed. Since $\widetilde{\boldsymbol{y}}_{\mathsf{ave},i}$ has finite second moment, from the dominated convergence theorem (Theorem D.6) we have that $g(\delta) \to 0$ as $\delta \to 0$. Using this result in (F.81) and accounting for (F.75) yields

$$\limsup_{\delta \to 0} \sum_{i=1}^{\infty} \left| \varphi_{\mathsf{ave}}(\zeta_i s) - 1 + \frac{\zeta_i^2 s^2}{2} \right| \le \varepsilon\, \frac{|s|^3}{6}. \tag{F.83}$$

Due to the arbitrariness of ε, we conclude that the first term on the RHS of (F.74) vanishes as $\delta \to 0$. Since we already proved that the second term vanishes, we conclude that (F.73) holds. And since we already showed that the second term on the RHS of (F.70) vanishes, we conclude that $\varphi_{\widetilde{z}}(s)$ converges to $e^{-s^2/2}$ as $\delta \to 0$. We have therefore shown that $\widetilde{\boldsymbol{z}}(\delta)$ in (F.66) converges in distribution to a standard Gaussian variable as $\delta \to 0$, and this completes the proof.

■

We conclude this appendix by examining the asymptotic properties of the LMGF of $\boldsymbol{z}(\delta)$ in (F.38). Preliminarily, it is useful to establish the following auxiliary lemma.

Lemma F.8 (Limiting property of a useful sum). Let $\mathsf{f}(s)$ be a function twice continuously differentiable on $\mathbb{R}$, with $\mathsf{f}(0) = 0$. Define the interval

$$\mathcal{J}_s = \begin{cases} [0, s] & \text{if } s \geq 0, \\ [s, 0] & \text{otherwise,} \end{cases} \tag{F.84}$$

and introduce the auxiliary functions

$$\mathsf{g}(s) = \begin{cases} \dfrac{\mathsf{f}(s)}{s} & \text{if } s \neq 0, \\ \\ \mathsf{f}'(0) & \text{if } s = 0, \end{cases} \tag{F.85}$$

and[4]

$$\mathsf{h}(s) \triangleq \frac{s^2}{2} \max_{\varsigma \in \mathcal{J}_s} \left| \mathsf{g}'(\varsigma) \right|. \tag{F.86}$$

Then, for all $s \in \mathbb{R}$,

$$\sum_{i=1}^{\infty} \mathsf{f}\left(s\,\delta(1-\delta)^{i-1} \right) = \frac{1}{\delta} \int_0^{s\,\delta} \frac{\mathsf{f}(\varsigma)}{\varsigma} d\varsigma + \mathsf{r}(s, \delta), \tag{F.87}$$

where $\mathsf{r}(s, \delta)$ is a function that satisfies the following bound:

$$|\mathsf{r}(s, \delta)| \leq \frac{\mathsf{h}(s\,\delta)}{2 - \delta}. \tag{F.88}$$

Proof. To begin with, observe from the properties of $\mathsf{f}(s)$ that the function $\mathsf{h}(s)$ is well defined and satisfies, for all $s \in \mathbb{R}$, the inequalities $0 \leq \mathsf{h}(s) < \infty$. Now, for the case $s = 0$, Eq. (F.87) is trivially verified with the choice $\mathsf{r}(0, \delta) = 0$. Let us consider the case $s > 0$; the proof for $s < 0$ is similar. We introduce the following infinite partition of $(0, \delta s]$:

$$s_i = s\,\delta(1-\delta)^{i-1}, \qquad i \in \mathbb{N}. \tag{F.91}$$

Let us introduce the function

$$G(s) = \int_s^{s_i} \mathsf{g}(\varsigma) d\varsigma, \tag{F.92}$$

where g is defined by (F.85). A second-order Taylor expansion of $G(s)$ around the point s_i gives

$$G(s_{i+1}) = \mathsf{g}(s_i)(s_i - s_{i+1}) - \frac{1}{2} \mathsf{g}'(\bar{s}_i)(s_i - s_{i+1})^2 \tag{F.93}$$

---

[4]The derivative $\mathsf{g}'(\varsigma)$ appearing in (F.86) is well defined. Indeed, for $\varsigma \neq 0$,

$$\mathsf{g}'(\varsigma) = \frac{\mathsf{f}'(\varsigma)\,\varsigma - \mathsf{f}(\varsigma)}{\varsigma^2}. \tag{F.89}$$

In addition, by applying L'Hôpital's rule [144], from (F.89) we obtain

$$\lim_{\varsigma \to 0} \mathsf{g}'(\varsigma) = \frac{\mathsf{f}''(0)}{2}, \tag{F.90}$$

which implies that $\mathsf{g}'(0) = \mathsf{f}''(0)/2$ — see footnote 3 in Appendix E.

for a certain $\bar{s}_i \in (s_{i+1}, s_i)$. Observing from (F.91) that $s_i - s_{i+1} = s_i\,\delta$ and using (F.85), Eq. (F.93) can be rewritten as

$$G(s_{i+1}) = \delta \mathsf{f}(s_i) - \frac{\delta^2}{2}\,\mathsf{g}'(\bar{s}_i)\,s_i^2. \tag{F.94}$$

Consider now an index $n > 0$ and observe from (F.91) that $s_1 = s\,\delta$ and $s_{n+1} = s\,\delta(1-\delta)^n$. Therefore, from (F.92) and the definition of integration we have

$$\int_{s\,\delta(1-\delta)^n}^{s\,\delta} \mathsf{g}(\varsigma)d\varsigma = \int_{s_{n+1}}^{s_1} \mathsf{g}(\varsigma)d\varsigma = \sum_{i=1}^{n}\int_{s_{i+1}}^{s_i} \mathsf{g}(\varsigma)d\varsigma = \sum_{i=1}^{n} G(s_{i+1}). \tag{F.95}$$

Then, from (F.94) we obtain

$$\int_{s\,\delta(1-\delta)^n}^{s\,\delta} \mathsf{g}(\varsigma)d\varsigma = \delta\sum_{i=1}^{n} \mathsf{f}(s_i) - \frac{\delta^2}{2}\sum_{i=1}^{n} \mathsf{g}'(\bar{s}_i)\,s_i^2, \tag{F.96}$$

which, exploiting (F.91) and (F.85), is equivalent to

$$\begin{aligned}\sum_{i=1}^{n} \mathsf{f}\Big(s\,\delta(1-\delta)^{i-1}\Big) &= \frac{1}{\delta}\int_{s\,\delta(1-\delta)^n}^{s\,\delta} \frac{\mathsf{f}(\varsigma)}{\varsigma}d\varsigma \\ &\quad + \frac{\delta}{2}\sum_{i=1}^{n}(1-\delta)^{2(i-1)}(s\,\delta)^2\mathsf{g}'(\bar{s}_i).\end{aligned} \tag{F.97}$$

Now, from the definition of integration we have

$$\lim_{n\to\infty}\int_{s\,\delta(1-\delta)^n}^{s\,\delta} \frac{\mathsf{f}(\varsigma)}{\varsigma}d\varsigma = \int_{0}^{s\,\delta} \frac{\mathsf{f}(\varsigma)}{\varsigma}d\varsigma, \tag{F.98}$$

which shows that, as $n \to \infty$, the first term on the RHS of (F.97) agrees with the first term on the RHS of (F.87).

Consider then the second term on the RHS of (F.97). Since $\bar{s}_i \in (s_{i+1}, s_i)$, and since $(s_{i+1}, s_i) \subset [0, s\,\delta]$ in view of (F.91), we have

$$\frac{1}{2}(s\,\delta)^2\,|\mathsf{g}'(\bar{s}_i)| \le \frac{1}{2}(s\,\delta)^2 \max_{\varsigma\in[0,s\,\delta]} |\mathsf{g}'(\varsigma)| = \mathsf{h}(s\,\delta), \tag{F.99}$$

where the equality follows from the definition of $\mathsf{h}(s)$ in (F.86). Using (F.99) we can write

$$\frac{\delta}{2}\sum_{i=1}^{n}(1-\delta)^{2(i-1)}(s\,\delta)^2|\mathsf{g}'(\bar{s}_i)| \le \delta\,\mathsf{h}(s\,\delta)\sum_{i=1}^{\infty}(1-\delta)^{2(i-1)} = \frac{\mathsf{h}(s\,\delta)}{2-\delta}, \tag{F.100}$$

which implies that the last summation in (F.97) is absolutely convergent as $n \to \infty$, allowing us to introduce the series

$$\mathsf{r}(s,\delta) \triangleq \frac{\delta}{2}\sum_{i=1}^{\infty}(1-\delta)^{2(i-1)}(s\,\delta)^2\mathsf{g}'(\bar{s}_i). \tag{F.101}$$

Letting $n \to \infty$ in (F.97), and applying (F.98) and (F.101), we obtain the representation in (F.87), with the function $\mathsf{r}(s,\delta)$ satisfying (F.88) in view of (F.100).

■

We are now ready to characterize the LMGF of $\boldsymbol{z}(\delta)$ in the small-δ regime.

Lemma F.9 (Limiting LMGF). Consider the setting described in Definition F.2. Assume that each entry $\boldsymbol{y}_{k,n}$ of the vector $\boldsymbol{y}_n$ has LMGF finite everywhere:

$$\Lambda_{y_k}(s) \triangleq \log \mathbb{E} \exp\left\{s\, \boldsymbol{y}_{k,n}\right\} < \infty \quad \forall s \in \mathbb{R}. \tag{F.102}$$

Let $\Lambda_{\mathsf{ave}}(s)$ be the LMGF of the average variable $\boldsymbol{y}_{\mathsf{ave},n}$ defined by (F.39), and let $\Lambda_\delta(s)$ be the LMGF of $\boldsymbol{z}(\delta)$ in (F.38). Then

$$\lim_{\delta \to 0} \delta \Lambda_\delta(s/\delta) = \int_0^s \frac{\Lambda_{\mathsf{ave}}(\varsigma)}{\varsigma} d\varsigma. \tag{F.103}$$

Proof. Let

$$\Lambda_{z_n}(s) \triangleq \log \mathbb{E} \exp\left\{s\, \boldsymbol{z}_n(\delta)\right\} \tag{F.104}$$

be the LMGF of the random variable $\boldsymbol{z}_n(\delta)$ in (F.33). Since the LMGF of the sum of independent random variables is equal to the sum of the LMGFs of the random variables, by exploiting the independence across i of the random variables $\alpha_i^\top \boldsymbol{y}_i$ that appear in (F.33), we get

$$\Lambda_{z_n}(s) = \sum_{i=1}^n \log \mathbb{E} \exp\left\{s\, \delta(1-\delta)^{i-1} \alpha_i^\top \boldsymbol{y}_i\right\}. \tag{F.105}$$

It is convenient to introduce the *multivariate* LMGF of the *vector* $\boldsymbol{y}_i$ [59, 159]:

$$\Lambda_y(u) \triangleq \log \mathbb{E} \exp\left\{u^\top \boldsymbol{y}_i\right\}, \qquad u \in \mathbb{R}^K, \tag{F.106}$$

from which (F.105) can be rewritten as

$$\begin{aligned}
\Lambda_{z_n}(s) &= \sum_{i=1}^n \Lambda_y\left(s\, \delta(1-\delta)^{i-1} \alpha_i\right) \\
&= \sum_{i=1}^n \Lambda_y\left(s\, \delta(1-\delta)^{i-1} \alpha\right) \\
&+ \sum_{i=1}^n \left[\Lambda_y\left(s\, \delta(1-\delta)^{i-1} \alpha_i\right) - \Lambda_y\left(s\, \delta(1-\delta)^{i-1} \alpha\right)\right] \\
&= \sum_{i=1}^n \Lambda_{\mathsf{ave}}\left(s\, \delta(1-\delta)^{i-1}\right) \\
&+ \sum_{i=1}^n \left[\Lambda_y\left(s\, \delta(1-\delta)^{i-1} \alpha_i\right) - \Lambda_y\left(s\, \delta(1-\delta)^{i-1} \alpha\right)\right],
\end{aligned} \tag{F.107}$$

where, to justify the last equality, we recall that $\Lambda_{\mathsf{ave}}(s)$ is the LMGF of the average random variable $\boldsymbol{y}_{\mathsf{ave},i} = \alpha^\top \boldsymbol{y}_i$, which, in view of (F.106), yields

$$\begin{aligned}\Lambda_{\mathsf{ave}}\Big(s\,\delta(1-\delta)^{i-1}\Big) &= \log \mathbb{E} \exp\Big\{s\,\delta(1-\delta)^{i-1}\,\boldsymbol{y}_{\mathsf{ave},i}\Big\} \\ &= \log \mathbb{E} \exp\Big\{s\,\delta(1-\delta)^{i-1}\alpha^\top \boldsymbol{y}_i\Big\} \\ &= \Lambda_y\Big(s\,\delta(1-\delta)^{i-1}\alpha\Big). \end{aligned} \tag{F.108}$$

We are interested in evaluating the limits of the summations on the RHS of (F.107) as $n \to \infty$. Regarding the first summation, we can apply Lemma F.8 with the choice $\mathsf{f}(s) = \Lambda_{\mathsf{ave}}(s)$,[5] obtaining

$$\sum_{i=1}^{\infty} \Lambda_{\mathsf{ave}}\Big(s\,\delta(1-\delta)^{i-1}\Big) = \frac{1}{\delta}\int_0^{s\,\delta} \frac{\Lambda_{\mathsf{ave}}(\varsigma)}{\varsigma}d\varsigma + \mathsf{r}(s,\delta), \tag{F.110}$$

where $\mathsf{r}(s,\delta)$ is the remainder term introduced in Lemma F.8.

Let us now focus on the second term on the RHS of (F.107). In view of (F.102), we know that the multivariate LMGF is finite for all $u \in \mathbb{R}^K$.[6] As a consequence, it is infinitely differentiable on $\mathbb{R}^K$ [59, 159]. In particular, we can use a first-order Taylor

[5] Recalling that $\Lambda_{\mathsf{ave}}(s)$ is the LMGF of the random variable defined by (F.39), we have

$$\begin{aligned}\Lambda_{\mathsf{ave}}(s) = \log \mathbb{E} \exp\left\{s\sum_{k=1}^{K}\alpha_k \boldsymbol{y}_{k,n}\right\} &\le \log \mathbb{E}\sum_{k=1}^{K}\alpha_k \exp\Big\{s\,\boldsymbol{y}_{k,n}\Big\} \\ &= \log\left(\sum_{k=1}^{K}\alpha_k \exp\{\Lambda_{y_k}(s)\}\right) < \infty, \end{aligned}\tag{F.109}$$

where the first inequality is an application of Jensen's inequality (see Theorem C.5 and in particular (C.10)) to the term $\exp\Big\{s\sum_{k=1}^{K}\alpha_k \boldsymbol{y}_k\Big\}$, accounting for the fact that the exponential function is convex and the weights $\{\alpha_k\}$ are nonnegative and add up to 1. The second inequality follows from (F.102). We conclude that $\Lambda_{\mathsf{ave}}(s)$ is finite for all $s \in \mathbb{R}$. This implies that it is infinitely differentiable on $\mathbb{R}$ (see Appendix E.1.2), thus fulfilling the hypotheses of Lemma F.8.

[6] If u has all zero entries, the multivariate LMGF is equal to 0. Thus, we consider the case where u has at least one nonzero entry, and let

$$\sigma_u = \sum_{k=1}^{K}|u_k|, \qquad q_k = \frac{|u_k|}{\sigma_u}. \tag{F.111}$$

We have the following chain of relations:

$$\begin{aligned}\exp\Big\{u^\top \boldsymbol{y}_i\Big\} = \exp\left\{\sum_{k=1}^{K} u_k \boldsymbol{y}_{k,i}\right\} &= \exp\left\{\sum_{k=1}^{K} q_k \mathrm{sign}(u_k)\sigma_u \boldsymbol{y}_{k,i}\right\} \\ &\le \sum_{k=1}^{K} q_k \exp\Big\{\mathrm{sign}(u_k)\sigma_u \boldsymbol{y}_{k,i}\Big\}, \end{aligned}\tag{F.112}$$

where the last step follows from Jensen's inequality (see Theorem C.5 and in particular (C.10)), which can be used because the exponential function is convex and the weights $\{q_k\}$ are nonneg-

expansion of Λ_y around the point $s\,\delta(1-\delta)^{i-1}\alpha$:

$$\Lambda_y\Big(s\,\delta(1-\delta)^{i-1}\alpha_i\Big) = \Lambda_y\Big(s\,\delta(1-\delta)^{i-1}\alpha\Big) + s\,\delta(1-\delta)^{i-1}(\alpha_i-\alpha)^\top \nabla\Lambda_y(\bar{u}), \tag{F.114}$$

where $\nabla\Lambda_y(u)$ is the gradient of $\Lambda_y(u)$ taken with respect to u, and where $\bar{u}$ is a point lying in the open line segment that joins the points $s\,\delta(1-\delta)^{i-1}\alpha_i$ and $s\,\delta(1-\delta)^{i-1}\alpha$. In particular, by introducing the hypercube $\mathcal{H}_s$ defined as

$$\mathcal{H}_s = \begin{cases} [0,s]^K & \text{if } s \geq 0, \\ [s,0]^K & \text{otherwise,} \end{cases} \tag{F.115}$$

we know that $\bar{u}$ is surely contained in $\mathcal{H}_{s\delta}$, because $(1-\delta)^{i-1} \leq 1$ and all the entries of the vectors α_i and α are nonnegative and bounded by 1. In view of (F.114), we can write

$$\begin{aligned} &\left|\Lambda_y\Big(s\,\delta(1-\delta)^{i-1}\alpha_i\Big) - \Lambda_y\Big(s\,\delta(1-\delta)^{i-1}\alpha\Big)\right| \\ &= s\,\delta(1-\delta)^{i-1}\left|(\alpha_i-\alpha)^\top \nabla\Lambda_y(\bar{u})\right| \leq \kappa\, s\,\delta\,\xi^i(1-\delta)^{i-1}\sum_{k=1}^K \left|\frac{\partial\Lambda_y}{\partial u_k}(\bar{u})\right|, \end{aligned} \tag{F.116}$$

where, in the last step, we used (F.37). In addition, since $\nabla\Lambda_y(u)$ is continuous on $\mathbb{R}^K$, we can write

$$\sum_{k=1}^K \left|\frac{\partial\Lambda_y}{\partial u_k}(\bar{u})\right| \leq \mathsf{h}_2(s\,\delta), \tag{F.117}$$

where we have defined the auxiliary function

$$\mathsf{h}_2(s) \triangleq \max_{u\in\mathcal{H}_s} \sum_{k=1}^K \left|\frac{\partial\Lambda_y}{\partial u_k}(u)\right|. \tag{F.118}$$

Letting

$$\mathsf{r}_2(s,\delta) \triangleq \sum_{i=1}^{\infty}\Big(\Lambda_y\left(s\,\delta(1-\delta)^{i-1}\alpha_i\right) - \Lambda_y\left(s\,\delta(1-\delta)^{i-1}\alpha\right)\Big), \tag{F.119}$$

from (F.116) and (F.117) we get

$$|\mathsf{r}_2(s,\delta)| \leq \kappa\,\xi\,s\,\delta\,\mathsf{h}_2(s\,\delta)\sum_{i=1}^{\infty}[\xi(1-\delta)]^{i-1} = \frac{\kappa\,\xi\,s\,\delta}{1-\xi(1-\delta)}\,\mathsf{h}_2(s\,\delta). \tag{F.120}$$

---

ative and add up to 1. Using (F.112), and (F.106), we can write

$$\begin{aligned} \Lambda_y(u) = \log\mathbb{E}\exp\left\{u^\top \boldsymbol{y}_i\right\} &\leq \log\left(\sum_{k=1}^K q_k\,\mathbb{E}\exp\left\{\text{sign}(u_k)\sigma_u \boldsymbol{y}_{k,i}\right\}\right) \\ &= \log\left(\sum_{k=1}^K q_k\exp\left\{\Lambda_{y_k}\left(\text{sign}(u_k)\sigma_u\right)\right\}\right) \\ &< \infty, \end{aligned} \tag{F.113}$$

where the last inequality follows from (F.102). We conclude that $\Lambda_y(u)$ is finite for all $u \in \mathbb{R}^K$.

Combining (F.107), (F.110), and (F.119), we arrive at the representation

$$\Lambda_\delta(s) = \lim_{n\to\infty} \Lambda_{z_n}(s) = \frac{1}{\delta}\int_0^{s\,\delta} \frac{\Lambda_{\mathsf{ave}}(\varsigma)}{\varsigma} d\varsigma + \mathsf{r}(s,\delta) + \mathsf{r}_2(s,\delta), \tag{F.121}$$

where, in the first equality, the LMGF of the *limiting* random variable $\boldsymbol{z}(\delta)$ in (F.38) appears because, from the continuity theorem *for moment generating functions* [55], it is legitimate to evaluate the LMGF of $\boldsymbol{z}(\delta)$ as the limit of the LMGF of the partial sum $\boldsymbol{z}_n(\delta)$. Exploiting (F.121) we can further write

$$\delta\,\Lambda_\delta(s/\delta) = \int_0^{s} \frac{\Lambda_{\mathsf{ave}}(\varsigma)}{\varsigma} d\varsigma + \delta\,\mathsf{r}(s/\delta\,,\,\delta) + \delta\,\mathsf{r}_2(s/\delta\,,\,\delta). \tag{F.122}$$

To prove (F.103), we now show that the last two terms in (F.122) vanish as $\delta \to 0$. Regarding the first term, from (F.88) we get

$$|\delta\,\mathsf{r}(s/\delta\,,\,\delta)| \leq \frac{\delta}{2-\delta}\,\mathsf{h}(s) \stackrel{\delta\to 0}{\longrightarrow} 0, \tag{F.123}$$

where $\mathsf{h}(s)$ is the auxiliary function introduced in (F.86). Likewise, regarding the second term, from (F.120) we can write

$$|\delta\,\mathsf{r}_2(s/\delta\,,\,\delta)| \leq \delta \frac{\kappa\,\xi\,s\,\mathsf{h}_2(s)}{1-\xi(1-\delta)} \stackrel{\delta\to 0}{\longrightarrow} 0, \tag{F.124}$$

which concludes the proof.

■

Appendix G

Rademacher Complexity

In the social machine learning problem examined in Chapter 12, each agent learns a decision statistic chosen from some admissible family. The degree of complexity of the decision statistic is an important parameter that is related to the performance achievable in the learning process. In statistical learning, one way to quantify the complexity of a family of functions is the *Rademacher complexity* [30, 130, 155], originally introduced as Rademacher penalty in [103].

G.1 General Case

We start with the definition of the Rademacher complexity.

Definition G.1 (Rademacher complexity). Let $\mathcal{G}$ be a family of real-valued functions

$$g : \mathcal{X} \mapsto \mathbb{R} \tag{G.1}$$

and consider a sequence of samples $x_n \in \mathcal{X}$, for $n = 1, 2, \ldots, E$. The sequence of these samples will be compactly denoted by

$$X = \{x_1, x_2, \ldots, x_E\}. \tag{G.2}$$

The *empirical* Rademacher complexity of the family $\mathcal{G}$ associated with a particular sequence X is[1]

$$\mathcal{R}(\mathcal{G}; X) \triangleq \mathbb{E} \sup_{g \in \mathcal{G}} \left| \frac{1}{E} \sum_{n=1}^{E} r_n g(x_n) \right|, \tag{G.3}$$

[1] Following [13, 30], we are defining the empirical Rademacher complexity in (G.3) and the Rademacher complexity in (G.4) with the absolute value. Other definitions are without the absolute value, and the two definitions coincide if the family $\mathcal{G}$ is symmetric, i.e., if for any function $g(x) \in \mathcal{G}$, the function $-g(x)$ also belongs to $\mathcal{G}$.

where the sequence $\{\boldsymbol{r}_n\}$ is formed by independent and identically distributed Rademacher random variables, i.e., binary variables taking on values ± 1 with equal probability.
Assume now that $\boldsymbol{X}$ is random, with the individual samples $\boldsymbol{x}_n$ being iid and also independent of the Rademacher variables. We define the Rademacher complexity as

$$\mathcal{R}(\mathcal{G}) \triangleq \mathbb{E}\mathcal{R}(\mathcal{G};\boldsymbol{X}) = \mathbb{E}\sup_{g\in\mathcal{G}}\left|\frac{1}{E}\sum_{n=1}^{E}\boldsymbol{r}_n g(\boldsymbol{x}_n)\right|, \tag{G.4}$$

where the expectation is taken over all the involved random quantities, i.e., $\boldsymbol{r}_n$ and $\boldsymbol{x}_n$.

The summations in (G.3) and (G.4) are a measure of the correlation between the functions $g \in \mathcal{G}$ and the Rademacher variables $\boldsymbol{r}_n$. As a result, the Rademacher complexity measures on average how well the function family $\mathcal{G}$ correlates with random noise. The capability to emulate random noise describes the richness (hence, the complexity) of the chosen family, and can also be seen as a measure of *overfitting* during training [12, 130].

The following known property of the Rademacher complexity is useful for some of our derivations [30, 109]. We recall that a function $\mathscr{Q} : \mathbb{R} \mapsto \mathbb{R}$ is Lipschitz-continuous with constant $\mathscr{L}$, also referred to as $\mathscr{L}$-Lipschitz, when

$$|\mathscr{Q}(z_1) - \mathscr{Q}(z_2)| \leq \mathscr{L}|z_1 - z_2| \qquad \forall z_1, z_2 \in \mathrm{dom}(\mathscr{Q}). \tag{G.5}$$

Lemma G.1 (Contraction principle). Let $\mathscr{Q} : \mathbb{R} \mapsto \mathbb{R}$ be an $\mathscr{L}$-Lipschitz function with $\mathscr{Q}(0) = 0$. Let $\mathcal{G}$ be a family of real-valued functions, and consider a sequence of samples $X = \{x_1, x_2, \ldots, x_E\}$. The empirical Rademacher complexity satisfies the following property:

$$\mathcal{R}(\mathscr{Q}\circ\mathcal{G}; X) \leq \mathscr{L}\,\mathcal{R}(\mathcal{G}; X), \tag{G.6}$$

where we denote by $\mathscr{Q}\circ\mathcal{G}$ the family generated by the composition of functions $\mathscr{Q}\circ g$, for $g \in \mathcal{G}$.

Proof. From (G.3) we can write the empirical Rademacher complexity of the composition of functions $\mathscr{Q}\circ g$ as

$$\mathcal{R}(\mathscr{Q}\circ\mathcal{G}; X) = \mathbb{E}\sup_{g\in\mathcal{G}}\left|\frac{1}{E}\sum_{n=1}^{E}\boldsymbol{r}_n\mathscr{Q}\big(g(x_n)\big)\right|. \tag{G.7}$$

Since $\mathscr{Q}$ is $\mathscr{L}$-Lipschitz and $\mathscr{Q}(0) = 0$, from (G.5) we can write

$$\big|\mathscr{Q}\big(g(x_n)\big)\big| = \big|\mathscr{Q}\big(g(x_n)\big) - \mathscr{Q}(0)\big| \leq \mathscr{L}\,\big|g(x_n)\big|, \tag{G.8}$$

for $n = 1, 2, \ldots, E$. By defining

$$u_n \triangleq \begin{cases} \dfrac{\mathscr{Q}\big(g(x_n)\big)}{\mathscr{L}g(x_n)} & \text{if } g(x_n) \neq 0, \\ \\ \text{an arbitrary number in } [-1, 1] & \text{if } g(x_n) = 0, \end{cases} \tag{G.9}$$

we can also write

$$\mathscr{Q}\big(g(x_n)\big) = \mathscr{L}\, u_n g(x_n), \tag{G.10}$$

which, when used in (G.7), yields

$$\mathcal{R}(\mathscr{Q} \circ \mathcal{G}; X) = \mathscr{L}\, \mathbb{E} \sup_{g \in \mathcal{G}} \left| \frac{1}{E} \sum_{n=1}^{E} \boldsymbol{r}_n u_n\, g(x_n) \right|. \tag{G.11}$$

Observe from (G.8) and (G.9) that

$$|u_n| \leq 1 \quad \text{for } n = 1, 2, \ldots, E. \tag{G.12}$$

Consider now the function

$$f(u) = \mathbb{E} \sup_{g \in \mathcal{G}} \left| \frac{1}{E} \sum_{n=1}^{E} \boldsymbol{r}_n u_n\, g(x_n) \right|, \tag{G.13}$$

with $u = [u_1, u_2, \ldots, u_E] \in [-1, 1]^E$. Applying the triangle inequality and exploiting the subadditivity of the supremum, it is readily seen that $f(u)$ is a convex function of u. As a result, over the hypercube $[-1, 1]^E$ (which is a compact convex set), the maximizers of $f(u)$ must be extreme points of this hypercube, i.e., vectors with all entries equal to ± 1.[2] Let $u^\star$ be one maximizer, then we can write

$$\mathcal{R}(\mathscr{Q} \circ \mathcal{G}; X) \leq \mathscr{L}\, \mathbb{E} \sup_{g \in \mathcal{G}} \left| \frac{1}{E} \sum_{n=1}^{E} \boldsymbol{r}_n u_n^\star\, g(x_n) \right|. \tag{G.16}$$

We want to evaluate explicitly the expectation over the Rademacher variables $\boldsymbol{r}_n$ in (G.16). To this end, we introduce the random vector $\boldsymbol{r} = [\boldsymbol{r}_1, \boldsymbol{r}_2, \ldots, \boldsymbol{r}_E]$ and observe

[2] Observe that $\mathcal{C} = [-1, 1]^E$ is a compact convex set, and denote by $\mathcal{E}_\mathcal{C}$ the set of extreme points of $\mathcal{C}$, which are defined as points that do not lie in any open line segment joining two points in $\mathcal{C}$ [146]. Accordingly, the extreme points are vectors with all entries equal to ± 1, yielding $|\mathcal{E}_\mathcal{C}| = 2^E$. We denote these extreme points by y_m, for $m = 1, 2, \ldots, 2^E$. From the Krein-Milman theorem [146], it follows that $\mathcal{C}$ is equal to the closed convex hull of the extreme points. This result implies that any $u \in \mathcal{C}$ can be written in the form

$$u = \sum_{m=1}^{2^E} q_m y_m, \qquad y_m \in \mathcal{E}_\mathcal{C}, \tag{G.14}$$

where $\{q_m\}$ are nonnegative weights that add up to 1. Then, using (G.14) and the convexity of the function f defined by (G.13), we get

$$f(u) = f\left(\sum_{m=1}^{2^E} q_m y_m \right) \leq \sum_{m=1}^{2^E} q_m f(y_m) \leq \max_{m \in \{1,2\ldots,2^E\}} f(y_m), \tag{G.15}$$

which shows that f is maximized at some extreme point(s).

that, since the Rademacher variables $\boldsymbol{r}_n$ are iid with $\mathbb{P}[\boldsymbol{r}_n = 1] = \mathbb{P}[\boldsymbol{r}_n = -1] = 1/2$, the vector $\boldsymbol{r}$ takes all possible values in the set $\{-1,1\}^E$ (i.e., in the E-fold Cartesian product of binary sets $\{-1,1\}$) with equal probability 2^{-E}. As a result, Eq. (G.16) can be rewritten as

$$\begin{aligned}
\mathcal{R}(\mathscr{Q} \circ \mathcal{G}; X) &\leq \mathscr{L} \times \sum_{r \in \{-1,1\}^E} \frac{1}{2^E} \sup_{g \in \mathcal{G}} \left| \frac{1}{E} \sum_{n=1}^{E} r_n u_n^\star \, g(x_n) \right| \\
&= \mathscr{L} \times \sum_{r' \in \{-1,1\}^E} \frac{1}{2^E} \sup_{g \in \mathcal{G}} \left| \frac{1}{E} \sum_{n=1}^{E} r'_n \, g(x_n) \right| \\
&= \mathscr{L} \, \mathbb{E} \sup_{g \in \mathcal{G}} \left| \frac{1}{E} \sum_{n=1}^{E} \boldsymbol{r}_n g(x_n) \right| = \mathscr{L} \, \mathcal{R}(\mathcal{G}; X),
\end{aligned} \tag{G.17}$$

where the first equality holds because, irrespective of the particular value of $u^\star$, when r spans the set $\{-1,1\}^E$, the vector $r' = [r_1 \, u_1^\star, r_2 \, u_2^\star, \ldots, r_E \, u_E^\star]$ spans the same set. Comparing (G.17) with (G.6), we see that the proof is complete.

■

G.2 Multilayer Perceptrons

It is desirable to relate the Rademacher complexity to the system parameters that characterize a particular problem, e.g., the depth and weights of a neural network, the feature space, the size of the training set, and so on. In this section we establish a useful relation for the multilayer perceptron (MLP) from Example 12.2, considering the binary classification case with $H = 2$. This relation will reveal in particular that with bounded features, the Rademacher complexity of norm-constrained MLPs (where the weight matrices are bounded) scales as $1/\sqrt{E}$ with the number of samples E. The result is illustrated in the next lemma, which is adapted from [13, 137].

Lemma G.2 (Rademacher complexity of norm-constrained MLP). Assume that each entry $x(i)$ of the feature vector $x \in \mathbb{R}^d$ is bounded, namely,

$$\max_{i \in \{1,2,\ldots,d\}} |x(i)| \leq x_{\mathsf{max}} < \infty. \tag{G.18}$$

Consider the MLP represented in Figure 12.2 for the binary case $H = 2$. This MLP has L layers (excluding the final softmax layer). For $l = 1, 2, \ldots, L$, the number of nodes at layer l is denoted by n_l. The output of the Lth layer has one node (because $H = 2$) and computes a decision statistic

$$h(x) = h(x; \theta_1). \tag{G.19}$$

Assume that the MLP is characterized by: *i)* an activation function σ_a that is $\mathscr{L}_\sigma$-Lipschitz and satisfies $\sigma_a(0) = 0$; and *ii)* a weight matrix at layer l, denoted by W_l, fulfilling the bounded-norm condition

$$\|W_l\|_1 = \max_{m\in\{1,2,\ldots,n_l\}} \sum_{i=1}^{n_{l-1}} \left|w_{im}^{(l)}\right| \le w_{\max} < \infty \quad \text{for } l = 1,2,\ldots,L, \tag{G.20}$$

where $\|W_l\|_1$ is the maximum absolute column sum norm of the matrix W_l and $w_{im}^{(l)}$ denotes the (i,m) entry of the same matrix.
Let $\mathcal{H}$ denote the family of possible functions h generated at the Lth layer by the considered MLP. Then, the empirical Rademacher complexity of this family of functions obeys the following bound:

$$\mathcal{R}(\mathcal{H};X) \le \frac{2\, w_{\max}\, x_{\max}}{\sqrt{E}} (w_{\max}\, \mathscr{L}_\sigma)^{L-1} \sqrt{\log(2d)}. \tag{G.21}$$

Proof. We recall that, according to the notation introduced in Example 12.2, $g_m^{(l)}$ denotes the function computed by node m belonging to layer l of the MLP. We start by showing that all functions $g_m^{(l)}$ corresponding to the same layer belong to a common function family $\mathcal{G}_l$. That is, $g_m^{(l)} \in \mathcal{G}_l$ for $m = 1,2,\ldots,n_l$.[3] We prove this result for any layer $l \ge 2$, with the reasoning being similar for the first layer described by (12.27).

In view of (12.26), for any layer $l \ge 2$ we have

$$g_m^{(l)}(x) = \sum_{i=1}^{n_{l-1}} w_{im}^{(l)}\, \sigma_a\left(g_i^{(l-1)}(x)\right). \tag{G.22}$$

Examining (G.22), we see that the admissible functions generated at nodes $m = 1,2,\ldots,n_l$ are linear combinations of a nonlinear transformation of the functions generated at the previous layer $l-1$. In view of (G.20), the weights $w_{im}^{(l)}$ of this linear combination must obey the condition, for $m = 1,2,\ldots,n_l$,

$$\sum_{i=1}^{n_{l-1}} \left|w_{im}^{(l)}\right| \le w_{\max}. \tag{G.23}$$

Since this condition is the same for all nodes m, we conclude that the admissible functions generated at all nodes of the lth layer belong to the same family, which we denote by $\mathcal{G}_l$. Note that the decision statistic we are interested in is the function computed by the single node of layer L, which means that we have the identity $\mathcal{H} = \mathcal{G}_L$.

To evaluate the (empirical) Rademacher complexity of $\mathcal{G}_L$, we construct a recursion over the number of MLP layers, similarly to what was done in [137]. First, we establish that the empirical Rademacher complexity of one layer is upper bounded by the empirical Rademacher complexity of the previous layer, scaled by a suitable constant. Second, we apply recursively the obtained bound from the last to the first layer. Third, we derive an upper bound on the Rademacher complexity of the first layer. The combination of the three steps yields an upper bound on the Rademacher complexity of the last layer, i.e., of the MLP.

Consider then the lth layer and denote by $\mathcal{W}_l$ the family of vectors $w = [w_i] \in \mathbb{R}^{n_{l-1}}$ that have L_1 norm bounded by $w_{\max}$. In view of (G.22), for $l \ge 2$ the Rademacher

[3]Actually, for layer L there is nothing to show since $n_L = 1$.

complexity of the family $\mathcal{G}_l$ can be expressed as

$$\mathcal{R}\left(\mathcal{G}_l; X\right) = \mathbb{E} \sup_{\substack{w \in \mathcal{W}_l \\ g_1 \in \mathcal{G}_{l-1} \\ g_2 \in \mathcal{G}_{l-1} \\ \vdots \\ g_{n_{l-1}} \in \mathcal{G}_{l-1}}} \left| \frac{1}{E} \sum_{n=1}^{E} \boldsymbol{r}_n \sum_{i=1}^{n_{l-1}} w_i \sigma_a\big(g_i(x_n)\big) \right|, \tag{G.24}$$

where $\boldsymbol{r}_n$ are iid Rademacher random variables and where we see that *each* function g_i, for $i = 1, 2, \ldots, n_{l-1}$, is selected from the family $\mathcal{G}_{l-1}$. By applying the triangle inequality, we have

$$\begin{aligned}
\left| \frac{1}{E} \sum_{n=1}^{E} \boldsymbol{r}_n \sum_{i=1}^{n_{l-1}} w_i \sigma_a\big(g_i(x_n)\big) \right| &= \left| \sum_{i=1}^{n_{l-1}} w_i \frac{1}{E} \sum_{n=1}^{E} \boldsymbol{r}_n \sigma_a\big(g_i(x_n)\big) \right| \\
&\le \sum_{i=1}^{n_{l-1}} |w_i| \times \left| \frac{1}{E} \sum_{n=1}^{E} \boldsymbol{r}_n \sigma_a\big(g_i(x_n)\big) \right|.
\end{aligned} \tag{G.25}$$

Taking the supremum over $w \in \mathcal{W}_l$ and $g_i \in \mathcal{G}_{l-1}$, and exploiting the subadditivity of the supremum (relative to $g_i \in \mathcal{G}_{l-1}$), we further get

$$\begin{aligned}
&\sup_{\substack{g_1 \in \mathcal{G}_{l-1} \\ g_2 \in \mathcal{G}_{l-1} \\ \vdots \\ g_{n_{l-1}} \in \mathcal{G}_{l-1}}} \sum_{i=1}^{n_{l-1}} |w_i| \times \left| \frac{1}{E} \sum_{n=1}^{E} \boldsymbol{r}_n \sigma_a\big(g_i(x_n)\big) \right| \\
&\le \sup_{w \in \mathcal{W}_l} \sum_{i=1}^{n_{l-1}} |w_i| \times \sup_{g_i \in \mathcal{G}_{l-1}} \left| \frac{1}{E} \sum_{n=1}^{E} \boldsymbol{r}_n \sigma_a\big(g_i(x_n)\big) \right| \\
&= w_{\mathsf{max}} \sup_{g_i \in \mathcal{G}_{l-1}} \left| \frac{1}{E} \sum_{n=1}^{E} \boldsymbol{r}_n \sigma_a\big(g_i(x_n)\big) \right|.
\end{aligned} \tag{G.26}$$

Using (G.25) and (G.26) in (G.24) we obtain

$$\mathcal{R}\left(\mathcal{G}_l; X\right) \le w_{\mathsf{max}} \mathbb{E} \sup_{g_i \in \mathcal{G}_{l-1}} \left| \frac{1}{E} \sum_{n=1}^{E} \boldsymbol{r}_n \sigma_a\big(g_i(x_n)\big) \right| = w_{\mathsf{max}} \mathcal{R}\left(\sigma_a \circ \mathcal{G}_{l-1}; X\right), \tag{G.27}$$

which, applying Lemma G.1, yields the following recursion relating the empirical Rademacher complexities at layers l and $l-1$:

$$\mathcal{R}\left(\mathcal{G}_l; X\right) \le w_{\mathsf{max}} \mathscr{L}_\sigma \mathcal{R}\left(\mathcal{G}_{l-1}; X\right). \tag{G.28}$$

Iterating (G.28) from the last layer L to the first layer, we obtain

$$\mathcal{R}\left(\mathcal{H}; X\right) = \mathcal{R}\left(\mathcal{G}_L; X\right) \le \left(w_{\mathsf{max}} \mathscr{L}_\sigma\right)^{L-1} \mathcal{R}\left(\mathcal{G}_1; X\right). \tag{G.29}$$

It remains to bound the empirical Rademacher complexity relative to each node of the first layer, which has a simpler structure implementing a linear combination of the feature vector entries. To characterize the complexity of such a structure, we can directly use [137, Lemma 15], applied with the choices $p = 1$, $\gamma = w_{\mathsf{max}}$, and $\|x_n\|_\infty = \max_{i\in\{1,2,\ldots,d\}} |x_n(i)| \leq x_{\mathsf{max}}$, obtaining

$$\mathcal{R}(\mathcal{G}_1; X) = \mathbb{E} \sup_{w\in\mathcal{W}_1} \left| \frac{1}{E} \sum_{n=1}^{E} r_n \sum_{i=1}^{d} w_i x_n(i) \right| \leq \frac{2\, w_{\mathsf{max}} x_{\mathsf{max}}}{\sqrt{E}} \sqrt{\log(2d)}, \tag{G.30}$$

which, when used in (G.29), yields the final result.

■

References

[1] D. Acemoglu, M. A. Dahleh, I. Lobel, and A. Ozdaglar (2011). "Bayesian learning in social networks". *The Review of Economic Studies* 78.4, pp. 1201–1236.

[2] D. Acemoglu and A. Ozdaglar (2011). "Opinion dynamics and learning in social networks". *Dynamic Games and Applications* 1.1, pp. 3–49.

[3] D. Acemoglu, A. Ozdaglar, and A. ParandehGheibi (2010). "Spread of (mis)-information in social networks". *Games and Economic Behavior* 70.2, pp. 194–227.

[4] J. Aczél and Z Daróczy (1975). *On Measures of Information and their Characterizations*. Academic Press.

[5] J. Alcock (2009). *Animal Behavior: An Evolutionary Approach*. Sinauer Associates.

[6] D. Alistarh, D. Grubic, J. Li, R. Tomioka, and M. Vojnovic (2017). "QSGD: Communication-efficient SGD via gradient quantization and encoding". *Proc. Neural Information Processing Systems* (NIPS), pp. 1707–1718.

[7] R. B. Ash and C. A. Doléans-Dade (2000). *Probability and Measure Theory*. Academic Press.

[8] R. R. Bahadur and R. R. Rao (1960). "On deviations of the sample mean". *The Annals of Mathematical Statistics* 31.4, pp. 1015–1027.

[9] D. Bajović, D. Jakovetić, J. M. F. Moura, J. Xavier, and B. Sinopoli (2012). "Large deviations performance of consensus+innovations distributed detection with non-Gaussian observations". *IEEE Transactions on Signal Processing* 60.11, pp. 5987–6002.

[10] A. Bandura (1977). *Social Learning Theory*. Prentice Hall.

[11] A.-L. Barabási and Z. N. Oltvai (2004). "Network biology: Understanding the cell's functional organization". *Nature Reviews Genetics* 5, pp. 101–113.

[12] P. L. Bartlett, S. Boucheron, and G. Lugosi (2002). "Model selection and error estimation". *Machine Learning* 48.1, pp. 85–113.

[13] P. L. Bartlett and S. Mendelson (2002). "Rademacher and Gaussian complexities: Risk bounds and structural results". *Journal of Machine Learning Research* 3, pp. 463–482.

[14] M. Basseville and I. V. Nikiforov (1993). *Detection of Abrupt Changes: Theory and Application*. Prentice Hall.

[15] M. F. Bear, B. W. Connors, and M. A. Paradiso (2006). *Neuroscience: Exploring the Brain*. Lippincott Williams & Wilkins.

[16] M. A. Beauchamp (1965). "An improved index of centrality". *Behavioral Science* 10.2, pp. 161–163.

[17] A. Beck and M. Teboulle (2003). "Mirror descent and nonlinear projected subgradient methods for convex optimization". *Operations Research Letters* 31.3, pp. 167–175.

[18] T. Berger, Z. Zhang, and H. Viswanathan (1996). "The CEO problem [multiterminal source coding]". *IEEE Transactions on Information Theory* 42.3, pp. 887–902.

[19] R. H. Berk (1966). "Limiting behavior of posterior distributions when the model is incorrect". *The Annals of Mathematical Statistics* 37.1, pp. 51–58.

[20] J. M. Bernardo and A. F. M. Smith (2000). *Bayesian Theory.* John Wiley & Sons.

[21] P. Billingsley (2008). *Probability and Measure.* John Wiley & Sons.

[22] B. Bollobás (1998). *Modern Graph Theory.* Springer.

[23] V. Bordignon, M. Kayaalp, V. Matta, and A. H. Sayed (2023a). "Social learning with non-Bayesian local updates". *Proc. European Signal Processing Conference* (EUSIPCO), pp. 1878–1882.

[24] V. Bordignon, V. Matta, and A. H. Sayed (2020). "Social learning with partial information sharing". *Proc. IEEE International Conference on Acoustics, Speech and Signal Processing* (ICASSP), pp. 5540–5544.

[25] — (2021). "Adaptive social learning". *IEEE Transactions on Information Theory* 67.9, pp. 6053–6081.

[26] — (2023). "Partial information sharing over social learning networks". *IEEE Transactions on InformationTheory* 69.3, pp. 2033–2058.

[27] — (2024). "Socially intelligent networks: A framework for decision making over graphs". *IEEE Signal Processing Magazine* 41.4, pp. 20–39.

[28] V. Bordignon, S. Vlaski, V. Matta, and A. H. Sayed (2021). "Network classifiers based on social learning". *Proc. IEEE International Conference on Acoustics, Speech and Signal Processing* (ICASSP), pp. 5185–5189.

[29] — (2023b). "Learning from heterogeneous data based on social interactions over graphs". *IEEE Transactions on Information Theory* 69.5, pp. 3347–3371.

[30] S. Boucheron, O. Bousquet, and G. Lugosi (2005). "Theory of classification: A survey of some recent advances". *ESAIM: Probability and Statistics* 9, pp. 323–375.

[31] S. Boucheron, G. Lugosi, and O. Bousquet (2003). "Concentration inequalities". *Summer School on Machine Learning.* Ed. by O. Bousquet, U. von Luxburg, and G. Rätsch. Springer, pp. 208–240.

[32] S. Boyd, P. Diaconis, P. Parrilo, and L. Xiao (2009). "Fastest mixing Markov chain on graphs with symmetries". *SIAM Journal on Optimization* 20.2, pp. 792–819.

[33] S. Boyd and L. Vandenberghe (2004). *Convex Optimization.* Cambridge University Press.

[34] L. M. Bregman (1967). "The relaxation method of finding the common point of convex sets and its application to the solution of problems in convex programming". *USSR Computational Mathematics and Mathematical Physics* 7.3, pp. 200–217.

[35] L. Breiman (1992). *Probability.* SIAM.

[36] S. Bubeck (2015). "Convex optimization: Algorithms and complexity". *Foundations and Trends in Machine Learning* 8.3-4, pp. 231–357.

[37] P. S. Bullen (2003). *Handbook of Means and Their Inequalities.* Vol. 560. Springer.

[38] G. Buzsaki (2011). *Rythms of the Brain.* Oxford University Press.

[39] S. Camazine, J. L. Deneubourg, N. R. Franks, J. Sneyd, G. Theraulaz, and E. Bonabeau (2003). *Self-Organization in Biological Systems.* Princeton University Press.

[40] M. Carpentiero, V. Matta, and A. H. Sayed (2023). "Distributed adaptive learning under communication constraints". *IEEE Open Journal of Signal Processing* 5, pp. 321–358.

[41] — (2024). "Compressed regression over adaptive networks". *IEEE Transactions on Signal and Information Processing over Networks.*

[42] C. Chamley, A. Scaglione, and L. Li (2013). "Models for the diffusion of beliefs in social networks: An overview". *IEEE Signal Processing Magazine* 30.3, pp. 16–29.

[43] C. P. Chamley (2004). *Rational Herds: Economic Models of Social Learning.* Cambridge University Press.

[44] H. Chernoff (1952). "A measure of the asymptotic efficiency of tests of a hypothesis based on a sum of observations". *The Annals of Mathematical Statistics* 23, pp. 493–507.

[45] B. D. Choi and S. H. Sung (1987). "Almost sure convergence theorems of weighted sums of random variables". *Stochastic Analysis and Applications* 5.4, pp. 365–377.

[46] M. Cirillo, V. Bordignon, V. Matta, and A. H. Sayed (2023). "Memory-aware social learning under partial information sharing". *IEEE Transactions on Signal Processing* 71, pp. 2833–2848.

[47] M. Cirillo, V. Matta, and A. H. Sayed (2023). "Estimating the topology of preferential attachment graphs under partial observability". *IEEE Transactions on Information Theory* 69.2, pp. 1355–1380.

[48] N. D. Condorcet (1785). *Essai sur l'Application de l'Analyse à la Probabilité des Décisions Rendues à la Pluralité des Voix.* Imprimerie Royale.

[49] J. Conlisk (1996). "Why bounded rationality?" *Journal of Economic Literature* 34.2, pp. 669–700.

[50] C. Cortes, M. Mohri, and U. Syed (2014). "Deep boosting". *Proc. International Conference on Machine Learning* (ICML), pp. 1179–1187.

[51] I. D. Couzin (2009). "Collective cognition in animal groups". *Trends in Cognitive Sciences* 13, pp. 36–43.

[52] T. M. Cover and J. A. Thomas (1991). *Elements of Information Theory.* John Wiley & Sons.

[53] H. Cramér (1938). "Sur un nouveau théorème-limite de la théorie des probabilités". *Proc. Colloque Consacré à la théorie des probabilités, Actualités Scientifiques et Industrielles.* 736, pp. 5–23.

[54] I. Csiszár (1975). "I-divergence geometry of probability distributions and minimization problems". *The Annals of Probability* 3.1, pp. 146–158.

[55] J. H. Curtiss (1942). "A note on the theory of moment generating functions". *The Annals of Statistics* 13.4, pp. 430–433.

[56] D. M. Cvetković, M. Doob, and H. Sachs (1980). *Spectra of Graphs: Theory and Applications.* Academic Press.

[57] Z. Daróczy and L. Losonczi (1967). "Über die erweiterung der auf einer punktmenge additiven funktionen". *Publicationes Mathematicae Debrecen* 14, pp. 239–245.

[58] M. H. DeGroot (1974). "Reaching a consensus". *Journal of the American Statistical Association* 69.345, pp. 118–121.

[59] A. Dembo and O. Zeitouni (1998). *Large Deviations Techniques and Applications.* Springer.

[60] F. Den Hollander (2000). *Large Deviations.* American Mathematical Society.

[61] J. A. Deri and J. M. F. Moura (2016). "New York city taxi analysis with graph signal processing". *Proc. IEEE Global Conference on Signal and Information Processing* (GlobalSIP), pp. 1275–1279.

[62] L. Devroye, L. Györfi, and G. Lugosi (2013). *A Probabilistic Theory of Pattern Recognition.* Springer.

[63] X. Dong, D. Thanou, M. Rabbat, and P. Frossard (2019). "Learning graphs from data: A signal representation perspective". *IEEE Signal Processing Magazine* 36.3, pp. 44–63.

[64] L. A. Dugatkin (2009). *Principles of Animal Behavior.* W. W. Norton & Company.

[65] R. Durrett (2019). *Probability: Theory and Examples.* Cambridge University Press.

[66] C. Efthimiou (2011). *Introduction to Functional Equations.* Mathematical Sciences Research Institute.

[67] E. O. Elliott (1963). "Estimates of error rates for codes on burst-noise channels". *The Bell System Technical Journal* 42.5, pp. 1977–1997.

[68] R. S. Ellis (1984). "Large deviations for a general class of random vectors". *TheAnnals of Probability* 12.1, pp. 1–12.

[69] P. Erdős (1939). "On a family of symmetric Bernoulli convolutions". *American Journal of Mathematics* 61.4, pp. 974–976.

[70] W. Feller (2008). *An Introduction to Probability Theory and Its Applications, vol. 2.* John Wiley & Sons.

[71] D. A. Freedman (1963). "On the Asymptotic Behavior of Bayes' Estimates in the Discrete Case I". *The Annals of Mathematical Statistics* 34.4, pp. 1386–1403.

[72] — (1965). "On the Asymptotic Behavior of Bayes' Estimates in the Discrete Case II". *The Annals of Mathematical Statistics* 36.2, pp. 454–456.

[73] Y. Freund, R. Schapire, and N. Abe (1999). "A short introduction to boosting". *Journal of Japanese Society of Artificial Intelligence* 14.5, pp. 771–780.

[74] K. Friston, J. Kilner, and L. Harrison (2006). "A free energy principle for the brain". *Journal of Physiology-Paris* 100.1-3, pp. 70–87.

[75] A. Fujita, J. R. Sato, H. M. Garay-Malpartida, R. Yamaguchi, S. Miyano, M. C. Sogayar, and C. E. Ferreira (2007). "Modeling gene expression regulatory networks with the sparse vector autoregressive model". *BMC Systems Biology* 1.39, pp. 1–11.

[76] F. Galton (1907). "Vox populi (the wisdom of crowds)". *Nature* 75.7, pp. 450–451.

[77] F. R. Gantmacher (1959). *The Theory of Matrices*, 2 volumes. AMS Chelsea Publishing.

[78] A. Gärtner (1977). "On large deviations from the invariant measure". *Theory of Probability and its Applications* 22.1, pp. 24–39.

[79] C. Genest (1984). "A characterization theorem for externally Bayesian groups". *The Annals of Statistics* 12.3, pp. 1100 –1105.

[80] C. Genest, K. J. McConway, and M. J. Schervish (1986). "Characterization of externally Bayesian pooling operators". *The Annals of Statistics* 14.2, pp. 487 –501.

[81] G. B. Giannakis, Y. Shen, and G. V. Karanikolas (2018). "Topology identification and learning over graphs: Accounting for nonlinearities and dynamics". *Proceedings of the IEEE* 106.5, pp. 787–807.

[82] E. N. Gilbert (1960). "Capacity of a burst-noise channel". *The Bell System Technical Journal* 39.5, pp. 1253–1265.

[83] B. Golub and E. Sadler (2017). "Learning in social networks". *SSRN, available at https://ssrn.com/abstract=2919146*.

[84] B. Golub and M. O. Jackson (2010). "Naïve learning in social networks and the wisdom of crowds". *American Economic Journal: Microeconomics* 2.1, pp. 112–49.

[85] G. Grimmett and D. Stirzaker (2020). *Probability and Random Processes.* Oxford University Press.

[86] J. D. Hamilton (1994). *Time Series Analysis.* Princeton University Press.

[87] J. Z. Hare, C. A. Uribe, L. Kaplan, and A. Jadbabaie (2020a). "Non-Bayesian social learning with uncertain models". *IEEE Transactions on Signal Processing* 68, pp. 4178–4193.

[88] — (2021). "A general framework for distributed inference with uncertain models". *IEEE Transactions on Signal and Information Processing over Networks* 7, pp. 392–405.

[89] J. Z. Hare, C. A. Uribe, L. M. Kaplan, and A. Jadbabaie (2020b). "Communication constrained learning with uncertain models". *Proc. IEEE International Conference on Acoustics, Speech and Signal Processing* (ICASSP), pp. 8609–8613.

[90] T. Hastie, R. Tibshirani, and J. Friedman (2009). *The Elements of Statistical Learning.* Springer.

[91] J. Hazla, A. Jadbabaie, E. Mossel, and M. A. Rahimian (2021). "Bayesian decision making in groups is hard". *Operations Research* 69.2, pp. 632–654.

[92] O. Hlinka, O. Slučiak, F. Hlawatsch, P. M. Djurić, and M. Rupp (2012). "Likelihood consensus and its application to distributed particle filtering". *IEEE Transactions on Signal Processing* 60.8, pp. 4334–4349.

[93] R. A. Horn and C. R. Johnson (2013). *Matrix Analysis.* Cambridge University Press.

[94] P. Hu, V. Bordignon, S. Vlaski, and A. H. Sayed (2022). "Optimal combination policies for adaptive social learning". *Proc. IEEE International Conference on Acoustics, Speech and Signal Processing* (ICASSP), pp. 5842–5846.

[95] — (2023). "Optimal aggregation strategies for social learning over graphs". *IEEE Transactions on Information Theory* 69.9, pp. 6048–6070.

[96] A. Jadbabaie, P. Molavi, A. Sandroni, and A. Tahbaz-Salehi (2012). "Non-Bayesian social learning". *Games and Economic Behavior* 76.1, pp. 210–225.

[97] S. T. Jose and O. Simeone (2021). "Free energy minimization: A unified framework for modeling, inference, learning, and optimization [Lecture Notes]". *IEEE Signal Processing Magazine* 38.2, pp. 120–125.

[98] B. H. Junker and F. Schreiber (2008). *Analysis of Biological Networks.* John Wiley & Sons.

[99] M. Kayaalp, V. Bordignon, S. Vlaski, and A. H. Sayed (2022). "Hidden Markov modeling over graphs". *Proc. IEEE Data Science and Learning Workshop* (DSLW), pp. 1–6.

[100] W. Kocay and D. L. Kreher (2005). *Graphs, Algorithms and Optimization.* Chapman & Hall/CRC Press.

[101] G. Koliander, Y. El-Laham, P. M. Djurić, and F. Hlawatsch (2022). "Fusion of probability density functions". *Proceedings of the IEEE* 110.4, pp. 404–453.

[102] A. Koloskova, S. Stich, and M. Jaggi (2019). "Decentralized stochastic optimization and gossip algorithms with compressed communication". *Proc. International Conference on Machine Learning* (ICML), pp. 3478–3487.

[103] V. Koltchinskii (2001). "Rademacher penalties and structural risk minimization". *IEEE Transactions on Information Theory* 47.5, pp. 1902–1914.

[104] M. Kuczma (1978). "Functional equations on restricted domains". *Aequationes Mathematicae* 18, pp. 1–34.

[105] S. Kullback (1988). "[Optimal information processing and Bayes's theorem]: Comment". *The American Statistician* 42.4, pp. 282–283.

[106] A. Lalitha, T. Javidi, and A. D. Sarwate (2018). "Social learning and distributed hypothesis testing". *IEEE Transactions on Information Theory* 64.9, pp. 6161–6179.

[107] C. C. Leang and D. H. Johnson (1997). "On the asymptotics of M-hypothesis Bayesian detection". *IEEE Transactions on Information Theory* 43.1, pp. 280–282.

[108] Y. LeCun, C. Cortes, and C. J. Burges (2010). MNIST handwritten digit database. Available at http://yann.lecun.com/exdb/mnist.

[109] M. Ledoux and M. Talagrand (1991). *Probability in Banach Spaces: Isoperimetry and Processes.* Springer.

[110] E. L. Lehmann and G. Casella (1998). *Theory of Point Estimation.* Springer.

[111] M. P. Lévy (1931). "Sur les séries dont les termes sont des variables éventuelles indépendantes". *Studia Mathematica* 3, pp. 119–155.

[112] R. Liégeois, A. Santos, V. Matta, D. Van de Ville, and A. H. Sayed (2020). "Revisiting correlation-based functional connectivity and its relationship with structural connectivity". *Network Neuroscience* 4.4, pp. 1235–1251.

[113] M. Loève (1951). "On almost sure convergence". *Proc. Second Berkeley Symposium on Mathematical Statistics and Probability*, pp. 279–303.

[114] J. M. Lucas and M. S. Saccucci (1990). "Exponentially weighted moving average control schemes: Properties and enhancements". *Technometrics* 32.1, pp. 1–12.

[115] S. Mahdizadehaghdam, H. Wang, H. Krim, and L. Dai (2016). "Information diffusion of topic propagation in social media". *IEEE Transactions on Signal and Information Processing over Networks* 2.4, pp. 569–581.

[116] S. Marano, V. Matta, T. He, and L. Tong (2013). "The embedding capacity of information flows under renewal traffic". *IEEE Transactions on Information Theory* 59.3, pp. 1724–1739.

[117] G. Mateos, S. Segarra, A. Marques, and A. Ribeiro (2019). "Connecting the dots: Identifying network structure via graph signal processing". *IEEE Signal Processing Magazine* 36.3, pp. 16–43.

[118] V. Matta, V. Bordignon, A. Santos, and A. H. Sayed (2020). "Interplay between topology and social learning over weak graphs". *IEEE Open Journal of Signal Processing* 1, pp. 99–119.

[119] V. Matta, P. Braca, S. Marano, and A. H. Sayed (2016a). "Diffusion-based adaptive distributed detection: Steady-state performance in the slow adaptation regime". *IEEE Transactions on Information Theory* 62.8, pp. 4710–4732.

[120] — (2016b). "Distributed detection over adaptive networks: Refined asymptotics and the role of connectivity". *IEEE Transactions on Signal and Information Processing over Networks* 2.4, pp. 442–460.

[121] V. Matta, A. Santos, and A. H. Sayed (2020). "Graph learning under partial observability". *Proceedings of the IEEE* 108.11, pp. 2049–2066.

[122] — (2022). "Graph learning over partially observed diffusion networks: Role of degree concentration". *IEEE Open Journal of Signal Processing* 3, pp. 335–371.

[123] V. Matta and A. H. Sayed (2018). "Estimation and detection over adaptive networks". *Cooperative and Graph Signal Processing.* Ed. by P. M. Djurić and C. Richard. Academic Press, pp. 69–106.

[124] — (2019). "Consistent tomography under partial observations over adaptive networks". *IEEE Transactions on Information Theory* 65.1, pp. 622–646.

[125] C. McDiarmid (1989). "On the method of bounded differences". *Surveys in Combinatorics* 141.1, pp. 148–188.

[126] C. D. Meyer (2000). *Matrix Analysis and Applied Linear Algebra.* SIAM.

[127] J. Mills, J. Hu, and G. Min (2020). "Communication-efficient federated learning for wireless edge intelligence in IoT". *IEEE Internet of Things Journal* 7.7, pp. 5986–5994.

[128] A. Mitra, S. Bagchi, and S. Sundaram (2020). "Event-triggered distributed inference". *Proc. IEEE Conference on Decision and Control* (CDC), pp. 6228–6233.

[129] A. Mitra, J. A. Richards, S. Bagchi, and S. Sundaram (2021). "Distributed inference with sparse and quantized communication". *IEEE Transactions on Signal Processing* 69, pp. 3906–3921.

[130] M. Mohri, A. Rostamizadeh, and A. Talwalkar (2018). *Foundations of Machine Learning.* MIT Press.

[131] P. Molavi, A. Tahbaz-Salehi, and A. Jadbabaie (2018). "A theory of non-Bayesian social learning". *Econometrica* 86.2, pp. 445–490.

[132] E. Mossel and O. Tamuz (2017). "Opinion exchange dynamics". *Probability Surveys* 14, pp. 155–204.

[133] G. V. Moustakides (1986). "Optimal stopping times for detecting changes in distributions". *The Annals of Statistics* 14.4, pp. 1379–1387.

[134] R. Nassif, S. Vlaski, M. Carpentiero, V. Matta, M. Antonini, and A. H. Sayed (2023). "Quantization for decentralized learning under subspace constraints". *IEEE Transactions on Signal Processing* 71, pp. 2320–2335.

[135] A. Nedić, A. Olshevsky, and C. A. Uribe (2017). "Fast convergence rates for distributed non-Bayesian learning". *IEEE Transactions on Automatic Control* 62.11, pp. 5538–5553.

[136] A. S. Nemirovski and D. B. Yudin (1983). *Problem Complexity and Method Efficiency in Optimization.* John Wiley & Sons.

[137] B. Neyshabur, R. Tomioka, and N. Srebro (2015). "Norm-based capacity control in neural networks". *Proc. Conference on Learning Theory* (COLT), pp. 1376–1401.

[138] B. L. Partridge (1982). "The structure and function of fish schools". *Scientific American* 246.6, pp. 114–123.

[139] M. S. Pinsker (1964). *Information and Information Stability of Random Variables and Random Processes.* Holden-Day.

[140] P. C. Pinto, P. Thiran, and M. Vetterli (2012). "Locating the source of diffusion in large-scale networks". *Physical Review Letters* 109, pp. 068702–1–068702–5.

[141] H. V. Poor and O. Hadjiliadis (2008). *Quickest Detection.* Cambridge University Press.

[142] Y. Ritov (1990). "Decision theoretic optimality of the CUSUM procedure". *The Annals of Statistics* 18.3, pp. 1464–1469.

[143] S. W. Roberts (1959). "Control chart tests based on geometric moving averages". *Technometrics* 1.3, pp. 239–250.

[144] W Rudin (1964). *Principles of Mathematical Analysis.* McGraw-Hill.

[145] — (1987). *Real and Complex Analysis.* McGraw-Hill.

[146] W. Rudin (1991). *Functional Analysis.* McGraw-Hill.

[147] H. Salami, B. Ying, and A. H. Sayed (2017). "Social learning over weakly connected graphs". *IEEE Transactions on Signal and Information Processing over Networks* 3.2, pp. 222–238.

[148] H. Salami, B. Ying, and A. H. Sayed (2021). "Belief control strategies for interactions over weakly-connected graphs". *IEEE Open Journal of Signal Processing* 2, pp. 265–279.

[149] R. Salhab, A. Ajorlou, and A. Jadbabaie (2020). "Social learning with sparse belief samples". *Proc. IEEE Conference on Decision and Control* (CDC), pp. 1792–1797.
[150] A. Santos, V. Matta, and A. H. Sayed (2020). "Local tomography of large networks under the low-observability regime". *IEEE Transactions on Information Theory* 66.1, pp. 587–613.
[151] A. H. Sayed (2014a). "Adaptation, learning, and optimization over networks". *Foundations and Trends in Machine Learning* 7.4–5, pp. 311–801.
[152] — (2014b). "Adaptive networks". *Proceedings of the IEEE* 102.4, pp. 460–497.
[153] — (2014c). "Diffusion adaptation over networks". *Academic Press Library in Signal Processing*, vol. 3. Ed. by R. Chellappa and S. Theodoridis. Academic Press, pp. 323–454.
[154] A. H. Sayed (2008). *Adaptive Filters.* John Wiley & Sons.
[155] A. H. Sayed (2022). *Inference and Learning from Data*, 3 volumes. Cambridge University Press.
[156] T. D. Seeley, R. A. Morse, and P. K. Visscher (1979). "The natural history of the flight of honey bee swarms". *Psyche* 86, pp. 103–114.
[157] S. Shahrampour, A. Rakhlin, and A. Jadbabaie (2015). "Distributed detection: Finite-time analysis and impact of network topology". *IEEE Transactions on Automatic Control* 61.11, pp. 3256–3268.
[158] C. E. Shannon (1948). "A mathematical theory of communication". *The Bell System Technical Journal* 27.3, pp. 379–423.
[159] J. Shao (2003). *Mathematical Statistics.* Springer.
[160] V. Shumovskaia, M. Kayaalp, M. Cemri, and A. H. Sayed (2023). "Discovering influencers in opinion formation over social graphs". *IEEE Open Journal of Signal Processing* 4, pp. 188–207.
[161] H. A. Simon (1990). "Bounded rationality". *Utility and Probability.* Ed. by J. Eatwell, M. Milgate, and P. Newman. Springer, pp. 15–18.
[162] O. Sporns (2010). *Networks of the Brain.* MIT Press.
[163] A. Tartakovsky, M. Basseville, and I. Nikiforov (2015). *Sequential Analysis: Hypothesis Testing and Changepoint Detection.* CRC Press.
[164] M. T. Toghani and C. A. Uribe (2022). "Communication-efficient distributed cooperative learning with compressed beliefs". *IEEE Transactions on Control of Network Systems* 9.3, pp. 1215–1226.
[165] A. B. Tsybakov (2009). *Introduction to Nonparametric Estimation.* Springer.
[166] A. W. van der Vaart (1998). *Asymptotic Statistics.* Cambridge University Press.
[167] V. N. Vapnik and A. Y. Chervonenkis (2015). "On the uniform convergence of relative frequencies of events to their probabilities". *Measures of Complexity.* Ed. by V. Vovk, H. Papadopoulos, and A. Gammerman. Springer, pp. 11–30.
[168] P. Venkitasubramaniam, T. He, and L. Tong (2008). "Anonymous networking amidst eavesdroppers". *IEEE Transactions on Information Theory* 54.6, pp. 2770–2784.
[169] H. Viswanathan and T. Berger (1997). "The quadratic Gaussian CEO problem". *IEEE Transactions on Information Theory* 43.5, pp. 1549–1559.
[170] S. Vlaski, L. Vandenberghe, and A. H. Sayed (2022). "Regularized diffusion adaptation via conjugate smoothing". *IEEE Transactions on Automatic Control* 67.5, pp. 2343–2358.

[171] M. J. Wainwright and M. I. Jordan (2008). "Graphical models, exponential families, and variational inference". *Foundations and Trends in Machine Learning* 1.1–2, pp. 1–305.

[172] L. Xiao and S. Boyd (2004). "Fast linear iterations for distributed averaging". *System & Control Letters* 53.1, pp. 65–78.

[173] B. Ying and A. H. Sayed (2016). "Information exchange and learning dynamics over weakly connected adaptive networks". *IEEE Transactions on Information Theory* 62.3, pp. 1396–1414.

[174] A. Zellner (1988). "Optimal information processing and Bayes's theorem". *The American Statistician* 42.4, pp. 278–280.

[175] X. Zhao and A. H. Sayed (2012). "Learning over social networks via diffusion adaptation". *Proc. Asilomar Conference on Signals, Systems and Computers*, pp. 709–713.

[176] Q. Zou, S. Zheng, and A. H. Sayed (2010). "Cooperative sensing via sequential detection". *IEEE Transactions on Signal Processing* 58.12, pp. 6266–6283.

About the Authors

Vincenzo Matta is a Full Professor in Telecommunications at the Department of Information and Electrical Engineering and Applied Mathematics, University of Salerno, Italy. An author of nearly 150 articles published in reputed journals and proceedings of international conferences, his research interests include adaptation and learning over networks, social learning, statistical inference on graphs, and security in communication networks. Dr. Matta has served IEEE in multiple capacities, including as a member of the editorial boards of several journals.

Virginia Bordignon received the Ph.D. degree in electrical engineering in 2022 from École Polytechnique Fédérale de Lausanne (EPFL), Switzerland, for which she was awarded the 2023 Best Dissertation Award from the IEEE Signal Processing Society. She served as a post-doctoral scholar with the Adaptive Systems Laboratory at EPFL until early 2024. Her research interests include statistical inference, distributed learning, and information processing over networks.

Ali H. Sayed is Dean of Engineering at EPFL, Switzerland, where he also directs the Adaptive Systems Laboratory. He served before as Distinguished Professor and Chair of Electrical Engineering at UCLA. He is a member of the US National Academy of Engineering and The World Academy of Sciences. He served as President of the IEEE Signal Processing Society in 2018 and 2019. An author of over 650 scholarly publications and 9 books, his research involves several areas including adaptation and learning theories, statistical inference, and multi-agent systems. His work has been recognized with several major awards including the 2022 IEEE Fourier Technical Field Award and the 2020 IEEE Wiener Society Award. He is a Fellow of IEEE, EURASIP, and the American Association for the Advancement of Science.

www.ingramcontent.com/pod-product-compliance
Lightning Source LLC
LaVergne TN
LVHW051051220926
848485LV00001BA/71
9781638284727